ACS SYMPOSIUM SERIES **467**

Water-Soluble Polymers

Synthesis, Solution Properties, and Applications

Shalaby W. Shalaby, EDITOR
Clemson University

Charles L. McCormick, EDITOR
The University of Southern Mississippi

George B. Butler, EDITOR
University of Florida–Gainesville

Developed from a symposium sponsored
by the Division of Polymer Chemistry, Inc.
at the 198th National Meeting
of the American Chemical Society,
Miami Beach, Florida,
September 10–15, 1989

American Chemical Society, Washington, DC 1991

Library of Congress Cataloging-in-Publication Data

American Chemical Society. Meeting (198th : 1989 : Miami Beach, Fla.)
  Water-soluble polymers : synthesis, solution properties, and
applications /Shalaby W. Shalaby, editor; Charles L. McCormick, editor;
George B. Butler, editor.

    p.   cm.—(ACS symposium series, ISSN 0097–6156; 467)

  "Developed from a symposium sponsored by the Division of Polymer
Chemistry at the 198th National Meeting of the American Chemical
Society, Miami Beach, Florida, September 10–15, 1989."

  Includes bibliographical references and index.

  ISBN 0–8412–2101–4

  1. Water-soluble polymers—Congresses.

  I. Shalaby, Shalaby W.   II. McCormick, Charles L., 1946–   .
III. Butler, George B., 1916–   . IV. American Chemical Society.
Polymer Chemistry Division. V. Title. VI. Series.

QD382.W3A43   1991
668′.9—dc20   91–751
                                            CIP

The paper used in this publication meets the minimum requirements of American National Standard for Information Sciences—Permanence of Paper for Printed Library Materials, ANSI Z39.48–1984.   ∞

PRINTED IN THE UNITED STATES OF AMERICA

ACS Symposium Series

M. Joan Comstock, *Series Editor*

1991 ACS Books Advisory Board

Foreword

THE ACS SYMPOSIUM SERIES was founded in 1974 to provide a medium for publishing symposia quickly in book form. The format of the Series parallels that of the continuing ADVANCES IN CHEMISTRY SERIES except that, in order to save time, the papers are not typeset, but are reproduced as they are submitted by the authors in camera-ready form. Papers are reviewed under the supervision of the editors with the assistance of the Advisory Board and are selected to maintain the integrity of the symposia. Both reviews and reports of research are acceptable, because symposia may embrace both types of presentation. However, verbatim reproductions of previously published papers are not accepted.

Contents

Preface..ix

POLYMERS AND INTERMEDIATES

1. Structural Design of Water-Soluble Copolymers............................... 2
   Charles L. McCormick

2. Synthetic Methods for Water-Soluble Monomers and Polymers:
   Nonpolyelectrolytes and Polyelectrolytes... 25
   George B. Butler and Nai Zheng Zhang

3. Water-Soluble Polymer Synthesis: Theory and Practice................... 57
   D. N. Schulz

4. Chemical Modifications of Natural Polymers and Their
   Technological Relevance 74
   Shalaby W. Shalaby and Kishore R. Shah

POLYMER SYNTHESIS AND MODIFICATION

5. Mechanism and Kinetics of the Persulfate-Initiated
   Polymerization of Acrylamide 82
   D. Hunkeler and A. E. Hamielec

6. Persulfate-Initiated Polymerization of Acrylamide at High
   Monomer Concentration...105
   D. Hunkeler and A. E. Hamielec

7. Copolymers of Acrylamide and a Novel Sulfobetaine Amphoteric
   Monomer..119
   Luis C. Salazar and Charles L. McCormick

8. Synthesis and Solution Characterization of Pyrene-Labeled
   Polyacrylamides 130
   Stephen A. Ezzell and Charles L. McCormick

9. Comblike Cyclopolymers of Alkyldiallylamines
   and Alkyldiallylmethylammonium Chlorides 151
   George B. Butler and Choon H. Do

10. New Fluorocarbon-Containing Hydrophobically Associating
    Polyacrylamide Copolymer 159
    Y.-X. Zhang, A.-H. Da, Thieo E. Hogen-Esch,
    and George B. Butler

11. Hydrophobically Associating Ionic Copolymers of Methyldiallyl-
    (1,1-dihydropentadecafluorooctoxyethyl)ammonium Chloride 175
    Sridhar Gopalkrishnan, George B. Butler, Thieo E.
    Hogen-Esch, and Nai Zheng Zhang

12. Peptide Graft Copolymers from Soluble Aminodeoxycellulose
    Acetate 189
    William H. Daly and Soo Lee

PHYSICOCHEMICAL ASPECTS OF AQUEOUS SOLUTIONS

13. Sodium-Ion Interactions with Polyions in Aqueous Salt-Free
    Solutions by Diffusion 202
    Paul Ander

14. Aqueous-Solution Behavior of Hydrophobically Modified
    Poly(acrylic Acid) 218
    T. K. Wang, I. Iliopoulos, and R. Audebert

15. Role of Labile Cross-Links in the Behavior of Water-Soluble
    Polymers: Hydrophobically Associating Copolymers 232
    Philip Molyneux

16. Probes of the Lower Critical Solution Temperature
    of Poly(N-isopropylacrylamide) 249
    Howard G. Schild

17. Complexation Between Poly(N-isopropylacrylamide) and
Sodium n-Dodecyl Sulfate: Comparison of Free and
Covalently Bound Fluorescent Probes 261
Howard G. Schild

18. Determination of Molecular-Weight Distribution
of Water-Soluble Macromolecules by Dynamic Light
Scattering 276
Michael J. Mettille and Roger D. Hester

19. Photophysical and Rheological Studies of the Aqueous-Solution
Properties of Naphthalene-Pendent Acrylic Copolymers 291
Mark D. Clark, Charles L. McCormick, and Charles E. Hoyle

20. Hydrophobic Effects on Complexation and Aggregation
in Water-Soluble Polymers: Fluorescence, pH,
and Dynamic Light-Scattering Measurements 303
Curtis W. Frank, David J. Hemker, and Hideko T. Oyama

21. Roles of Molecular Structure and Solvation on Drag
Reduction in Aqueous Solutions 320
Charles L. McCormick, Sarah E. Morgan,
and Roger D. Hester

22. Rheological Properties of Hydrophobically Modified
Acrylamide-Based Polyelectrolytes 338
John C. Middleton, Dosha F. Cummins,
and Charles L. McCormick

BIOMEDICAL AND INDUSTRIAL APPLICATIONS

23. Bioadhesive Drug Delivery 350
Sau-Hung S. Leung and Joseph R. Robinson

24. Heparin, Heparinoids, Synthetic Polyanions, and Anionic
Dyes: Opportunities and New Developments 367
William Regelson

25. Amine-Functionalized, Water-Soluble Polyamides as Drug
Carriers 394
Eberhard W. Neuse and Axel G. Perlwitz

26. Hydrosoluble Polymeric Drug Carriers Derived from Citric
Acid and L-Lysine 405
J. Huguet, M. Boustta, and M. Vert

27. New Polyethylene Glycols for Biomedical Applications 418
J. Milton Harris, Julian M. Dust, R. Andrew McGill,
Patricia A. Harris, Michael J. Edgell, Reza M.
Sedaghat-Herati, Laurel J. Karr, and Donna L. Donnelly

28. Biomedical Applications of High-Purity Chitosan: Physical,
Chemical, and Bioactive Properties 430
Paul A. Sandford and Arild Steinnes

29. Viscosity Behavior and Oil Recovery Properties of Interacting
Polymers 446
John K. Borchardt

ADVANCES IN LESS CONVENTIONAL SYSTEMS

30. Hydrophilic–Hydrophobic Domain Polymer Systems 468
Kishore R. Shah

31. Enzyme-Degradable Hydrogels: Properties Associated
with Albumin-Cross-Linked Polyvinylpyrrolidone Hydrogels 484
Waleed S. W. Shalaby, William E. Blevins, and Kinam Park

32. Comparison of Solution and Solid-State Structures of Sodium
Hyaluronan by ^{13}C NMR Spectroscopy 493
Joan Feder-Davis, Daniel M. Hittner, and Mary K. Cowman

33. Thermoreversible Gels 502
Shalaby W. Shalaby

Author Index 507

Affiliation Index 508

Subject Index 508

Preface

THE CONTENTS OF THIS BOOK are based primarily on the proceedings of a symposium on water-soluble polymers held during the ACS National Meeting in Miami in September 1989. This was organized as a response to a highly noted interest in this class of polymers for their distinct relevance to theoretical and applied research in biotechnology, controlled delivery of active agents, and protection of the environment. To meet the call for a comprehensive coverage of this area in one volume, authoritative investigators other than those who participated in the symposium were invited to contribute to this publication.

Contents of the volume are organized to provide the readers with fundamental information on the organic and physical chemistry of synthetic and natural polymers, as well as recent developments in polymer synthesis and modification (the first three sections). New and non-traditional applications of water-soluble polymers — particularly in the health care industry — are dealt with in the fourth section. In concert with the growing emphasis on interdisciplinary research, the editors elected to conclude the book with a brief, highly focused section on advances in less conventional systems. These pertain to topics which bridge high technology research areas in chemistry and biology, medicine, and engineering.

It is hoped that the readers of this volume will find it not only a valuable source of information, but also an inspiring lead to recognize and pursue exciting new areas of research.

SHALABY W. SHALABY
Clemson University
Clemson, SC 29634

CHARLES L. MCCORMICK
The University of Southern Mississippi
Hattiesburg, MS 39406

GEORGE B. BUTLER
University of Florida
Gainesville, FL 32611

April 1991

POLYMERS AND INTERMEDIATES

Chapter 1

Structural Design of Water-Soluble Copolymers

Charles L. McCormick

Department of Polymer Science, The University of Southern Mississippi, Hattiesburg, MS 39406–0076

Macromolecules exhibiting solubility in aqueous solutions represent a diverse class of polymers ranging from biopolymers that direct life processes to synthetic systems with enormous commercial utility. In this chapter discussion will be directed toward fundamental concepts of structural arrangement, functionality, synthesis, and solution behavior. Subsequent chapters will detail advanced synthetic methods, polymerization kinetics, solution behavior including that of polyelectrolytes, polyampholytes, and polysoaps, and a number of selected technological applications.

Solution properties and ultimate performance of water-soluble polymers are determined by specific structural characteristics of the solvated macromolecular backbone. Primary structure depends upon the nature of the repeating units (bond lengths and valence bond angles) as well as composition, location, and frequency. Macromolecules may be linear or branched with repeating units arranged in random, alternating, block, or graft fashion (Figure 1).

Secondary structure in water-soluble polymers is related to configuration, conformation, and intramolecular effects such as hydrogen bonding and ionic interactions. Tertiary structure is defined by intermolecular interactions while quaternary structure is governed by multiple-chain complexation.

The key to water solubility lies in positioning sufficient numbers of hydrophilic functional groups along the backbone or side chains. Figure 2 lists some of the major substituents that possess sufficient polarity, charge, or hydrogen bonding capability for hydration. Some biopolymers and synthetic co- and terpolymers are not totally water-soluble, but instead exist in microheterogeneous states in aqueous media. Reversible associations may be controlled to yield technologically important responses.

Synthetic Methods

Water-soluble copolymers are prepared by step-growth or chain-growth polymerization of appropriate monomers or by post-reaction procedures. Distribution of the units along the backbone or on the side chain may be accomplished in a number of ways. In nearly all procedures, proper sequencing

0097–6156/91/0467–0002$06.75/0

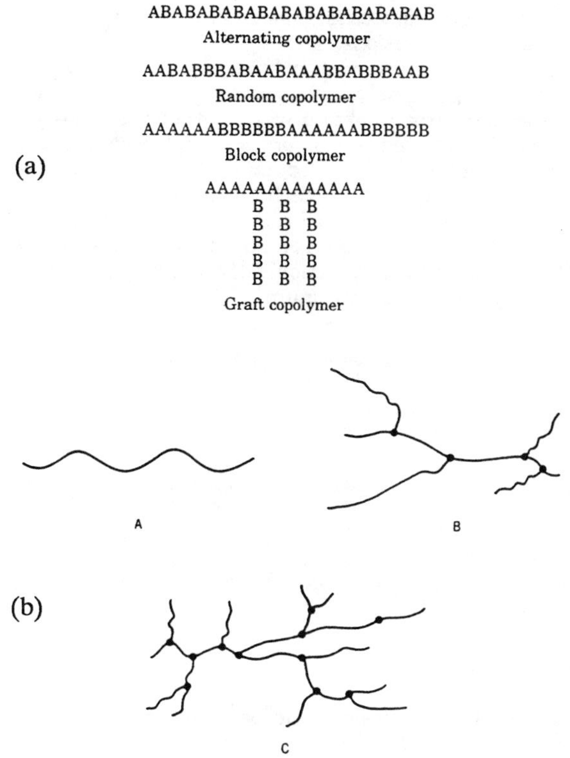

ABABABABABABABABABABABAB
Alternating copolymer

AABABBBABAABAAABBABBBAAB
Random copolymer

AAAAAABBBBBBAAAAAABBBBBB
Block copolymer

(a)

AAAAAAAAAAAA
   B B B
   B B B
   B B B
   B B B
   B B B
Graft copolymer

(b)

A   B

C

Figure 1. (a) Monomer distributions; (b) segmental disposition in soluble copolymers: A, linear; B, branched—long branches; C, branched—branches protruding from branches give a dendritic structure.

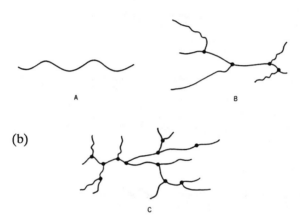

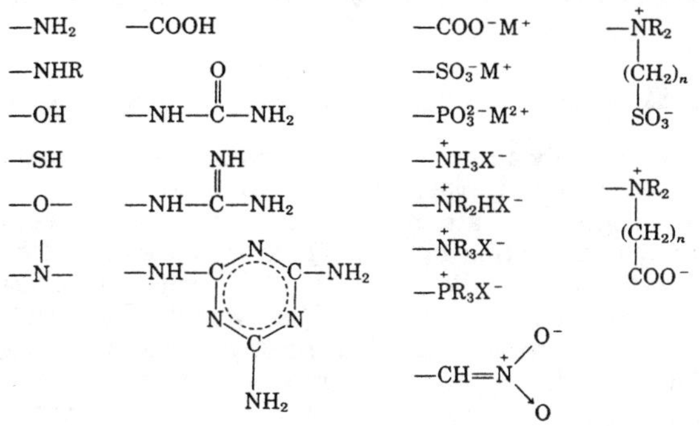

—NH$_2$	—COOH	—COO$^-$M$^+$	—$\overset{+}{N}$R$_2$
—NHR	$\overset{O}{\underset{\|}{}}$	—SO$_3^-$M$^+$	(CH$_2$)$_n$
—OH	—NH—C—NH$_2$	—PO$_3^{2-}$M^{2+}	SO$_3^-$
—SH	NH	—$\overset{+}{N}$H$_3$X$^-$	
—O—	—NH—$\overset{\|}{C}$—NH$_2$	—$\overset{+}{N}$R$_2$HX$^-$	—$\overset{+}{N}$R$_2$
		—$\overset{+}{N}$R$_3$X$^-$	(CH$_2$)$_n$
—N—	—NH—C$\underset{\diagdown}{\diagup}$C—NH$_2$	—$\overset{+}{P}$R$_3$X$^-$	COO$^-$
	NH$_2$	—CH=$\overset{+}{N}\diagdown^{O^-}_{O}$	

Figure 2. Functional groups imparting water solubility.

can be obtained by carefully controlling monomer reactivity, concentration, order of addition, and reaction conditions.

Step growth condensation reactions may be conducted in solution, bulk, interfacially, microheterogeneously, or on solid supports. Often active esters are employed at low temperatures to obtain water-soluble polyesters or polyamides. Sequential addition of protected monomers onto polymer supports is employed for preparation of synthetic polypeptides, polynucleotides, and polysaccharides.

Scheme 1 illustrates the major mechanisms for preparation of commercial synthetic water-soluble polymers by direct chain growth or ring opening of functionalized alkenes, carbonyl monomers, or strained ring systems. Initiation may be accomplished free radically, anionically, cationically, or by coordination catalysis depending on monomer structure.

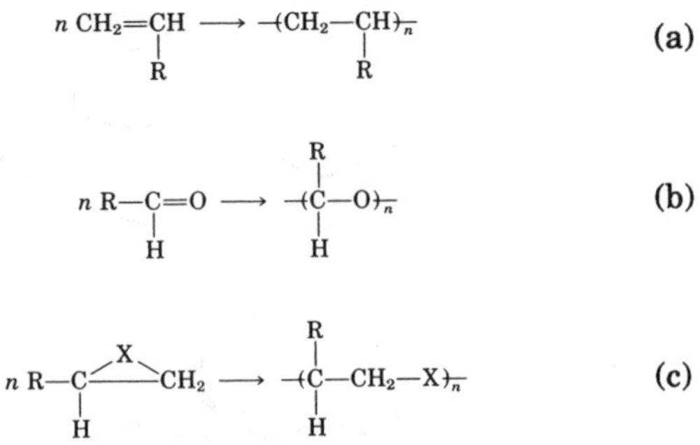

$$n\ CH_2{=}\underset{R}{CH} \longrightarrow {\leftarrow}(CH_2{-}\underset{R}{CH}){\rightarrow}_n \qquad \text{(a)}$$

$$n\ R{-}\underset{H}{C}{=}O \longrightarrow {\leftarrow}(\overset{R}{\underset{H}{C}}{-}O){\rightarrow}_n \qquad \text{(b)}$$

$$n\ R{-}\underset{H}{C}{\overset{X}{\underset{}{\diagup\diagdown}}}CH_2 \longrightarrow {\leftarrow}(\overset{R}{\underset{H}{C}}{-}CH_2{-}X){\rightarrow}_n \qquad \text{(c)}$$

Scheme 1. Major synthetic pathways for preparing
water-soluble polymers.

A number of synthetic procedures utilizing aqueous solutions, dispersions, suspensions, or emulsions are particularly useful for commercial production. For example, water-soluble monomers are easily copolymerized in water-in-oil emulsions or suspensions. For copolymerization of hydrophobic and hydrophilic monomers, microemulsions or aqueous solutions of high surfactant concentration are often required. The large number of variables that control polymer microstructure and reaction kinetics are discussed in subsequent chapters of this book.

Hydrodynamic Volume

Solution behavior of polymers can best be predicted by considering chemical structure and hydrodynamic volume (HDV) or that volume occupied by the solvated chain. Theoretical attempts to relate dimensions of macromolecular chains in solution to repeating unit structure were pioneered by Flory (1). Of particular importance to the extension of such theories to aqueous polyelectrolytes were the contributions of Morawetz (2), Tanford (3), and Strauss (4). Some of the conceptually simple aspects of these theories, useful to the organic chemist in tailoring polymers, are discussed below.

Initially, a freely jointed chain was described having an end-to-end distance, r, related to the number of bonds, n, and the length of each bond, l, by Equation 1, Figure 3a. Valence bond angle (θ) and conformational angle (φ) restrictions were introduced (Equation 2, Figure 3b) to impose directionality to the chain. Often the latter terms are replaced with a stiffness parameter.

Despite the chain stiffening introduced by valence bond and conformational restrictions, Equation 2 greatly underestimates the experimentally determined end-to-end distances even for uncharged polymers. In a macromolecular chain, additional expansion is predicted since segments cannot spatially occupy sites filled by other segments. Additionally in thermodynamically "good" solvents, the chain is markedly expanded. Reasonably good approximations of the end-to-end distance can be obtained by including an expansion factor (α) shown in Equation 3.

In addition to the excluded volume and solvation effects, longer range molecular interactions, charge-charge repulsions (or attractions), hydrogen bonding, etc., may contribute to the value of α. A number of theoretical models (2,3) incorporate additional terms to specifically deal with both short- and long-range repulsive effects schematically shown in Figure 3c. Recently, a number of powerful computer programs have been developed to spatially represent macromolecules in solution. In some limited cases—most frequently well behaved homopolymers of low molecular weight in organic solvents—models agree with experimental observations. Water-soluble polymers, particularly those with microheterogeneous phase behavior, have not been successfully modeled to date.

Synthetic Strategies for Controlling HDV

An examination of Equation 3 reveals potential synthetic approaches for increasing average end-to-end distance and thus HDV of macromolecules. The number of bonds, n, can be increased by increasing the degree of polymerization (DP). Choice of appropriate monomers and mechanisms (free radical, anionic, cationic, coordination, or template) can lead to high DP. Step growth polymers usually have relatively low DP and rely on other interactions for reasonable values of HDV. Notable exceptions are found in biopolymers.

The effective bond length, l, may be increased by appropriate choice of monomers. Note that r varies directly with l, and with $n^{0.5}$ in Equation 3. Introduction of cyclic rings (polysaccharides), double or triple bonds increase rigidity by effecting valence bond angle (θ) changes. Likewise, rotational bond angles (ϕ) might be changed by introducing steric bulk along the backbone. This must, however, be accomplished without compromising DP and thus n. Temperature, of course, determines the availability of various rotational states (i.e., the number of trans conformations) and thus HDV.

Other effective ways to increase hydrodynamic volume are shown in Figure 4. Adjacent mer units may act in concert if associated by hydrogen bonding (partially hydrolyzed acrylamide) or ionic charge interactions to yield longer effective bond lengths with restrictive rotations. In some instances configurational restrictions lead to chain-stiffening by helix formation (polynucleotides and polypeptides).

Finally, chain expansion may be realized by increasing the polymer solvent interaction or the "goodness" of the solvent and by introduction of like charges along or pendent to the macromolecular backbone. Extremely large HDV's are attainable for flexible polyelectrolytes in deionized water; addition of simple electrolytes dramatically reduces HDV. This typical polyelectrolyte behavior and the contrasting behavior of polyampholytes will be addressed later.

$$\langle r^2 \rangle = nl^2 \qquad (1)$$

$$\langle r^2 \rangle = nl^2 \left(\frac{1 - \cos\theta}{1 + \cos\theta} \right) \left(\frac{1 + \overline{\cos\phi}}{1 - \overline{\cos\phi}} \right) \qquad (2)$$

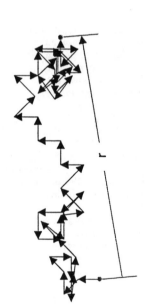

a

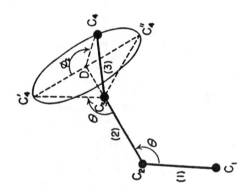

b

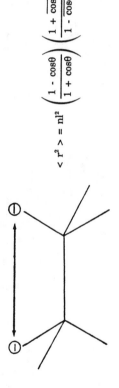

$$\langle r^2 \rangle = nl^2 \left(\frac{1 - \cos\theta}{1 + \cos\theta} \right) \left(\frac{1 + \overline{\cos\phi}}{1 - \overline{\cos\phi}} \right) \alpha^2 \qquad (3)$$

Figure 3. (a) Vectorial representation in two dimensions of a freely jointed chain. A random walk of fifty steps. (b) Spatial representation of a simple singly bonded carbon chain. (c) Intramolecular charge–charge repulsion.

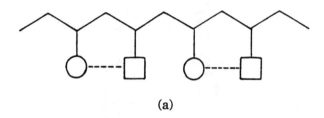

(a)

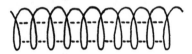

(b)

Figure 4.   Intramolecular interactions; (a) adjacent groups, (b) helix formation.

Role of Water in Solvation

The shape and size of the hydrated coil in water-soluble polymers is determined by placement of charged groups, hydrophobic moieties, hydrogen bonds, chiral centers, or restrictive rings. However, water also plays a vital role in determining solution properties, macroscopic organization, and phase behavior. Hydration may involve simple interactions with polar or hydrogen-bonding sites in uncharged random coils, charged sites on polyelectrolytes or more complex association in microheterogeneous hypercoils, micelles, and vesicles (Figure 5). Solution behaviors of these various types are quite diverse. Nonionic water-soluble polymers and polyelectrolytes are often used for rheology control and formulation. Extended rods such as cellulose derivatives may exhibit lyotropic liquid crystalline behavior. Amphiphilic molecules are utilized in personal care formulation, drug delivery, and phase transfer catalysis.

In amphiphilic polymers such as those illustrated in Figure 5 (d-f), water structure around the hydrophobic portion of the chain is considered more "ordered" than that in the bulk; conversely, hydrophilic portions "disorder" water structure. The unusual behavior of amphiphilic polymers continues to be of great interest and controversy and is the subject of several papers, books, and reviews (5-10).

Although universal acceptance of the role of water structuring is unlikely in the near term, polymer solubility and phase behavior is usually rationalized in terms of entropy- or enthalpy-dominated events. Some polymers such as poly(acrylic acid) or polyacrylamide precipitate from aqueous solutions when cooled (normal solubility behavior) whereas others such as poly(ethylene oxide), poly(propylene oxide), and poly(methacrylic acid) precipitate when heated (inverse solubility behavior). Solution turbidimetry is often used to determine upper and lower critical solution temperatures for phase separation at fixed conditions. Changes in ionic strength, concentration, molecular weight, and addition of cosolvents or structure-breakers affect the shapes of the phase behavior curves.

Viscosity and Rheology

As discussed earlier, chemical structure and hydrodynamic volume are the key determinants of polymer behavior in solution. Viscosity, which is directly related to HDV, is an easily measured property which can yield valuable insight into structure/performance relationships. Dilute solution measurements can yield the intrinsic viscosity, $[\eta]$, an indication of the hydrodynamic volume of an isolated polymer chain in solution. The Mark-Houwink-Sakurada (MHS) expression can then be used to relate that parameter to molecular weight, M.

$$[\eta] = KM^a \qquad (4)$$

The constants K and a are characteristic of a particular polymer chain under specified conditions of solvent and temperature. Of significance is the extension of the coil in dilute solution indicated by the value of a which can range from 0.5 for random coils in theta conditions to nearly 2.0 for extended rods.

In dilute solution flexible polyelectrolytes are more extended than nonionic polymers at low ionic strength due to charge-charge repulsions. Estimates of $[\eta]$ under such conditions are made by the Fuoss (11) relationship (Equation 5). η_{sp} is the specific viscosity, c polymer concentration, B and k are characteristic constants. The traditionally used Huggins relationship (Equation 6) shows a dramatic upturn in the η_{sp}/c vs c plot at low polymer concentrations

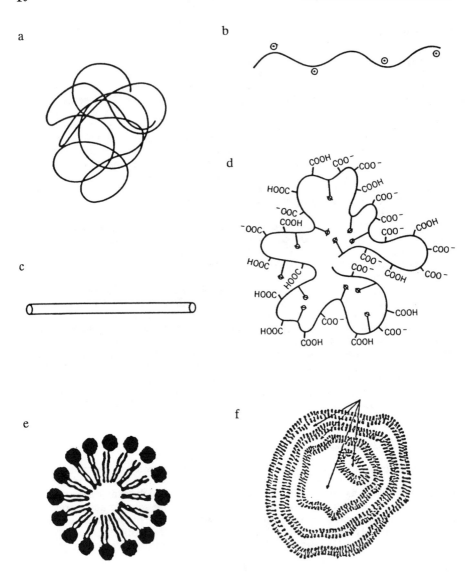

Figure 5. Structural dependence of molecular shapes of copolymers in aqueous media. (a) Solvated random coil; (b) extended coil; (c) rod-like polymer; (d) hypercoil; (e) polymeric micelle; (f) vesicle.

for polyelectrolytes (Figure 6), making accurate determination of $[\eta]$ impossible. However, addition of a sufficient quantity of a simple electrolyte suppresses charge-charge repulsion allowing use of Equation 6.

$$\frac{\eta_{sp}}{c} = [\eta] / (1 + B\sqrt{c}) \tag{5}$$

$$\frac{\eta_{sp}}{c} = [\eta] + kc \tag{6}$$

$$\eta \propto c^b M^d \tag{7}$$

In semidilute and concentrated solutions, macromolecules are no longer isolated. Intermolecular interactions above a critical concentration c*, often termed the "overlap" concentration, lead to increased values of apparent viscosity. Apparent viscosity is related to concentration c and molecular weight M through Equation 7 in which b and d are scaling constants.

Plots of $\ln \eta$ vs $\ln c$ at constant molecular weight (Figure 7a) or $\ln \eta$ vs $\ln M$ at constant concentration (Figure 7b) are made to determine the onset of associations. Onset rarely occurs sharply due to sample polydispersity. Viscometric or rheometric measurements for the above plots are made at constant shear rate or shear stress, temperature, pH, and solvent conditions. Sufficient aging of samples during dissolution and between measurements is particularly important for highly viscous water-soluble polymers.

Rheological characteristics of aqueous solutions are dictated by molecular structure, hydrodynamic volume, solvation, and associations within and between chains. In many cases, segmental interactions must be evaluated not only in regard to the number of statistical encounters but also as to free energy changes due to enthalpic interactions and entropically driven hydrophobic associations. Additionally, the time dependency of these phenomena under ambient conditions must be considered.

Some typical rheological responses of fluids are illustrated in Figure 8. Figure 8a represents shear thickening (dilatant), Newtonian, and shear thinning (pseudoplastic) behavior. Figure 8b demonstrates time dependence on viscosity for rheopectic and thixotropic polymer solutions.

Biopolymers
‾‾‾‾‾‾‾‾‾‾

Water-soluble or dispersable biopolymers represent a immensely diverse class of polymers including polynucleotides, polypeptides, and polysaccharides. Also included are the lipids and glycoproteins with enormous potential in pharmaceuticals and health care. Recombinant techniques, enzyme- assisted in vitro techniques and solid support syntheses of nucleic acids and proteins will have a great technological impact during the next decade.

While a detailed discussion of the molecular structure of biopolymers is beyond the scope of this chapter, a brief look at some representative features of commercial importance is instructive. Structural details can be found in Reference 5. Figure 9 illustrates the diversity of the water soluble natural polymers: DNA—a highly ordered antiparallel double-stranded helix with charged, alternating deoxyribose phosphate repeating units, intermolecularly bound by pendent base hydrogen bonding around a hydrophobic core; a globular protein with micelle-like properties, a hydrophilic exterior and hydrophobic

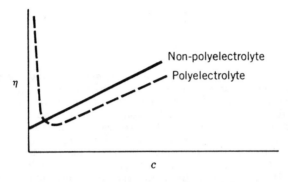

Figure 6. Viscosity response vs concentration below C* (other variables held constant) for a charged and a neutral polymer.

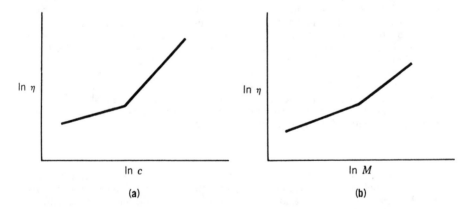

Figure 7. Relationship between apparent viscosity and (a) concentration at constant molecular weight; (b) molecular weight at constant concentration.

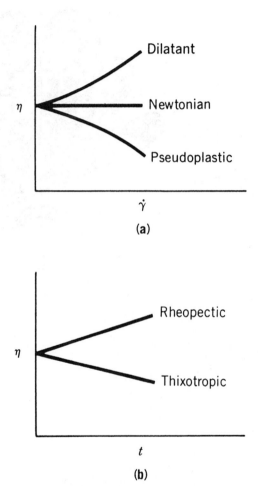

Figure 8.   Viscosity response as a function of (a) shear rate (other variables held constant), three general cases; (b) time (other variables held constant), two cases.

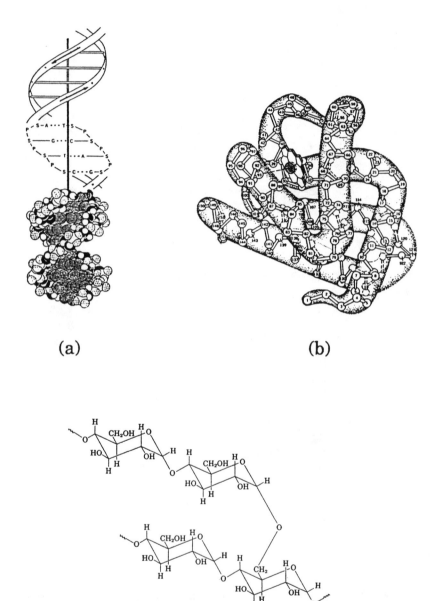

(a)   (b)

(c)

Figure 9. Naturally occurring water-soluble polymers; (a) DNA, (b) globular protein, (c) amylopectin. [(b) Reprinted with permission from ref. 5. Copyright 1989 John Wiley & Sons, Inc.]

interior; and a storage polysaccharide with anhydroglucose mers polymerized through the 1,4 positions and having branching through the 1,6 position.

Nucleic Acids. Polynucleotides are biopolymers that carry genetic information involved in the processes of replication and protein synthesis. The primary structural units of the nucleic acids are five-carbon cyclic sugars with a phosphate ester at C-5. Attached at C-1 are heterocyclic amine bases (purine or pyrimidine). Deoxyribonucleic acids DNAs carry the bases adenine, guanine, thymine, and cytosine while as do RNA's with the exception of uracil being substituted for thymine. Both polynucleotides are polymerized through the 3' and 5' position on the sugar. DNA does not possess a 2' hydroxyl group.

The biosynthetic process allows organization of DNA into an antiparallel double stranded helix with pendent pairing of adenine with thymine and guanine with cytosine on adjacent chains. The strong hydrogen bonding of the complementary pairs in the hydrophobic regions orients the charged phosphate group on each repeating unit outwardly. The balance of hydrophobic and hydrophilic forces and the presence of divalent ions such as $Mg++$ are also responsible for helix stabilization. Molecular weights of DNAs vary depending on source. Human chromosomes are reported to have up to 240 million base pairs corresponding to a molecular weight of 1.6×10^{11}.

Some DNA's are linear; others are cyclic or supercoiled. Most bacteria contain cyclic DNA's, called plasmids, in addition to chromosomal DNA. Recombinant technology has been developed to insert genes from other species into isolated plasmids and then back into the host cell. The recombinant DNA segments then serve as templates for RNA transcription and eventual synthesis of specific proteins with pharmacological or agricultural applications.

Ribonucleic acids are synthesized from the DNA template through transcription. Unlike DNAs, RNAs are single stranded although hairpin twisting in an intramolecular fashion can occur in order to match complementary base pairs. The three major types-messenger RNA, transfer RNA, and ribosomal RNA-are involved in protein synthesis (12,13).

During the past decade rapid advancements have been made in nucleotide sequencing and polynucleotide synthesis. Kornberg and coworkers (14) first prepared synthetic DNAs from purified mononucleotides utilizing isolated DNA polymerase enzymes. Shortly thereafter RNAs were prepared in a similar fashion utilizing RNA polymerases (15). In neither case was a template utilized and only oligomeric homopolynucleotides were prepared. Currently polynucleotides with programmed sequences may be prepared on solid state supports (16). Such sequences may be joined by appropriate use of restriction enzymes and ligases.

Polypeptides and Proteins. Naturally occurring polypeptides are synthesized biologically in a complex sequence of events involving template assembly of selected amino acids (17). Each protein has a unique structure and precise molecular weight. The 20 most common amino acids appear in various microstructural combinations, sequence lengths, and total chain lengths as specified by triplet code sequences along mRNA templates (18).

Most polypeptides are water-soluble or water-dispersible due to sufficient numbers of polar and hydrogen-bonding sites along the chain. The four major types of pendent groups on the amino acids which dictate secondary and tertiary structure are: hydrophobic, polar, basic, and acidic. Solubilities of proteins vary considerably based on composition and the pH and ionic strength of the surrounding medium.

Transport function and enzyme activity are usually dependent on conformational changes which may be triggered by pH, solute concentration, and electrolyte changes. Among the many functions of water-soluble polypeptides are oxygen transport, digestion, nerve response, nutrient storage, muscle movement, hormonal regulation, and gene expression.

Synthetic polypeptides can be made through sequential addition of protected amino acids onto a solid support. The procedure, pioneered by Merrifield (19,20), has been automated and hundreds of polypeptides have been synthesized. Large numbers of proteins have also been prepared by expression from cloned bacterial and yeast cells utilizing recombinant techniques (21).

Commercial applications of proteins are growing rapidly. In addition to the traditional roles in detergent additives, hydrolysis of polysaccharides and proteins, isomerization, beer and wine making, leather, and mineral recovery, genetically engineered proteins such as insulin, growth hormones, interferon, and antibodies are dramatically impacting medicine.

Polysaccharides.   Water-soluble polysaccharides comprise a large class of biopolymers formed by enzyme-directed step-growth condensation reactions of various activated cyclic sugar molecules (22). More than 100 sugars substituted variously by alcohol, sulfate, phosphate, acetate, acetamide, carboxylate, pyruvate, malonate, or other functional groups serve as possible structural repeating units.   Covalent linkages between repeating units may occur at different ring positions in various configurational and isomeric fashions. Linear or branched structures with single or multiple monomers may be formed.

Aqueous solution properties such as solubility, phase behavior, and viscosity are highly dependent on the degree of branching, stereochemistry along the cyclolinear backbone, and the chemical microstructure.   The availability of acidic or basic functional groups results in pH-, electrolyte-, and temperature-dependent behavior.   Degrees of polymerization for various polysaccharides depend upon source.

Industrial polysaccharides have traditionally been extracted from renewable resources such as plant seeds (starch), fruit (pectin), and algae (carrageenan and algin). Recently polysaccharides such as dextran, curdlan, pullulan, xanthan, scleroglucan, schizophyllan, and some pharmacologically active oligosaccharides have been commercially produced from microbial sources. These polymers have found use largely in the food and pharmaceutical industries although several have been tested in petroleum recovery processes.

A number of mucopolysaccharides consisting of amino sugars with N-acetyl or O-sulfate functionality are found in the connective tissues of higher animals (5).   Functions include wound healing, ion binding, calcification, metabolite control, and water structuring. Polymers such as hyaluronic acid, chondroitin sulfate, dermatin sulfate, and heparin are used widely in biomedical, cosmetic, and personal care applications.

Derivatized Polysaccharides.   The three most abundant naturally-occurring polymers cellulose, chitin, and starch have been derivatized to yield water solubility. Common derivatives include hydroxyethyl-, hydroxypropyl-, methyl-, carboxymethyl-, sulfate, and phosphate. Cationic, anionic, and zwitterionic substituents have been reported for a large number of applications.

Inorganic Polymers

There are relatively few inorganic polymers which have been studied in detail. Among the most important are poly(metaphosphoric acid) and poly(silicic acid) and their salts. The latter are raw materials for window glass and fiber glass (23,24).

Synthetic Water-soluble Polymers

Nonionic Polymers.   Polar, nonionic functional groups can impart water-solubility if present in sufficient numbers along or pendent to the backbone. Figure 10 lists major commercial polymers of this type. Acrylamide has an unusually high rate of polymerization in aqueous solutions leading to high molecular weight.   Acrylamide polymers and copolymers have widespread application in rheology control, flocculation, and adhesion.   Poly(ethylene oxide) prepared by ring opening polymerization is currently used for more applications than any other water-soluble system (5).   Major functions include adhesion, binding, coating, sizing, dispersion, flocculation, lubrication, water retention, and viscosification.   Poly(vinyl alcohol)-poly(vinyl acetate) copolymers are used in papermaking, emulsification, thickening, and cosmetic formulation.   Poly(N-vinylpyrrolidinone) and poly(hydroxyethyl acrylate) have been widely used in biomedical, pharmaceutical, cosmetic, and personal care products.

Charged Polymers.   Polymers possessing charges along or pendent to the molecular backbone can be classified into two main groups based on behavior in aqueous electrolyte solutions.   Polyelectrolytes, polyanions (negative charges) or polycations (positive charges) with their associated counterions, normally collapse to smaller hydrodynamic dimensions with addition of electrolytes while polyampholytes (both positive and negative charges along the macromolecular chain) expand in dilute solution.   Electrostatic effects, counterion binding, solvation, and local dielectric effects determine phase behavior and solubility (1,2,5).   Molecular structure of polyelectrolytes can be tailored to allow significant conformational changes with pH, temperature, or added electrolytes. Parameters that influence behavior include number, type, and distribution of charged mers, hydrophobe/hydrophile balance, spacing from the backbone, and counterion type.   Polyelectrolytes are prepared by homo- or copolymerization of appropriate monomers or by modification of functional polymers.

Figure 11 presents a representative group of industrially significant monomers for preparing anionic polyelectrolytes including acrylic acid, methacrylic acid, maleic anhydride (an easily hydrolyzed comonomer), p-styrene carboxylic and p-styrene sulfonic acids, vinyl sulfonic acid, 2-acrylamido-2-methylpropane sulfonic acid, and 3-acrylamido-3-methylbutanoic acid. The large number of applications in coatings formulation, dispersion, flocculation, and as oil field chemicals are reviewed in reference 4.

Cationic polyelectrolytes are usually prepared by copolymerization of ammonium, sulfonium, or phosphonium salts with appropriate comonomers or by polymer modification reactions. Figure 12 lists structures and acronyms for a number of industrially important quaternary ammonium salts (5).   Of particular interest from a structural standpoint is the cyclopolymerization of DADMAC first studied by Butler and coworkers (25).   Representative of polymer modification reactions are alkylation reactions of poly(2-vinyl pyridine) or poly(4-vinyl pyridine) to yield polycations.   Cationic polyelectrolytes have been utilized extensively in water treatment, papermaking, mineral processing, and petroleum recovery.

Amphoteric water-soluble polymers contain zwitterions on the same monomer (i.e., betaines) or along the same backbone (ampholytes) as illustrated in Figure 13.   In some cases water solubility can be attained by preparing

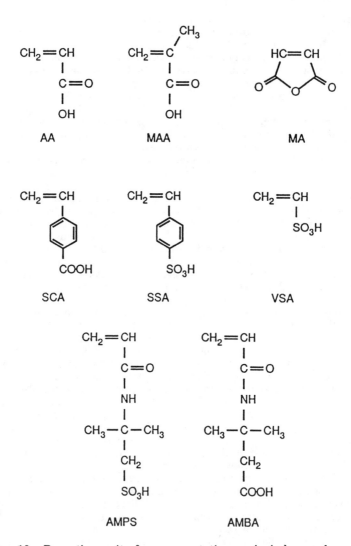

Figure 10. Repeating units for representative nonionic homopolymers.

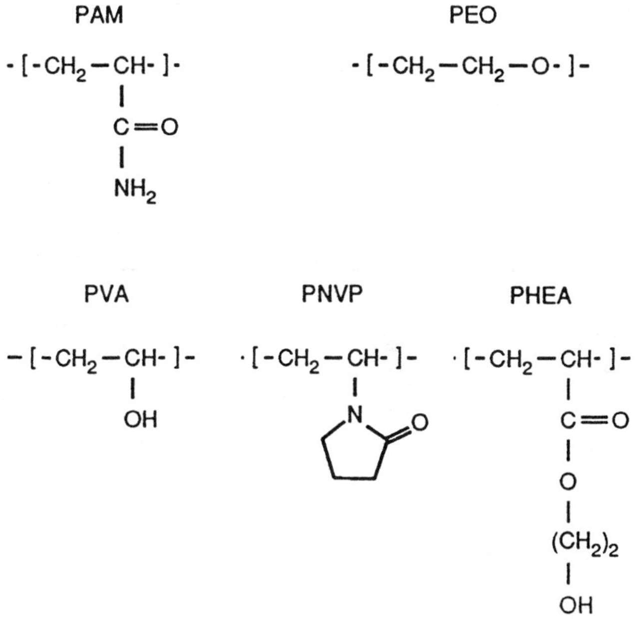

PAM

$-[-CH_2-CH-]-$
$\qquad\quad |$
$\qquad\quad C=O$
$\qquad\quad |$
$\qquad\quad NH_2$

PEO

$-[-CH_2-CH_2-O-]-$

PVA

$-[-CH_2-CH-]-$
$\qquad\quad |$
$\qquad\quad OH$

PNVP

$\cdot[-CH_2-CH-]-$
$\qquad\qquad |$
$\qquad\qquad N$

PHEA

$\cdot[-CH_2-CH-]-$
$\qquad\qquad |$
$\qquad\qquad C=O$
$\qquad\qquad |$
$\qquad\qquad O$
$\qquad\qquad |$
$\qquad\qquad (CH_2)_2$
$\qquad\qquad |$
$\qquad\qquad OH$

Figure 11. Representative anionic monomers for synthetic polyelectrolytes.

Name	Abbreviation	Structure
diallyldimethylammonium chloride	DADMAC	$(CH_3)_2\overset{+}{N}(CH_2-CH=CH_2)_2Cl^-$
diallyldiethylammonium chloride	DADEAC	$(C_2H_5)_2\overset{+}{N}(CH_2-CH=CH_2)_2Cl^-$
diethylaminoethyl methacrylate	DEAEMA	$(C_2H_5)_2\overset{+}{N}CH_2CH_2O-\overset{\overset{O}{\|}}{C}-\overset{\overset{CH_3}{\|}}{C}=CH_2$
dimethylaminoethyl methacrylate	DMAEMA	$(CH_3)_2\overset{+}{N}CH_2CH_2O-\overset{\overset{O}{\|}}{C}-\overset{\overset{CH_3}{\|}}{C}=CH_2$
methacryloyloxyethyltrimethylammonium sulfate	METAM	$CH_2=\overset{\overset{CH_3}{\|}}{C}-COO(CH_2)_2\overset{+}{N}(CH_3)_3OSO_3^-$
methacryloyloxyethyltrimethylammonium chloride	METAC	$CH_2=\overset{\overset{CH_3}{\|}}{C}-COO(CH_2)_2\overset{+}{N}(CH_3)_3Cl^-$
3-(methacrylamido)propyltrimethylammonium chloride	MAPTAC	$CH_2=\overset{\overset{CH_3}{\|}}{C}-CO-NH(CH_2)_3\overset{+}{N}(CH_3)_3Cl^-$

Figure 12. Cationic monomers, quaternary ammonium salts.

Polymer type	Generic name	Example
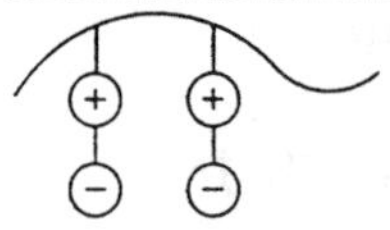	polybetaines	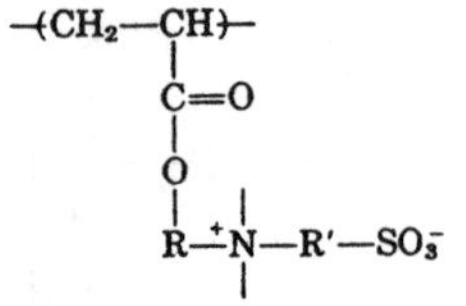 $-\!(CH_2\!-\!CH)\!-$ $C\!=\!O$ O $R\!-\!\overset{+}{N}\!-\!R'\!-\!SO_3^-$
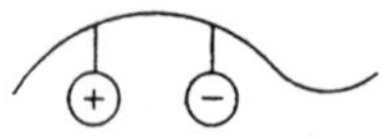	polyampholytes	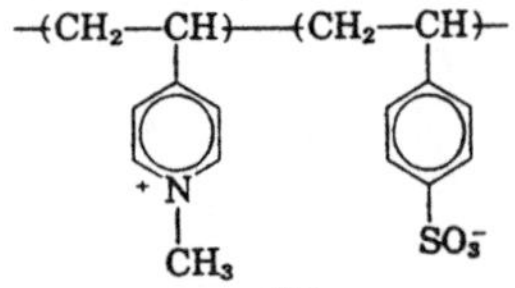 $-\!(CH_2\!-\!CH)\!-\!(CH_2\!-\!CH)\!-$ pyridinium $\overset{+}{N}$ CH_3 SO_3^-
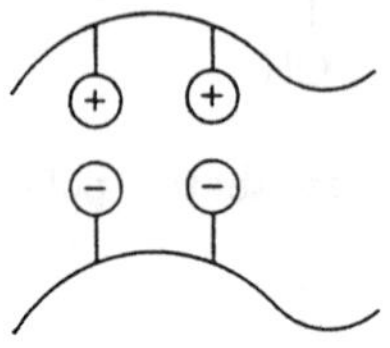	interpolymer complexes	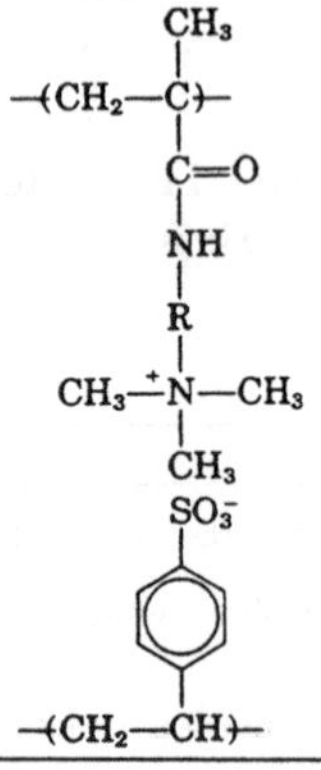 CH_3 $-\!(CH_2\!-\!C)\!-$ $C\!=\!O$ NH R $CH_3\!-\!\overset{+}{N}\!-\!CH_3$ CH_3 SO_3^- $-\!(CH_2\!-\!CH)\!-$

Figure 13. Amphoteric polymers.

interpolymer complexes of polyanions and polycations. Amphoteric polymers are generally prepared by copolymerization of betaine monomers or ampholytic monomer pairs with appropriate comonomers (26-32; McCormick, C. L. and Johnson, C. B., Polymer, in press; McCormick, C. L. and Johnson, C. B., J. Macromol. Sci. Chem., in press).

Polyampholytes and polybetaines have unusual phase behavior and solution properties. Unlike typical polyelectrolytes, amphoteric polymers are more soluble with significantly higher viscosities in aqueous salt solutions than in deionized water. For this reason, these polymers have been called "anti-polyelectrolytes". Typically, a diminution of the upper critical solution temperature upon addition of salt is observed; the degree to which this effect is apparent is determined by a number of molecular parameters as well as the effectiveness of solvation of the added ions.

The commercial utility of polybetaines, polyampholytes, and interpolymer complexes is largely unexplored to date. However, the unusual properties may find utility in brine viscosification, superabsorbency, enhanced oil recovery, and drag reduction (5,33,34).

Hydrophobically Modified Polymers. Synthetic copolymers containing nonpolar groups which aggregate in polar media were first studied as models to mimic conformational behavior of proteins. Strauss and coworkers (4,35-38) demonstrated a clear transition from conventional polyelectrolyte behavior to "polysoap" behavior by hydrophobic modification of polycations or polyanions. These copolymers, prepared from n-dodecylation of poly(2-vinylpyridine) or hydrolysis of alternating copolymers of maleic anhydride with alkylvinyl ethers, possess surfactant-like properties.

Hydrophobically-modified water-soluble polymers can be placed into two categories: those that form intramolecular associations and those that form intermolecular associations. Both, in reality, form microheterogeneous solutions in water. Intra- or intermolecular associations or domains are responsible for solution properties including phase behavior, viscosity, etc. The unusual behavior is based on hydrophobic interactions and associated water ordering discussed earlier in this chapter. Structural parameters including concentration, distribution of hydrophobes, remaining chemical microstructure, hydrophilic segment length, fixed macroions and counterions are among the many which can be changed to yield desired properties. Intra- and interpolymer associations may also be manipulated by controlling polymer concentration, molecular weight, ionic strength, pH, and temperature.

Hydrophilic cellulose derivatives were among the first to be utilized as associative thickeners (39-45). Some possible reaction pathways using long chain alkyl halides, acyl halides, anhydrides, isocyanates, and epoxides are shown in Scheme 2. Chain growth polymers with similar associative properties have been prepared by copolymerization of hydrophilic monomers with low molar concentrations of hydrophobic monomers (46-49). Dissolution problems have been overcome by introducing charged functional groups which can, after dissolution, be triggered by pH to induce associations (50,51).

Perhaps the most useful feature of the rheology of associating polymers is the unique response to shear and solvent conditions (5). Viscosity can be independent (Newtonian), increase (dilatant), or decrease (pseudoplastic) with shear rate depending upon polymer microstructure and ionic strength. Rheological characteristics have been utilized in enhanced oil recovery processes, coatings and personal care formulation, controlled release, and phase transfer.

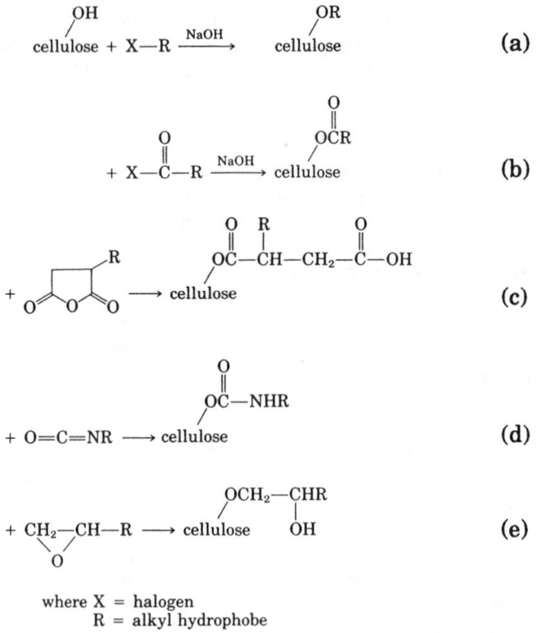

Scheme 2. Synthetic pathways for preparing water-soluble hydrophobically modified cellulose derivatives.

Conclusions

In this chapter we have introduced general concepts relating molecular structure to behavioral characteristics of water-soluble polymers. Polymers have been grouped into the categories of biopolymers, modified polymers, synthetic nonionics, polyelectrolytes, amphoteric polymers, and hydrophobically associating polymers. Structure and solution properties have been discussed with particular emphasis placed on those with present or future industrial significance.

Literature Cited

1. Flory, P. J. Principles of Polymer Chemistry; Cornell University Press: Ithaca, NY, 1953; p 399.
2. Morawetz, H. Macromolecules in Solution, 2nd ed.; John Wiley & Sons, Inc.: New York, 1975.
3. Hopfinger, A. J. Conformational Properties of Macromolecules; Academic Press, Inc.: Orlando, FL, 1973.
4. Strauss, U. P.; Jackson, E. G. J. Polym. Sci. 1951, 6, 649.
5. McCormick, C. L.; Bock, J.; Schulz, D. N. In Encyclopedia of Polymer Science and Engineering, 2nd ed.; John Wiley & Sons, Inc.: New York, 1989; Vol. 17, pp 730-784.
6. Molyneaux, P. Water-Soluble Synthetic Polymers. Properties and Behavior; CRC Press, Inc.: Boca Raton, FL, 1984; Vol. 1, Chapter 1.

7. Tanford, C. The Hydrophobic Effect: Formation of Micelles and Biological Membrane; John Wiley & Sons, Inc.: New York, 1973.
8. Saenger, W. Ann. Rev. Biophys. Chem. 1987, 16, 93.
9. Edsall, J. T.; McKenzie, H. A. Adv. Biophys. 1978, 16, 53.
10. Finney, J. L. J. Mol. Biol. 1978, 119, 415.
11. Fuoss, R. M. J. Polym. Sci. 1948, 3, 603.
12. Itakura, K.; Rossi, J. J.; Wallace, R. B. Annu. Rev. Biochem. 1984, 53, 323.
13. Freifelder, D. The DNA Molecule—Structure and Properties; W. H. Freeman & Company: San Francisco, 1978.
14. Kornberg, A. DNA Replication; W. H. Freeman & Company: San Francisco, 1980.
15. Chamberlin, M. J. RNA Polymerases: An Overview; Cold Spring Harbor Laboratory: Cold Spring Harbor, New York, 1976.
16. Oligonucleotide Synthesis, A Practical Approach; Gait, M. J., Ed.; IRL Press: Oxford, U.K., 1984.
17. Lehninger, A. L. Principles of Biochemistry; Worth Publishers: New York, 1982; pp 127-137.
18. Fersht, A. Enzyme Structure and Mechanism; W. H. Freeman & Company: San Francisco, 1977.
19. Merrifield, R. B. J. Am. Chem. Soc. 1963, 85, 2149.
20. Merrifield, R. B. Adv. Enzymol. 1969, 32, 221.
21. Freifelder, D. Recombinant DNA; W. H. Freeman & Company: San Francisco, 1978.
22. Ref. 17, pp 572-577.
23. Elias, H. G. Macromolecules 1977, 2, 1113.
24. Molyneaux, P. Water-Soluble Synthetic Polymers; CRC Press, Inc.: Boca Raton, FL, 1984; Vol. 2, p 31.
25. Butler, G. B.; Ingley, F. L. J. Am. Chem. Soc. 1951, 73, 895.
26. Hart, R.; Timmerman, D. J. J. Polym. Sci. 1958, 28, 118.
27. Salamone, J. C.; Volksen, J. C.; Olson, A. P.; Israel, S. C. Polymer 1978, 19, 1157.
28. Monray Soto, V. M.; Galin, J. S. Polymer 1984, 25(121), 254.
29. Schulz, D. N.; Peiffer, D. G.; Agarwal, P. K.; Larabee, J. J.; Kaladas, L. S.; Handwerker, B.; Garner, R. T. Polymer 1986, 27, 1734.
30. Salamone, J. C.; Tsai, C. C.; Olson, A. P.; Watterson, A. C. J. Polym. Sci. Polym. Chem. Ed. 1980, 18, 2983.
31. McCormick, C. L.; Johnson, C. B. Macromolecules 1988, 21, 686.
32. McCormick, C. L.; Johnson, C. B. Macromolecules 1988, 21, 694.
33. Johnson, C. B. Ph.D. Thesis, The University of Southern Mississippi, Hattiesburg, MS, 1987.
34. Morgan, S. E. Ph.D. Thesis, The University of Southern Mississippi, Hattiesburg, MS, 1988.
35. Dubin P.; Strauss, U. P. J. Phys. Chem. 1973, 77(11), 1427.
36. Dubin P.; Strauss, U. P. J. Phys. Chem. 1967, 71, 2757.
37. Dubin P.; Strauss, U. P. J. Phys. Chem. 1967, 74, 2842.
38. Strauss, U. P.; Vesnaver, G. J. Phys. Chem. 1975, 79(15), 1558.
39. Landoll, L. M. U.S. Patent 4 352 916 A, 1982.
40. Landoll, L. M. U.S. Patent 4 228 277, 1980.
41. Landoll, L. M. J. Polym. Sci. Polym. Chem. Ed. 1982, 20, 443.
42. Hoy, K. L.; Hoy, R. C. (to Union Carbide) U.S. Patent 4 426 485, 1984.
43. Evani, S.; Lalk, R. H. U.S. Patent 3 779 970, 1973.
44. Evani, S.; Rose, G. D. Polym. Matl. Sci. Eng. 1987, 57, 477.
45. Evani, S.; Corson, F. P. U.S. Patent 3 963 684, 1976.

46. Turner, S. R.; Siano, D. B.; Bock, J. (to Exxon Research and Engineering).
     U.S. Patent 4 520 182, 1985.
47. Emmons, W. D.; Stevens, T. E. (to Rohm and Haas). U.S. Patent 4 395
     524, 1983.
48. Evani, S. (to Dow Chemical) U.S. Patent 4 432 881, 1984.
49. McCormick, C. L.; Johnson, C. B.; Nonaka, T. Polymer 1987, 29(4), 731.
50. Bock, J.; Siano, D. B.; Turner, S. B. (to Exxon Research and Engineering).
     U.S. Patent 4 694 046, 1987.
51. McCormick, C. L; Middleton, J. C. Polym. Matl. Sci. 1987, 57, 700.

RECEIVED June 4, 1990

Chapter 2

Synthetic Methods for Water-Soluble Monomers and Polymers

Nonpolyelectrolytes and Polyelectrolytes

George B. Butler and Nai Zheng Zhang

Center for Macromolecular Science and Engineering, University of Florida, Gainesville, FL 32611-2046

This review is based on a lecture that has been presented over the past several years by one of us (GBB) in connection with certain well-established short courses. Although the major water-soluble non-polyelectrolytes are covered briefly in this review, the major emphasis is placed on well-studied and either commercially available polyelectrolyte monomers or such monomers which have the potential for commercial development. The latter judgment is based on availability of raw materials, monomer and polymer yields, and/or quality of derivable polymers or copolymer. In many cases, only the monomers are emphasized, since methods for their polymerization generally parallel each other, all being predominantly initiated and polymerized via free radical mechanisms. No distinction is generally made among the several polymerization techniques such as aqueous or other solution processes, suspension or emulsion processes. In many cases all types have been studied. In most of the others, success in the application of the several techniques to the polymerization process can reasonably be predicted.

Organic compounds having hydrophilic groups are soluble in water provided the hydrophobic-hydrophilic balance is favorable. Alcohols of low molecular weight are soluble in water; however, those of higher molecular weight are no longer soluble. The same principles apply to polymers. Table I includes the more common functional groups which impart water-solubility to organic compounds as well as to polymers.

Non-polyelectrolytes

Aside from the naturally occurring water-soluble polysaccharides or their synthetically modified derivatives, the major contributors to the non-polyelectrolyte group of water-soluble polymers are: poly(acrylamide), poly(acrylic) or (methacrylic acids), poly(ethylene oxides), poly(methyl vinyl ether) and poly(vinyl alcohols).

Polyacrylamide (PAM). Acrylamide polymerizes easily in aqueous solution in presence of a wide variety of free-radical initiators (1). Many added salts affect polymerization rate and molecular weight. Often, commercially produced poly(acrylamides) may contain a fraction that appears to be insoluble in water, a property which presents a problem for many uses. Much attention

0097–6156/91/0467–0025$09.00/0

TABLE I. Some Functional Groups Which Impart Water-Solubility to Organic Polymers

$-NH_2$	HO- Alcohols	$-CH = N - O^-$ $\quad\quad\quad\;\; O$	$-NH\diagdown_{C\diagup N\diagup C}\diagup^{NH_2}$
$-NHR$	Phenols		$N\diagdown_{C}\diagdown N$
$-NR_2$	Enols	$-PO_3^=$	NH_2
$\overset{+}{-NR_2H}\; X^-$	$\underset{OH}{-C=O}$	$-SO_3^-$	
$\overset{+}{-NR_3}\; X^-$	$\underset{O^-M^+}{-C=O}$	$-NH - \underset{NH}{\overset{\|}{C}} - NH_2$	Water-Soluble Polymers
$\overset{+}{-PR_3}\; X^-$	$\underset{NH_2}{-C=O}$	$-NH - \underset{O}{\overset{\|}{C}} - NH_2$	$-(-CH_2 - \underset{OH}{\overset{\|}{CH}}-)_n-$
$\overset{+}{-OR_2}\; X^-$	$\underset{NHR}{-C=O}$	$OH\diagdown_{C\diagup N\diagup C}\diagup OH$	$-(-CH_2 - \underset{OCH_3}{\overset{\|}{CH}}-)_n-$
$-SR_2\; X^-$	$\underset{NHCH_2OH}{-C=O}$	$N\diagdown_{C}\diagdown N$ $\quad\quad OH$	$-(-CH_2 - CH_2-O-)_n-$

has been given to water-in-oil emulsion polymerization where a concentrated solution is polymerized as the dispersed phase and the continuous phase consists of a liquid hydrocarbon.

Poly(acrylic acid) and Poly(methacrylic acid). Linear polymers of acrylic and methacrylic acids can be prepared by use of free-radical initiation, generally in aqueous solution (2). However, nonaqueous media, inverse phase emulsion or suspension techniques may also be used. Salts of acrylic and methacrylic acids can also be polymerized in aqueous solutions. The rate of polymerization is dependent on pH, being high at low pH, dropping to a minimum at pH 6-7 for both monomers, and rising again to a maximum of pH 10 for acrylic acid and near pH 12 for the methacrylic acid.

Poly(methyl vinyl ether). The only vinyl ether polymer which is completely miscible with water is methyl vinyl ether (3). The vinyl ethers are generally converted to polymers via use of Friedel-Crafts type (or cationic) initiators. The most important commercial materials derived from vinyl ethers are their copolymers with maleic anhydride. These copolymers are hydrolyzable to the corresponding polycarboxylic acids.

Poly(vinyl alcohol) (PVA). Polyvinyl alcohol is the largest volume synthetic water-soluble polymer produced in the world. Vinyl alcohol does not exist to any substantial extent in the free state, since it is the enol tautomer of acetaldehyde. Consequently, PVA is commercially produced by hydrolysis of poly(vinyl acetate) (4).

Polymers of Ethylene Oxide. Polymeric derivatives of ethylene oxide are divided into classes which are defined by molecular weight. Low mol. wt. polymers of average mol. wts. less than 20,000 are generally defined as poly-(ethylene glycols), while polymers with av. mol. wt. up to five million are generally referred to as ethylene oxide polymers. The distinction is based on limitations on mol. imposed by the method of synthesis. All appear to be miscible with water at room temperature in all proportions (5).

Polyelectrolytes

Cationic Polymers

Quaternary Ammonium: The most important and extensively used cationic polymers are the quaternary ammonium polymers. The quaternary ammonium function is highly hydrophilic, and organic compounds containing this group are extensively hydrated.   The same is true for the polymers. Among the first water-soluble quaternary ammonium polymers was that shown in Equation I (6-7). This polymer, as the chloride salt, is now manufactured under the trade name "Cat-Floc" and was also the first such polymer to be approved by the Food and Drug Administration for use in potable waters. It is extensively used as a flocculating agent, but has many other uses.

(1)

The monomer, dimethyldiallylammonium chloride, known as DMDAC or DADMAC, can be synthesized by either of the methods shown in Equation 2, as well as by several other methods.   The second method is the most economical and is used commercially.   The monomer generally is not isolated, but is purified and polymerized as a 62-70% solution.

The mechanism for polymerization of a non-conjugated diene such as DADMAC is generally referred to as cyclopolymerization.   This process does not conform to the theory of Staudinger (8) which postulates that non-conjugated dienes will produce cross-linked polymers on polymerization.   Structure I shows two alternative structures for the polymer.   Structure Ia is predicted by the theory of Flory (9) that states that the more stable intermediate will predominate in vinyl polymerization, which in this case is the secondary radical and would lead to Structure Ib.   It has been shown (10) that Structure Ia is the almost exclusive structure obtained in polymerization of DADMAC which requires the less stable primary radical to be the predominant propagating species, contradictory to the theory of Flory.

Monomer Synthesis:

1).

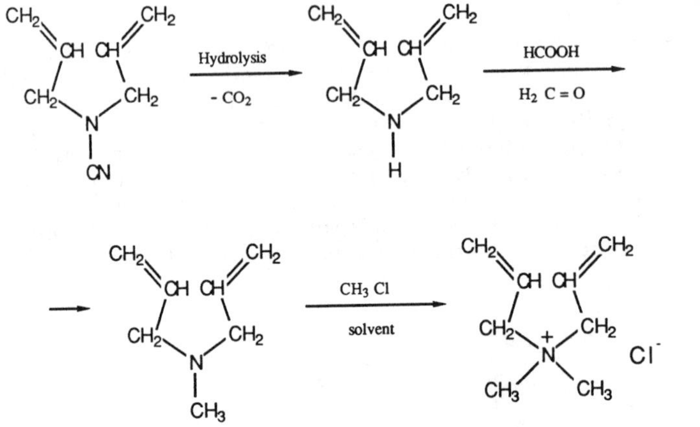

$$\text{Hydrolysis} \quad -CO_2 \quad\quad HCOOH \quad H_2C=O$$

$$CH_3\,Cl \quad \text{solvent} \qquad\qquad (2)$$

2).

$$(CH_3)_2NH \quad + \quad CH_2 = CHCH_2Cl \xrightarrow[\text{aq.}]{NaOH} CH_2 = CHCH_2N(CH_3)_2$$

$$\xrightarrow{CH_2 = CHCH_2Cl}$$

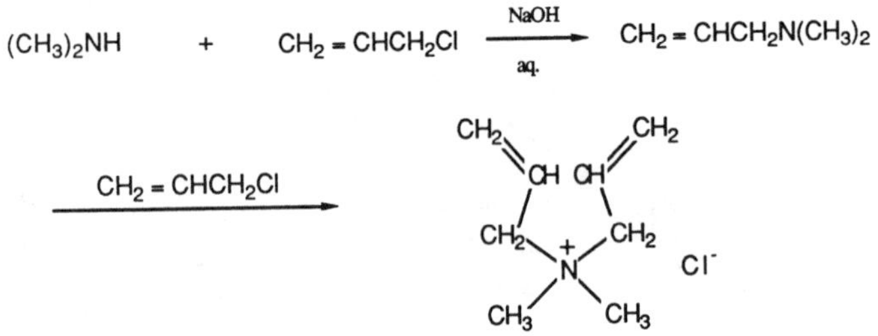

a   b

Str. 1

Figure 1 confirms that poly(DADMAC) contains essentially one nitrogen structure, and thus predominantly one ring size (11).

Poly(DADMAC) has been fractionated and its solution properties studied. Figure 2 shows the dependence of intrinsic viscosity of various fractions of the polymer upon temperature at a constant 0.5 ionic strength of the medium (12-13).

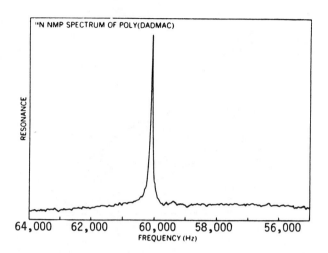

Figure 1. ^{14}N-NMR spectrum of poly(DADMAC) in water in the scan range 56000–63000 Hz at 25 °C. (Reprinted with permission from ref. 11. Copyright 1979 John Wiley & Sons, Inc.)

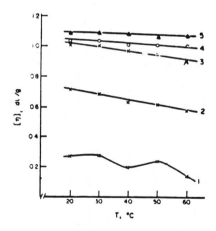

Figure 2. Dependence of intrinsic viscosity upon temperature and molecular weight for an ionic strength of 0.5. Curve 1, $\overline{M}_w = 0.0835 \times 10^6$; curve 2, $\overline{M}_w = 0.426 \times 10^6$; curve 3, $\overline{M}_w = 0.981 \times 10^6$; curve 4, $\overline{M}_w = 1.943 \times 10^6$; curve 5, $\overline{M}_w = 0.701 \times 10^6$. (Reprinted with permission from ref. 12. Copyright 1977 Pergamon.)

Figure 3 shows the dependence of intrinsic viscosity of a sample of constant molecular weight on ionic strengths of medium varying from 0.1 to 0.5. From Mark-Houwink plots, the following equations which relate intrinsic

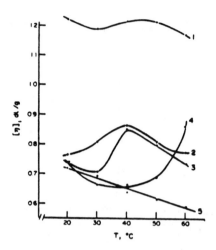

Figure 3. Dependence of intrinsic viscosity of a sample with $\overline{M}_w$= 0.0835 X 10⁶, on ionic strength (i): 1. i=0.1; 2. i=0.2; 3. i=0.3; 4. i=0.4; and 5, i=0.5. (Reprinted with permission from ref. 13. Copyright 1977 Pergamon.)

viscosity to weight average molecular weight of the polymer at 20°C for the various ionic strengths were obtained:

(0.1 M NaCl solution)
$$[\eta] = 3.98 \times 10^{-4} \, \overline{M}_w^{0.660}$$

(0.2 M NaCl solution)
$$[\eta] = 2.512 \times 10^{-4} \, \overline{M}_w^{0.638}$$

(0.3 M NaCl solution)
$$[\eta] = 2.510 \times 10^{-4} \, \overline{M}_w^{0.620}$$

(0.4 M NaCl solution)
$$[\eta] = 1.390 \times 10^{-4} \, \overline{M}_w^{0.542}$$

(0.5 M NaCl solution).
$$[\eta] = 3.982 \times 10^{-4} \, \overline{M}_w^{0.566}$$

Poly(DADMAC) can be fractionated and characterized by size exclusion chromatography (SEC) as shown in Figure 4. The column substrate in this case was a chloromethylated and quaternized (with trimethylamine) sample of the ion exchange resin, Amberlite 900 (14).

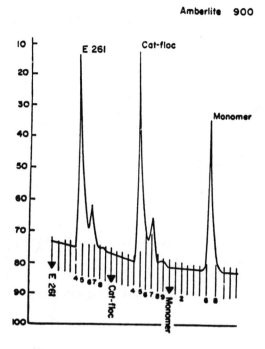

Figure 4. Fractionation of poly-(diallyldimethylammonium chloride) on chloromethylated and quaternized styrene-divinylbenzene gels. (Reprinted from ref. 14.)

The polymer can also be fractionated and characterized by SEC by use of a quaternized silicious substrate as shown in Figure 5 (15).

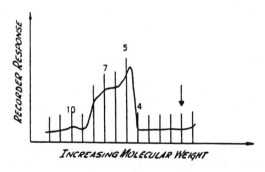

Figure 5. GPC fractionation of a high MW poly-(diallyldimethylammonium chloride) on a quaternized silicious support. (Reprinted from ref. 15.)

Diallylmethyl β-propionamido ammonium chloride monomer and homo-polymers have been prepared (16). The polymers as shown in Equation 3 were found to be useful as primary coagulants and in many flocculation applications. Copolymers with acrylamide have excellent flocculation properties (17).

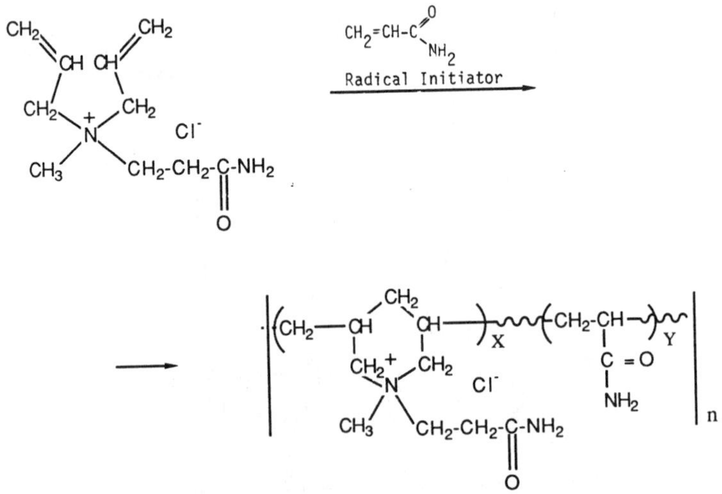

(3)

Structure 2 shows some readily polymerizable tertiary amino monomers as well as the corresponding quaternized monomers. The Dow monomers are probably a mixture of the m- and p-isomers. The three amine monomers polymerize most effectively in aqueous solution at the pH of the hydrochloride salt. The "Tsuruta" monomer can readily be prepared by addition of one mole of dimethyl amine to p- divinylbenzene.

Str. 2

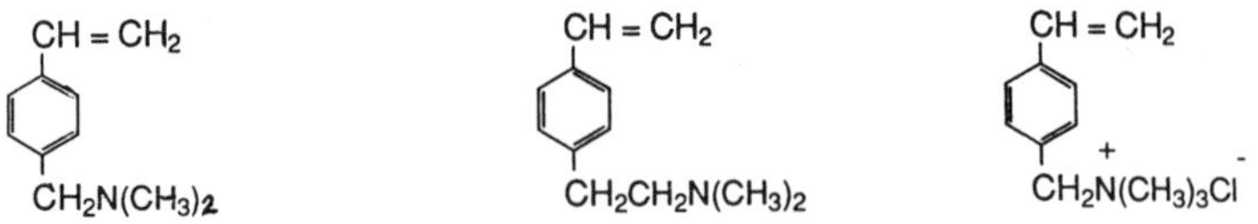

| p-Dimethylaminomethyl | p-Dimethylaminoethyl | Trimethyl(p-Vinylbenzyl) |
| Styrene (Dow) | Styrene (Tsuruta) | Ammonium Chloride (Dow) |

Structure 3 shows types of monomers available by the Mannich reaction of acrylamide with formaldehyde and dimethylamine. The methylol monomer is useful as a post-cross-linking site. The corresponding polymer or copolymer when treated with acid converts methylol containing chains to the methylene-bis-acrylamide cross-links (18).

$$CH_2 = CHC = O$$
$$NHCH_2N(CH_3)_2$$

$$CH_2 = CHC = O$$
$$NHCH_2OH$$

Str. 3

Dimethylaminomethyl

Acrylamide

N-Methylolacrylamide

(Available Commercially)

Structure 4 shows some tertiary amino acrylate monomers which are available (19-20). However, the ester linkage is not sufficiently stable to hydrolysis to make this monomer too attractive. The amide monomers are more stable and are extensively used. All can be quaternized.

$$CH_2 = CHC = O$$
$$OCH_2CH_2N(CH_3)_2$$

β-Dimethylaminoethyl

Acrylamide

Str. 4

$$CH_2 = CHC = O$$
$$NHCH_2CH_2CH_2N(CH_3)_2$$

γ -Dimethylaminopropyl

Acrylamide

$$CH_2 = CHC = O$$
$$NHCH_2CH_2N(CH_3)_2$$

β –Dimethylaminoethyl

Acrylamide

Structures 5 and 6 show some interesting tertiary amino monomers which are synthesized via the Ritter reaction, first developed in 1951 (21). Diacetone acrylamide can be synthesized from the relatively inexpensive raw materials, acetone and acrylonitrile (22-29).

$$CH_2 = CHC = O$$
$$NHC(CH_3)_2CH_2COCH_3$$

Str. 5

Diacetone Acrylamide

$$CH_2 = CHC = O \quad CH_3$$
$$NH-----C--CH_2CH_2N(CH_3)_2$$
$$CH_3$$

Str. 6

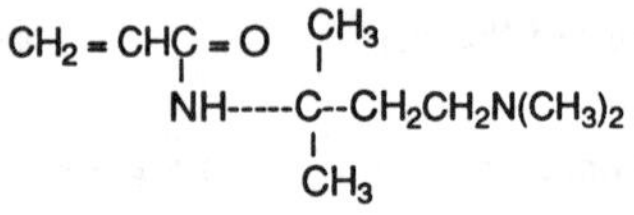

N-(1,1-Dimethyl-3-dimethylamino-

propyl) Acrylamide

Cationic quaternary ammonium monomers, as shown in Structures 7-9, and their polymers have attained considerable importance because of their exceptionally wide range of applications. The N-substituted acrylamides offer significant advantages over the N-alkyl acrylate esters because of their increased reactivity and hydrolylic stability.

Cationic (Quaternary) Monomers

Str. 7

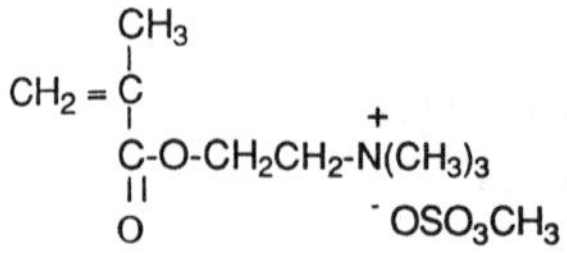

$$CH_3$$
$$CH_2 = C$$
$$C-O-CH_2CH_2-N(CH_3)_3$$
$$O \qquad \text{-}OSO_3CH_3$$

2-Methacryloyloxy Ethyl

Trimethyl Ammonium

Methosulfate

METAMS

Str. 8

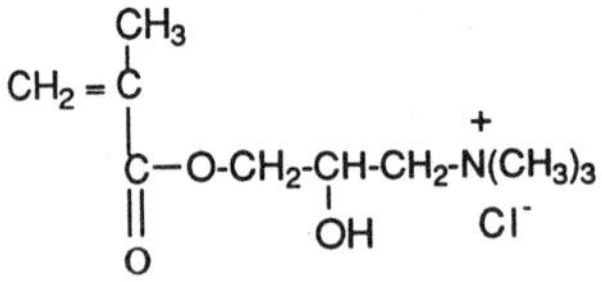

$$CH_3$$
$$CH_2 = C$$
$$C-O-CH_2-CH-CH_2-N(CH_3)_3$$
$$O \qquad OH \quad Cl^-$$

3-Methacryloyloxy-

2-hydroxy propyl

Trimethyl Ammonium Chloride

G-MAC

Str. 9

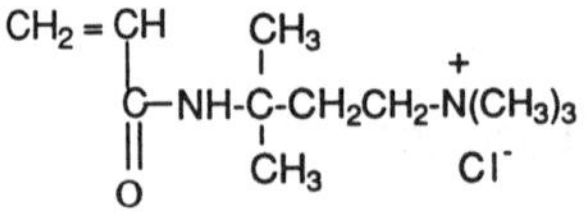

$$CH_2 = CH \quad CH_3$$
$$C-NH-C-CH_2CH_2-N(CH_3)_3$$
$$O \quad CH_3 \quad Cl^-$$

3-Acrylamido-3-methyl

Butyl Trimethyl Ammonium

Chloride

AMBTAC

The N-(dialkylaminoalkyl)acrylamides have been known for many years; however, the only convenient synthesis of these materials, the reaction of a suitable N,N-dialkyldiamine with acryloyl chloride, is expensive. The synthesis of AMBTAC is shown in Equation 4.

$$(CH_3)_2NH \ + \ CH_2 = \overset{\overset{\displaystyle CH_3}{|}}{C} - CH = CH_2 \ \xrightarrow{[Na]} \ (CH_3)_2N - CH_2 - CH = \overset{\overset{\displaystyle CH_3}{|}}{\underset{\underset{\displaystyle CH_3}{|}}{C}}$$

Dimethylammination of Isoprene   75% $(4\text{-}1)$

$$\underset{(4\text{-}1)}{(CH_3)_2N - CH_2 - CH = \overset{\overset{\displaystyle CH_3}{|}}{\underset{\underset{\displaystyle CH_3}{|}}{C}}} \ + \ CH_2 = \overset{\underset{\displaystyle CN}{|}}{CH} \ \xrightarrow[\text{Heat}]{H_2SO_4/H_2O} \ CH_2 = \overset{|}{CH}$$

Ritter Reaction on Amino Olefin

$$\underset{\underset{\displaystyle O}{\|}}{C}-NH-\overset{\overset{\displaystyle CH_3}{|}}{\underset{\underset{\displaystyle CH_3}{|}}{C}}-CH_2CH_2-N(CH_3)_2 \quad (4)$$

$(4\text{-}2)$

$$\xrightarrow{CH_3Cl} \quad CH_2 = \overset{|}{CH}$$

$$\underset{\underset{\displaystyle O}{\|}}{C}-NH-\overset{\overset{\displaystyle CH_3}{|}}{\underset{\underset{\displaystyle CH_3}{|}}{C}}-CH_2CH_2-\overset{+}{N}(CH_3)_3 \quad Cl^- \quad (4\text{-}3)$$

AMBTAC

3-Acrylamido-3-methylbutyltrimethylammonium chloride (AMBTAC) has been synthesized by a novel synthetic route which utilizes the sodium catalyzed reaction of dimethylamine with isoprene, to give N,N-3-trimethyl-2-butenylamine) and other by-products. The reaction mixture consists of about 75% of N,N-3-trimethyl-2-butenylamine (4-1). The Ritter reaction with acrylonitrile and sulfuric acid yield 85-90% of N-(1,1-dimethyl-3-dimethyl-aminopropyl) acrylamide (4-2) (30-31).

This amine (4-2) can be homo- or copolymerized and then subsequently quaternized or quaternized directly to give 3-acrylamido-3-methylbutyl trimethylammonium chloride (AMBTAC) (4-3).

Polyquaternary ammonium compounds with the ammonium ion integral to the backbone of the polymer chain, have been known for about 40 years since the work of Marvel (32-36). Other researchers (37-41) have used the Menshutkin reaction to form linear, unsymmetrical polyquaternary ammonium compounds by reacting ditertiary amines with dihalides.

The generic name "Ionenes" was proposed (42) for this class of compounds. This term has been widely accepted and is generally used to describe this type of polymer. They can be derived either from reaction of A-A/B-B monomers (ditertiary amines and dihalides) or by the reaction of an A-B monomer (aminoalkyl halide) with itself to form the linear polyquaternary ammonium compounds.

Investigators in this field have elucidated the mechanism (43-58) involved in the step growth polymerizations of both the A-A/B-B reaction sequence and the A-B reaction to form ionenes of well defined character. The highest charge density and the highest molecular weight ionene polymer is the

3-3-ionene. This was first synthesized by Marvel from 3-dimethylamino-n-propyl bromide. Polymerization of dimethylamino propyl chloride in a concentrated aqueous solution (59) yields a reasonably high molecular weight polymer which has properties as a flocculant (Equation 5). Cyclic intermediates have been observed in mechanistic studies of these reactions (Equation 6).

3-Ionene

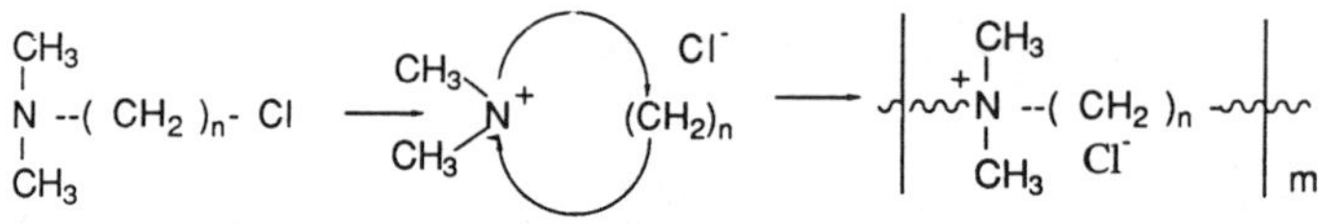

$$
\begin{array}{c}
CH_3 \\
| \\
N --(CH_2)_n\text{-} Cl \\
| \\
CH_3
\end{array}
\longrightarrow
\begin{array}{c}
CH_3 \\
\diagdown N^+ \\
CH_3 / \quad (CH_2)_n
\end{array} Cl^-
\longrightarrow
\left[
\begin{array}{c}
CH_3 \\
+ | \\
\sim\!\!\sim\!\!N --(CH_2)_n \sim\!\!\sim \\
| \quad Cl^- \\
CH_3
\end{array}
\right]_m
\qquad (5)
$$

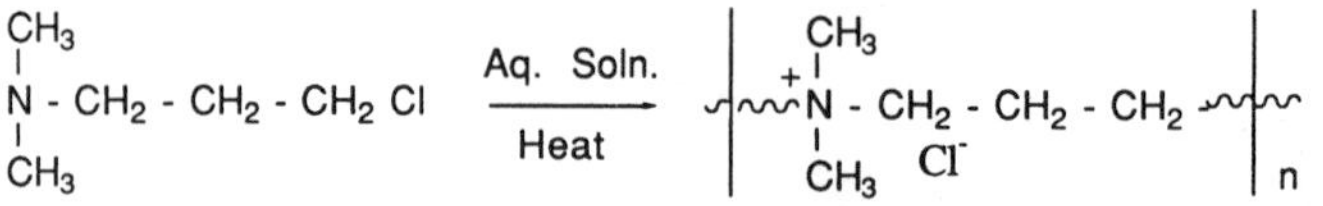

$$
\begin{array}{c}
CH_3 \\
| \\
N - CH_2 - CH_2 - CH_2 Cl \\
| \\
CH_3
\end{array}
\xrightarrow[\text{Heat}]{\text{Aq. Soln.}}
\left[
\begin{array}{c}
CH_3 \\
+ | \\
\sim\!\!\sim\!\!N - CH_2 - CH_2 - CH_2 \sim\!\!\sim \\
| \quad Cl^- \\
CH_3
\end{array}
\right]_n
\qquad (6)
$$

3-Chloropropyl   3-Ionene

Dimethylamine

A novel approach to quaternary ammonium polymers and copolymers is shown in Equation 7. This group in Japan pioneered in the alternating copolymers of DADMAC monomer with sulfur dioxide (60-61).

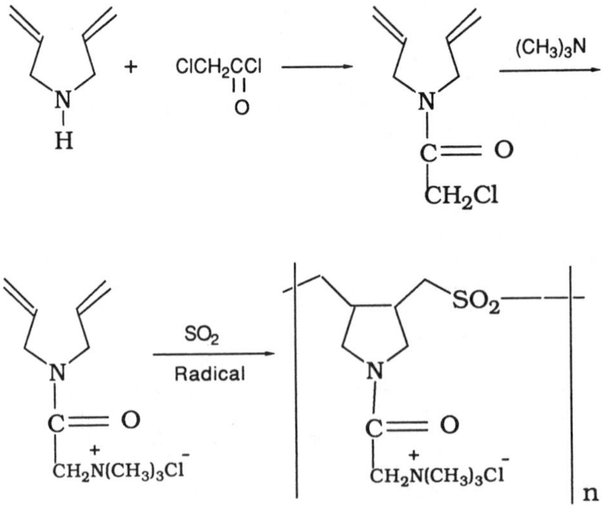

(7)

A novel monomer and its polymer are described in Equation 8. N-Hydroxysuccinimide has extremely useful properties as a "catalyst" for low temperature esterification reactions. The potential for the polymer described is obvious (62).

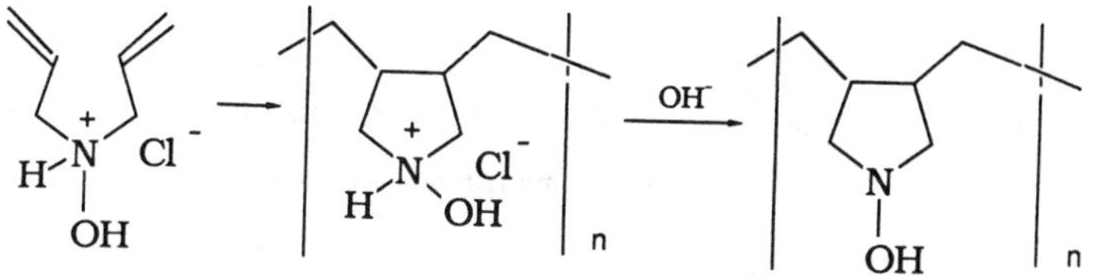

(8)

Equation 9 illustrates the synthesis of linear and crystalline polyethyleneimine. Polyethylenimine (PEI), produced by the ring opening cationic polymerization of ethylenimine monomer (aziridine), has a highly branched structure. The secondary amine groups in the main polymer chain cause the monomer to "graft" onto the already formed polymer. The preparation of linear PEI from aziridine monomer has yet to be accomplished. Synthesis of linear, crystalline polyethylenimine (9-2) has been accomplished (63-64) by an isomerization polymerization of 2-oxazoline, and alkaline hydrolysis of the intermediate, poly(N-formylethylenenimine) (9-1).

Linear Crystalline Polyethylenimine

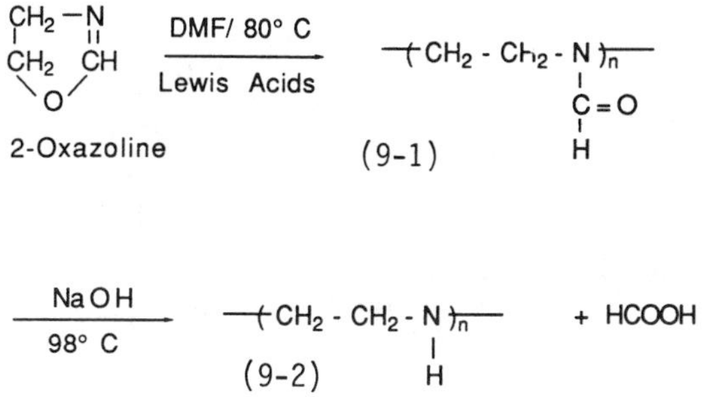

(9)

Previously, Fuhrmann (65) had prepared poly(N-acylethylenimines) from 2-substituted-2-oxazolines but did not report the hydrolysis of these materials to the free polyethylenimine. Continued work (66-69) has resulted in elucidation of the mechanism of 2-oxazoline polymerization.

By DSC analysis the melting point of this polyethylenimine (9-2) was shown to be 58.5°C and the glass transition point was -23.5°C. The polymer is insoluble in water at room temperature but soluble in hot water. This solubility behavior is quite different from that of the conventional polyethylenimine which is readily soluble in cold water.

Extensive investigations on the poly(2- and 4-vinylpyridinium salts) have been carried out. Several monomeric quaternary ammonium salts are known to spontaneously polymerize in aqueous solutions when their concentration is above about 25% by weight.

Both 2- and 4-vinylpyridine have been reported (70-82) to undergo spontaneous polymerization either upon attempted quaternization with alkyl halides or protonation with mineral acids. Further discussion of the extensive mechanistic studies reported on this problem is beyond the scope of this summary (Equation 10-14).

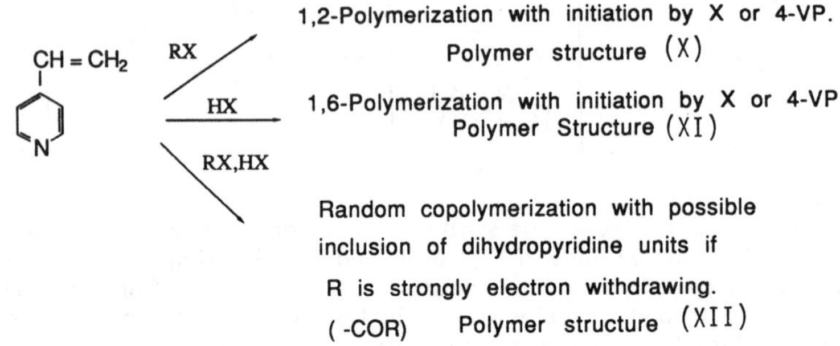

1,2-Polymerization with initiation by X or 4-VP. (10)

Polymer structure (X)

1,6-Polymerization with initiation by X or 4-VP
Polymer Structure (XI)   (11)

Random copolymerization with possible inclusion of dihydropyridine units if R is strongly electron withdrawing. (-COR)   Polymer structure (XII)   (12)

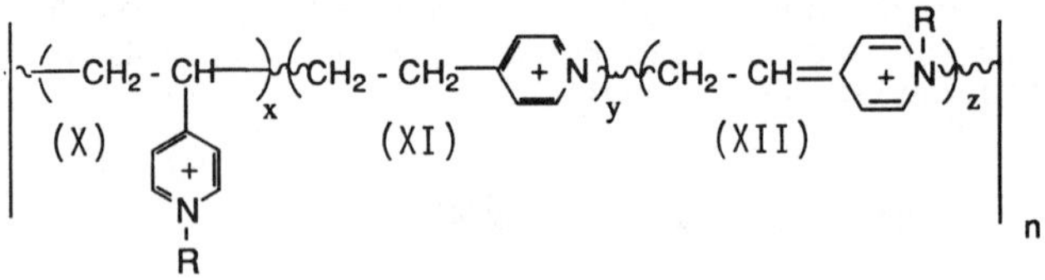

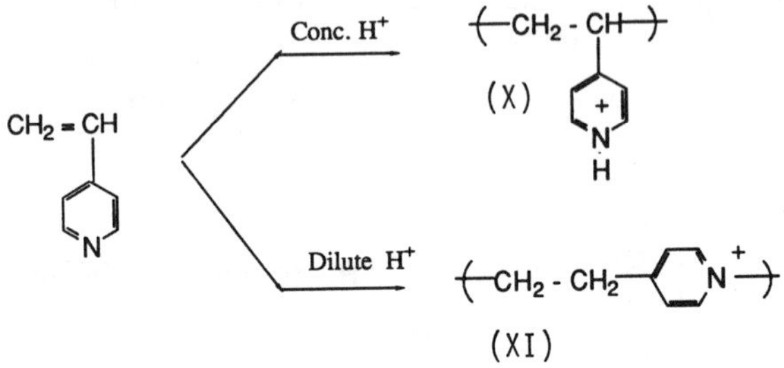

$$\cdot\sim\!\!\left(\!\!-CH_2-CH\!\!-\!\!\right)_x\!\!\left(\!CH_2-CH_2-\langle +N\rangle\!\!\right)_y\!\!\left(\!CH_2-CH=\langle +N\rangle\!\!\right)_z\!\!\sim$$

(X)   (XI)   (XII)   n

$$CH_2=CH \xrightarrow{\text{Conc. } H^+} \left(\!-CH_2-CH\!-\right) \quad (13)$$

(X)

$$CH_2=CH \xrightarrow{\text{Dilute } H^+} \left(\!-CH_2-CH_2-\langle N^+\rangle\!-\right)$$

(XI)   (14)

Equation 15 shows some recently claimed methacryloyl urea derivatives containing quaternary ammonium groups and their copolymers with acrylamide (83).

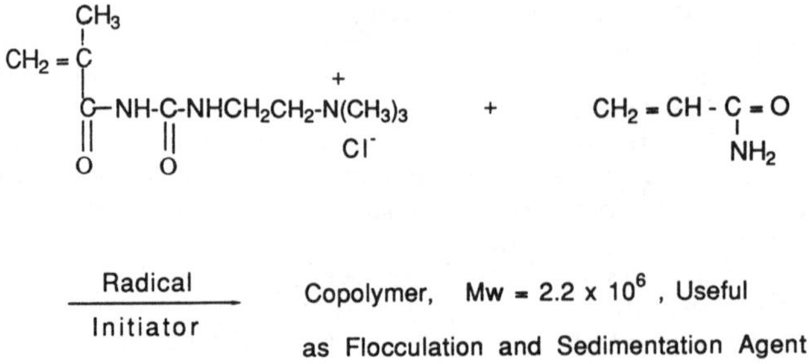

$$\begin{array}{c} CH_3 \\ | \\ CH_2 = C \\ | \\ C-NH-C-NHCH_2CH_2-N(CH_3)_3 \\ \| \quad \| \\ O \quad O \quad Cl^- \end{array} \quad + \quad \begin{array}{c} CH_2 = CH - C = O \\ | \\ NH_2 \end{array} \quad (15)$$

$$\xrightarrow[\text{Initiator}]{\text{Radical}}$$ Copolymer, Mw = 2.2 x 10^6 , Useful as Flocculation and Sedimentation Agent

A polymeric tertiary amine is available by cationic ring-opening polymerization of 1-azabicyclo[4.2.0]octane (the so-called condinine) (Equation 16). The synthesis and homopolymerization of this monomer was first published in 1960. Several authors studied the mechanism (Equation 16a,b) and kinetics of the polymerization (84-86).

$$RX + \quad \longrightarrow \quad X^- \tag{16a}$$

$$\tag{16b}$$

$$\Big| \quad CH_3 I$$

$$\tag{16c}$$

The initiation of the polymerization proceeds by an alkylation and the alkylation agent is fixed at the chain end. Therefore, using a polymeric initiator - for instance, a polystyrene with a bromoacetylester end group (Equation 17a) - a diblock copolymer is obtained (Equation 17b) (87) which can be quaternized (Equation 17c).

$$-[CH_2-CH-]_m-(CH_2)_2-O-C-CH_2-Br \quad + \tag{17a}$$
$$\Big| \quad 60°\ C$$

$$-[CH_2-CH-]_m-(CH_2)_2-O-C-CH_2-\tag{17b}$$
$$\Big| \quad CH_3 I$$

$$-[CH_2-CH-]_m-(CH_2)_2-O-C-CH_2-\tag{17c}$$

The second monomer shown as Structure 10 provides an excellent route to polyvinylamine. Many efforts have been made over the years to find an economical source of this polyamine of low repeat unit; however, most have been disappointing. N-Vinylacetamide is synthesized as follows (Equation 18) (88):

Str. 10 $CH_2 = CH - NH - \underset{\underset{CH_3}{|}}{C} = O$ N-Vinylacetamide

$$CH_3CH = O \ + \ 2\,CH_3 - \underset{\underset{NH_2}{|}}{C} = O \quad \xrightarrow{H_2SO_4} \quad CH_3CH(NH-\underset{\underset{O}{\|}}{C}-CH_3)_2$$

$$\xrightarrow[\text{30-40 Torr Celite}]{175\text{-}95°\text{ C}} \quad CH_2 = CH - NH - \underset{\underset{NH_2}{|}}{C} = O \quad \xrightarrow{AIBN} \quad \begin{array}{c} \text{Polymer} \\ \text{80\% from} \\ CH_3 - \underset{\underset{NH_2}{|}}{C} = O \end{array} \qquad (18)$$

$$\xrightarrow{\text{Hydrol.}} \quad \text{Polyvinylamine}$$

Sulfonium Monomers and Polymers: 2-Methylene-3-butenyl ammonium and sulfonium monomers and their polymers have been synthesized from 2-chloromethylbutadiene by reaction with a tertiary amine or sulfide, followed by polymerization (Equation 19) (89-91). Polymer modification as shown in Equation 20 can also lead to water-soluble (20-2) polymers.

Dialkyl-(2-Methylene-3-Butenyl) Sulfonium Chloride

$$CH_2 = \underset{\underset{CH_2 - Cl}{|}}{C} - CH = CH_2 \quad \xrightarrow{R_2 S} \quad \sim\!\!\left(\!\!\sim CH_2 - \underset{\underset{\underset{19\text{-}2}{\overset{+}{CH_2\text{-}SR_2}}}{|}}{C} = CH - CH_2 \sim\!\!\right)_n \overset{}{Cl^-} \qquad (19)$$

$$-\!\!\left(\, O\text{-}CH_2 - \underset{\underset{CH_2Cl}{|}}{CH} \,\right)_x \longrightarrow -\!\!\left(\, O\text{-}CH_2 - \underset{\underset{CH_2Cl}{|}}{CH} \,\right)_{x-y}\!\!\left(\, O\text{-}CH_2 - \underset{\underset{CH_2}{|}}{CH} \,\right)_y \qquad (20)$$

$$\text{20-1} \qquad\qquad\qquad \text{20-2} \quad \underset{\underset{CH_3}{\overset{}{CH_3}\diagdown\overset{|}{N}\diagup}}{\overset{+}{N}}\ Cl^-$$

The corresponding cationic polymers (19-2, or 20-2) could be obtained by either polymerization of the ammonium or sulfonium monomers or by reaction of poly(2-chloromethyl-1,3-butadiene) or polymer 20-1 with an appropriate tertiary amine or sulfide (92). These quaternary monomers can be homopolymerized or copolymerized for use as flocculants, dispersants, electroconductive paper coatings and wet strength additives for paper. The sulfonium polymers are water soluble and, upon heating or treatment with a base, tend to decompose and crosslink and become hydrophobic and water insoluble. Chiu (93) disclosed their use in this regard as durable antistatic coatings.

Water soluble monomers containing the sulfonium cation have been extensively studied. The hydrolytic instability of the sulfonium group, however, makes the polymeric materials unstable. This instability has been turned to good use in that these polymers can be applied, e.g., as coatings and the sulfonium function subsequently decomposed to yield hydrophobic and thus water resistant coatings. A new class of coatings has been developed on this basis (Equations 21 and 22) (94-96). A detailed discussion of these monomers and their polymerization was presented by Schmidt and co-workers (97).

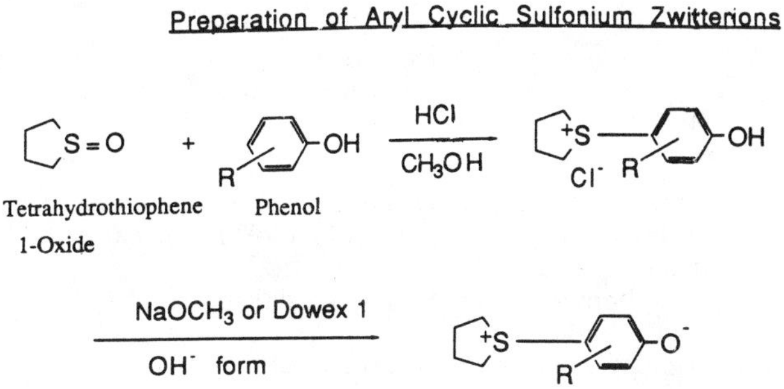

Preparation of Aryl Cyclic Sulfonium Zwitterions

Tetrahydrothiophene   Phenol
1-Oxide

(21)

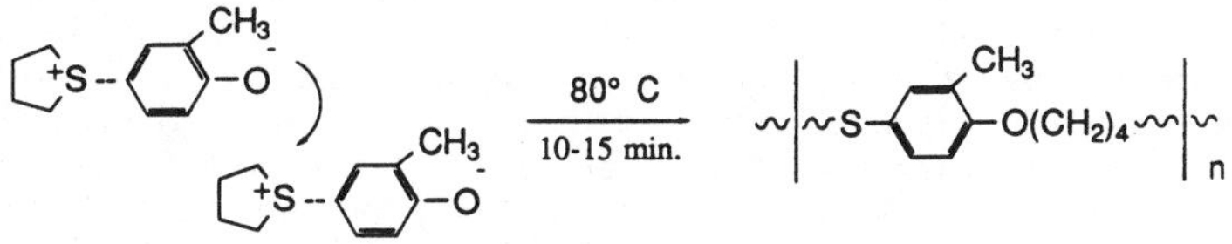

(22)

Polymerization of these cyclic sulfonium zwitterions proceeds by a novel mechanism involving both ring opening and a loss of charge to form nonionic polymers.

Phosphonium Monomers and Polymers: A variety of phosphonium polymers were synthesized via cyclopolymerization of a variety of diallylphosphonium salts (98). The viscosity data (Figure 6) for the poly-phosphonium

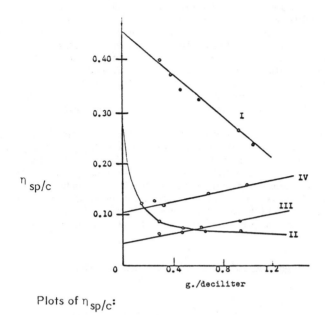

Plots of $\eta_{sp/c}$:

I. Poly-(diallyldiphenylphosphonium bromide) in 95% ethanol.
II. Poly-(diallyldiphenylphosphonium chloride) in a 1.0 M 1:1 ethanol-water solution of potassium chloride.
III. Poly-(diallyldiphenylphosphonium bromide) in 1:1 ethanol-water 0.1 M in potassium bromide.
IV. Poly-(diallylphenylphosphine oxide) in 95% ethanol.

Figure 6. Intrinsic viscosity determinations of phosphonium polymers. (Reprinted from ref. 98. Copyright 1969 American Chemical Society.)

salts were typical of the data for polyelectrolytes in general. The viscosity curves ($\eta_{sp/c}$ vs. c) of polyelectrolytes are strongly concave upward, in contrast to the behavior of uncharged linear polymers. This is due to the dissociation of the ionic bond in the solution which leads to large repulsive forces between the positively charged groups remaining on the chain. These forces give rise to greatly expanded configurations and very large intrinsic viscosities. At high concentrations, the molecules in the polyelectrolyte are not significantly ionized and tend to partially overlap; thus, they are not appreciably expanded. As the solution is diluted, the molecules no longer fill all the space and some of the halide ions leave the region of the chain. This causes a development of charge and extension of the chains. The addition of a strong electrolyte suppresses the loss of halide ion and the viscosity behavior becomes more normal. However, the intrinsic viscosity is dependent on the volume of the polymer in the solution and added electrolyte compresses the hydrodynamic unit corresponding to the polyelectrolyte. The compression changes the shape of the polymer and causes a marked decrease in the viscosity of the solution. Thus, the term intrinsic viscosity, for a polyelectrolyte can not be used in the same sense as for an uncharged polymer since it is dependent largely on the concentration of added salt. This type of behavior was exhibited by the poly-phosphonium compounds and is illustrated in Figure 6. Line I, represents poly-(diallyldiphenylphosphonium bromide) dissolved in ethanol, and as was expected, $\eta_{sp/c}$ tended to increase with dilution. Addition of a strong electrolyte, such as a potassium halide, provided a common ion effect which repressed the ionization of the polymer. At this time the viscosity behaved similar to a linear uncharged polymer, as shown by line III. Line II represents poly-diallyldiphenylphosphonium chloride rather than the bromide; but demonstrates an intermediate change between line I and III where insufficient common ion was present to completely repress ionization. Line IV represents the viscosity of the phosphine oxide, which was produced from the same sample of diallyldiphenylphosphonium bromide used for the determination of line III.

Oxonium Monomers and Polymers: Of the several possibilities of introducing cations into polymers, perhaps the pyrylium function has been given least attention. However, in order to use most organic photoconductors as the photosensitive layer in electrophotography and electrographic copying processes, it is necessary to increase their sensitivity (e.g., the speed of electrical response) and to extend their photosensitive region to the longer wavelength end of the visible spectrum. To meet these requirements many sensitizers have been investigated and typical materials found useful in this regard are triphenylmethane dyestuffs such as crystal violet and rhodamine, pyrylium salts, benzopyrylium salts, and electron acceptor compounds such as anthraquinone, chloranil and TCNQ (99-101).

Interest in polymers containing this function followed naturally. A synthesis (102) of vinylpyrylium salts and polymeric benzopyrylium salts for applications in both photosensitizers and photopolymerizable monomers is now available (Equation 23).

(23)

Anionic Polymers.

Polysulfonic Acids and Their Salts: We now turn our attention to some acidic monomers capable of polymerizing to anion containing polymers rather than cation containing polymers as are derived from the quaternary ammonium monomers. Four such monomers are shown as Structures 11-14 (103-116). AMPSA is derived via the Ritter reaction as illustrated in Equation 24.

Str. 11

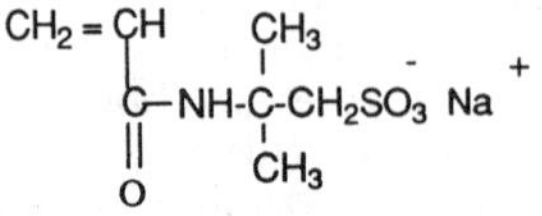

$$CH_2=CH \quad CH_3$$
$$C-NH-C-CH_2SO_3^- \ Na^+ \qquad [AMPSA]$$
$$\| \quad CH_3$$
$$O$$

Sodium 2-Acrylamido-2-methylpropane Sulfonate

2-Acrylamido-2-methyl Propane Sulfonic Acid

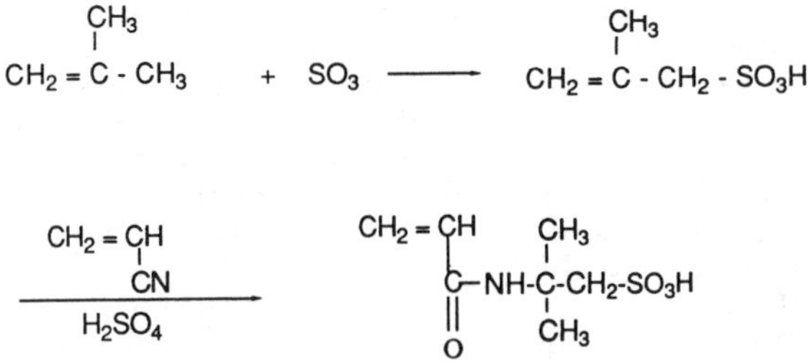

$$CH_3 \qquad\qquad\qquad CH_3$$
$$CH_2=C-CH_3 \ + \ SO_3 \longrightarrow CH_2=C-CH_2-SO_3H \qquad (24)$$

$$CH_2=CH \qquad CH_2=CH \quad CH_3$$
$$\xrightarrow[\ H_2SO_4\ ]{CN} \qquad C-NH-C-CH_2-SO_3H$$
$$\| \quad CH_3$$
$$O$$

(AMPSA)

Extensive development work has been done (105-106) on 2-sulfoethyl methacrylate (Structure 12) which has proven to be of limited commercial value due to the hydrolytic instability of the ester linkage. This same hydrolytic instability has been a serious problem with 3-sulfo-2-hydroxypropyl methacrylate reported by Schaper (107) (Structure 13).

Str. 12

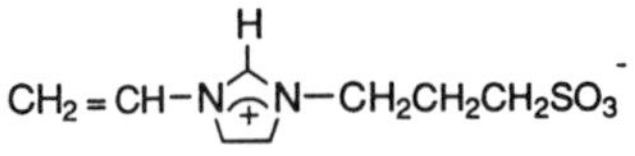

$$CH_2=CH-N\overset{H}{\underset{\oplus}{N}}-CH_2CH_2CH_2SO_3^-$$

Vinyl Imidazolium Sulfobetaine

Str. 13

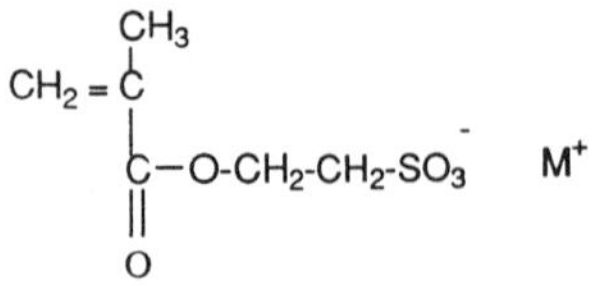

$$CH_3$$
$$CH_2=C \qquad\qquad\qquad \text{2-Methacryloyloxy Ethyl}$$
$$C-O-CH_2-CH_2-SO_3^- \quad M^+ \qquad \text{Sulfonate}$$
$$\|$$
$$O \qquad\qquad\qquad\qquad (SEM)$$

AMPSA was reported by Murfin and Miller (108-109) who prepared this monomer by the reaction of SO_3 with isobutylene followed by the Ritter reaction with acrylonitrile to yield AMPSA (Structure 14, Equation 24).

Str. 14

$$CH_2 = \overset{\overset{\displaystyle CH_3}{|}}{\underset{\underset{\displaystyle O}{\|}}{C}} \\ C-O-CH_2-CH-CH_2-SO_3^- \;\; M^+ \\ \qquad\quad \underset{\displaystyle OH}{|}$$

3-Methacryloyloxy-
2-hydroxypropyl
Sulfonate

(SHPM)

AMPSA is highly reactive in both homo- and copolymerizations and can be incorporated into many polymer systems either by homogeneous, solution, or emulsion polymerization techniques. It is used in several industrial polymers and copolymers which have many applications, including improving the affinity for basic dyes in acrylonitrile fibers (110), improving emulsion stability (111), flocculants (112), improving dry strength in paper (113), sludge dispersant in boiler water treatment (114), silt control in cooling water systems (115), and secondary oil recovery (116). AMPSA homopolymers form excellent polysalts with polycations.

Polyphosphonic Acids and Their Salts: Equation 25 shows an example of polymer modification to yield ion containing products. It has been shown (117-118) that bubbling oxygen through a solution of polyethylene in phosphorus trichloride, followed by hydrolyzing the product leads to a series of polymers containing pendant phosphonic acid groups. The original work indicated that solubility in aqueous sodium hydroxide solutions could be attained at only 12 molar % of phosphonic acid substitution.

Phosphonated Polyethylene

$$\sim\!\!\!(-CH_2 - CH_2 -)_n\!\!\!\sim \quad \xrightarrow[\;2)\; H_2O\;]{1)\; O_2/PCl_3} \quad \left| \sim\!\!\cdot(-CH_2 - \underset{\underset{\underset{\displaystyle O^-\;\;^-O}{\diagdown}}{\overset{\displaystyle P=O}{|}}}{CH} -)_x\!\!\!\sim\!\!\!(-CH_2 - CH_2 -)_y \right|_n \cdot \tag{25}$$

The mechanical and dielectric relaxation behavior as well as thermal characteristics of a series of ethylene-phosphonic acid copolymers have been reported (119-120).

The use of the Mannich reaction in modification of acrylamide for commercial production of useful polymers has been described earlier (Structure 3). A similar reaction is utilized in the second case of Equation 26 to produce novel phosphonic acid polymers (121).

$$\underset{\underset{\displaystyle H}{|}}{N}\diagup\!\!\!\diagup \quad + \; HCHO \; + \; H_3PO_3 \quad \longrightarrow \quad \underset{\underset{\underset{\displaystyle P(O)(OH)_2}{|}}{CH_2}}{N}\diagup\!\!\!\diagup \quad \xrightarrow{\text{Radical}} \quad \text{Polymer} \tag{26}$$

98%

Some problems associated with the exceptionally high molecular weight polymers necessary for satisfacory flocculation characteristics are high viscosities at low concentrations, and times required for dissolution of the solid polymers. Both of these problems can be overcome by synthesis as water-in-oil emulsions followed by inversion and rapid dissolution of the polymer. The details of this procedure are beyond the scope of this review;

however, a few of the many references dealing with this topic are given (122-125).

Amphoteric Polymers.

   Quaternary Ammonium-Carboxylates: Polyampholytes have been synthesized (126) by terpolymerizing dimethyldiallylammonium chloride with acrylamide and acrylic acid. The polymers were used as dry strength resins in paper (Structure 15) (127).

Str. 15

where x is 0.1 - 40 mole %

   Amineimides: Structure 16 illustrates a novel monomer type which imparts interesting properties to polymers. Structure 16 is an amphoteric monomer, a derivative of hydrazine, and has been discussed in a review of water-soluble polymers (103).

Str. 16

$$CH_2 = CH - C = O$$

$$N^- - N^+(CH_3)_3$$

Amineimides and Ammonium Imides

   The aminimides represent a highly functional class of organic compounds which have been exploited for their unique properties in polymers. An aminimide moiety is both a di-polar ion and an isocyanate precursor. The di-polar ion characteristics contribute toward their water solubility, and as isocyanate precursors, they offer many opportunities for incorporation into polymers where this serves as a latent crosslinking or branching site and can provide a desirable route to adhesion.
   A convenient synthesis of the monomeric methacrylimides was first disclosed by Slagel (128) (Equation 27). A variety of aminimide monomers have been prepared and their polymerization characteristics reported. Culbertson and Slagel (129-131) first reported the polymerization of trimethylamine-4-vinylbenzimide. Culbertson, Sedor and Slagel (132) also reported the preparation and polymerization of 1,1-dimethyl-1-(2-hydroxypropyl) amine methacrylimide (I) and other derivatives. An excellent review of the chemistry of aminimides was recently prepared by McKillip, et al. (133).
   Culbertson, et al. (134-136) have shown that the hydrochloride salt (the hydrazinium form) of 1,1-dimethyl-1-(2-hydroxypropyl) amine methacrylimide more readily copolymerizes with styrene to produce soluble polymers containing pendent quaternary ammonium groups. These polymers can then be treated with base to provide modified polystyrenes containing pendent aminimide residues. The latter polymers can then be thermolyzed in solution or in a solid phase to produce modified polystyrenes containing pendent isocyanate functions. If the thermolysis is carried out in the presence of isocyanate reactive moieties, high molecular weight crosslinked polymers can be synthesized. Several methacrylimide monomers find valuable industrial uses (137).

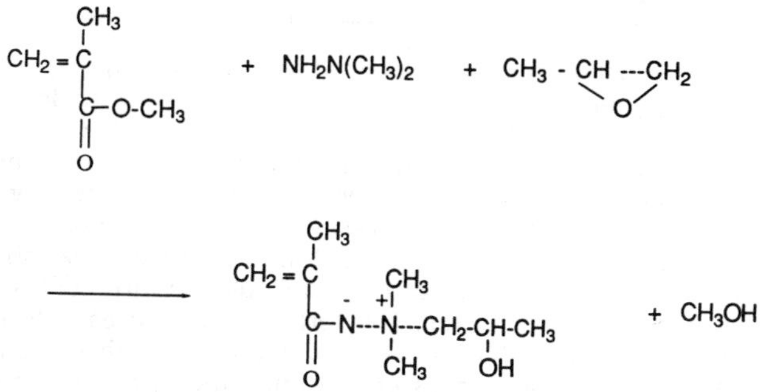

$$
\begin{array}{c}
\underset{\underset{\underset{O}{\parallel}}{\overset{CH_3}{\underset{|}{C}}}}{CH_2 = C} \\
\underset{C-O-CH_3}{}
\end{array}
\quad + \quad NH_2N(CH_3)_2 \quad + \quad CH_3 - CH\cdots CH_2 \atop \diagdown O \diagup
$$

(27)

$$
\xrightarrow{\hspace{2cm}}
\quad
CH_2 = \underset{\underset{O}{\parallel}}{\overset{CH_3}{\underset{|}{C}}} \quad
\underset{C-N\cdots N\cdots CH_2\text{-}CH\text{-}CH_3}{\overset{CH_3 \atop +|}{}} \quad + \quad CH_3OH
$$

These monomers are generally used in copolymer systems as isocyanate precursors since they thermolyze to the isocyanate group upon heating. They are very effective in crosslinking and bonding reactions for use in polymeric coatings on metal, rubber (such as for tire cord adhesives) and textiles (such as in shrink-proofing of wool) (138).

Regular Polyampholytes: Regular polyampholytes are polymers which contain equal concentrations of both basic and acidic functional groups. There are many reports in the literature on the copolymerization of acid-containing monomers with base-containing monomers, e.g., acrylic acid and vinylpyridine. However, recent studies have shown that some of the original work in this field did not yield polyampholytes at all under the experimental conditions used, but rather homopolymers of one of the ionic monomers with the other present as the counterion. Although many interesting methods (139-142) have been used to synthesize regular polyampholytes, one of the most interesting involves cyclopolymerization of the monomer shown in Equation 28 (143).

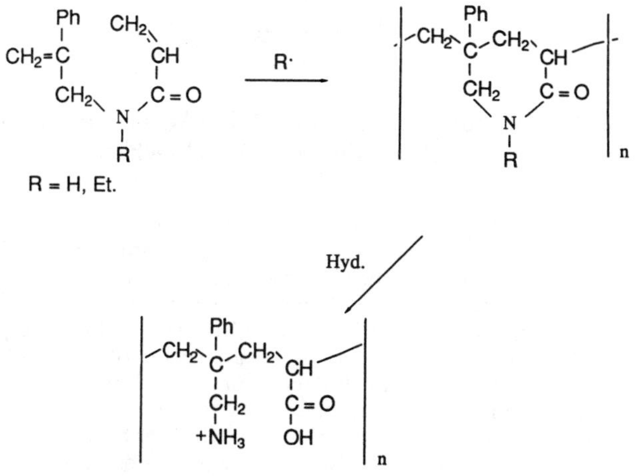

(28)

Some Applications of Water-Soluble Polymers

Flocculation: Flocculation of suspended solids from aqueous solutions represents an important area of application of polyelectrolytes. In a laboratory method used to measure the flocculation characteristics of polymers based on that developed by Vilaret (144) a specially prepared kaolimite clay suspension was treated with varying concentrations of polymer, and measurements were made of the equilibrium residual turbidity. Figure 7 shows the results obtained with two polymers of different intrinsic viscosity (145). Residual turbidity is plotted as per cent of initial turbidity (T/T_i X 100) against dosage in micrograms per liter. The turbidity decreases with increasing dosage, reaches a minimum, then increases because of redispersion caused by overtreatment. From these curves one obtains the optimum dosage for coagulation (in this system only) for a particular polymer sample. In the cases shown, the sample with an intrinsic viscosity of 1.36 dl/g has an optimum dosage of 60 µ g/liter and the sample with an intrinsic viscosity of 0.36 dl/g has an optimum dosage of 125 µ g/liter.

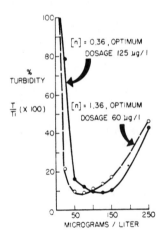

Figure 7. % Residual turbidity vs. poly-DMDAAC dosage. (Reprinted with permission from ref. 145. Copyright 1970 Marcel Dekker, Inc.)

Sludge Dewatering: A series of DMDAC-acrylamide copolymers were evaluated for sludge dewatering (145-150). The method of evaluating these polymers was the standard Buchner Funnel Vacuum Filtration Test used in the sewage treatment field. Figure 8 shows the dewatering curves for the copolymers. On the left, volume of filtrate in mililiters is plotted against time in seconds. On the right are the various weight ratios of DMDAAC to acrylamide in the monomer feed. The bottom curve is that obtained with 100% polyacrylamide, which gave identical results to that obtained with no treatment. The third curve from the bottom is that obtained with 100% poly-DMDAAC. Several mixtures of 100% polyacrylamide and poly-DMDAAC were evaluated and in no case were the results as good as that obtained from 100% poly-DMDAAC alone. However, all of the copolymers having >10 mole % of DADMAC were more effective.

The dosage used to obtain these curves was the optimum for each sample and varied from 250 to 350 ppm. The dosage for the top three curves was 250 ppm, the next two 300, and the rest 350. The curves clearly demonstrate that there is an optimum charge density for a polyelectrolyte used in dewatering digested sewage sludge.

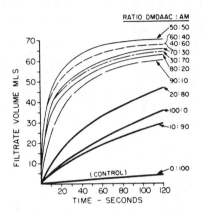

Figure 8. Sewage sludge dewatering curves. (Reprinted with permission from ref. 145. Copyright 1970 Marcel Dekker, Inc.)

Sedimentation of Suspended Polymers: Goodman, et al. (151-152) and others (153), on the basis of a plant-scale study, concluded that the application of a cationic copolymer to the biological oxidation process at a rate of 0.1 lb./ton of secondary effluent dried suspended solids increased the overall BOD removal efficiency of the plant to 95% and decreased the loss of suspended solids in secondary effluent by 99%. At the beginning of this study (153-154) (Figure 9), the secondary effluent suspended solids were 700 mg/l.

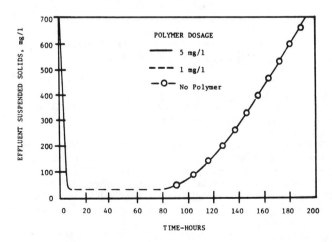

Figure 9. Effect of a cationic polyelectrolyte on secondary sedimentation basin performance. (Reprinted with permission from ref. 150. Copyright 1973 Plenum Press.)

Initially, the copolymer was applied at a treatment dosage of 5 mg/l. Due to the increased capture of suspended solids in the secondary sedimentation basin, the polymer treatment was reduced to 1.0 mg/l and was fed for another 60 hrs. At this treatment dosage, the suspended solids in the secondary effluent remained at 25 mg/l. When the polymer treatment study was stopped at 100 hours, the suspended solids in the secondary effluent gradually increased to 700 mg/l in 90 hours.

     Vacuum Filtration: Figure 10 illustrates the vacuum filtration of primary sludges. The ferric chloride and lime treatment dosage required to dewater primary sludge is between 3% and 20% (60-100 lb/ton) which results in filtration production rates from a low of 3 lb/ft^2/hr to a high of 7 lb/ft^2/hr. The required cationic polyelectrolyte dosage to dewater primary sludge is between 0.2% to 1/2% (4-24 lb/ton), which results in filtration production rates between 6 and 20 lb/ft^2/hr. (150,154-156).

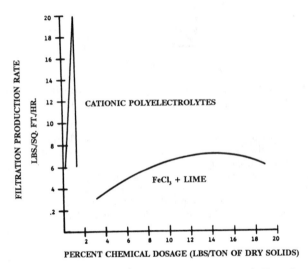

Figure 10. Vacuum filtration of primary sludges. (Reprinted with permission from ref. 150. Copyright 1973 Plenum Press.)

   Electroconductivity in Electrophotography. The technology associated with electrophotography began with the work of Carlson (157) who developed xerography by utilizing the physics and chemistry of photoconductivity, conductivity, and electrostatics.

   An alternative process, developed at RCA (158), involves phenomena similar to xerography, except they all occur on a sheet of specially coated paper. One of the key developments related to this process is the development of commercial electroconductive polymers by several companies. Poly(DADMAC) has been one of the most widely used electroconductive polymers for this purpose. The details of the process are beyond the scope of this review; however, since the conductive polymers conduct via ion transport rather than by an electronic mechanism, the nature of the counter ion affects conductivity. For a series of anions of poly(DADMAC), it was found (Table II) that, in general, the halides are the most conductive and that of these, chloride is the most effective (159). Fortunately, the chloride form of these polymers, typically, is the most economical to produce (160).

TABLE II. Surface Resistivity as a Function of Polymer Anion Form

Anion	Surface Resistivity at 20% RH 0.5 lb/3000 Ft^2 coat weight, ohms
F^-	7.4×10^{11}
Cl^-	1.1×10^{10}
Br^-	8.9×10^{10}
OH^-	3.6×10^{12}
NO_2^-	4.1×10^{10}
Ac^-	2.5×10^{10}
HSO_4^-	8.2×10^{12}
$H_2PO_4^-$	4.5×10^{13}

Figure 11 shows the relative effectiveness of various types of polymers for this application.

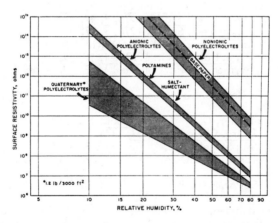

Figure 11. Surface resistivity vs. humidity for polyelectrolytes. (Reprinted with permission from ref. 160. Copyright 1968 Technical Association of the Pulp and Paper Industry.)

Literature Cited

1. Thomas, W. M.; Wang, D. W. Encyclo. Polym. Sci., and Eng., 1988, 1, 169-211; J. I. Kroschwitz, Exec. Ed.; J. Wiley and Sons, Inc.
2. Nemec, J. W.; Bauer, W., Jr. Encycl. Polym. Sci. & Eng., 1988, 1, 211-234; J. I. Kroschwitz, Exec. Ed.; J. Wiley and Sons, Inc.

3. Biswas, M.; Mazumdar, A.; Mitra, P. Encycl. Polym. Sci. and Eng.,
    1988, 17, 446-468; J. I. Kroschwitz, Exec. Ed.; J. Wiley and Sons, Inc.
4. Marten, F. L. Encyclo. Polym. Sci. and Eng., 1989, 17, 167-198; J. I.
    Kroschwitz, Exec. Ed., J. Wiley and Sons, Inc.
5. Clinton, N.; Matlock, P. Encycl. Polym. Sci. and Eng., 1989, 17, 225-
    273; J. I. Kroschwitz, Exec. Ed.; J. Wiley and Sons, Inc.
6. Butler, G. B.; Angelo, R. J. J. Am. Chem. Soc., 1957, 79, 3128.
7. Butler, G. B. U.S. Pat. 3, 288,770 (Nov. 28, 1966).
8. Staudinger, H.; Heuer, W. Ber., 1934, 57, 1169.
9. Flory, P. J. J. Am. Chem. Soc., 1937, 59, 241.
10. Lancaster, J. E.; Baccei, L.; Panzer, H. P. Polymer Letters, 1976, 14,
    549.
11. Bowman, L. M.; Cha, C. Y. J. of Polym. Sci.: Polym. Letts. Ed., 1979,
    17, 167-173.
12. Maxim, S.; Dumitriu, E.; Ioan, S.; Carpov, A. Euro. Polym. J., 1977,
    13, 105.
13. Ioan, S.; Dumitriu, E.; Maxim, S.; Carpov, A. Euro. Polym. J., 1977,
    13, 109.
14. Butler, G. B.; Wu, C. U.S. Pat. 3,982,206, June 8, 1976.
15. Talley, C. F.; Bradley, G. M.; Guliana, R. T. U.S. Pat. 4,118,316,
    Oct. 3, 1973.
16. Hoover, M. F.; Schaper, R. J.; Boothe, J. E. U.S. Pat. 3,412,019 (Nov.
    19, 1968) to Calgon Corp.
17. Schaper, R. J. Can. Pat. 820,379 (Aug. 12, 1969) to Calgon Corp.
18. Ger. Pat. 1,102,157; Makromol. Chem., 57, 27 (1962).
19. From -Propiolactone: U.S. 2,649,438; C.A. 47, 10863E (1953).
20. From Acrylolyl Chloride: U.S. 2,567,836; C.A. 45, 10669H (1951).
21. Plot, H.; Ritter, D. J. J. Am. Chem. Soc., 1951, 73, 4070.
22. Hoke, Donald I. Macromol. Synth., 1974, 5, 87.
23. Hoke, Donald I. Macromol. Synth., 1974, 5, 87.
24. Hoke, Donald I.; Surbey, Donald L.; Oviatt, William R. Macromol.
    Synth. 1976, 6, 95; J. Polym. Sci., 1972, A1, 10, 595.
25. Bork, John F.; Wyman, D. P.; Coleman, L. E. J. Appl. Polym. Sci.,
    1963, 7, 451.
26. Coleman, L. E.; Bork, J. F.; Wyman, D. P.; Hoke, D. I. J. Polym. Sci.,
    1965, A3, 1601.
27. Hoover, M. F.; Butler, G. B. J. Polym. Sci., 1974, C45, 1-37.
28. Slagel, R. C. J. Org. Chem., 1988, 33, 1374.
29. Culbertson, R. M.; Slagel, R. C. J. Polym. Sci., 1968, A1, 363.
30. Hoke, D .I.; Surbey, D. L.; Oviatt, W. R. J. Poly. Sci., Part A-1, 1972,
    10(2), 595-604.
31. Hoke, D. I., U.S. Pat. 3,666,810 (May 30, 1972), The Lubrizol Corp.
32. Littmann, E. R.; Marvel, C. S. JACS, 1930, 52, 287-293.
33. Gibbs, C. F.; Littmann, E. R.; Marvel, C. S. JACS 1933, 55, 573-7.
34. Lehman, M. R.; Thompson, C. D.; Marvel, C. S. JACS, 1933, 55,
    1977-81.
35. Gibbs, C. F.; Marvel, C. S. JACS 1934, 56, 725-27.
36. Gibbs, C. F.; Marvel, C. S. JACS, 1935, 57, 1137-39.
37. Kern, W.; Brenneisen, E. J. Prakt. Chem., 1941, 159, 194-217.
38. Ritter, D. M. U.S. Pat. 2,261,002 (Oct. 28, 1941).
39. Searle, N. E. U.S. Pat. 2,271,378 (Jan. 27, 1942).
40. Kirby, J. E. U.S. Pat. 2,375,853 (Mar. 15, 1945).
41. Kirby, J. E.; Lontz, J. F. U.S. Pat. 2,388,614 (Nov. 6, 1945).
42. Rembaum, A.; Baumgartner, W.; Eisenberg, A. J. Poly. Sci., 1968,
    B6(3), 159-171.

43. Casson, D.; Rembaum, A. Macromol., 1972, 5(1), 75-81.
44. Casson, D.; Rembaum, A. J. Poly. Sci. (Poly. Letters) 1970, 8, 773-83.
45. Eisenberg, A.; Tokoyama, T. Polymer Preprints, 1968, 9(1), 617-22.
46. Hadek, V.; Noguchi, H.; Rembaum, A. Macromol., 1971, 4(4), 494-499.
47. Noguchi, H.; Rembaum, A. J. Poly. Sci., 1969, B7(3), 383-394.
48. Noguchi, H.; Rembaum, A. Macromol., 1972, 5(3), 253-260.
49. Rembaum, A. J. Macromol. Sci.-Chem., 1969, A3(1), 87-99.
50. Rembaum, A.; Casson, D.; Yen, S. P. S. Polymer Preprints, 1972, 13(1), 521.
51. Rembaum, A.; Hermann, A. M. J. Poly. Sci., 1969, B7, 507-514.
52. Rembaum, A.; Hermann, A. M.; Stewart, F. E.; Gutmann, F. J. Phys. Chem., 1969, 73(3), 513-520.
53. Rembaum, A.; Noguchi, H. Macromol., 1972, 5(3), 261-269.
54. Rembaum, A.; Rile, H.; Samoano, R. J. Poly. Sci. (Polymer Letters), 1970, 8, 457-466.
55. Rembaum, A.; Singer, S.; Keyzer, J. J. Poly. Sci. (Polymer Letters), 1969, 7, 395-402.
56. Rembaum, A.; Yen, S. P. S.; Landel, R. F.; Shen, M. J. Macromol. Sci.-Chem., 1970, A4(3), 715-738.
57. Samoano, R.; Yen, S. P. S.; Rembaum, A. J. Poly. Sci. (Polymer Letters), 1970, 8, 467-479.
58. Yen, S. P.S.; Rembaum, A. J. Biomed. Mater. Res. Symp., 1970, 1, 83-97.
59. Yen, S. P. S.; Casson, D.; Rembaum, A. Polymer Preprints, 1972, 32(2), 293; Poly. Sci. & Tech., 2, "Water Soluble Polymers", 291, N. M. Bikales, Ed., Plenum Press (1973).
60. Harada, S.; Kato, T. Japan Kokai, Dec. 24, 1973, 73, 102,900; C.A. 1974, 80, 109091H.
61. Harada, S.; Kato, T. Japan Kokai, 1973, 73, 103,700; C.A. 1974, 81, 121391J.
62. Nagasawa, K.; Kuroiwa, K.; Narita, K. Japan, April 6, 1974, 74 14,351; C.A. 1975, 82, 5851A.
63. Saegusa, T.; Ikeda, H.; Fujii, H. Poly. Jour., 1972, 3(1), 35-39.
64. Saegusa, T.; Ikeda, H.; Fujii, H. Macromol., 1972, 5(1), 108.
65. Fuhrmann, R.; et al., U.S. Pat. 3,373,197 (March 12, 1968), Allied Chemical Corp.
66. Saegusa, T.; Ikeda, H.; Fujii, H. Poly. Jour., 1972, 3(2), 176-180.
67. Saegusa, T.; Ikeda, H.; Fujii, H. Poly. Jour., 1973, 4(1), 87-92.
68. Saegusa, T.; Ikeda, H.; Fujii, H. Macromol., 1972, 5(4), 59-62.
69. Saegusa, T.; Ikeda, H.; Fujii, H. Macromol., 1973, 6(3), 15-319.
70. Kabanov, V. A.; Aliev, K. V.; Patrikeeva, T. I.; Kargina, O. V.; Kargin, V. A. J. Poly. Sci., 1967, C16, 1079.
71. Kabanov, V. A.; Aliev, K. V.; Kargin, V. A. Polym. Sci. USSR, 1968, 10, 1873.
72. Salamone, J. C.; Snider, B.; Fitch, W. L. J. Poly. Sci. (Polymer Letters), 1971, 9, 13-17.
73. Salamone, J. C.; Ellis, E. J. ACS Preprints, 1972, 32(2), 294-299.
74. Salamone, J. C.; Snider, B.; Fitch, W. L. Macromol., 1970, 3(5), 707-9.
75. Salamone, J. C.; Snider, B.; Fitch, W. L. J. Poly. Sci., 1971, A9(1), 1493-1504.
76. Salamone, J. C.; Ellis, E. J.; Israel, S. C. J. Poly. Sci., 1972, 10, 605-613.

77. Kabanov, V. A.; Patrikeeva, T. I.; Kargina, O. V.; Kargin, V. A. J. Poly. Sci., 1968, C23, 357.
78. Kargina, O. V.; Ul'yanova, M. V.; Kabanov, V. A.; Kargin, V. A. Polym. Sci. USSR, 1967, 9, 380.
79. Salamone, J. C.; Snider, B.; Fitch, W. L.; Ellis, E. J.; Dholakia, P. L. Vol. I and II Macromol. Preprints, (XXIII), Inter. Cong. of Pure & Appl. Chem., Boston (July 1971).
80. Salamone, J. C.; Ellis, E. J.; Wilson, C. R.; Bardoliwalla, D. F. Macromol., 1973, 6(3), 475-6.
81. Mielke, I.; Ringsdorf, H. Helsinki IUPAC Symposium on Macromolecules (July 1972). (See also Ibid. J. Poly. Sci., C9, 1-12 (1971).
82. Moore, J. A.; Goldstein, J. J. Poly. Sci., 1972, A1(10), 2103-2113.
83. Kuester, E.; Dahmen, K.; Barthell, E. Ger. Offen. 2,857,432, July 3, 1980; C.A. 1980, 93, 168899U.
84. Lavagnino, E. R.; Chanvette, R. R.; Cannon, W. N.; Cornfeld, E. L. J. Am. Chem. Soc., 1960, 82, 2609.
85. Troy, M. S.; Price, C. C. J. Am. Chem. Soc., 1960, 82, 2613.
86. C.A. 110:213390d. Schulz, R. C. M.; Schwarsenbach, E.; Zoeller, J. Inst. Org. Chem., Ur D-6500 Mainz, Fed. Rep. Ger.; Makromol. Chem., Symp. 1989, 26, 221-31 (Eng.).
87. Schulz, R. C.; Muhlbach, K.; Perner, Th.; Ziegler, P. Polymer Preprints, 1986, 27(2), 25; C.A. 110:213390d, Schulz, R. C. M.; Schwarzenbach, E.; Zoeller, J. Inst. Org. Chem., Ur D-6500 Mainz, Fed. Rep. Ger.; Makromol. Chem., Symp. 1989, 26, 221-31 (Eng.).
88. Dawson, D. J.; Glass, R. P.; Wingard, R. E., Jr., J. Am. Chem. Soc. 1976, 98, 5996.
89. Jones, G. D.; Geyer, G. R.; Hatch, M. J. U.S. Pat. 3,494,965 (Feb. 10, 1970), The Dow Chemical Co.
90. Jones, G. D.; Geyer, G. R.; Hatch, M. J. U.S. Pat. 3,544,532 (Dec. 1, 1970), The Dow Chemical Co.
91. Jones, G. D.; Geyer, G. R.; Hatch, M. J. U.S. Pat. 3,673,164 (June 27, 1972), The Dow Chemical Co.
92. Jones, G. D.; Meyer, W. C.; Tefertiller, N. B.; MacWilliams, D. C. J. Poly. Sci., Part A-I, 1970, 8, 2123-2138.
93. Chiu, T. T. U.S. Pat. 3,726,821 (April 10, 1973), The Dow Chemical Co.
94. Hatch, M. J.; Yoshimine, M.; Schmidt, D. L.; Smith, H. B. JACS, 1971, 93, 4617.
95. Hatch, M. J.; et al., U.S. Pat 3,636,052 (Jan. 18, 1972), The Dow Chemical Co.
96. Hatch, M. J.; et al., U.S. Pat. 3,660,431 (May 2, 1972), The Dow Chemical Co.
97. Schmidt, D. L.; Smith, H. B.; Yoshimine, M.; Hatch, M. J. J. Poly. Sci., (Poly. Chem. Ed.), 1972, 10, 2951-2966.
98. Butler, G. B.; Skinner, D. L.; Bond, W. C., Jr.; Rogers, C. L. Polym. Prepr., 1969, 10(2), 923.
99. Morimoto, K.; Murakami, Y. Appl. Optics, Proceedings of International Conference on Electrophotography, pp. 50-54 (Sept. 6, 1968).
100. Murakami, Y.; Morimoto, K. Japan 69 07, 351, March 31, 1969; C.A. 71:125473d.
101. Strzelecki, L. C. R. Acad. Sci., Paris, Ser. C, 265(20), 1094-6 (1967); C.A. 68:40155q.
102. Petropoulos, C. C. J. Poly. Sci., 1972, A10(1), 957-974.
103. Hoover, M. F.; Butler, G. B. J. Polym. Sci., 1974, C45, 1-37.

104. Overberger, C. G.; Pacansky, T. J. J. Polym. Sci., Symp. 45, 1974, 39-50.
105. Kangas, D. A. J. Poly. Sci., Part A-I, 1970, 8, 1813-1821.
106. Kangas, D. A. J. Poly. Sci., Part A-I, 1970, 8, 3543-3555.
107. Schaper, R. J. U.S. Pat. 3,541,059 (Nov. 17, 1970), Calgon Corporation.
108. Murfin, D. L.; Miller, L. E. U.S. Pat. 3,478,091 (Nov. 11, 1969), The Lubrizol Corp.
109. Miller, L. E.; Murfin, D. L. U.S. Pat. 3,506,707 (April 14, 1970), The Lubrizol Corp.
110. Arlt, D.; Glabisch, D.; Suling, C. U.S. Pat. 3,547, 899 (Dec. 15, 1970), Farbenfabriken Bayer.
111. LaCombe, E. M.; Miller, W. P. U.S. Pat. 3,332,904 (July 25, 1967), Union Carbide Corp.
112. Hoke, D. I. U.S. Pat. 3,692,673 (Sept. 19, 1972), The Lubrizol Corp.
113. Slagel, R. C.; Sinkovitz, G. D. U.S. Pat. 3,709,780 (Jan. 9, 1973), Calgon Corp.
114. Boothe, J. E.; Cornelius, T. E., III, U.S. Pat. 3,709,816 (Jan. 9, 1973), Calgon Corp.
115. Walker, J. L.; Boothe, J. E. U.S. Pat. 3,709,816 (Jan. 9, 1973), Calgon Corp.
116. Kaufman, P. R. U.S. Pat 3,679,000 (July 25, 1972), The Lubrizol Corp.
117. Johnson, R. G. L.; Delf, B. W.; MacKnight, W. J. J. Poly. Sci., (Poly Phys. Ed.), 1973, 11, 571-585.
118. Schroeder, J. P.; Sopchak, W. P. J. Poly. Sci., 1960, 47, 417-433.
119. Phillips, P. J.; Emerson, F. A.; MacKnight, W. J. Macromol., 1970, 3(6), 767-777.
120. Phillips, P. J.; Emerson, F. A.; MacKnight, W. J. Macromol., 1970, 3(6), 771-777.
121. Haruta, M.; Kageno, K.; Soeta, M.; Harada, S. Japan Kokai, 75 72,987, June 16, 1975; C.A. 1975, 83, 1648554.
122. Vanderhoff, J. W.; Wiley, R. M. U.S. Pat. 3,284,393, Nov. 8, 1966.
123. Dixon, K. W. U.S. 4,225,445, Sept. 30, 1980.
124. Morgan, J. E.; Boothe, J. E. U.S. 3,968,037, July 6, 1975.
125. Shibahara, Y.; Okada, M.; Tominaga, Y.; Noda, K.; Osuga, Y. Jpn. Kokai Tokkyo Koho 79 74,841; C.A. 1979, 91, 194147A.
126. Schuller, W. H.; Moore, S. T.; House, R. R. U.S. Pat. 2,884,058.
127. Hoover, M. F. J. Macromol. Sci.-Chem., 1970, A4, 1327.
128. Slagel, R. C. J. Org. Chem,, 1968, 33(4), 1374-8.
129. Culbertson, B. M.; Slagel, R. C. J. Poly. Sci., 1968, A6(1), 363-373.
130. Culbertson, B. M.; Sedor, E. A.; Dietz, S.; Freis, R. E. J. Polym. Sci., 1968, A6(1), 2197-2207.
131. Culbertson, B. M.; et al., U.S. Pat. 3,641,145 (Feb. 8, 1972), Ashland Chemical Co.
132. Culbertson, B. M.; Sedor, E. A.; Slagel, R. C. Macromol., 1968, 1, 254-160.
133. McKillip, W. J.; et al., Chemical Reviews, 1973, 73(3), 255-281.
134. Culbertson, B. M.; Freis, R. E. Macromol., 1970, 3, 715.
135. Culbertson, B. M.; Randen, N. A. J. Appl. Poly. Sci., 1971, 15, 2609-2621.
136. Slagel, R. C.; Culbertson, B. M. U.S. Pat. 3,715,343 (Feb. 6, 1973), Ashland Chemical Co.
137. Ashland Chemical Technical Information Bulletin, "Ashland Aminimides - A New Organic Functional Group" (1972).

138. McKillip, W. J.; et al., J. Textile Chemists and Colorists, 1970, 2(18), 25/329.
139. Marvel, C. S.; Moyer, W. W., Jr., JACS, 1957, 79, 4990-4994.
140. Marvel, C. S.; DeTommaso, G. L. J. Org. Chem., 1960, 25(2), 2207-2209.
141. Allison, J. P.; Marvel, C. S. J. Poly. Sci., 1965, 3A, 137-144.
142. Neufeld, C. H. H.; Marvel, C. S. J. Poly. Sci., 1967, 5(A1), 537-543.
143. Panzik, H. L.; Mulvaney, J. E. J. Poly. Sci. (Poly. Chem. Ed.), 1972, 10, 3469-3487.
144. Vilaret, M. R., Ph.D. Dissertation, University of Florida.
145. Boothe, J. E.; Flock, H. G.; Hoover, M. F. J. Macromol. Sci.-Chem., 1970, 4, 1419.
146. Butler, G. B.; Ingley, F. L. J. Amer. Chem. Soc., 1951, 73, 895.
147. Butler, G. B.; Angelo, R. J. J. Amer. Chem. Soc., 1957, 79, 3128.
148. Butler, G. B., U.S. Pat. 3,288,770 (Nov. 29, 1966).
149. Suen, T. J.; Schiller, A. M., U.S. Pat. 3,171,805 (Mar. 2, 1965).
150. Flock, H. G.; Rausch, E. G. in Water-Soluble Polymers, N. M. Bikales, Ed., Plenum Press, New York, 1973, p. 42.
151. Goodman, B. L. Water and Wastes Engineering, 1966, 3(2), 62-64.
152. Goodman, B. L.; Mikkelson, K. A. Advanced Wastewater Treatment, Chemical Engineering/Deskbook Issue, April 27, 1970, 77(9).
153. Anon., Water and Sewage Works, Jan. 1964, 111, 64-66.
154. Schepman, B. A. Wastes Engineering, Apr. 1956, 162-165.
155. Brown, J. M. Water and Sewage Works, 1960, 107(5), 193-195.
156. Trubnick, E. H. Water and Sewage Works, 1960, 107, R-287-R-292.
157. Carlson, C. F. U.S. Pat. 2,297,691 (Oct. 1942); 2,357,809 (Sept. 1944).
158. Young, C. F.; Grieg, A. A. RCA Rev., 1954, 15, 471.
159. Dolinski, R. J.; Dean, W. R. Chem. Tech. (May 1971), pp. 307-309.
160. Hoover, M. F.; Carr, H. E. Tappi, 1968, 51, 552.

RECEIVED July 2, 1990

Chapter 3

Water-Soluble Polymer Synthesis
Theory and Practice

D. N. Schulz

Exxon Research and Engineering Company, Route 22E, Annandale, NJ 08801

This paper reviews the theory and practice of water soluble synthesis, as exemplified by the polymerization of acrylamide and N-vinyl pyrrolidone. It combines basic kinetics information with practical polymerization recipes and tips. Various polymerization techniques (e.g. solution, inverse emulsion, inverse microemulsion, bulk, suspension, precipitation) for homo and copolymerization are covered. In addition, the special techniques used in the preparation of hydrophobically associating polymers are presented.

This is a review of the fundamental/practical aspects of water soluble polymer synthesis. It looks at the major polymerization mechanism (e.g. free radical) and the various polymerization techniques (e.g. solution, inverse emulsion, inverse microemulsion, bulk, suspension, precipitation) for two well-known water soluble monomers (i.e. acrylamide and N-vinyl pyrrolidone). This paper will combine basic kinetics information with practical polymerization recipes and tips. Both homopolymerization and copolymerization are considered. In addition, the special synthetic methods for preparing hydrophobically associating polymers are covered.

Kinetics of Free Radical Polymerization

Free radical polymerization involves the well known initiation, propagation, transfer, and termination events (Scheme A). Table I shows the practical consequences of these kinetic steps. Thus, to achieve a high product Degree of Polymerization, $\overline{DP}$, (or molecular weight), one should maximize the $k_p/k_t^{1/2}$ and $[M]/[I]^{1/2}$ ratios, as well as minimize the concentration of chain transfer agent $[CT]$. To increase the rate of polymerization, one should maximize the product $[M] \cdot [I]^{1/2}$. Moreover, increasing reaction

0097–6156/91/0467–0057$06.00/0

temperature increases R_p and tends to decrease $\overline{DP}$. Since $k_p/k_t^{1/2}$ is an intrinsic function of the monomer itself, only the quantities [M], [I], [CT] and temperature are easily controlled by the synthetic chemist.

SCHEME A

FREE RADICAL POLYMERIZATION

- KINETIC SCHEME

INITIATION   TRANSFER

I ---K_D--> 2R· M_N· + CT-H ---K_{CT}--> M_NH + CT·
(K_D OFTEN RATE CONTROLLING)

R· + CH$_2$ = CH ---K_I--> R-CH$_2$-CH·   CT· + M ------> C$_T$-M ETC.

PROPAGATION   TERMINATION

M_1· + M ---K_p--> M_2· M_N·+ M_N· ---K_T--> M_N - M_N

M_2· + M ---K_p--> M_3· M_N· + M_N ------> M_N + M_N

M_N· + M ------> M_{N+1}·

I   = Initiator
K_D = Rate of constant for decomposition of initiator
K_I = Initiation rate constant
K_p = Propagation rate constant
K_{CT} = Chain transfer rate constant
CT-H = Chain transfer agent
CT· = Chain transfer radical
K_T = Chain termination constant

TABLE I. PRACTICAL CONSEQUENCES OF KINETICS SCHEME

KINETICS EQS.	CONSEQUENCES
I. Degree of PZN ($\overline{DP}$) $MW = \overline{DP} = \dfrac{k_p[M]}{[fk_dk_t]^{1/2}[I]^{1/2}}$	I. To Get High MW: • High k_p/k_t $^{1/2}$ (function of monomer) • High $[M]/[I]$ $^{1/2}$
II. Rate of PZN Rate of PZN $= R_p = k'$ [m] • $[I]^{1/2}$	II. To Get Fast PZN • Maximize [M] • $[I]^{1/2}$
III. Chain Transfer $\dfrac{1}{\overline{DP}} = \dfrac{k_t}{k_p^2} \cdot \dfrac{R_p}{[M]^2} + C_{ct}\dfrac{[CT-H]}{[M]}$ $C_{ct} = \dfrac{k_{ct}}{k_p}$ (chain transfer constant)	III. Effects: • R_p – No Effect • $\overline{DP}$ (MW) dec. as [CT-H] inc. • Alcohols – High C_{ct}
IV. Temperature $k = A$ exp $-E/RT$	IV. Effects: • R_p inc. as temp inc. • $\overline{DP}$ dec. as temp inc.

ACRYLAMIDE

Solution Polymerization. Acrylamide is a unique hydrophlic monomer because it is easily purified and has an exceptionally high $k_p/k_t^{1/2}$ ratio and a low chain transfer constant to monomer. Acrylamide also has the highest $k_p/k_t^{1/2}$ ratio within the family of acrylamide monomers, e.g. methacrylamide, N,N' dimethylacrylamide (Tables II, III) (1). As such, acrylamide is capable of being polymerized to very high DP's (molecular weights).

TABLE II TYPICAL KINETIC PARAMETERS

Monomer	$k_p/k_t^{1/2}$
Styrene	0.01
Methyl acrylate	0.35
Methyl methacrylate	0.04
Vinyl acetate	0.12
Acrylamide	3.5
Vinyl Pyrrolldone	?

TABLE III RATE CONSTANTS FOR PZN. OF ACRYLAMIDES

Monomer	$k_p \times 10^{-4}$	$k_t \times 10^{-6}$	$k_p/k_t^{1/2}$	Temp°C	$C_m \times 10^4$
AM	1.8	1.45	4.2	25	2
Me AM	0.08	1.65	0.2	25	540
NN'di me AM	1.1	3.8	1.78	50	15

SOURCE: Reprinted with permission from ref. 1. Copyright 1983 Plenum.

TABLE IV EFFECT OF PH ON KINETIC PARAMETERS

pH	kp x 10^{-4}	kt x 10^{-6}	kp/kt$^{1/2}$	Temp (°C)
1.0	1.72	16.3	4.3	25
5.5	0.6	3.3	3.3	25
13.0	0.4	1.0	4.0	25

SOURCE: Reprinted with permission from ref. 1. Copyright 1983 Plenum.

The kinetics constants for the free radical solution polymerization of acrylamide are medium dependent. Thus, these parameters are subject to pH, polarity, and salt effects. Table IV (1) shows the effects of pH on k_p and k_t. The high k_p at low pH is presumably caused by protonation of the acrylamide radical and electron localization (Equation 1).

$$\sim CH_2 - \overset{|}{\underset{|}{C}}H\cdot \; + \; H\cdot \; \longleftrightarrow \; \sim CH_2 - \overset{|}{\underset{|}{C}}H\cdot \qquad (1)$$

On the other hand, the low k_p at high pH is thought to be the result of charge repulsion of partially hydrolyzed acrylamide groups (Equation 2).

$$-CH_2-CH\cdot \; + \; CH_2{=}CH \quad \longrightarrow \qquad (2)$$

TABLE V POLYMERIZATION PARAMETERS
AM IN H_2O/DMSO

H_2O:DMSO	$kp \times 10^{-3}$	$kt \times 10^{-7}$
100:0	66.6	55.1
65:35	14.1	6.5
38:62	5.38	8.8
0:100	2.5	16.7

- Corresponds to a maximum in η
  - Suppression of kt
  - Tromsdorf effect

SOURCE: Reprinted with permission from ref. 1. Copyright 1983 Plenum.

    Changes in solvent polarity can also have a huge effect on k_p and k_t. Table V (1) shows how changes in H_2O/DMSO ratios affect k_p and especially k_t. In fact, k_t goes through a minimum around a H_2O/DMSO ratio of 50/50. This solvent ratio also corresponds to a maximum in polymerization viscosity, which suppresses k_t and leads to higher polymer molecular weights (Tromsdorf effect).

    Medium effects may also affect the reactivity ratios for the copolymerization of acrylamide with ionogenic monomers (e.g. acrylic acid), as shown in Table VI. Thus, r_1 (acrylic acid) decreases with increasing pH, presumably because of charge repulsion effects, caused by the ionization of the acrylic acid moieties. When salts are added, the r_1 values decrease as the ionic radius of the cation increases (2). However, other workers (3) believe that solvent effects in copolymerization are artifactual.

TABLE VI. MEDIUM EFFECTS IN COPZN OF AM
WITH IONOGENIC COMONOMERS

Monomer 1	Monomer 2	Rx Media[a]	r_1	r_2
Acrylic Acid	Acrylamide	No neutralization	1.43	0.60
		pH 5.2 (OH-)[b]	0.35	1.10
		ph 2 (OH-)	0.92	0.25
		pH 4 (OH-)	0.32	0.57
		pH 6 (OH-)	0.33	0.85
		pH 8 (OH-)	0.63	0.12
		pH 4 (OH-) + 1 M NaCl	0.92	0.68
		pH 6 (OH-) + 1 M NaCl	0.28	0.85
Li salt		pH (7.1)	0.37	0.66
Na salt		pH (7.1)	0.30	0.94
K salt		pH (7.1)	0.24	1.45

a 30°C
b 60°C

SOURCE: Reprinted with permission from ref. 2. Copyright 1981 Marcel Dekker, Inc.

A typical solution polymerization recipe for polyacrylamide is shown below (4):

	Polymerization Charge, g
Acrylamide	32
Deionized Water	268
Sodium Bromate	0.0032
Dry Ice	5.3
Sulfuric Acid	0.165

SOURCE: Reprinted with permission from ref. 4. Copyright 1973 Plenum.

The recipe for copolymerization is similar except for the addition of the second monomer. Some experimental "tricks" for obtaining high MW, gel-free acrylamide polymers include the use of low temperature initiation, high $[M]/[I]^{1/2}$ ratios, and the addition of an antioxidant (e.g. 2-mercaptobenzimadazole) to the polymerization or drying steps (5).

Inverse   Emulsion   (w/o)   Polymerization.   Inverse   emulsion
polymerization of acrylamide has a number of advantages over
solution methods, i.e. higher solids level, lower solution
viscosities, as well as better heat removal and mixing.
A schematic diagram of inverse emulsion polymerization is
shown in Figure 1 (6). The polymerization mixture consists of
aqueous miceles, aqueous monomer-polymer particles, and aqueous
monomer droplets swimming around in a continuous hydrophobic
phase. Initiation can take place either in the hydrophobic or
aqueous phases. The kinetics of inverse (w/o) emulsion
polymerization is complex and not a mirror image of convential
emulsion (o/w) polymerization. So far, no unified quantitative
theory exists for the w/o systems. The kinetics of each system is
dependent upon the level and type of initiator, emulsifier, and
monomer.
    Thus, Demonie (7) found inverse suspension polymerization
kinetics for low concentrations of aqueous $K_2S_2O_8$ of
initiator. Kurevkov (8) used another water soluble initiator and
found that R_p was proportional to [E] up to 20% conversion and
independent of [E] at >20% conversion; MW was found to be inversely
proportional to [I] and [E]. On the other hand, Baade and Reichart
(8) found R_p to be proportional to [I] and $[E]^{0.2}$ for an oil
soluble initiator. Vanderhoff (10-12) used a tetronic emulsifier
and benzoyl peroxide initiator and found R_p to be proportional to
$[I]^2$ and [E], and $\eta(MW)$ to be proportional to $1/[I]^{0.3}$ and
$1/[E]^{0.5}$.
    A typical recipe for inverse emulsion copolymerization of
acrylamide and acrylic acid is shown below (13):

		Polymerization Charge, g
A.	Monomer/Aqueous Phase	
	70:30 Acrylamide/Acrylic Acid	302.7
	Deionized Water	173.5
	EDTA, Sodium Salt	.6
	Ammonium Ferrous Sulfate	.6
B.	Oil Phase	
	50:50 (HLB = 5.6)/HLB = 10.0	33.0
	Paraffinic Hydrocarbon	206.3

    The surfactant is dissolved in the hydrocarbon phase. Then, A
is added to B to form a w/o emulsion. 1.10% t-butyl hydroperoxide
is added. The mixture is purged with N_2 and sodium bisulfite is
added. Polymerization follows.

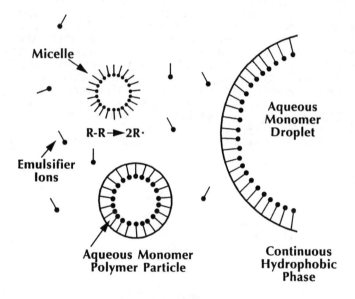

Figure 1. Schematic diagram of inverse emulsion (w/o) polymerization. (Reproduced from Ref. 6. Copyright 1962 American Chemical Society.)

Polymer isolation from the inverse w/o emulsion can be achieved by precipitation or phase inversion. In the case of the phase inversion method, a low surfactant HLB package is used for polymerization and a high HLB surfactant package is used to invert the w/o emulsion to a o/w emulsion (14-16). The advantage of the phase inversion method is that it yields a rapidly redissolvable product. The disadvantage of the phase inversion method is that inversion must be complete or else polymer (activity) is lost.

Inverse w/o Microemulsion Polymerization. Candau and coworkers (17-19) have prepared water soluble polymers by use of inverse (w/o) microemulsion polymerization. A microemulsion is an emulsion of very small particle size (e.g. 50-100 A), which is optically transparent. It forms spontaneously and is thermodynamically stable when formed with an appropriate surfactant package. No external stirring is needed. Candau found that the R_p for inverse microemulsion polymerization was 10-200X faster than convential inverse emulsion polymerization because of the high number of small particles. Moreover, the kinetic order of [AM], [surfactant] depends upon the type of initiators and surfactants that are used. Thus, for the AIBN system, $R_p \propto [AM]^{1.1}$; $R_p \propto [AIBN]^{0.1}$; $R_p \propto [AOT]^{-0.55}$ while $Mv \propto [AM]^{1.1}$, $[AIBN]^{-0.03}$, $[AOT]^{-0.8}$.
A typical inverse microemulsion recipe is below (17-19):

		Polymerization Charge, g
	Acrylamide	106
	Aqueous solution (containing 106 g AM, 15.9 g NaOAC)	250
	Isoparafin fraction	300
HLB = 9.3	(Sorbitan Sesquioleate	22
	(Solbitan Hexaoleate	135
	AIBN	.32

The above mixture is irradiated by UV at 19°C for 15 minutes. The resulting product is a stable inverse microlatex.

Precipitation Polymerization. In the precipitation polymerization
of acrylamide, a solvent is used which is a good solvent for the
monomer and a poor solvent for the polymer. The advantages of this
type of polymerization are the medium never gets viscous and the
polymer is easy to dry. A typical precipitation polymerization
recipe for acrylamide (19) involves a 40-55% solution of monomer in
t-butanol (with optional salt).

Synthesis of Hydrophobically Associating PAM's. Hydrophobically
associating polyacrylamides have interesting property profiles,
namely increased viscosification efficiency, shear stability, shear
thickening rheology and improved salt tolerance (21-26). However,
hydrophobically associating polymers are difficult to synthesize
because of the difficulty in combining oil soluble and water
soluble polymers in the same pot. These difficulties are overcome
by the use of micellar (or microemulsion) copolymerization or
surfactant macromonomers.

Equation 3 shows the use of micellar polymerization for the
synthesis of copolymers of acrylamide and alkyl acrylamide (RAM
polymers). In this method, the oil soluble acrylamide is
solubilized in the aqueous phase by suitable surfactant(s). The
main problem with this method is the need for substantial amounts
of external surfactant for the solubilization process.

Equation 4 shows the synthesis of hydrophobically associating
polyacrylamides from surfactant macromonomers (PAM-surf polymers).
Here, water soluble (or dispersable) surfactant acrylates are
copolymerized with acrylamide. Since both the surfactant and
hydrophobe (R) are built into the macromonomer, no external
surfactants are needed to solubilize the hydrophobic-R-group. A
comparison of typical recipies for synthesizing RAM and Pan-Surf
polymers is shown in Table VII.

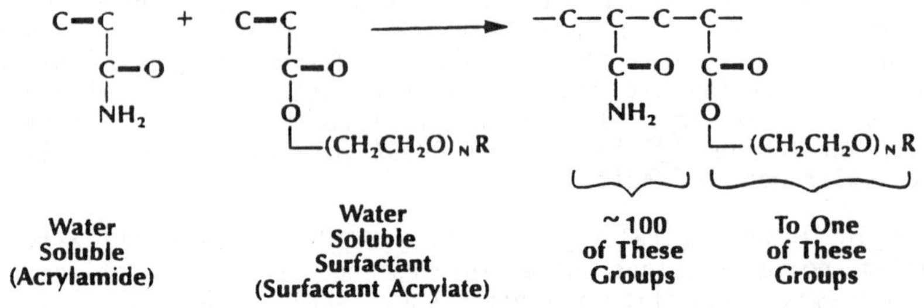

$$(4)$$

| Water Soluble (Acrylamide) | Water Soluble Surfactant (Surfactant Acrylate) | ~100 of These Groups | To One of These Groups |

TABLE VII

SYNTHESIS OF HYDROPHOBICALLY ASSOCIATING PAM'S

TYPICAL RECIPES

RAM (MICELLAR PZN)(A)	PAM-SURF. (SURFACTANT MACROMONOMER)(B)
• 470.7 G H2O 15.85 G SODIUM DODECYL SULFATE 14.76 G ACRYLAMIDE 0.288 G OCTYL ACRYLAMIDE • ABOVE HEATED TO 50°C • 0.01 G K2S2O8	• 500 ML H2O 30.0 G ACRYLAMIDE 1-5 G NONYL PHENOXY (ETHEROXY) ACRYLATE • ABOVE HEATED TO 50° • 0.01 G K2S2O8 ADDED

(A) S. R. TURNER, D. B. SIANO, J. BOCK. U.S. PATENT 4,528,348
(ASSIGNED TO EXXON) JULY 9, 1985.

(B) D. N. SCHULZ, J. J. MAURER, J. BOCK. U.S. PATENTS 4,463,151
AND 4,463,152 (ASSIGNED TO EXXON) JULY 31, 1984.

N-Vinyl Pyrrolidone (NVP)

Unlike acrylamide, NVP is not easily purified. Consequently, polymerization of NVP monomer has been plagued by nonreproducible results and low $\overline{DP}$ products. Typical monomer impurities are acetaldehyde, 1-methylpyrrolidone, and 2-pyrrolidone. Many purification methods have been tried, e.g. zone melting, vacuum distillation, and even pre-polymerization, but with only limited success.

Bulk Polymerization. In the bulk polymerization of NVP, AIBN is the preferred initiator, while peroxides are acceptable. However, persulfates react with the monomer and should be avoided. A typical bulk polymerization of acrylamide is shown below (27):

	Polymerization Charge, g
NVP (freshly distilled)	5.0
AIBN	0.1
Temp.	-60°C
Time	72 hours

Table VIII shows r, r_2 values for the bulk phase copolymerization of NVP with various monomers. NVP is a mildly electron donating monomer. Consequently, it has an alternating tendency with electron withdrawing monomers, such as maleic anhydride and acrylonitrile. However, it shows very little copolymerization with styrene (27).

TABLE VIII. NVP (M1) COPOLYMERIZATIONS
BULK

Monomer	r^1	r^2
Acrylonitrile	0.06	0.18
Allyl Alcohol	1.0	0.0
Butyl Methacrylate	0.23	1.16
Crotonic Acid	0.85	0.02
Maleic Anhydride	0.16	0.08
Styrene	0.045	15.7
trichloroethylene	0.54	<0.01
vinyl chloride	0.38	0.53
vinyl phenyl ether	4.43	0.22

SOURCE: Reprinted with permission from ref. 27. Copyright 1977 Academic Press.

<u>Solution   Polymerization</u>.   One   of   the   curiosities   of   NVP
polymerization is that a polymerization with 50% H_2O polymerizes
faster than the bulk phase (Figure 2) (<u>28</u>).   Presumably, the NVP
monomer forms a complex with water that is more reactive than the
monomer itself.
    NVP copolymerizes with various simple ionic and zwitterionic
sulfonated   monomers   in   solution   (Equation 5)   (<u>29</u>).   The
copolymerization of NVP and NaAMPS shows some signs of having
donor–acceptor   character   because   (1) it   is   an   azeotropic
polymerization; (2) r_1 and r_2 are less than 1.0 and r_1 times
r_2 is very low; (3) there is a rate enhancement near the 50/50
NVP/NaAMPS ratio; (4) the two monomers can copolymerize in the
absence of initiation.

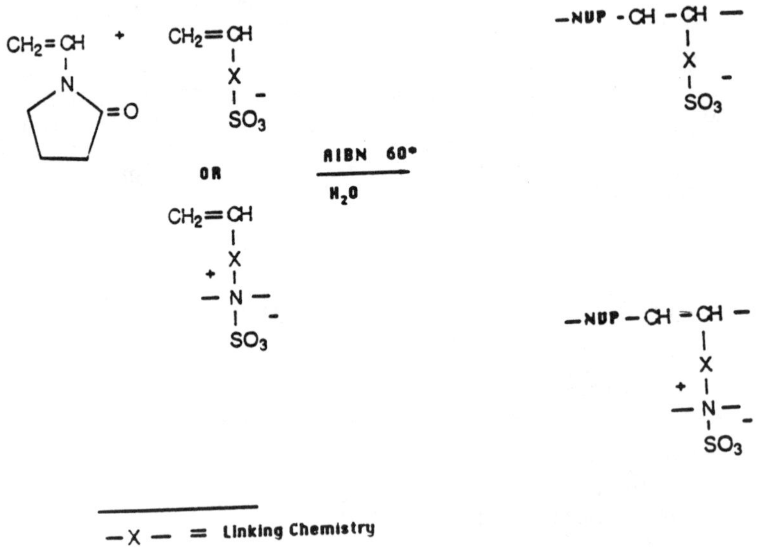

(5)

_X_ = Linking Chemistry

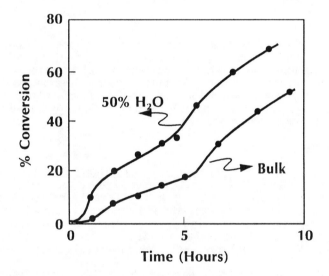

Figure 2. Kinetics of NVP polymerization. (Reproduced with permission from Ref. 28. Copyright 1975 John Wiley & Sons, Inc.)

Suspension Polymerization. NVP can be polymerized in suspension in n-heptane (30). A typical recipe is shown below:

In reactor	In cylinder
2.12 g NVP	375 g NVP
425 g Dist. H_2O	1 g Vazo 64
50 g Suspending agent	
5,000 g n-heptane	

The contents of the reactor are brought to 75°C and the contents of the cylinder (initiator dissolved in NVP) are added.

Summary

This paper reviews the free radical polymerization of two well known water soluble monomers, acrylamide and N-vinyl pyrrolidone. The polymerizations are examined from fundamental kinetics, as well as practical points of view. Solution, bulk, inverse emulsion, inverse microemulsion, suspension, and precipitation polymerization techniques, as well as special methods for hydrophobically associating polymers are presented.

Literature Cited

1. Thomson, R.A.M. In Chemistry and Technology of Water Soluble Polymers, Finch, C.A., Ed., Plenum Press, pp. 31-70, and references therein; Thomson, R.A.M., "Water Soluble Polymers - Polymerization", presented at a Residential School of the Royal Society of Chemistry on Water Soluble Polymers, Cambridge, England, September 14-18, 1981.
2. Plochka, K., J. Macromol. Sci.-Dev. Macromol. Chem., C20 (1) 67 (1981).
3. Park, K.Y.; Santee, E.R.; Harwood, H.J., ACS Polym. Prepr. 27, (2), 81 (1986).
4. Bikales, N.M. In Water Soluble Polymers, Bikales, N.D. Ed., Polymer Science and Technology, Vol. 2, Plenum Press, New York, 1973, pp. 213-255.
5. Yoshida, K.; Ogawa, Y.; Handa, R.; Hosoda, J.; Kurashige, N.; Furino, A., U.S. Patent 4,306,045 (assigned to NiHo Co.), December 15, 1981.
6. Vanderhoff, J.W.; Bradford, E.B.; Tarkowski, H.L.; Shaffer, J.B.; Wiley, R.M., Adv. Chem. Ser. 34, 32 (1962).
7. Dimonie, M.V.; Boghina, G.M.; Marinescu, N.N.; Cincu, C.J., Opescu, O.G. Eur. Polym. J., 18, 639 (1982).
8. Kurenkov, V.F.; Osipova, T.M.; Kuzuetsov, E.V.; Myagchenkov, V.A. Vysokomol. Soldin. Ser. A20 (11), 2608 (1978).
9. Baade, V.; Reichert K.H.; Eur. Polym. J. 20 (5), 505 (1984).

10. Vanderhoff, J.W.; Distefano, F.V.; El-Aasser, M.S.; O'Leary, R.; Shaffer, O.M.; Visioli, D., J. Disper. Sci. Technol. 5, 323 (1984).
11. Visioli, D.L.; El-Aasser, M.S.; Vanderhoff, J.W., ACS Div. of Polym. Mat'l. Sci. & Eng. Prepr. 51, 258 (1984).
12. Vanderhoff, J.W.; Visioli, D.L.; El-Aasser, M.S., ACS Div. of Polym. Mat'l. Sci. & Eng. Prepr. 54, 375 (1986).
13. U.S. Patent 4,022,731 (assigned to American Cyanamid).
14. Frisque, A.J., "Inverse Emulsion Polymers: Their Properties and Applications", presented at the 1984 Short Course on Water Soluble Polymers, New Paltz, NY, August 1984.
15. Frisque, A.J., U.S. Patent 3,624,019 (assigned to Nalco Chemical Co.) (Nov. 1971).
16. Anderson, D.P.; Frisque, A.J., U.S. Patent Re. 28,576 (assigned to Nalco Chemical Co.) (October 21, 1975; U.S. Patent Re. 28,474 (assigned to Nalco Chemical Co.) (July 8, 1984).
17. Durand, J.P.; Candau, F.; Nicholas, D.; Ger. Offen. DE 3,524,969 (Jan. 16, 1986).
18. Candau, F.; Leong, Y.S.; Fitch, R.M.; J. Polym. Sci. Polym. Chem. 23 193 (1985).
19. Candau, F.; Zehuini, Z.; Heatley, F.; Fronta, E., Colloid Polym. Sci., 264 (8), 676 (1986).
20. Monagle, D.J.; Shyluk, W.P., U.S. Patent 3,336,269 (assigned to Hercules) (August 15, 1967).
21. Turner, S.R.; Siano, D.B.; Bock, J., U.S. Patents 4,520,182 (assigned to Exxon Res. & Eng. Co.); 4,521,580 (1985); 4,528,348 (1985).
22. Bock, J.; Siano, D.B.; Schulz, D.N.; Turner, S.R.; Valint, Jr., P.L.; Pace, S.J., ACS Div. Polym. Matl. Sci. & Eng. Prepr. 55,355 (1986).
23. Schulz, D.N.; Kaladas, J.J.; Mauer, J.; Bock, J.; Pace, S.J.; Schulz, W.W., Polymer 28 2110 (1987).
24. Evani, S.; U.S. Patent 4,432,881 (assigned to Dow Chemical Co.) (February 21, 1984).
25. McCormick, C.L.; Johnson, C.B., ACS Div. Polym. Matl. Sci. & Eng. Prepr. 55, 366 (1986).
26. Bock, J.; Valint, P.L. Jr.; Pace, S.J.; Siano, D.B.; Schulz, D.N.; Turner, S.R. In Water Soluble Polymers for Petroleum Recovery, Stahl, G.A. and Schulz, D.N. Eds., Plenum Publishing Co., New York, 1988, pp. 147-160.
27. Sandler, S.R.; Karo, W., Polymer Synthesis, Volume II, Academic Press, New York, 1977, Ch. 8.
28. Senogles, E.; Thomas, R., J. Polym. Sci. Symp. 49, 203 (1975).
29. Schulz, D.N.; Kitano, K.; Danik, J.A.; Kaladas, J.J. in Polymers in Aqueous Media - Performance through Association, Advances in Chemistry Series, No. 223, American Chemical Soc., Washington, D.C., 1989, Ch. 9.
30. U.S. Patent 4,190,718.

RECEIVED June 4, 1990

Chapter 4

Chemical Modifications of Natural Polymers and Their Technological Relevance

Shalaby W. Shalaby[1] and Kishore R. Shah[2,3]

[1]Bioengineering Department, Clemson University, Clemson, SC 29634–0905
[2]Ethicon, Inc., Somerville, NJ 08876

Natural polymers which are water-soluble or can be rendered water-soluble by chemical modification have been the subject of extensive technical reviews and original research reports, over the past fifty years. Early interests in this area were associated with the food, leather, paper and textile industries and to a lesser extent, the cosmetics and pharmaceutical industries. Proteins, such as collagen and gelatin, as well as polysaccharides, such as cellulose derivatives and starch, were dominant among all natural polymers and their derivatives. Over the past 20 years and until recently, the polymer and allied industries have focused on synthetic polymers and limited to moderate level of efforts were directed toward the modification of natural polymers. With the new interests in the biomedical and pharmaceutical industries in these polymers over the past few years, impressive activities on the modification of natural polymers to meet the growing needs are being recognized and noted in recent reviews. (Ford, 1986; Glass, 1986; Molyneux, 1982; Pariser and Lombardi, 1989; Skjak-Braek and Sanford, 1989; Shalaby, 1988). It was, therefore, decided to focus the discussion in this chapter on the most recent advances pertinent to the chemical modification of natural polymers, which are of particular interest to those in the biomedical, pharmaceutical and allied industries.

Chitosan

Chitosan, a partially deacetylated product of the natural polysaccharide chitin, is based on glucosamine and acetylated glucosamine units. Depending on the free amine content of the chain, chitosan exhibits variable degrees of solubility in dilute aqueous media. In the most common grade of chitosan the mole ratio of acetylated to deacetylated amine groups is 30/70. Not only does this ratio affect the polymer solubility but also it determines its susceptibility to enzyme and hence its biodegradability. Thus, a good fraction of the research activities on chitosan modification pertained to modification of the substituted and unsubstituted amine side groups of the polysaccharide chain as well as the remaining two hydroxyl

[3]Current address: ConvaTec, Bristol-Myers Squibb Co., CN9254, Princeton, NJ 08543

functionalities (Pariser and Lombardi, 1989, Sanford 1989). Given below is an outline of major research activities on such chemical modifications of chitosan.

Recognizing the ability of chitosan to adsorb cupric ions, Koyama and Taniguchi (1986) have prepared homogeneously crosslinked chitosan, using glutaraldehyde, toward enhancing its uptake of the transition metal ions. A second approach to enhance chitosan adsorption of metal ions entailed its phosphorylation (Nishi, et al. 1987). The interaction of blood with chitosan derivatives has been addressed by Kurita (1986) and Hirano, et al. 1986) who were interested in cyanoethylated and sulfated chitosan, respectively. The cyanoethyl chitosan was studied in conjunction with cellulose nitrate membranes for microfiltration, while chitosan sulfate was evaluated for its anticoagulant activity. Kurita (1986) has studied several other derivatives of chitosan including N-alkyl, N-carboxybenzyl, N-carboxymethyl, and carboxyacyl chitosan. He also studied the complete deacetylation, succinylation and reductive amination of chitosan.

Properties and biological activities of chitosan derivatives having modulated solubility and ionic behavior has been treated by a few recent investigators. The amphoteric N-carboxymethyl and N-carboxybutyl chitosan have been reported by Muzzarelli (1989) to be more effective than chitosan as chelating agents of metals ions. They were also noted as being more effective bacteriostatic agents than other chitosans. Hydroxypropylation of chitosan and the effect of the ratio of O-/N-substitution and the overall degree of substitution on the solubility and related properties of resulting derivatized chitosan was studied by Maresch, Clausen and Lang (1989), N,O-carboxymethyl chitosan was prepared and its properties were studied by Davies, Elson and Hayes (1989) as a new water-soluble polymer with potential use in a few applications. N-carboxymethyl chitosan, on the other hand, was reported by Bioagini, et al. (1989) to induce neovascularization. The authors related their findings to the understanding of healing mechanisms for chitosan-treated wounds.

Cellulose

Among the recent studies on the modification of cellulose toward increasing its solubility in organic solvents and not necessarily water were those dealing with its cyanoethylation and subsequent reduction by borane-tetrahydrofuran to aminopropyl cellulose followed by grafting amino acid N-carboxyanhydrides (Miyamoto, et al. 1986; Hasegawa, et al. 1988). The resulting cellulose/polypeptide graft copolymers were shown to be soluble in common organic solvents. Similar grafts can be made starting with the commercially available aminoethyl cellulose (Miyamoto, et al. 1986; Miyamoto, et al. 1987). Although the technological importance of these polymers is not fully recognized, Miyamoto, et al. (1986), have described members of this family of graft copolymers to have excellent antithrombogenic properties.

Starch

An extensive review of new developments and potential applications of starch products and other polysaccharides in biotechnology, diagnostics, and pharmaceuticals has been recently published (Yalpani, M., 1987). The use of hydroxyethyl starch for regulation of osmotic pressure in blood substitutes based on artifical perfluorohexamethylenetetramine and synthetic phospholipids has been claimed (Dandliker, et al., 1988). The pharmacokinetics of hydroxyethyl starch and dextran, the most commonly used plasma expanders, has been recently reviewed (Klotz and Kroemer, 1987). The complex compositions of these polysaccharides and biphasic time-dependent decline of their plasma concentrations render meaningful pharmacokinetic analysis difficult. Epichlorhydrin crosslinked starch hydrogels, conjugated to proteolytic enzymes (chymotrypsin, trypsin, etc.) and lysozyme, have been reported for use as biologically active wound shield (Zamek, et al., 1988). Similarly, hydrogel wound dressing, comprising of proteolytic enzymes immolized in glutaraldehyde crosslinked oxidized starch, has been described (Aliaga, et al., 1988). Kulicke, et al. (1989) have studied dynamic mechanical properties of hydrogels based on starch, oxidized starch, and amylopectin, all crosslinked with epichlorhydrin.

Pectin

The major use of pectins, which are made up primarily of α-1,4-poly(galacturonic acid) partially esterified with methanol, is in the food industry. The degree of esterification affects both the aqueous solubility and the gel-forming ability. The use of polyelectrolyte complex, formed by interaction of pectin with diethylamino-ethyl dextran, as an anticholesteremic agent has been claimed by Kito, et al., (1986). Complexation of selected model polypeptides with potassium pectates and pectinates of various esterification degrees has been reported by Bystricky, et al. (1988). It was observed that the complex-forming efficacy continuously decreases with the increase in the degree of esterification of pectin.

Alginate

Alginate is a collective term for a family of copolymers containing segments of 1,4-linked β-D-manuronic and α-L-guluronic acid residues (M-block and G-block, respectively) in varying proportions and sequential arrangements (Martinsen, Skjak and Smidsrod, 1989). Most of the recent work on alginates deals with correlating the chain composition with the properties of the alginate salts and particularly those with calcium ions, for their demonstrated importance in the biomedical, pharmaceutical and textile industries. It has been documented earlier that calcium alginate forms gels and the gel-forming properties are strongly correlated with the proportion and lengths of the blocks of contiguous L-guluronic acid residues (G-blocks) in the polysaccharide chain (Smidsrod, Haug and Wittington, 1972). In a recent study (Martinsen, Skjak and Smidsrod, 1989), it was shown that calcium alginate beads, made from alginate with 70% G-blocks having an average degree of polymerization of at least 15, exhibit the highest mechanical strength and lowest shrinkage. One of the important applications of the calcium alginate beads is the encasing of living cells and immobilization of enzymes. For this reason purity of the soluble alginate is considered critical to the success of the calcium alginate beads as carriers of enzymes or living cells. Polyphenolic contaminants are found in commercial alginates and their detection and removal was addressed in a recent report by Skiak-Braek, Murano and Paoletti (1989).

Being a polyanion, alginate can complex with the polycationic chitosan. A recent patent has described the utility of such complex using modified alginates (Daly, Keown and Knorr, 1989). Thus, capsules, for use as carriers of enzymes or cells, with modulated properties, were claimed to have been prepared from chitosan alginate wherein the alginate component is 10 to 60% esterified with 1,2-propylene glycol. This esterification was pursued to increase capsule permeability. As usual, the capsule hardening is achieved by allowing the chitosan alginate to react with calcium chloride. McKnight, et al. (1988) have reported modification of calcium alginate beads by formation of a polyelectrolyte complex membrane with chitosan. The effect of the chitosan molecular weight and the degree of deacetylation upon the membrane wall forming characteristics was studied.

Hyaluronic Acid

Modification of this natural, tissue lubricant to explore new applications, was made possible through its availability in sufficient quantities as a fermentation product. In two patent applications, Della Valle (1987, 1988) disclosed two key modifications of hyaluronic acid to produce biodegradable materials for use by the biomedical and pharmaceutical industries. In one instance, the carboxyl groups of the polymer were esterified with monohydric alcohols or reacted with basic drugs and the products were geared primarily to pharmaceutical applications (Della Valle, 1987). Hunt, et al. (1990) have studied transport properties of thin films formed from alkyl esters of hyaluronic acid having varying degrees of alkyl group chain lengths. Small, neutral and positively charged molecules showed relatively high permeability through the films. Whereas, smaller permeability values were observed for negatively charged molecules and for solutes of molecular weights greater than 3000. Toward the preparation of biodegradable plastics primarily for sanitary and surgical articles Della Valle (1988) used polydric alcohols to produce a crosslinked hyaluronic acid (1988). Malson and Debelder (1986, 1988) have described crosslinking of hyaluronic acid with polyfunctional reagents, such as diepoxides, to produce water-swellable and biodegradable materials for use as surgical implants and as adjuvants for the prevention of postsurgical adhesions between body tissues.

Proteins

This section focuses on modification of proteins such as enzyme and growth factors which are intended, primarily, to increase their stability without impairing their biological activities. With the new development in recombinant DNA products, growth factors can be obtained in sufficient purity to study their chemical modification. In a recent study of the human epidermal growth factor (EGF) Njieha and Shalaby (1989) have shown that acylation of the primary amino-group of this simple protein increases its stability without compromising its biological activities. This was supported by data based on receptor binding and mitotic assays.

This is perhaps one of the earliest studies on the direct modification of growth factors. On the other hand, modification of enzymes has been known for many years. However, recent focus on this technology is to seek not only an increase in enzyme stability but also to transform the native enzymes to achieve new catalytic activities. Some attempts made toward this goal have been noted by Kabanov, Levashov and Martinek (1987). An interesting approach to enzyme modification was dealt with during the preparation of semisynthetic enzymes. As in the case of flavopapain (Kaiser, 1987). In general, this approach entails taking existing enzymes with appropriately reactive functional groups in their active sites and modifying these groups with appropriate coenzyme analogs in such a way that the particularly altered substrate binding sites are still accessible to a range of substrate molecules. (Kaiser, 1987). A second approach to this preparation of semisynthetic enzyme entails that conformational modification of protein (Keyes, Albert and Sarawathi, 1987). Although the success in the development of totally synthetic enzyme has been limited, recent studies on polymers containing chiral cavities for racemic resolution appear to hold some promise (Sellergren, Lepisto and Moshbach, 1988; Wulff, 1986). Most of the recent efforts were directed toward the synthesis of enzyme-analogue built polymers for the resolution of racemic α-amino acids and the ultimate synthesis of optically active amino acids from glycine (Wulff and Vietmeir, J. 1989).

Literature Cited
1. Aliaga, I.; Monsan, P.; Fauran, F.; Couzinier, J., Fr. Demande, FR 2,600,897 (Jan. 8, 1988).
2. Biagini, G., Pugnaloni, A. Frongia, G., Gazzanelli, G., Lough, C. and Muzzarelli, R.A.A., in "Chitin, and Chitosan:Sources, Chemistry, Biochemistry Physical Properties and Applications"(Skjak-Braek, G. and Sanford, P.A., Eds.) Elsevier Appl. Sci. New York, 1989, p.671.
3. Bystricky, S.; Malovikova, A.; Sticzay, T.; and Blaha, K., Collect. Czech. Chem. Commun., 53,2833(1988)
4. Daly, M.M., Keown, R.W. and Knorr, D.W., U.S. Pat. (to University of Delaware) 4,808,707(1989).
5. Dandliker, W.; Watson, K.; and Drees, T., Europ. Pat. Appl. EP 261,802 (30 March, 1988).
6. Davies, D.H., Elson, C.M. and Hayes, E.R., in "Chitin and Chitosan:Sources, Chemistry, Biochemistry, Physical Properties and Applications" (Skjak-Braek, G. and Sanford, P.A., Eds.) Elsevier Appl. Sci. New York, 1989, p.467.
7. Debelder, A.N. and Malson, T., EP 190,215 (Aug. 13, 1986).
8. Della Valle, F., Eur. Pat. Appl. (to Fidia, SpA) 0,216,453-A$_2$ (1987).
9. Della Valle, F., Eur. Pat. Appl. (to Fidia, SpA.) 0,265,116-A$_2$ (1988).
10. Ford, W.T., Ed. "Polymeric Reagents and Catalysts", ACS Symp. Series #308, Am. Ch. Soc., Washington, D.C., 1986.

11. Glass, J.E., Ed. "Water-Soluble Polymers-Beauty with Performance" Adv. Chem. Series Vol. #213, Amer. Chem. Soc. Washington, D.C., 1986.
12. Hasegawa, O., Takahashi, S., Zuzuki, H. and Miyamoto, T., Bull. Inst. Chem. Res., Kyoto Univ. 66,93(1988).
13. Hirano, S., Kinugawa, J. andNishioka, A., in "Chitin in Nature and Technology" (Muzzarelli, R., Jeuniaux C., and Gooday, C. W., Eds.) Plenum Press, New York, 1986, p-461.
14. Hunt, J.A., Joshi, H.N., Stella, V.J., and Topp, E.M., J. Control. Rel., 12, 159(1990).
15. Kabanov, A.V., Leashov, A.V. and Martinek, K., Ann. N.Y. Acad. Sci., 501, 63(1987).
16. Kaiser, E.T., Ann. N.Y. Acad. Sci., 501, 14,(1987).
17. Keyes, M.H., Albert, D.E. and Saraswathi, S. Ann. N.Y. Acad. Sci., 501, 63(1987).
18. Kito, K.; Ogawa, T.; Vemida, H; Tanahaski, E.; Ito, Y.; and Fujii, M., Jpn. Kokgi Tokkyo Koho JP 61,106,602 (24 May, 1986).
19. Klotz, U. and Kroemer, H., Clin. Pharmacokinetic, 12, 123(1987).
20. Koyama, Y. and Taniguchi, A., J. Appl. Polym. Sci. 31, 1951(1986).
21. Kulicke, W. M.; Aggour, Y. A.; Nottelmann, H.; and Elaabee, M., Starch/Staerke, 41, 140(1989).
22. Kurita, K. in "Chitin in Nature and Technology" (R. Muzzarelli, C. Jeuniaux and G. W. Gooday, Eds.), Plenum Press, New York, 1986, p-287.
23. McKnight, C.A., Ku, A., Goosen, M.F.A., Sun, D. and Penney, C., J. Bioact. and Comp. Polym. 3, 334(1988).
24. Malson, T., EP 272,300 (June 29, 1988).
25. Maresch, G., Claussen, T. and Lang, G. in "Chitin and Chitosan:Sources, Chemistry, Biochemistry, Physical Properties and Applications" (Skjak, Braek, G. and Sanford, P.A., Eds.) Elsevier Appl. Sci. New York, 1989, p-389.
26. Martinsen, A., Skjak-Braek, G. and Smidsrod, O., Biotech. Bioeng., 33, 79(1989).
27. Miyamoto, T., Takahashi, H., Tsuji, S., Inagaki, H. and Noishiki, J. Appl. Polym. Sci., 31, 2303(1986).
28. Miyamoto, T., Ito, H., Takahashi, S., Inagaki, H. and Noishiki, Y., in "Wood and Cellulosics" (Kennedy, J.F., Phillips, G.O. and Williams, P.A., Eds.) E. Horwood Ltd., Chichester, U.K. 1987, Chap. 53.
29. Molyneux, P., "Water-Soluble Polymers:Properties and Behavior" Volumes I and II., CRC Press, Boca Raton, Fl. 1982.
30. Muzzarelli, R.A.A. in "Chitin and Chitosan:Sources, Chemistry, Biochemistry, Physical Properties and Applications" (Skjak-Braek, G. and Sanford, P.A., Eds.) Elsevier Appl. Sci., New York 1989, p-87.
31. Nishi, N., Mekita, Y., Nishimura, S., Int. J. Biol. Macromol., 9, 109(1987).
32. Njieha, F.K. and Shalaby, S.W., U.S. Pat. Applic (to Ethicon, Inc.) Doc.-S-761, Ser.#383-518, filed 7/24/89.

33. Pariser, E.R. and Lombardi, D.P. "Chitin Sourcebook:A Guide to the Research Literature", John Wiley, New York, 1989.
34. Sellergren, B., Lepisto, M. and Mosbach, K., J. Amer. Chem. Soc., 110, 5853(1988).
35. Shalaby, S.W. in "Encyclopedia of Pharmaceutical Technology" Vol. 1, Marcel Dekker, New York, 1988, p-465.
36. Skjak-Braek, G., Murano, E., and Paoletti, S., Biotech. Bioeng., 33, 90(1989).
37. Skjak-Braek, G. and Sanford, P.A., Eds. "Chitin and Chitosan:Sources, Chemistry, Biochemistry, Physical Properties and Applications, Elsevier Appl. Sci., New York, 1989.
38. Smidsrod, O., Haug, A. and Wittington, S., Acta. Chem. Scand., 26, 2563(1972).
39. Wulff, G. in "Polymeric Reagents and Catalysts" (W.T. Ford, Ed.) ACS Symposium Series #308, Washington, 1986, p.186.
40. Wulff, G. and Vietmeir, J., Makromol. Chem., 190, 1717(1989).
41. Yalpani, M., Prog. Biotechnol., 3, 311(1987).
42. Zamek, J.; Jurkstovic, T.; and Kumiak, L. Czech. Patent CS 250,018 (May 15, 1988).

RECEIVED July 24, 1990

POLYMER SYNTHESIS
AND MODIFICATION

Chapter 5

Mechanism and Kinetics of the Persulfate-Initiated Polymerization of Acrylamide

D. Hunkeler and A. E. Hamielec

Department of Chemical Engineering, Institute for Polymer Production Technology, McMaster University, Hamilton, Ontario L8S 4L7, Canada

A mechanism has been developed for the polymerization of acrylamide initiated with potassium persulfate. It has been found that donor-acceptor interactions between the monomer and initiator cause the formation of a charge-transfer complex which acts as a secondary source of initiation. This "weakly bonded complex" will be shown to be equivalently represented as a "diffuse monomer-swollen cage". That is, the complex and cage treatments are non-discriminating models of the same physical phenomena as far as amide-persulfate interactions are concerned. This analogy will enable the development of a kinetic model which can predict the observed high rate orders with respect to monomer concentration without the thermodynamic inconsistencies characteristic of previous mechanisms. The efficiencies of thermal and monomer-enhanced decomposition will also be included. The mechanism will be generalized for nonionic and ionogenic acrylic water soluble monomers in polar solvents.

Polyacrylamide homopolymers derive their utility from their long chain lengths and expanded configuration in aqueous solutions. As such they are used primarily for water modification purposes. For example, drag reduction agents function by transferring energy from the eddies to provide a laminar flow regime and decrease the hydrolytic resistance. Polyacrylamides are also applied as thickening agents (1), cutting fluids (2), soil stabilizers (3) and to a lesser extent in gel electrophoresis (4), soaps (5) and textile applications (6). They are also used in emulsion or microemulsion form as cleaners and in enhanced oil recovery (7). Recently, hydrophobic modifications have expanded the market for polyacrylamides. For commercial applications polyacrylamide quality is derived from its moisture insensitivity, oxidative stability (8) and rapid dissolution in water.

The amide substituent groups are capable of undergoing most of the reactions characteristic of their small molecule counterpart. By comparison the polymer backbone is relatively inert, although it is susceptible to attack from strong oxidizing agents, such as persulfates and peroxides (9). Polymer degradation can occur through excessive high speed agitation (10) or attack from hydroxy radicals (11). The latter results in chain scission only in the presence of dissolved oxygen (12). Ferric ions and EDTA, often part of redox initiation systems and residual to the product, can also enhance degradation (13). This effect is most severe at elevated temperatures (14).

0097–6156/91/0467–0082$06.75/0

Polyacrylamide prepared with free radical initiators is atactic in nature with the sequence length distribution conforming to Bernoullian statistics (15), as is generally observed for the addition polymerization of vinyl monomers. Isotactic polyacrylamide has, however, been prepared indirectly (16) by reacting polyphenolacrylate and ammonia in dimethyl sulfoxide.

Aqueous solutions of high molecular weight polyacrylamides ($\geq$ one million daltons) experience a decrease in intrinsic viscosity with standing time. This "aging" phenomena has been interpreted as evidence of an intramolecular hydrogen bond rearrangement to a less extended structure (17). The flocculation efficiency is however unaffected by the decrease in molecular size (18,19).

Review of Polymerization

Acrylamide is most often polymerized by a radical addition mechanism in aqueous solution. These reactions proceed at moderate temperatures (40-60°C) in order to generate a high molecular weight linear product. At more extreme temperatures (140-160°C) or in strongly acidic media, crosslinking occurs via an intra-intermolecular imidization process (20,21). The onset of polymer degradation occurs at 180°C (22). Low temperatures can also be employed, in conjunction with redox couples, which are operative down to 0°C.

The enhanced decomposition of initiators by reducing agents or "activators" was inadvertently discovered by Bacon in 1946 (23) while testing inhibitors for persulfate initiated polymerization of acrylonitrile. This new initiation process was termed "Reduction Activation", although it is presently referred to as Reduction Oxidation. Since this discovery there have been several investigations of acrylamide polymerization with redox pairs (Table I). Other initiation methods such as irradiation by [60]Co γ-rays (38-40), electrolytic initiation (41,42) and photosensitized polymerization (43-47) have been studied. Acrylamide can also be polymerized through an anionic mechanism to yield poly-β-alanine (Nylon 3) (48), and in the solid state (49). The latter were the first investigators to utilize ESR to measure radical concentrations in the acrylamide-polyacrylamide lattice.

Dainton and co-workers (38,43,50,51) were the first to perform a detailed kinetic investigation on acrylamide polymerization. They applied the rotating sector method to measure propagation and termination rate constants, and found the former to be much larger than for any other commercial monomer. They postulated that termination was principally due to combination, although subsequent investigations have determined that disproportionation is predominant (52).

Cavell (53) was the first to employ azo initiators, and determined that molecular weight was controlled by transfer to monomer. Currie (54) determined the effect of pH on propagation and termination and found both to decrease by an order of magnitude as pH rose from 1 to 13. However, the rate parameter $k_p/k_t^{1/2}$ did not significantly vary over the entire pH range. This was later confirmed by Gromov (40). Ishige and Hamielec (55) were the first to investigate the kinetics at moderate monomer concentrations. They concluded that diffusion limitations on termination reactions existed for polymer concentrations beyond approximately 7 wt.%. Later Kim and Hamielec (56) would quantify this, and suggest the primary cause of the gel effect was entanglement between the polymer chains.

Effect of Solvents. The effect of organic additives such as methanol (57,58), ethanol (37,59), dimethly formamide (46,60) and dimethyl sulfoxide (61) reduce the rate of polymerization and the molecular weight. At high levels organic solvents also cause the polymer to precipitate (40). Chapiro (62) conducted an extensive investigation of the solvent effects on acrylamide polymerization and reported:

$$R_{p,\,water} > R_{p,\,acetic\,acid} > R_{p,\,methanol} > R_{p,\,DMF} \approx R_{p,\,dioxane} \approx R_{p,\,toluene} > R_{p,\,acetonitrile}$$

Table I: Acrylamide Polymerizations with Redox Couples

Redox System	Reference
Ammonium persulfate/Sodium thiosulfate	Morgan, 1946 (24)
Ammonium persulfate/Sodium metabisulfite	Rodriguez, 1961 (25)
Ammonium persulfate/Sodium metabisulfite/EDTA	Hoover, 1966 (26)
Ammonium persulfate/Sodium bisulfite/Cu^{++}	Hoover, 1967 (27)
Sodium chlorate/Sodium sulfite	Suen, 1958 (28)
Potassium persulfate/Sodium metabisulfite	Jursich, 1969 (29)
Potassium persulfate/Sodium metabisulfite	Kurenkov, 1987 (30)
Potassium persulfate/Sodium metabisulfite/Fe^{++}	Sackis, 1967 (31)
Potassium persulfate/Sodium thiosulfate	Riggs and Rodriguez, 1967 (32)
Potassium persulfate/Sodium thiosulfate	Pohl and Rodriguez, 1980 (33)
Ceric ammonium sulfate/mercaptoethanol	Hussain and Gupta, 1977 (34)
Cerium (IV)/2-chloroethanol	Gupta, 1987 (35)
Peroxydiphosphate/sodium thiosulfate	Lenka, 1984 (36)
Permanganate/Glycolic acid	Behari, 1986 (37)

These strong effects are due to the polarity and hydrogen bonding affinity of acrylamide. Gromov (63) studied acrylamide polymerization in water, DMSO and THF and concluded that as the polarity of the solvent rises its ability to donate protons to the carbonyl rises. This results in a positive charge on the amide and an increased electron localization on the α-carbon. Bune et al. (64) have confirmed, by proton and ^{13}C NMR, that hydrogen bonding occurs predominantly through the carbonyl group ($\delta_{c=o}$ shifts to weaker field positions while δ_{NH2} is essentially unchanged) They also correlated the electron accepting ability of the carbonyl with an increased electron density on the conjugated double bond. Furthermore, they observed that the downfield displacement of the olefinic carbons, upon increasing solvent polarity, was proportional to the rate of polymerization (or $k_p/k_t^{1/2}$) in that solvent.

The rate is also affected by the dielectric constant of the medium, which influences the stability of the growing macroradical.

Inhibition. Free radical polymerization of acrylamide is very sensitive to residual oxygen, which efficiently scavenges primary radicals even at concentrations as low as 1 ppm. Ghosh and George (65,66) have quantified the effect of oxygen on the polymerization of acrylamide in water and ethanol. They have concluded that O_2 also reacts with macroradicals, and the polyperoxides produced terminate through a unimolecular reaction. Kishore (67) has postulated that polyperoxide radicals can also participate in monomer addition reactions. This would produce backbone O-O bonds which would be susceptible to scission at elevated temperatures, although this has not been directly observed. Recently, Vaskova (68) has shown that 4-hydroxy-2,2,6,6-tetramethyl piperdin-1-oxyl functions strictly as an inhibitor, rendering $k_p/k_t^{1/2}$ unchanged.

Review of Kinetics

In 1967, Riggs and Rodriguez (69) observed an unusual rate dependence for aqueous acrylamide polymerizations initiated with potassium persulfate:

$$R_p = k[M]^{1.25}[I]^{0.5}$$

Over the past two decades, twenty two investigations have confirmed a monomer dependency exceeding first order while maintaining that termination occurs predominantly through a bimolecular macroradical reaction (Table II). Riggs and Rodriguez (32) interpreted the high rate order as evidence of monomeric influences on the rate of initiation. This had previously been postulated by Jenkins (87) to account for similar observations made while polymerizing styrene in toluene with benzoyl peroxide as an initiator. Morgan (24) had taken the inference a step further, suggesting his sesqui-molecular order was attributable to secondary initiation caused by the monomer-enhanced decomposition of peroxide. The credibility of this hypothesis has been substantiated through experimental work performed by Dainton and co-workers (38,43,50,51). They observed the rate dependence to revert to unity in the absence of chemical initiators (Polymerizations were initiated with 60Co-γ rays, which generated H and OH radicals, but left propagation, termination and transfer reactions unchanged).

To account for high rate orders with respect to monomer three mechanisms have been proposed: the cage-effect theory (Matheson, 1946) (88), the complex theory (Gee and Rideal, 1936, 1939) (89,90) and the solvent-transfer theory (Burnett, 1955; Allen, 1955) (91,92). The latter assumes that the solvent acts as a unimolecular terminating agent. A necessary consequence is that the corresponding transfer radical is unreactive. The solvent-transfer mechanism is not applicable in aqueous media where the hydroxy radical is very unstable and is capable of initiating olefinic monomers (Noyes, 1955) (93). Chambers (94) quantified the initiation by hydroxy radicals and found the rate constant

Table II: Kinetic Investigations of Acrylamide Polymerizations in Aqueous Media

Authors	Year	"a" in $R_p \propto M^a$	Initiator
Morgan (24)	1946	1.50	potassium persulfate
Riggs and Rodriguez (69)	1967	1.25	potassium persulfate
Gromov (40)	1967	1.60	ammonium persulfate
Friend and Alexander (70)	1968	1.25	potassium persulfate
Geczy (71)	1971	1.50	potassium persulfate/ascorbic acid
Ishigie and Hamielec (55)	1973	1.24	4,4'-azobis-4-cyanovaleric acid
Hussain and Gupta (34)	1977	1.50	Ce(IV)-mercaptoethanol
Trubitsyna (72)	1977	1.50	potassium persulfate
Osmanov (73,74)	1978	1.20 in H_2O 1.30 in H_2O/ DMSO (1:1) 1.50 in DMSO	potassium persulfate
Trubitsyna (75,76)	1978	1.60	potassium persulfate
Kurenkov (77)	1978	1.70	potassium persulfate
Singh, Manickam and Venkataroo (78)	1979	1.50	potassium persulfate
Osmanov (79,80)	1979, 1980	1.70 in LiBr sol. 1.90 in LiCl sol. 2.60 in $CaCl_2$ sol.	UV irradiation with AIBN photosensitizer
Pohl and Rodriguez (33)	1980	1.53	potassium persulfate/sodium bisulfite
Kurenkov (81)	1981	1.30	ammonium persulfate
Lenka (36)	1984	1.12	peroxydiphosphate/sodium thiosulfate
Khokhrin (82)	1985	1.47	alkylbenzene sulfonato-bis acetylacetonate
Kurenkov (83)	1986	1.10	potassium persulfate/sodium bisulfite
Baade (84)	1986	1.10	2.2'-azobis[N-2-hydroxyethyl butra-amidine]
Fouassier (47)	1987	1.50 1.60	benzophenone/triethylamine benzophenone
Kurenkov (30)	1987	1.37	potassium persulfate/sodium bisulfite
Chen and Kuo (85)	1988	1.38	Co(III)/triethylene tetramine
Rafi'ee Fanood (86)	1988-I	1.16	4,4'-azobis-4-cyanovaleric acid

for the reaction with acrylamide to be $1.1 \cdot 10^{11}$ L/mol $\cdot$ min, far in excess of that for sulfite radicals formed via persulfate decomposition.

The complex theory assumes a reversible association complex is formed between the monomer and initiator. This decomposes to produce a primary radical and a macroradical of length 1:

1.
$$I + M \; \underset{}{\overset{K_c}{\rightleftharpoons}} \; I - M$$

2.
$$I - M \; \overset{k_i}{\rightarrow} \; R_1^{\bullet} + R_{in}^{\bullet}$$

Matheson's alternative explanation for the initiation mechanism assumes that as two fragments of a dissociated molecule are produced they are contained in a "cage" of solvent molecules. This radical pair may combine several times before diffusing out of the cage. This hypothesis is based on Eyring's (1940) observation (95) for benzene at room temperature, where a molecule made 10^{10} movements in its equilibrium position per second but underwent $10^{13\text{-}14}$ collisions in the same period. For the persulfate initiated polymerization of acrylamide the cage-effect theory can be written as:

1.
$$I \; \underset{k_1'}{\overset{k_1}{\rightleftharpoons}} \; (R_{in}^{\bullet} \quad {}^{\bullet}R_{in})$$

2.
$$(R_{in}^{\bullet}) \; \overset{k_D}{\rightarrow} \; R_{in}^{\bullet}$$

3.
$$2R_{in}^{\bullet} \; \overset{k_r}{\rightarrow} \; I$$

4.
$$(R_{in}^{\bullet} \; {}^{\bullet}R_{in}) + M \; \overset{k_{MI}}{\rightarrow} \; R_1^{\bullet} + R_{in}^{\bullet}$$

Although based on very different premises both these mechanisms reduce to an identical rate equation:

$$R_p = \frac{k_p}{k_t^{1/2}} [I]^{1/2} [M]^{3/2} (2k_i)^{1/2} \left(\frac{K}{1 + K[M]} \right)^{1/2}$$

where k_i is the complex/initiator decomposition constant for the complex and cage theories, respectively, and K is the complex association constant (K_c) or the ratio k_{MI}/k_r, which represents the reltaive rate a caged radical undergoes propagation and recombination. These predict a decrease in the rate order from 1.5 to 1.0 as the conversion increases. While either mechanism can satisfactorily predict the conversion time development for acrylamide polymerizations (55), there has been no direct verification that the order is changing with monomer concentration. Furthermore, since both theories reduce to the same rate equation, kinetics cannot be used to discriminate between the mechanisms.

Evaluation of the Cage-Effect and Complex Theories. Flory (1953) has shown that for typical values of radical diffusivities (10^{-5} cm^2/s) the monomer cannot appreciably influence the events in the cage (96). That is, the enhanced decomposition of caged radicals is insignificant relative to diffusion out of the cage, unless, as Jenkins showed (87), the cage has enormous dimensions (10^4 Å radius), which seems improbable. The theory does however predict low efficiencies of initiation (97) and these have been reported for aqueous acrylamide polymerizations (f = 0.024), (34).

Noyes (93) invoked the concept of a hierarchal cage structure and defined the following:

Primary recombination: Between two molecular fragments that are
                                  separated by less than one molecular diameter.

Secondary recombination: Between two fragments of the same molecule that
                                  have diffused greater than one molecular diameter
                                  apart.

Tertiary recombination: Between two fragments from different initiator
                                  molecules.

Primary and secondary recombination occur in $\approx 10^{-13}$ and 10^{-9} seconds respectively. Since the time between diffusive displacements is approximately 10^{-11} seconds, monomer-enhanced decomposition cannot compete with primary recombination. However, if the scavenger (monomer) concentration is high, Noyes calculated that the fraction of radicals reacting with scavenger that would otherwise have undergone secondary recombination is, to the first approximation, proportional to $[M]^{1/2}$. That is:

$$R_I \alpha [M]^{1/2}$$

which accounts for the observed rate behavior for acrylamide polymerizations.

Noyes' theoretical calculation is equivalent to our kinetic approach if we assume the existence of two cage entities: "compact", where the radicals are separated by less than one molecular diameter and "diffuse" where they have diffused further apart. It is assumed that only the latter are susceptible to monomer attack. These calculations suggest that although monomer-cage interactions are insignificant for cages with short lifetimes, for the fraction of cages where radicals are significantly separated, the enhanced decomposition reactions are competitive. Furthermore, the long cage lifetimes necessary for enhanced decomposition suggest that if the cage theory is true, Flory and Jenkins have overestimated the macroradical diffusion coefficient.

Complex Theory. The complex theory has been historically criticized and rejected because experimental data indicate that the association constant rises with temperature. This is inconsistent with energetic predictions which indicate the complex is less favorable at higher temperatures. Further indirect evidence against amide-persulfate complexability was presented by Riggs and Rodriguez (69) who showed the overall activation energy for aqueous acrylamide polymerization (16, 900 cal/mol) was almost exclusively composed of the contribution from the thermal decomposition of potassium persulfate ($E_{kd/2} = 16, 800$ cal/mol), (complex formulation should reduce the activation energy). However, more recent experiments (75) indicate that the overall activation energy is appreciably lowered; to below 10 kcal/mol in the presence of monomer. In the same investigation ultraviolet spectrometry was used to identify new complexes produced when acrylonitrile and N-vinyl pyrrolidone; two nitrogen containing monomers were mixed with potassium persulfate. Furthermore, the optical intensity of the new bands reached a maximum at a time coincident with the induction period of the reaction. The authors concluded a donor-acceptor complex was produced between the nitrogen containing monomer and the persulfate (Trubitsyna, 1966) (98). These authors also attributed the colour change, noticeable immediately after mixing nitrogen containing monomers and persulfates, to the formation of a molecular complex. Further evidence to this end came from polymerizations with monomers that are stronger proton donors than acrylonitrile and N-vinyl pyrrolidone: styrene, methyl methacrylate, isoprene and methyl acrylate. When these are added to persulfate no polymerization occurs although iodometrically the concentration of potassium persulfate decreases by 70% in the first hour. These monomers are preferentially forming a complex with potassium persulfate and blocking out the nitrogen containing monomers. This is significant in three respects: it shows that the new bands in the UV spectrum cannot be attributable to non-monomeric speices, such as oxygen, secondly it demonstrates that proton donation from the monomer to the persulfate is a prerequisite for enhanced decomposition. It also implies that the complex is initiating polymerization, since potassium persulfate alone at 20°C is incapable of initiating the reaction.

Other peroxide initiators (benzoyl peroxide: Trubitsyna, 1965) (99) have also been shown to decompose at faster rates in the presence of nitrogen containing additives. In this case the enhanced decomposition was also accompanied by a decrease in the overall activation energy. This was observed by the same authors several years later for reactions between benzoyl peroxide and aminated polystyrene (100), indicating that the donor-acceptor interactions are not dependent on molecular architecture.

In 1978, Trubitsyna (76) used conductivity to monitor the charged particles produced from the interaction of acrylamide and potassium persulfate. These experiments found the onset of charged particle generation (80 minutes) corresponded to the induction period of the reaction and the onset of radical generation, as determined by ESR. Furthermore, they were performed below 20°C, where the thermal decomposition of potassium persulfate is negligible, indicating a secondary decomposition reaction was occurring. Based on these observations Trubitsyna (76) proposed the following electron donor mechanism, with concurrent radical and charge generation:

$$CH_2=C(H)C(O)NH_2 + K_2S_2O_8 \rightleftharpoons [(CH_2=C(H)C(O)NH_2)K_2S_2O_8]$$

$$CCT$$

$$CCT \rightleftharpoons [(CH_2=C(H)C(O)N^\bullet H_2)^+KSO_4^{-1} + KSO_4^\bullet$$

Ionic adduct

$$Ionic\ adduct \rightleftharpoons KHSO_4 + CH_2=C^\bullet C(O)NH_2$$

Morsi (101) observed that diphenyl amine enhanced the decomposition of benzoyl peroxide. He also attributed this to a donor-acceptor interaction between the amine and the peroxide, and suggested the interaction was caused by a modification of the peroxides dihedral angle. Further evidence that acrylamide complexes with potassium persulfate comes from Bekturov (102) who found SO_4^{2-} salted out polyvinyl pyrrolidone, but could not precipitate polyacrylamide, presumably because it was neutralized by a reaction with the amide side chains. The reaction of amides with persulfate is not surprising in light of NMR evidence (61) which shows that the carbonyl groups are hydrogen bonded to water but the amides are relatively free. Coleman (103) has confirmed that the carbonyl and amide substituents behave complementarily, in that the binding of one functional group is concurrent with the reactivity of the second.

Based on the literature review the following conclusions can be made:

1)   Highly electronegative monomers can form "associates" with persulfates which lead to donor-acceptor interactions between the monomer and the peroxide. These enhance the decomposition of, for example, potassium persulfate, at low temperatures, reducing the activation energy and extending the useful range of the initiator.

2)   The electrical environment of the monomer is the primary factor in its interaction with peroxide.

3)   The enhanced decomposition of complexed persulfate is caused by hindered recombination and not a greater frequency of fragment dissociations. When the initiator is bound to the monomer the dissociated radical pair cannot regenerate potassium persulfate by recombination. Either radicals or an inert recombination product of the form $^{-\bullet}OSO_3MO_3SO^{\bullet-}$ are produced. Therefore, in the presence of donor-acceptor interactions each radical separation or "transient dissociation" results in the consumption of one initiator molecule. In the absence of bound monomer, an initiator transiently decomposes and recombines 10^{2-3} times for each "permanent decomposition".

Elucidation of an Initiation Mechanism

Both the cage and complex theories are inapplicable in their present forms as they require a unit efficiency for monomer-enhanced initiation. Manickam (104) has also developed a mechanism where monomer enhanced decomposition is included outside the framework of cage or complex theories, but this also assumes a 100% efficiency for enhanced decomposition. Additional reactions must therefore be included to account for cage or complex destruction, either of which can potentially generate non-reactive products.

Complex/Cage Equivalence. Prior to monomer enhanced decomposition an intermediate "associate" must be formed. This allows the electron donating group to attack the peroxide for a sufficient period to be competitive with the rapid radical fragmentation and recombination reactions. This monomer-initiator associate can result from either the diffusive displacement of a monomer to the volume element of the peroxide (cage approach) or the formation of a molecular complex. Both of these phenomena can be represented by a general reaction that is non-specific to the forces drawing the monomer and initiator into close proximity (This modification must also be applied to Manickam's mechanism to successfully apply it):

$$\{R_{in}^{\bullet} \: {}^{\bullet}R_{in}\} + M \; \rightarrow \; \left\{ \frac{R_{in}^{\bullet} + R_{in}^{\bullet}}{M} \right\}$$

That is, the "associate" is a broadly defined concept encompassing both covalent and weak-bonding interactions. However, since experimental measurements (76) have found the associate to be irreversibly formed, uncharacteristic of covalently bonded molecules, the weak-bonding interaction must be responsible for monomer-initiator association. This "associate" can therefore be represented as a "weak complex", formed presumably due to hydrogen bonding, or equivalently, as a "diffuse monomer swollen cage", the difference being entirely semantic since in both models the monomer is physically contained within the three dimensional volume element of a diffuse radical pair. In other words, the complex and cage treatments are non-discriminating models of the same physical phenomena as far as amide-persulfate interactions are concerned.

The ambiguity in defining cage or complex structures allows us to combine the positive features of both mechanisms. Specifically, Noyes' hierarchical cage structure will be used as a precursor to charge transfer interactions. This allows the implementation of an electron donating mechanism without the inconsistency of association phenomena increasing with temperature. This will be presented in a subsequent section following a discussion of the existing mechanism.

Historically used Mechanism. Based on experimental observations the following mechanism has been proposed (76):

1.
$$S_2O_8^{2-} \; \xrightarrow{k_d} \; 2\,SO_4^{\bullet -}$$

2.
$$S_2O_8^{2-} + M \; \xrightarrow{k_a} \; I-M \quad \text{(Complex formation)}$$

where M designates acrylamide monomer.

3.
$$I-M \xrightarrow{k_b} CH_2C^\bullet CONH_2 + SO_4^{\bullet-}$$

4.
$$I-M \xrightarrow{k_c} Q \quad (\text{Inert products})$$

We have added step 4 to avoid the improbable situation of a unit efficiency for monomer-enhanced decomposition. Bunn (105) has shown that sulfate radicals can also react with water to produce hydroxy radicals, which are capable of initiation (94):

5.
$$SO_4^{\bullet-} + H_2O \xrightarrow{k_H} HSO_4^- + {}^\bullet OH$$

6.
$$SO_4^{\bullet-} + M \xrightarrow{k_{i1}} R_1^\bullet$$

7.
$${}^\bullet OH + M \xrightarrow{k_{i2}} R_1^\bullet$$

where $R_1^\bullet$ denotes a macroradical of length 1.

We must also include the propagation, termination and transfer reactions that have previously been identified for aqueous acrylamide polymerizations with water soluble initiators.

8.
$$R_r^\bullet + M \xrightarrow{k_p} R_{r+1}^\bullet$$

9.
$$R_r^\bullet + M \xrightarrow{k_{fm}} P_r + R_1^\bullet$$

10.
$$R_r^\bullet + R_s^\bullet \xrightarrow{k_{td}} P_r + P_s$$

Where $R_r^\bullet$ is a macroradical of length r and P_r is a dead polymer chain of length r.
Defining

$$[R^\bullet] = \sum_{r=1}^{\infty} [R_r^\bullet]$$

we can construct the balances on the reactive species:

$$\frac{d[S_2O_8^{2-}]}{dt} = -k_d[S_2O_8^{2-}] - k_a[S_2O_8^{2-}][M] \tag{1}$$

$$\frac{d[SO_4^{\bullet-}]}{dt} = 2k_d[S_2O_8^{2-}] + k_b(I-M) - k_{i1}[SO_4^{\bullet-}][M] - k_H[SO_4^{\bullet-}][H_2O] \approx 0 \tag{2}$$

$$\frac{d[^\bullet OH]}{dt} = k_H[SO_4^{\bullet-}][H_2O] - k_{i2}[^\bullet OH][M] \approx 0 \tag{3}$$

$$\frac{d[I-M]}{dt} = k_a[I][M] - (k_b + k_c)[I-M] \approx 0 \tag{4}$$

$$\frac{d[R^\bullet]}{dt} = k_{i1}[SO_4^{\bullet-}][M] + k_{i2}[^\bullet OH][M] + k_b[I-M] - k_{td}[R^\bullet]^2 \approx 0 \tag{5}$$

Assuming k_p is independent of chain length ($k_{i1} \approx k_{i2} = k_p$) equations (2-4) can be substituted into equation (5) to yield:

$$k_{td}[R^\bullet]^2 = 2f\,k_d[S_2O_8^{2-}] + \frac{2\,k_a[S_2O_8^{2-}][M]}{1 + \dfrac{k_c}{k_b}}$$

The macroradical concentration may be written as:

$$[R^\bullet] = \left(\frac{2f\,k_d[S_2O_8^{2-}]}{k_{td}} + \frac{2f_c\,k_a[S_2O_8^{2-}][M]}{k_{td}} \right)^{1/2}$$

Where the complex/cage efficiency is defined as:

$$f_c = \frac{1}{1 + \dfrac{k_c}{k_b}}$$

Where the first term represents chains initiated by thermal decomposition of the persulfate and the second monomer-enhanced decomposition.

The rate of polymerization (R_p) may be expressed as:

$$R_p = k_p[M]\left(\frac{2f\ k_d[S_2O_8^{2-}]}{k_{td}} + \frac{2f_c\ k_a[S_2O_8^{2-}]}{k_{td}}[M] \right)^{1/2}$$ (6)

This can be used to describe the kinetics for acrylamide polymerizations. However, if applied, two parameter inconsistencies occur: The complex decomposition efficiency is very low ($f_c = 0.06$–0.4) suggesting a cage process is responsible for the kinetics, and the association constant (k_a) increases with temperature.

Proposed Mechanism. The apparent increase in k_a with temperature and the low efficiency can be rationalized by considering the charge transfer reaction to be preceded by the generation of diffuse cage radical fragments which have large enough molecular separations to prevent recombination and provide access to proton donating monomers. The mechanism can therefore be expressed as a hybrid of the cage and complex theories as follows:

Initiator reactions

1.
$$S_2O_8^{2-} \underset{k_1}{\overset{k_1}{\rightleftharpoons}} (SO_4^{\bullet-}\ {}^{-\bullet}O_4S)$$

2.
$$(SO_4^{\bullet-}\ {}^{-\bullet}O_4S) \overset{k_2}{\rightarrow} \{SO_4^{\bullet-}\ {}^{-\bullet}O_4S\}$$

3.
$$\{SO_4^{\bullet-}\ {}^{-\bullet}O_4S\} \overset{k_d^*}{\rightarrow} 2SO_4^{\bullet-}$$

4.
$$\{SO_4^{\bullet-}\ {}^{-\bullet}O_4S\} \overset{k_r}{\rightarrow} S_2O_8^{2-}$$

Solid parentheses indicate a "compact cage" and braced parentheses signify a "diffuse cage".

Swollen cage formation and decomposition

5.
$$\{SO_4^{\bullet-}\ {}^{-\bullet}O_4S\}+M \overset{k_a^*}{\rightarrow} \left\{ \begin{matrix} SO_4^{\bullet-}\quad {}^{-\bullet}O_4S \\ M \end{matrix} \right\}$$

6.
$$\left\{ \begin{matrix} SO_4^{\bullet-}\quad {}^{-\bullet}O_4S \\ M \end{matrix} \right\} \overset{k_b}{\rightarrow} HSO_4^- + SO_4^{\bullet-} + CH_2 = C^{\bullet}C(O)NH_2$$

7.
$$\left\{ \begin{array}{cc} SO_4^{\bullet-} & {}^{-\bullet}O_4S \\ & M \end{array} \right\} \xrightarrow{k_c} Q \quad \text{(Inert products)}$$

Chain Initiation

8.
$$SO_4^{\bullet-} + M \xrightarrow{k_{i1}} R_1^{\bullet}$$

9.
$$SO_4^{\bullet-} + H_2O \xrightarrow{k_H} HSO_4^- + {}^{\bullet}OH$$

10.
$${}^{\bullet}OH + M \xrightarrow{k_{i2}} R_1^{\bullet}$$

Propagation

11.
$$R_r^{\bullet} + M \xrightarrow{k_p} R_{r+1}^{\bullet}$$

Transfer to Monomer

12.
$$R_r^{\bullet} + M \xrightarrow{k_{fm}} P_r + R_1^{\bullet}$$

Termination

13.
$$R_r^{\bullet} + R_s^{\bullet} \xrightarrow{k_{td}} P_r + P_s$$

At any time t, the persulfate is comprised of undissociated initiator ($S_2O_8{}^{2-}$), "compact caged fragments" ($SO_4^{\bullet-} \quad {}^{\bullet}O_4S$), and "diffuse cage fragments" $\{SO_4^{\bullet-} \quad {}^{\bullet}O_4S\}$. The total persulfate in the system (I_t) is:

$$I_t = [S_2O_8^{2-}] + (SO_4^{\bullet-} \quad {}^{-\bullet}O_4S) + \{SO_4^{\bullet-} \quad {}^{-\bullet}O_4S\}$$

where

$$[S_2O_8^{2-}] = \phi_0 I_t$$

$$(SO_4^{\bullet-} \quad {}^{-\bullet}O_4S) = \phi_1 I_t$$

$$\{SO_4^{\bullet-} \quad {}^{-\bullet}O_4S\} = \phi_2 I_t$$

and $\Phi_0 + \Phi_1 + \Phi_2 = 1.0$

The balances on all reactive species follow:

$$\frac{d[S_2O_8^{2-}]}{dt} = -k_1[S_2O_8^{2-}] + k_1'(SO_4^{\bullet-} \quad {}^{-\bullet}O_4S) + k_r\{SO_4^{\bullet-} \quad {}^{-\bullet}O_4S\} \tag{7}$$

$$\frac{d(SO_4^{\bullet-} \quad {}^{-\bullet}O_4S)}{dt} = k_1[S_2O_8^{2-}] - (k_1' + k_2)(SO_4^{\bullet-} \quad {}^{-\bullet}O_4S) \tag{8}$$

$$\frac{d\{SO_4^{\bullet-} \quad {}^{-\bullet}O_4S\}}{dt} = k_2(SO_4^{\bullet-} \quad {}^{-\bullet}O_4S) - k_d^*\{SO_4^{\bullet-} \quad {}^{-\bullet}O_4S\}$$

$$-k_r\{SO_4^{\bullet-} \quad {}^{-\bullet}O_4S\} - k_a^*\{SO_4^{\bullet-} \quad {}^{-\bullet}O_4S\}[M] \tag{9}$$

$$\frac{d\left\{ \begin{matrix} SO_4^{\bullet-} \quad {}^{-\bullet}O_4S \\ M \end{matrix} \right\}}{dt} = k_a^*\{SO_4^{\bullet-} \quad {}^{-\bullet}O_4S\}[M] - (k_b + k_c)\left\{ \begin{matrix} SO_4^{\bullet-} \quad {}^{-\bullet}O_4S \\ M \end{matrix} \right\} \approx 0 \tag{10}$$

$$\frac{d(SO_4^{\bullet-})}{dt} = 2k_d^*\{SO_4^{\bullet-} \quad {}^{-\bullet}O_4S\} + k_b\left\{ \begin{matrix} SO_4^{\bullet-} \quad {}^{-\bullet}O_4S \\ M \end{matrix} \right\}$$

$$-k_H[SO_4^{\bullet-}][H_2O] - k_{i1}[SO_4^{\bullet-}][M] \approx 0 \tag{11}$$

$$\frac{d[^\bullet OH]}{dt} = k_H[SO_4^{\bullet-}][H_2O] - k_{i2}[^\bullet OH][M] \approx 0 \tag{12}$$

$$\frac{d[R^\bullet]}{dt} = k_{i1}[SO_4^{\bullet-}][M] + k_{i2}[^\bullet OH][M] + k_b\left\{ \begin{matrix} SO_4^{\bullet-} \quad {}^{-\bullet}O_4S \\ M \end{matrix} \right\} - k_{td}[R^\bullet]^2 \approx 0 \tag{13}$$

From equation (10):

$$\left\{ \begin{matrix} SO_4^{\bullet-} \quad {}^{-\bullet}O_4S \\ M \end{matrix} \right\} = \left(\frac{k_a^*}{k_b + k_c} \right)\{SO_4^{\bullet-} \quad {}^{-\bullet}O_4S\}[M] \tag{10'}$$

Substituting equations 10′, 11 and 12 into equation 13 yields:

$$[R^\bullet] = \left(\frac{2f\, k_d^*\{SO_4^{\bullet-} \quad {}^{-\bullet}O_4S\}}{k_{td}} + \frac{2f_c\, k_a^*\{SO_4^{\bullet-} \quad {}^{-\bullet}O_4S\}[M]}{k_{td}} \right)^{1/2}$$

Where

$$\{SO_4^{\bullet-} \quad {}^{-\bullet}O_4S\} = \phi_2\,(S_2O_8^{2-})$$

and

$$f_c = \frac{1}{1 + \dfrac{k_c}{k_b}}$$

Where f_c can be defined equivalently as a cage-destruction or complex-decomposition efficiency.

The long chain approximation subsequently yields:

$$R_p = k_p[M]\left(\frac{2f(\phi_2 k_d^*)[S_2O_8^{2-}]}{k_{td}} + \frac{2f_c(\phi_2 k_a^*)[S_2O_8^{2-}]}{k_{td}}[M] \right)^{1/2} \tag{14}$$

<u>Discussion of the derived rate equation.</u> The decomposition constant (k_d) has been determined from measurements of the concentration of undecomposed initiator (106):

$$\frac{d[S_2O_8^{2-}]}{dt} = -k_d[S_2O_8^{2-}] \tag{15}$$

However, since only diffuse cage fragments can dissociate to produce unpaired free radicals equation (15) can more appropriately be written as:

$$\frac{d[S_2O_8^{2-}]}{dt} = k_d^*\{SO_4^{\bullet-} \quad {}^{-\bullet}O_4S\} \tag{16}$$

Expressing the concentration of these diffuse cages as a fraction of the overall initiator level, equations (15) and (16) can be combined to yield:

$$\frac{d[S_2O_8^{2-}]}{dt} = \phi_2 k_d^*[S_2O_8^{2-}]$$

Which provides the identity $k_d = \Phi_2 k_d^*$. By an analogous procedure it can be shown $k_a = \Phi_2 k_a^*$. Where k_a is the apparent (overall) association constant and k_a^* is the actual (specific) association constant.

In the "hybrid cage/complex" theory, the apparent association constant is experimentally observed to rise with temperature. This can be attributed to either an increase in the specific association parameter, which is entropically unfavorable, or an increase in the fraction of persulfate present as diffuse cages. The latter appears to be the correct explanation, since supplying additional energy to the system would increase the frequency of equilibrium displacements while reducing the probability of radical recombination. This important conclusion enables us to include donor-acceptor interactions between monomer and initiator without the thermodynamic inconsistencies of prior charge transfer mechanisms. The mechanism derived herein will be used in the following chapter for evaluation against experimental data.

The rate equation (14) is unique in a second regard. It suggests that the 1.25 order for acrylamide polymerization with persulfate is comprised of two initiation reactions thermal decomposition, providing first order kinetics, and monomer-enhanced decomposition which has a sesquimolecular rate order. Table III shows that the rate data generated from equation (14), with arbitrary parameter values, are virtually indistinguishable from kinetics generated by a rate equation with a single $5/4^{th}$ order term. This implies we cannot make inferences as to whether the observed order is caused by single or multiple initiation reactions. Furthermore, kinetics exclusively cannot disprove the hybrid mechanism, or distinguish complex and cage theories.

Generalization of the Initiation Mechanism to Other Water Soluble Monomers

The charge transfer complexes that form between persulfate and nitrogen containing monomers require free electrons on the donor (monomer) and a polar medium. For acrylamide polymerizations in dimethyl sulfoxide-water mixtures the rate order with respect to monomer increases as the fraction of DMSO in the solvent mixture rises (73,74). This is caused by the aprotic nature of dimethyl sulfoxide and its inability to hydrogen bond to the acrylamide carbonyl group. This less solvated monomer is therefore more readily complexed with the peroxide, and undergoes enhanced decomposition at an accelerated rate. The rate therefore approaches sesquimolecular order, the limit if monomer and initiator form 1:1 complexes, as the fraction of DMSO rises, in agreement with experimental observations.

This effect is also observed in other monomers with electronegative atoms, for example in acrylic acid(oxygen)-persulfate systems (104). The polarity and hydrogen bonding affinity of the hydroxy group exceed that of the corresponding amide, and the rate order again approaches the sesquimolecular limit. Indeed, the monomer's ability to form hydrogen bonds, through the side chain functional group, appears to be one of the contributing factors for the formation of monomer-initiator complexes.

For N, N, dimethyl acrylamide, which cannot hydrogen bond to either the peroxide or persulfate oxygens, due perhaps to steric interference of the bulky methyl substituents, a unimolecular rate order with respect to monomer is observed (57,107). Ergozhin (108) investigated the kinetics of a series of N-substituted amides and observed the rate order with respect to monomer concentration to decrease as the accessibility to the vinyl group was hindered. This confirms Trubitsyna's postulate (76) that the amide is responsible for the electron rearrangement leading to the monomer-initiator association. Haas (109) has observed that other amide containing monomers, for example acrylylglycinamide, enhance the decomposition of potassium persulfate, and have the same rate order with respect to monomer concentration as acrylamide. If methacrylamide replaces acrylamide as the monomer in a persulfate reaction a greater than first order dependence is again observed ($R_p \propto M^{1.13}$; Gupta, 1987) (35). The reduced order from the $5/4^{th}$ power may be an experimental anomaly, as methacrylamide has not been extensively investigated. However, it is more likely that the α-methyl substitution is affecting the electron rearrangement necessary to produce a monomer-initiator complex (110); (Methacrylamide radicals are present in a resonance stabilized structure where the β-carbon can more easily stabilize a radical than the α-methyl substituted carbon (111)).

The mechanism, therefore, in its present form can explain the rate behavior for several water soluble monomers (acrylamide, acrylic acid, N, N-dimethyl acrylamide, methacrylamide). The stronger the monomer polarity, or its proton donating ability, the greater the proportion of initiator that decomposes through a monomer-enhanced mechanism. Contrarily, steric interferences reduce the ability of charge transfer complexes to form, and increase the proportion of sulfate radicals produced via thermal decomposition of the peroxide

The sum of a first and sesquimolecular order initiation mechanism is therefore flexible enough to describe a broad array of kinetic observations for ionogenic and nonionic monomers.

Generalization to Cationic Monomers. Friend and Alexander (70) were the first to observe an interaction between persulfate and quaternary ammonium compounds. Trubitsyna (72,75,76) and Kurenkov (112) latter observed complexes and enhanced decomposition of persulfate due to cationic ammonium additives. Further, the magnitude of the enhanced decomposition was significantly greater than was observed for acrylamide. It has been shown (70) that charge transfer interactions are responsible for the formation of a 1:2 stoichiometric complex of the following type:

$$
R_4N^+ \quad {}^-O-\underset{\underset{O}{\|}}{\overset{\overset{O}{\|}}{S}}-O-O-\underset{\underset{O}{\|}}{\overset{\overset{O}{\|}}{S}}-O \quad {}^+NR_4
$$

These decompose to produce two macroradicals of length 1 (when 1:1 complexes are produced, one primary radical is liberated). Such a mechanism reduces to a second order rate dependence on monomer concentration, in agreement with experimental observations for diallyldimethylammonium chloride polymerization (Jaeger, 1984) (113). Jaeger also showed that the rate order was reduced by 1 when a non-complexing initiator (azobispentanoic acid) was used in place of potassium persulfate. Therefore, the high rate orders observed for polymerization of acrylic water soluble monomers in aqueous media initiated by persulfate are almost certainly due to the unique donor-acceptor interaction of the monomer-initiator pair. The strength in this interaction determines the deviation in order from unity. Table IV summarizes the observed kinetic relationships for several nonionic, anionic and cationic acrylic water soluble monomers. A correlation between the rate order with respect to monomer and the strength of the monomer-initiator complex is again observable.

Based on the preceding literature survey we can propose the following general initiation mechanism for acrylic water soluble monomers with persulfate:

1.
$$ S_2O_8^{2-} \leftrightarrows (SO_4^{\bullet-} \quad {}^-{}^\bullet O_4S) $$

2.
$$ (SO_4^{\bullet-} \quad {}^-{}^\bullet O_4S) \to \{SO_4^{\bullet-} \quad {}^-{}^\bullet O_4S\} $$

3.
$$ \{SO_4^{\bullet-} \quad {}^-{}^\bullet O_4S\} \to S_2O_8^{2-} $$

4.
$$ \{SO_4^{\bullet-} \quad {}^-{}^\bullet O_4S\} \to 2SO_4^{\bullet-} $$

Table III: Equivalence of two kinetic processes which provide 5/4th
power rate dependencies

Monomer Concentration [M]	$R_p = k_1^\dagger[M]^{1.25}$	$R_p = k_1^\dagger([M]+[M]^{1.5})/2$
0.5	0.42	0.43
1.0	1.0	1.0
2.0	2.38	2.41
3.0	3.95	4.10
4.0	5.66	6.00
5.0	7.48	8.09
6.0	9.39	10.34
7.0	11.39	12.76

$\dagger$ k_1 is assigned an arbitrary value of 1.0

Table IV: Rate Equations for Several Acrylic Water Soluble Monomers

Monomer	"a" in R_p & M^a	Reference
N,N,dimethylacrylamide	1.0	Kurenkov, 1980 (112)
Methacrylamide	1.13	Gupta, 1987 (35)
Acrylylglycinamide Acrylamide	1.22 1.25	Haas, 1970 (109) Riggs and Rodriguez, 1967 (69)
Acrylic acid	1.50	Manickam, 1979 (104)
Diallyl dimethyl ammonium chloride	2.0+	Jaeger, 1984 (113) Hahn, 1983 (114, 115)

+   A third order dependence has been reported. However, the monomeric salt influences propagation between charged radicals and monomer molecules. This results in a first order relationship between the propagation rate constant and the monomer concentration.

5
$$SO_4^{\bullet-} + H_2O \rightarrow HSO_4^{\bullet-} + {}^{\bullet}OH$$

6
$$SO_4^{\bullet-} + M \rightarrow R_1^{\bullet}$$

7.
$${}^{\bullet}OH + M \rightarrow R_1^{\bullet}$$

8.
$$\{SO_4^{\bullet-} \quad {}^{-\bullet}O_4S\} + xM \rightarrow \left\{ \begin{matrix} SO_4^{\bullet-} \quad {}^{-\bullet}O_4S \\ xM \end{matrix} \right\}$$

9.
$$\left\{ \begin{matrix} SO_4^{\bullet-} \quad {}^{-\bullet}O_4S \\ xM \end{matrix} \right\} \rightarrow (2-x)R_{in}^{\bullet} + xR_1^{\bullet}$$

10.
$$\left\{ \begin{matrix} SO_4^{\bullet-} \quad {}^{-\bullet}O_4S \\ xM \end{matrix} \right\} \rightarrow Q \quad \text{(Inert products)}$$

Where x is the stoichiometric ratio of monomer in the initiator complex. That is:

For anionic and nonionic monomers, $x = 1$ and $R_p \, \alpha \, k \cdot M + k' \cdot M^{3/2}$

For cationic monomers, $x = 2$ and $R_p \, \alpha \, k \cdot M + k'' \cdot M^2$

An oppositely charged monomer-initiator pair is therefore able to form higher stoichiometric complexes than if the initiator and monomer are of the same charge, or if one or both of the species are uncharged. This accounts for the higher order in rate with respect to monomer concentration for aqueous polymerizations of cationic monomers with persulfate.

Literature Cited

1.   Seymour, K.G., Harper, B.G., Ger. Pat., 1 197 272 (1965).
2.   Gramain, Ph., Myard, Ph., J. Coll. Interfac. Sci., 84, 114 (1981).
3.   Mowry, D.T., Hendrick, R.M., US Patent, 2 625 471 (1953).
4.   Bikales, N.M., in "Water Soluble Polymers", V2, N.M.Bikales, ed., Plenum Press, New York, 1973.
5.   Pye, D.J., US Patent, 3 072 536 (1963).
6.   Cyanamid, "Chemistry of Acrylamide", New Jersey, 1969.
7.   Meltzer, X.L., "Water Soluble Polymers", Noyes Data Corporation, Park Ridge, New Jersey, 1979.
8.   Stackman, R.W., Hurley, S.M., Polym. Mater. Sci. Eng., 57, 830 (1987).
9.   MacWilliams, D.C., in "Functional Monomers, Their Preparation, Polymerization and Application", V2, R.H. Yocum, E.B. Nyquist, ed., Dekker, New York, 1974.
10. Nagashiro, W., Tsunoda, T., Tanaka, M., Oikawa, M., Bull. Chem. Soc. Japan, 48, 2597 (1975).
11. Grollmann, U., Schnabel, W., Polym. Degradation and Stability, 4, 203 (1982).
12. Ramsden, D.K., McKay, K., Polym. Degradation and Stability, 14, 217 (1986).

13.   Ramsden, D.K., Fielding, S., Atkinson, N., Boota, M., Polym. Degradation and Stability, 17, 49 (1987).
14.   Klein, J., Westercamp, A., J. Polym. Sci. Chem., 19, 707 (1981).
15.   Lancaster, J.E., O'Connor, M.N., J. Polym. Sci. Letters, 20, 547 (1982).
16.   Alaya, A., Carriere, F., Monjal, P., Sekiguchi, H., Eur. Polym. J., 21, 663 (1985).
17.   Kulicke, W.-M., Kniewske, R., Klein, J., Prog. Polym. Sci., 8, 373 (1982).
18.   Hunt, J.A., Young, T.-S., Green, D.W., Willhite, G.P., SPE/DOE Fifth Symposium on Enhanced Oil Recovery, Tulsa, OK, April 20-23, 14949, 325, 1986.
19.   Henderson, J.M., Wheatley, A.D., J. Appl. Polym. Sci., 33, 669 (1987).
20.   Minsk, L.M., Kotlarchik, C., Meyer, G.N., Kenyon, W.O., J. Polym Sci. Chem., 12, 133 (1974).
21.   Guerrero, S.J., Boldarino, P., Zurinendi, J.H., J. Appl. Polym. Sci., 30, 955 (1985).
22.   Molyneux, P., "Water Soluble Synthetic Polymers: Properties and Behaviour", V1, CRC Press, Boca Raton, Florida (1983).
23.   Bacon, R.G.R., Trans. Farad. Soc., 42, 140 (1946).
24.   Morgan, L.B., Trans. Farad. Soc., 42, 169 (1946).
25.   Rodriguez, F., Givey, R.D., J.Polym.Sci., 55, 713 (1961).
26.   Hoover, M.F., Padden, J.J., US Patent, 3 442 803 (1966).
27.   Hoover, M.F., US Patent, 3 332 922 (1967).
28.   Suen, T.J., Jen, Y., Lockwood, J.V., J. Polym. Sci., 31, 481 (1958).
29.   Jursich, M.J., Randich, G.T., US Patent, 3 450 580 (1969).
30.   Kurenkov, V.F., Baiburdov, T.A., Stupen'kova, L.L., Vysokomol.soyed., A29, 348 (1987).
31.   Sackis, J.J., US Patent, 3 316 181 (1967).
32.   Riggs, J.P., Rodriguez, F., J.Polym.Sci., A1, 5, 3167 (1967).
33.   Pohl, K., Rodriguez, F., J. Appl. Polym. Sci., 26, 611 (1980).
34.   Hussain, M.M., Gupta, A., J. Macromol. Sci. Chem., A11, 2177 (1977).
35.   Gupta, K.C., Lai, M., Behari, K., Polymer Preprints 28(1), 118 (1987).
36.   Lenka, S., Nayak, P.L., Ray, S., J. Polym. Sci. Chem., 22, 959 (1984).
37.   Behari, K., Gupta, K.C., Verma, M., Vysokomol.soyed., A28, 1781 (1986).
38.   Collinson, E., Dainton, F.S., McNaughton, G.S., Trans. Farad. Soc., 53, 476 (1957).
39.   Collinson, E., Dainton, F.S., Smith, D.R., Trudel, G.J., Tuzuke, S., Disc. Farad. Soc., 29, 188 (1960).
40.   Gromov, V.F., Matveyeva, A.V., Khomikovskii, P.M., Abkin, A.D., Vysokomol.soyed., A9, 1444 (1967).
41.   Ogumi, Z., Tari, I., Takehara, Z., Yoshizawa, S., Bull. Chem. Soc. Japan, 47, 1843 (1974).
42.   Samal, S.K., Nayak, B., J. Polym. Sci. Chem., 26, 1035 (1988).
43.   Collinson, E., Dainton, F.S., McNaughton, G.S., Trans. Farad. Soc., 53, 489 (1957).
44.   Venkatarao, K., Santappa, M., J.Polym.Sci.A1, 8, 1785 (1970).
45.   Rodriguez, F., Chu, C.H., Chu, W.T.W.K., Rondinella, M.A., J. Appl. Polym. Sci., 30, 1629 (1985).
46.   Iwai, K., Vesugi, M., Takemura, F., Polymer J., 17, 1005 (1985).
47.   Fouassier, J.P., Lougnot, D.J., Zuchowicz, I., Green, P.N., Timpe, H.J., Kronfeld, K.P., Muller, U., J. Photochem., 36, 347 (1987).
48.   Ogata, N., J. Polym. Sci., 46, 271 (1960).
49.   Fadner, T.A., Morawetz, H., J. Polym. Sci., 45, 475 (1960).
50.   Dainton, F.S., Tordoff, M., Trans. Farad. Soc., 53, 499 (1957).
51.   Dainton, F.S., Tordoff, M., Trans. Farad. Soc., 53, 666 (1957).
52.   Suen, T.J., Rossler, D.F., J. Appl. Polym. Sci., 3, 126 (1960).
53.   Cavell, E.A.S., Makromol. Chem., 54, 70 (1962).
54.   Currie, D.J., Dainton, F.S., Watt, W.S., Polymer, 6, 451 (1965).
55.   Ishige, T., Hamielec, A.E., J. Appl. Polym. Sci., 17, 1479 (1973).
56.   Kim, C.J., Hamielec, A.E., Polymer, 25, 845 (1984).

57. Misra, G.S., Rebello, J.J., Makromol. Chem., 175, 3117 (1974).
58. Pantar, A.V., Eur. Polym. J., 22, 939 (1986).
59. Saini, G., Leoni, A., Franco, S., Makromol. Chem., 144, 235 (1971).
60. Chapiro, A., Perec, L., Eur. Polym. J., 7, 1335 (1971)
61. Zhuravleva, I.L., Zav'yalova, YE.N., Bogachev, Yu.S., Gromov, V.F., Vysokomol.soyed., A28, 873 (1986).
62. Chapiro, A., Perec-Spritzer, L., Eur. Polym. J., 11, 59 (1975).
63. Gromov, V.F., Galperina, N.I., Osmanov, T.O., Khomikovskii, P.M., Abkin, A.D., Eur. Polym. J., 16, 529 (1980).
64. Bune, YE.V., Zhuravleva, I.L., Sheinker, A.P., Bogachev, Yu.S., Teleshov, E.N., Vysokomol.soyed., A28, 1279 (1986).
65. Ghosh, A., George, M.H., Polymer, 19, 1057 (1978).
66. George, M.H., Ghosh, A., J. Polym. Sci. Chem., 16, 981 (1978).
67. Kishore, K., Bhanu, V.A., J. Polym. Sci. Chem., 24, 379 (1986).
68. Vaskova, V., Oremusova, D., Barton, J., Makromol.Chem., 189, 701 (1988)
69. Riggs, J.P., Rodriguez, F., J.Polym.Sci., A1, 5, 3151 (1967).
70. Friend, J.P., Alexander, A.E., J. Polym. Sci., A1, 6, 1833 (1968).
71. Geczy, I., Nasr, H.I., Acta Chim, 70, 319 (1971).
72. Trubitsyna, S.N., Ismailov, I, Askarov, M.A., Vysokomol.soyed., A19, 495 (1977).
73. Osmanov, T.O., Gromov, V.F., Komikovsky, P.M., Abkin, A.D., Vysokomol.soyed., B20, 263 (1978).
74. Osmanov, T.O., Gromov, V.F., Komikovsky, P.M., Abkin, A.D., Dokl. Akad Nauk SSSR, 240, 910 (1978).
75. Trubitsyna, S.N., Ismailov, I., Askarov, M.A., Vysokomol.soyed., A20, 1624 (1978).
76. Trubitsyna, S.N., Ismailov, I., Askarov, M.A., Vysokomol.soyed., A20, 2608 (1978).
77. Kurenkov, V.F., Osipova, T.M., Kuznetsov, E.V., Myagchenkov, V.A., Vysokomol.soyed., B20, 647 (1978).
78. Singh, U.C., Manickam, S.P., Venkatarao, K., Makromol. Chem., 180, 589 (1979).
79. Osmanov, T.O., Gromov, V.F., Komikovsky, P.M., Abkin, A.D., Vysokomol.soyed., A21, 1766 (1979).
80. Osmanov, T.O., Gromov, V.F., Komikovsky, P.M., Abkin, A.D., Vysokomol.soyed., A21, 668 (1980).
81. Kurenkov, V.F., Verihznikova, A.S., Kuznetsov, E.V., Myagchenkov, V.A., Izv. Vuz. SSSR Khimiya Khim.technol., 24 (1981).
82. Khokhrin, S.A., Shibalovich, V.G., Nikolayev, A.F., Chudnova, U.M., Vysokomol.soyed., A27, 1694 (1985).
83. Kurenkov, V.F., Verizhnikova, A.S., Myagchenkov, V.A., Vysokomol.soyed., A28, 488 (1986).
84. Baade, W., Ph.D. Dissertation, Technical University of Berlin, FRG, 1986.
85. Chen, C.-Y., Kuo, J.-F., J. Polym. Sci. Chem., 26, 1115 (1988).
86. Rafi'ee Fanood, M.H., George, M.H., Polymer, 29, 128 (1988).
87. Jenkins, A.D., J.Polym.Sci., 29, 245 (1958).
88. Matheson, M.S., J.Chem.Phys., 13, 584 (1945).
89. Gee, G., Rideal, E.K., Trans. Farad. Soc., 32, 666 (1936).
90. Cuthbertson, A.C., Gee, G., Rideal, E.K., Proc. Roy. Soc.(London), A 170, 300 (1939).
91. Burnett, G.M., Loon, L.D., Trans. Farad. Soc., 51, 214 (1955).
92. Allen, P.W., Merrett, F.M., Scalon, J., Trans. Farad. Soc., 51, 95 (1955).
93. Noyes, R.M., J. Am. Chem. Soc., 77, 2042 (1955).
94. Chambers, K., Collinson, E., Dainton, F.S., Seddon, W., Chem. Commun. (London), 15, 498 (1966).
95. Stearn, A.E., Irish, E.M., Eyring, H., J. Phys. Chem., 44, 981 (1940).
96. Flory, P.J., "Principals of Polymer Chemistry", Cornell University Press, Ithaca, New York, 1953.

97. Baer, M., Caskey, J.A , Fricke, A L., Makromol Chem., 158, 27 (1972).
98. Trubitsyna, S.N., Margaritova, M F., Prostakov, N.S., Vysokomol.soyed , 8, 532 (1966).
99. Trubitsyna, S.N., Margaritova, M.F , Medvedev, S.S., Vysokomol.soyed., 7, 2160 (1965).
100. Trubitsyna, S.N., Ruzmetova, Kh K., Askarov, M.A., Vysokomol.soyed., A13, 1950 (1971).
101. Morsi, S.E., Zaki, A.B., El-Shamy, T.M., Habib, A., Eur. Polym. J., 12, 417 (1976)
102. Bekturov, E.A., Khamzamulina, R.E., JMS- Rev. Macromol. Chem. Phys., C27, 253 (1987).
103. Coleman, M.M., Skrovanek, D.J., Hu, J., Painter, P.C., Polymer Preprints, 28(1), 19 (1987).
104. Manickam, S.P., Venkatarao, K., Subbaratnam, N.R., Eur. Polym. J., 15, 483 (1979).
105. Bunn, D., Trans. Farad. Soc., 42, 190 (1946).
106. Kolthoff, I.M., Miller, I.K., J. Am. Chem. Soc., 73, 3055 (1951).
107. Jacob, M., Smets, G., DeSchryver, F.C., J. Polym. Sci., B10, 669 (1972).
108. Ergozhin, E.E., Tausarova, B.R., Sariyeva, R.B., Makromol. Chem. Rapid Commun., 8, 171 (1987).
109. Haas, H.C., MacDonald, R.L., Schuler, A.N., J. Polym. Sci., A1, 8, 1213 (1970).
110. Manickam, S.P., Subbaratnam, N.R., Venkatarao, K., J. Polym. Sci. Chem., 18, 1679 (1980).
111. Burfield, D.R., Ng, S.C., Eur. Polym. J., 12, 873 (1976).
112. Kurenkov, V.F., Myagchenkov, V.A., Eur. Polym. J., 16, 1229 (1980).
113. Jaeger, W., Hahn, M., Seehaus, F., Reinisch, G., J. Macromol. Sci. Chem., A21, 593 (1984).
114. Hahn, M., Jaeger, W., Reinisch, G., Acta Polymerica, 34, 322 (1983).
115. Hahn, M., Jaeger, W., Reinisch, G., Acta Polymerica, 35, 350 (1983).

RECEIVED June 4, 1990

Chapter 6

Persulfate-Initiated Polymerization
of Acrylamide at High Monomer Concentration

D. Hunkeler and A. E. Hamielec

Department of Chemical Engineering, Institute for Polymer Production
Technology, McMaster University, Hamilton, Ontario L8S 4L7, Canada

Acrylamide was polymerized in inverse-microsuspension using potassium
persulfate as the initiator. Experiments were performed between 40 and
60°C with initial monomer concentrations of 25-50 wt % of the aqueous
phase. The rate of polymerization was determined to be proportional to the
monomer concentration to the 1.34 power, with 95% confidence limits of
±0.12. The rate order, therefore, does not significantly deviate from 1 25,
a dependence reported previously at low and moderate monomer levels.
This suggests the rate exponent is invariant to the acrylamide
concentration up to its solubility limit in water. Limiting conversions
have also been observed and have been attributed to the depletion of the
persulfate initiator. A reciprocal dependence between the limiting conversion and the initial acrylamide concentration implies that a secondary,
monomer-enhanced decomposition reaction is occurring. A hybrid
complex-cage mechanism, in which initiator-monomer association is a
necessary precursor to enhanced decomposition, will be shown to give good
quantitative predictions of the polymerization rate, monomer and initiator
consumption, and molecular weight. Further, it overcomes the
thermodynamic inconsistencies characteristic of unmodified cage or
complex theories.

In the preceding chapter a hybrid mechanism was developed in which a weakly bonded
monomer-initiator associate was formed via a caged precursor. The decomposition of this
associate proceeds by two independent reaction pathways, both of which have important
kinetic consequences. Either the amide and peroxide can combine, liberating active
sulfate and macroradicals which accelerate the polymerization rate, or they can form inert
species through a molecular rearrangement, and thereby effectively consume initiator.
The competition between the association phenomena and the two decomposition processes
permits the mechanism to describe the kinetic behaviour of polymerizations of acrylic
water soluble monomers without the free energy inconsistencies characteristic of the
unmodified complex or cage-effect theories. However, if correct, the hybrid mechanism
suggests that the rate order with respect to monomer concentration should be constant,
independent of the acrylamide level up to its solubility limit in water (≈ 50 wt%). In order
to verify this hypothesis a sequence of acrylamide polymerizations will be performed at
monomer concentrations between 25 and 50 wt%, varied at 5% increments. This will

0097–6156/91/0467–0105$06.00/0
© 1991 American Chemical Society

supplement the extensive kinetic data available at low and moderate monomer levels (≤ 30 wt%) (1-10).

The suitability of the hybrid mechanism to describe the kinetics at high conversions, and the polymer molecular weight, will also be investigated.

Review of Analytical Methods

Residual Monomer Concentration. The usual chemical methods for amide concentration determination are not applicable for polyacrylamides due to its limited solubility in organic solvents (11). Therefore, conversions are usually inferred from measurements of monomer concentration. Residual monomer levels have historically been determined by bromate-bromide or coulometric titrations (12). Chromatography (SEC: Ishige, (13); Kim, (10), LC: Husser, (11)) and UV absorption spectroscopy (14) have also been used. Spectrophotometric methods (NMR, IR) are occasionally employed however they are limited to relatively high levels of residual acrylamide.

Croll (15) has shown that Gas Chromatography is more sensitive to trace monomer levels than titrations, which fail below 0.1 wt.%. Croll (16) and Hashimoto (17) have developed a GC procedure suitable for the quantification of environmental samples, with a sensitivity below 0.1 ppb. The technique requires bromination of the unsaturated carbons followed by detection of the resulting α, β-dibromopropionamide.

For analysis of polymerized samples a liquid chromatography method has been described (18,19), which is applicable below 1 ppm with a 95% confidence limit of ± 0.25% This method, which was used in this investigation, offers a significant improvement in the accuracy for rapid LC measurement of acrylamide, which was previously limited to ± 2 7% (20). The additional accuracy is likely due to measurement (UV) at a wavelength (215 nm) closer to the peak absorption wavelength of acrylamide (201 nm). Previous investigations have detected acrylamide at 225 nm (11) and 240 nm (20), where the signal strength is more than an order of magnitude weaker. Researchers have also determined residual monomer levels by inferences from gravimetric or dilatometric observations (1, 2, 5, 21 24).

Molecular Weight Determination. Polyacrylamide molecular weight averages are usually determined indirectly from intrinsic viscosity measurements (2, 25-30). However, molecular weight estimates from this procedure are ambiguous because of uncertainty in the Mark-Houwink-Sakurada parameters, and the specificity of the correlation to the breadth of the molecular weight distribution. These produce estimates as divergent as ± 50%. By comparison light scattering offers reproducibilities below ± 10%, and the single point method (31) reduces this error twofold.

Molecular weight distributions have been measured by size exclusion chromatography (32). However, non-steric separation mechanisms are operative in aqueous solvents. The method is further limited to molecular weights below 10^7 g/mol, even for optimal mobile and stationary phase combinations. This is often insufficient for characterizing polymeric flocculants. Recently, SEC/LALLS have been coupled with reasonable success. However, the technique is limited by the column resolution, and often results in underprediction of the polydispersity (33). Holzworth (34) has coupled Band Sedimentation/LALLS with very good results. Band Sedimentation avoids some of the disadvantages of SEC, for example adsorption, while preclarifying the solvent before it passes through the light scattering detector. The procedure, which separates molecules based on their sedimentation velocity, has an upper limit of 10^8 g/mol. Furthermore, it has been shown to deviate by <5% from static light scattering measurements on DNA samples of 30 and 60 million daltons. Viscosity has also been coupled with SEC (35) and with LALLS/DRI/SEC (36), although it has not been used for polyacrylamide.

The molecular weight distribution of polyacrylamide has also been measured by electron microscopy in a solvent-nonsolvent mixture (13,37) and by turbidimetric titration (38,39).

Experimental

Conversions were inferred from measurements of the residual monomer concentration by High Performance Liquid Chromatography. A CN column (9% groups bonded to a μ-porasil (silica) substrate, Waters Assoc.) with a 8 mm I.D. and 4 μm particles was used as the stationary phase. The column was housed in a radial compression system (RCM-100, Waters) and was operated at a nominal pressure of 180 kg/cm². The HPLC system consisted of a degasser (ERC-3110, Erma Optical Works) a Waters U6K injector, a stainless steel filter and a CN precolumn (Waters). An ultra violet detector (Beckman 160) with a zinc lamp operating at a wavelength of 214 nm was used to measure the monomer absorption. A Spectra-Physics SP4200 integrator was used to compute peak areas.

The mobil phase was a mixture of 50 vol% acetonitrile (Caledon, Distilled in Glass, UV Grade) and 50 vol% double distilled deionized water, containing 0.005 mol/L dibutylamine phosphate. The flow rate was 2.0 mL/min. The peak separation was optimized by varying the acetonitrile-water ratio.

Molecular weights were measured using a Chromatix KMX-6 Low Angle Laser Light Scattering photometer, with a cell length of 15 mm and a field stop of 0.2. This corresponded to an average scattering angle of 4.8°.

A 0.45 μm cellulose-acetate-nitrate filter (Millipore) was used for polymer solutions. A 0.22 μm filter of the same type was used to clarify the solvent. Distilled deionized water with 0.02M Na₂SO₄ (BDH, analytical grade) was used as a solvent.

The refractive index increment of the "polyacrylamide solution" was determined using a Chromatix KMX-16 laser differential refractometer at 25°C and a wavelength of 632.8 nm. The dn/dc was found to be 0.1869.

Weight average molecular weights were calculated from measurements of the Rayleigh factor using the one-point method (31). This reduced the error in light scattering two fold over the conventional procedure.

For polymerizations solid acrylamide monomer (Cyanamid B.V., The Netherlands) was recrystallized from chloroform (Caledon, Reagent Grade) and washed with benzene (BDH, Reagent Grade). Potassium persulfate (Fisher Certified, assay 99.5%) was recrystallized from double distilled deionized water. Both reagents were dried in vacuo to constant weight and stored over silica gel in desiccators.

Method of Polymerization. Polymerization at high monomer concentrations in solutions require chain transfer additives to lower molecular weight, reduce viscosity and provide more efficient heat transfer. However, for this experimental set, chain transfer agents are undesirable since they can affect the initiation mechanism through redox coupling with persulfate. Therefore, inverse-microsuspension polymerization (40) was used. A prior investigation has demonstrated that inverse-microsuspension and solution polymerization are kinetically equivalent if a water soluble initiator is employed (Figure 1). In such instances, each isolated monomer droplet contains all reactive species and behaves like a microbatch reactor. Furthermore, the large particle diameter (≈10 μm) in this work minimizes interfacial effects.

Inverse-microsuspension polymerizations were performed using Isopar-K (Esso Chemicals) as the continuous phase and sorbitan monostearate (Alkaril Chemicals) as the emulsifier. Polymerizations were performed in a one-gallon stainless steel reactor, continuously agitated at 323 ± 1 RPM. The reactor was purged with nitrogen (Canadian Liquid Air, UHP Grade, 99.999% purity). A complete description of the experimental procedures is given in reference (19).

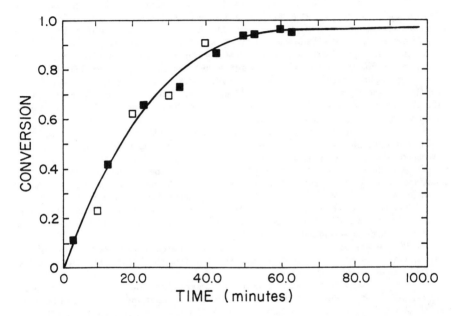

Figure 1: Conversion-time data for acrylamide polymerizations in solution (□) and inverse-microsuspension (■) under identical experimental conditions [Monomer] = 3.40 mol/L_w, [$K_2S_2O_8$] = 0.52 · 10^{-3} mol/L_w, Temperature = 60°C. For inverse-microsuspension: [Sorbitanmonooleate] = 0.168 mol/L_o, $\Phi_{w/o}$ = 1.42. Solution polymerization data are from Kim (1984).

Experimental Conditions.. Polymerizations were performed isothermally at 40, 50 and 60°C at monomer concentrations between 25 and 50 wt% of the aqueous phase. The latter corresponding to the solubility limit of acrylamide in water. Table I summarizes the experimental conditions for all polymerizations.

For all experiments reactor control was excellent, with thermal deviations never exceeding 1°C.

Results and Discussion

Rate Order with Respect to Monomer Concentration. The measured residual monomer concentrations were used to calculate the initial rates of polymerization. A series of six experiments were performed at 50°C with monomer concentrations between 25 and 50 wt.% of the aqueous phase, varied in 5% increments. From these data the following rate equation was estimated:

$$R_p \propto M^{1.34 \pm 0.12}$$

The 95% confidence limits were determined from a non-linear least squares estimation routine based on Marquardt's algorithm. The 95% confidence interval surrounds 1.25 and therefore we reject the hypothesis that at high monomer concentration the rate order deviates from the 5/4th power. This has mechanistic implications as it suggests the variable order rate models, the <u>unmodified</u> cage-effect and complex theories discussed in the previous chapter, are not applicable to aqueous persulfate initiated polymerization of acrylamide. The reliability of the 1.34 order is accentuated by Kurenkov's recent (1987) result: $R_p \propto M^{1.37}$ for monomer levels between 0.85 and 4.93 mol/L, which has been published after this work began.

Limiting Conversion. During these polymerizations, limiting conversions were observed for several reactions at high monomer or low initiator levels (Table II) (The usual reciprocal relationship between the limiting conversion and the rate of polymerization (41) has been found in this investigation). Incomplete monomer consumption is generally attributed to either a depletion of the initiator or isolation of the macroradicals. These phenomena can be distinguished by raising the temperature after the limiting conversion is reached. If residual initiator is present but is physically hindered from reaching the acrylamide monomer, increasing the thermal energy to the system will increase the diffusion of small molecules and increase the rate. Figure 2 shows such an experiment for this system. The temperature rise did not increase the conversion, indicating the initiator concentration had previously been exhausted. Therefore, the limiting conversions offer additional evidence of a second initiator decomposition reaction. Furthermore, the simultaneous occurrence of limiting conversions in polymerizations with high rate orders with respect to monomer concentration, which has historically been attributed to enhanced decomposition, suggest the following:

The unique kinetics for potassium persulfate initiated polymerization of acrylamide in aqueous media are due to the experimentally verifiable enhancement in the decomposition of peroxides by the proton donating monomer via a charge transfer complex interaction. Furthermore, the incorporation of this single modification to the mechanism is sufficient to explain all the unique phenomena for aqueous free radical polymerization of acrylamide and other acrylic water soluble monomers. Therefore, the hybrid mechanism is <u>not</u> refuted by the kinetic data.

Parameter Estimation. The conversion-time data were used to estimate two grouped parameters: $\Phi_2 k_a^*$ and f_c ($= 1/1 + k_c/k_b$), which are unique to the hybrid mechanism derived in the previous chapter. Additionally, measurements of weight average molecular weight were used to estimate the transfer to interfacial emulsifier parameter (Molecular

Table I: Polymerization Conditions

Temperature (°C)	[Acrylamide] (mol/L)	[K$_2$S$_2$O$_8$] (mmol/L)	Mass of Aqueous Phase (g)	Mass of Isopar-K (g)	Mass of SMS (g)
50	3.35	0.252	1000.0	1000.0	100.0
50	4 03	0.228	1000.0	1000.3	100.0
50	4.69	0.251	1000.0	999.9	100.0
50	5.37	0 248	1000.0	1000.0	100.0
50	6.04	0 250	1000.0	1002.1	99.9
50	6.41	0 238	1000.0	1000.3	100.0
40	6.70	1.573	1000.0	1000.2	100.0
60	6.70	0.0609	1000.0	1000.6	100.0

Table II: Limiting Conversion versus Monomer and Initiator Concentration

Limiting Conversion (X$_l$)	Acrylamide Concentration (mol/L)	Potassium persulfate Concentration (mmol/L)
0.92	6.41	0.238
0.95	6.04	0.250
0.96	5.37	0.248
0.996	4.69	0.251
0.996	4.03	0.228
0.997	3.35	0.252
0.76	6.70	0.0609
0.999	6.70	1.573

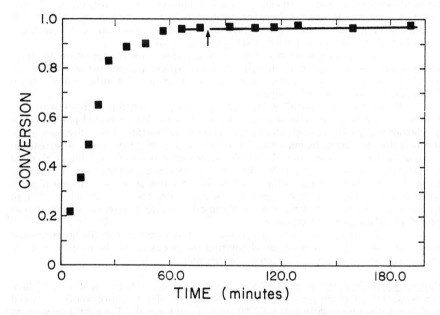

Figure 2: Conversion-time data (■) for an acrylamide polymerization. The experimental conditions were: [Monomer]: 5.37 mol/L_w, [$K_2S_2O_8$] = 2.44 × 10^{-3} mol/L_w, Temperature = 50°C. A limiting conversion of 0 97 is observed. Increasing the temperature to 60°C (arrow) did not lead to consumption of additional monomer.

weight development in inverse-microsuspension polymerization is detailed in reference (40)). The differential equations were solved with a variable order Runge-Kutta procedure with a step size of one minute.

At each temperature parameter estimates were obtained from a non-linear least squares regression routine based on Marquardt's procedure. These estimates were obtained utilizing residual monomer concentration and molecular weight data for all polymerizations at a given temperature. This is preferable to the common practice of estimating parameters from individual experiments, which is unable to identify inter-experimental data inconsistencies. This parameter overfitting leads to unreliable estimates which cannot be generalized to other reaction conditions.

Table III shows the values of parameters determined in this investigation. The activation energy for transfer to emulsifier was found to be −141 J/mol, typical of a termination reaction. It was not however, significantly smaller than zero at the 95% confidence level, and we can conclude that unimolecular termination with interfacial emulsifier is thermally invariant over the range investigated.

The complex/cage efficiency is observed to decrease with temperature, implying the monomer-swollen-cage preferentially forms inert species rather than active radicals. This is consistent with limiting conversion data which indicate that initiator deactivation is more favourable at higher temperatures.

The apparent association parameter ($k_a = \phi_2 k_a^*$) is found to increase with temperature according to an Arrhenius dependence. This has been reported previously for acrylamide polymerizations initiated by potassium persulfate (2). As was discussed in the previous chapter, this is the manifestation of two independent phenomena: a decrease in the specific association constant (k_a^*) which is entropically less favourable at elevated temperatures, and an increase in the fraction of potassium persulfate present as diffuse cages (ϕ_2). The latter, which is the only form of potassium persulfate capable of participating in monomer-enhanced decomposition reactions, are more abundant at high temperatures due to a greater frequency of radical diffusive displacements and a lower rate of radical fragment recombination.

All other rate parameters (k_p, k_{td}, k_{fm}, k_d) were obtained from the literature and have been summarized in reference (40) during the discussion of the modelling of the inverse-microsuspension polymerization of acrylamide.

Comparison of Kinetic Model to Experimental Data. Figures 3, 4 and 5 show representative conversion-time and weight average molecular weight-conversion data and model predictions for experiments at 60, 50 and 40°C, respectively. The hybrid mechanism is capable of predicting the initial polymerization rate and weight average molecular weight well over a range of temperatures, monomer concentrations and rates of initiation. The molecular weight behavior with conversion is typical of acrylamide polymerizations where transfer to monomer dominates. A slight decrease in molecular weight with conversion, and the increase with the initial monomer concentration, is evidence that a fraction of the chains are terminated through a bimolecular process. The limiting conversion is also predicted well.

Figure 6 illustrates the relative magnitudes of thermal and monomer-enhanced decomposition of potassium persulfate. At the onset of polymerization the majority of chains are initiated through a donor-acceptor interaction between the acrylamide and persulfate. This effect is most extreme at elevated temperatures. As the conversion rises both the monomer and initiator have depleted (Figure 7) and thermal bond rupture of the peroxide becomes the predominant initiation reaction. Figure 7 also shows that the rate of consumption of initiator is strongly dependent on the initial monomer concentration. Furthermore, for the conditions of the simulation, the potassium persulfate is exhausted before the reaction is completed for initial acrylamide concentrations exceeding 25 wt.%. For example, at 50 wt.% monomer, the radical generation ceases at 78% conversion, in agreement with experimental observations ($X_L = 0.76$).

Table III: Parameter Estimates

Parameter	Value	Units
f_c	40°C. 1.0 50°C: 0.372 60°C: 0.065	dimensionless
$(\phi_2 k_a^*)\dagger$	40°C: $3.17 \cdot 10^{-5}$ 50°C: $1.06 \cdot 10^{-3}$ 60°C: $1.93 \cdot 10^{-2}$	L/mol · min
k_{fE}	$2.433 \cdot 10^{-9}$	dm²/mol · min
A_0 + +	8.01	dimensionless
A_1 + +	$2.0 \cdot 10^{-2}$	K⁻¹ (Kim and Hamielec, 1984)

† $k_a = (\phi_2 k_a^*) = 8.77 \cdot 10^{41} \exp(-66,500/RT)$

+ +   A is a gel effect parameter in the expression:

$$\frac{k_{td}^o}{k_{td}} = \exp(A \cdot wp)$$

where $A = A_0 - A_1 T$

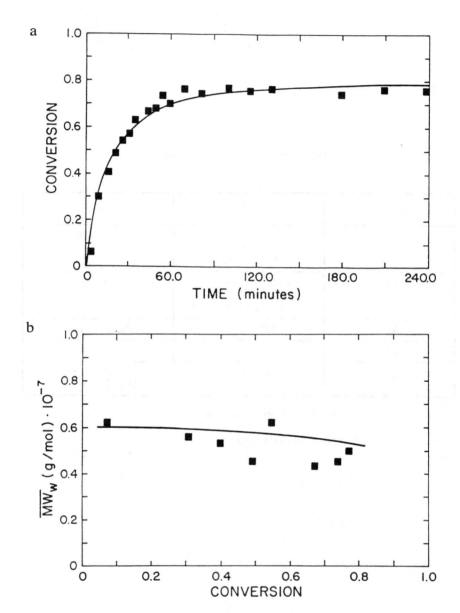

Figure 3: (a)   Conversion-time data (■) and kinetic model predictions (——————) for
an acrylamide polymerization at 60°C.   [Monomer] = 6.70 mol/L_w,
[$K_2S_2O_8$] = 0.0609 · 10^{-3} mol/L_w.

(b)   Weight average molecular weight-conversion data (■) and kinetic model
predictions (——————) for the same experiment.

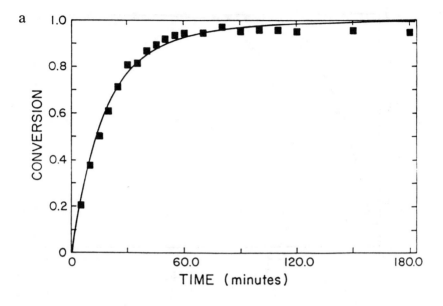

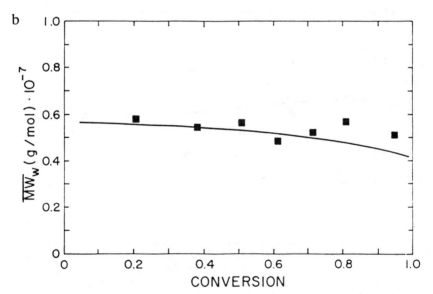

Figure 4: (a) Conversion-time data (▪) and kinetic model predictions (————) for
          an acrylamide polymerization at 50°C. [Monomer] = 6.04 mol/L_w,
          [$K_2S_2O_8$] = 0.250 · 10^{-3} mol/L_w.

   (b) Weight average molecular weight-conversion data (▪) and kinetic model
          predictions (————) for the same experiment.

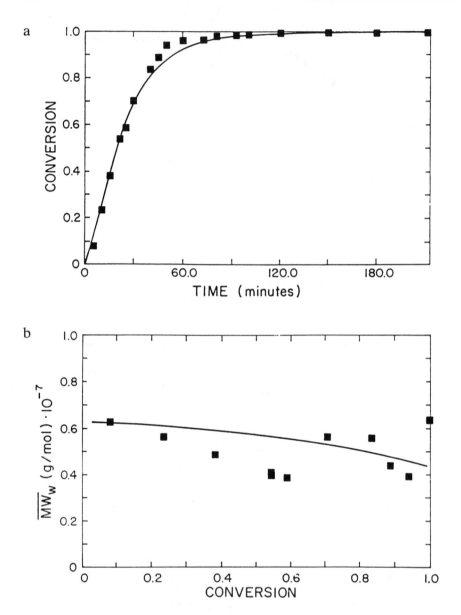

Figure 5: (a) Conversion-time data (■) and kinetic model predictions (——————) for
              an acrylamide polymerization at 40°C. [Monomer] = 6.70 mol/L$_w$,
              [K$_2$S$_2$O$_8$] = 1.573 · 10^{-3} mol/L$_w$.
         (b) Weight average molecular weight-conversion data (■) and kinetic model
              predictions (——————) for the same experiment.

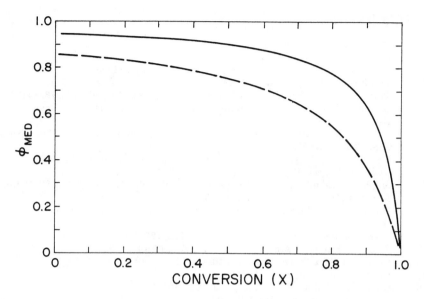

Figure 6: Fraction of polymer chains initiated by monomer-enhanced decomposition of potassium persulfate (Φ_{MED}) as a function of conversion. Simulations were performed at 40°C (-----) and 60°C (—————) with $[K_2S_2O_8] = 6.088 \cdot 10^{-5}$ mol/L_w, [Acrylamide] = 7.04 mol/L_w and $\Phi_{w/o} = 0.74$.

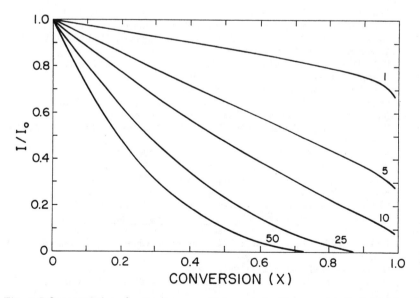

Figure 7: Concentration of potassium persulfate scaled with respect to its initial level (I/I_0) as a function of conversion. Simulations were performed at 60°C with $[K_2S_2O_8] = 2.4 \cdot 10^{-4}$ mol/L_w and initial acrylamide concentrations of 1, 5, 10, 25 and 50 wt.% based on the aqueous phase.

LITERATURE CITED

1.   Morgan, L.B., Trans. Farad. Soc., 42, 169 (1946)
2.   Riggs, J.P., Rodriguez, F., J.Polym.Sci., A1, 5, 3151 (1967).
3.   Friend, J.P., Alexander, A.E., J. Polym. Sci., A1, 6, 1833 (1968).
4.   Trubitsyna, S.N., Ismailov, I., Askarov, M.A., Vysokomol.soyed., A19, 495 (1977).
5.   Trubitsyna, S.N., Ismailov, I., Askarov, M.A., Vysokomol.soyed., A20, 1624 (1978)
6.   Osmanov, T.O., Gromov, V.F., Komikovsky, P.M., Abkin, A.D., Vysokomol.soyed., B20, 263 (1978).
7.   Osmanov, T.O., Gromov, V.F., Komikovsky, P.M., Abkin, A.D., Dokl. Akad Nauk SSSR, 240, 910 (1978).
8.   Kurenkov, V.F., Osipova, T.M., Kuznetsov, E.V., Myagchenkov, V.A, Vysokomol.soyed., B20, 647 (1978).
9.   Singh, U.C., Manickam, S.P., Venkatarao, K., Makromol. Chem., 180, 589 (1979).
10. Kim, C.J., Hamielec, A.E., Polymer, 25, 845 (1984).
11. Husser, E.R., Stehl, R.H., Prince, D.R., DeLap, R.A., Analytical Chemistry, 49, 155 (1977).
12. Suen, T.J., Rossler, D.F., J. Appl. Polym. Sci., 3, 126 (1960).
13. Ishige, T., Hamielec, A.E., J. Appl. Polym. Sci., 17, 1479 (1973).
14. Chatterjee, A.M., Burns, C.M., Canadian J. Chem., 49, 3249 (1971).
15. Croll, B.T., Analyst, 96, 67 (1971).
16. Croll, B.T., Simkins, G.M., Analyst, 97, 281 (1971).
17. Hashimoto, A., Analyst, 101, 932 (1976).
18. Hunkeler, D., Hamielec, A.E and Baade, W., "Polymerization of Cationic Monomers with Acrylamide", in Polymers in Aqueous Media: Performance through Association, J.E. Glass, ed., American Chemical Society, Washington, D.C., 1989.
19. Hunkeler, D., Ph.D. Dissertation, McMaster University, Hamilton, Canada (1990).
20. Ludwig, F.J., Sr., Besand, M.F., Analytical Chemistry, 50, 185 (1978).
21. Collinson, E., Dainton, F.S., McNaughton, G.S., Trans. Farad. Soc., 53, 476 (1957).
22. Gromov, V.F., Matveyeva, A.V., Khomikovskii, P.M., Abkin, A.D., Vysokomol.soyed., A9, 1444 (1967).
23. Baer, M., Caskey, J.A., Fricke, A.L., Makromol. Chem., 158, 27 (1972).
24. George, M.H., Ghosh, A., J. Polym. Sci. Chem., 16, 981 (1978).
25. Cavell, E.A.S., Makromol. Chem., 54, 70 (1962).
26. Shawki, S., Ph.D. Dissertation, McMaster University, Hamilton, Canada, 1978.
27. Baade, W., Ph.D. Dissertation, Technical University of Berlin, FRG, 1986.
28. Kurenkov, V.F., Baiburdov, T.A., Stupen'kova, L.L., Vysokomol.soyed., A29, 348 (1987).
29. Rafi'ee Fanood, M.H., George, M.H., Polymer, 29, 128 (1988).
30. Rafi'ee Fanood, M.H., George, M.H., Polymer, 29, 134 (1988).
31. Hunkeler, D., Hamielec, A.E., J. Appl. Polym. Sci., 35, 1603 (1988).
32. Klein, J., Westercamp, A., J. Polym. Sci. Chem., 19, 707 (1981).
33. Ouano, A.C., Kaye, W., J. Polym. Sci. Chem., 12, 1151 (1974).
34. Holzwarth, F., Soni, L., Schulz, D.N., Macromolecules, 19, 422 (1986).
35. Styring, M., Armonas, J.E., Hamielec, A.E., J. Liq. Chrom., 10, 783 (1987).
36. Tinland, B., Mazet, J., Rinaudo, M., Makromol. Chem. Rapid Commun., 9, 69 (1988).
37. Quayle, D.V., Polymer, 8, 217 (1967).
38. Omorodion, S.N.E., Masters Thesis, McMaster University, Hamilton, Canada, 1976.
39. Ramazanov, K.R, Klenin, S I., Klenin, V.I., Novichkova, L.M., Vysokomol.soyed., A26, 2052 (1984).
40. Hunkeler, D., Hamielec, A E. and Baade, W., Polymer, 30, 127 (1989).
41. Joshi, M.G., Rodriguez, F., J. Polym. Sci. Chem., 26, 819 (1988).

RECEIVED June 4, 1990

Chapter 7

Copolymers of Acrylamide and a Novel Sulfobetaine Amphoteric Monomer

Luis C. Salazar and Charles L. McCormick

Department of Polymer Science, The University of Southern Mississippi, Hattiesburg, MS 39406-0076

The free radical copolymerization of acrylamide (AM) with 3-(2-acrylamido-2-methylpropyldimethylammonio)-1-propanesulfonate (AMPDAPS) has been studied in the range from 25 to 90% AM in the feed. The value of r_1r_2 has been determined to be 0.60 for the AM-AMPDAPS pair. The copolymer compositions have been determined from elemental analysis and ^{13}C NMR. Molecular weights were determined using LALLS and were found to vary from 1.7 to 10.2 x 10^6g/mol. The copolymer microstructures, including run numbers and sequence distributions, were calculated from the reactivity ratios. The solution properties of the AM-AMPDAPS copolymers, as well as the AMPDAPS homopolymer, have been studied as a function of composition, pH, and added electrolytes. The solutions of these polymers show increased intrinsic viscosities in the presence of sodium chloride and/or calcium chloride. The solution behavior of the homopolymer of AMPDAPS is independent of pH. The observed properties are consistent with the charge density of the polymers and the sulfobetaine structure of the AMPDAPS monomer.

Polyampholytic polymers have been under investigation in our laboratories for the past several years (1,2). These polymers are polyions which have both positive and negative charges bound along the backbone (3). The solution properties of these polymers are of interest because their viscosities increase with the addition of electrolytes (4-7). This behavior has become known as the "antipolyelectrolyte" effect. Polymers which display this type of behavior have promise as absorbent materials and viscosifiers in the presence of electrolytes.
    The first polyampholytes studied in our laboratories consisted of copolymers made from cationic and anionic acrylamido monomers (8). These monomers were copolymerized in varying feed ratios to obtain copolymers of different compositions. Antipolyelectrolyte behavior was observed when the copolymers contained an equivalent amount of each monomer (9). We next studied polyampholytes with low-to-medium charge density (McCormick, C. L. and Johnson, C. B., Polymer, in press). The charge density could be varied by forming cationic-anionic monomer pairs and then polymerizing them with varying amounts of acrylamide (AM). These terpolymers exhibited

0097–6156/91/0467–0119$06.00/0
© 1991 American Chemical Society

antipolyelectrolyte behavior in solution when the ratio of each oppositely charged monomer was approximately one-to-one (McCormick, C. L. and Johnson, C. B., J. Macromol. Sci. Chem., in press).

In all cases it was important that the net charge of the polymer be zero regardless of the charge density. The antipolyelectrolyte effect was compromised if a charge imbalance existed. If the polymers contained an excess of either charged monomer, the solution behavior reflected that of a typical polyelectrolyte. The importance of having an equivalent amount of each cationic and anionic moiety in the polymer led us to our work with the amphoteric monomer 3-(2-acrylamido-2-methylpropyldimethylammonio)-1-propanesulfonate (AMPDAPS).

The use of an amphoteric monomer insures that a one-to-one ratio of positive to negative charges is maintained (10,11). A monomer with both charges can be incorporated in varying amounts into a copolymer to produce a series of polymers in which the charge density of polymer can be varied from low to high while the net charge remains zero. Other polyampholytes in which different monomers provide the negative and positive charges do not guarantee this. AMPDAPS has the added advantage of having charged groups which are stable to hydrolysis. The positive charge is provided by an ammonium quaternized by alkyl groups and the negative charge is provided by a sulfonate group which is difficult to protonate.

This chapter reports the synthesis and characterization of a series of copolymers of acrylamide and the new amphoteric monomer AMPDAPS. The methods used to synthesize these polymers and their dilute solution behavior will be discussed.

Experimental Details

Monomer Synthesis.   3-(2-Acrylamido-2-methylpropyldimethylammonio)-1-propanesulfonate (AMPDAPS) was synthesized by the ring opening reaction of 1,3-cyclopropanesultone (PS) with 2-acrylamido-2-methylpropanedimethylamine (AMPDA), Figure 1. 1,3-Cyclopropane-sultone was obtained from the Aldrich Chemical Co. and was used without further purification. The synthesis AMPDA has been previously reported (12). In a typical monomer synthesis, 0.144 mol AMPDA and 0.156 mol of PS were reacted in 500ml propylene carbonate (PGC) under N_2 at 55°C for 4 days. During this period the product formed as a white precipitate. This was then filtered and washed with diethyl ether until all the propylene carbonate was removed. Acrylamide (AM) from Aldrich Chemical Co. was recrystallized twice from acetone and vacuum-dried at room temperature. Potassium persulfate from J.T. Baker Co. was recrystallized twice from deionized water.

Copolymer Synthesis.   The homopolymer of AMPDAPS and the copolymers of AMPDAPS with AM, the DAPSAM series, were synthesized free radically in a 0.512-M NaCl aqueous solution under nitrogen at 30°C using 0.1 mol % potassium persulfate as the initiator. The monomer concentration was 0.45 M with a pH of 7.0 ± 0.1. A low conversion sample was analyzed to allow reactivity ratio studies. The reaction was terminated at <30% conversion due to the high viscosity of the reaction medium and as a precaution against copolymer drift. The polymers were precipitated in acetone, redissolved in deionized water, dialyzed using Spectra/Por 4 dialysis bags with molecular weight cutoffs of 12,000 to 14,000 daltons, and then lyophilized. When more than 40 mol % AMPDAPS was incorporated in the copolymers, swelling but not dissolution could be achieved in deionized water. These "hydrogels" were

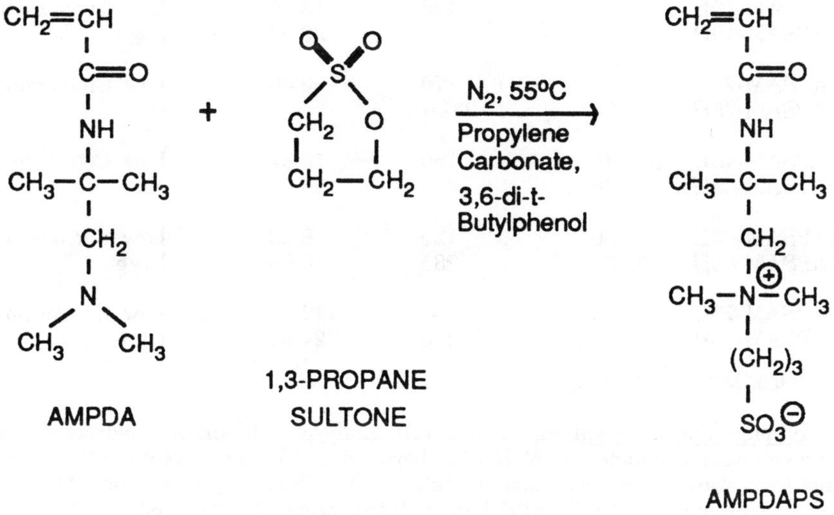

Figure 1. Synthesis of AMPDAPS monomer.

washed repeatedly with deionized water to remove any remaining salt or monomer and then lyophilized. Conversions were determined gravimetrically. Table I lists reaction parameters for the copolymerization of AMPDAPS with AM and the homopolymerization of AMPDAPS.

Table I. Reaction Parameters for the
Copolymerization of AMPDAPS with AM

AM + AMPDAPS ---------→ DAPSAM Copolymers

Copolymer	% AMPDAPS[a]	Time (min)	% Conversion	Comments
DAPSAM-10L	10	165	12.86	Low Conversion
DAPSAM-10H	10	270	22.07	High "
DAPSAM-25L	25	270	3.40	Low Conversion
DAPSAM-25H	25	23-hrs	--	High "
DAPSAM-40L	40	160	13.61	Low Conversion
DAPSAM-40H	40	300	26.63	High "
DAPSAM-60L	60	120	9.23	Low Conversion
DAPSAM-60H	60	285	16.68	High "
DAPSAM-75L	75	180	12.24	Low Conversion
DAPSAM-75H	75	420	24.66	High "

[a] % DAPSAM in the Feed

Copolymer Characterization. Elemental analyses for carbon, hydrogen, and nitrogen were conducted by M-H-W Laboratories of Phoenix AZ on both the low and high conversion copolymer samples. ^{13}C NMR spectra of the DAPSAM copolymers and the AMPDAPS homopolymer were obtained using 5-10 wt % aqueous (D_2O) polymer solutions. The procedure for quantitatively determining copolymer compositions from ^{13}C NMR has been discussed in detail elsewhere (13). FT-IR spectra for all materials synthesized were obtained using a Perkin-Elmer 1600 Series FT-IR spectrophotometer. Light scattering studies were performed on a Chromatix KMX-6 low-angle laser light scattering spectrophotometer and refractive index increments were obtained using a Chromatix KMX-16 laser differential refractometer. All measurements were conducted at 25°C in 0.512 M NaCl at a pH of 7.0 ± 0.1.

Viscosity Measurements. Stock solutions of sodium chloride (0.042, 0.086, 0.257, and 0.514-M NaCl) were prepared by dissolving the appropriate amount of salt in deionized water in volumetric flasks. Polymer stock solutions were then made by dissolving the appropriate amount of polymer in these salt solutions. The solutions were then diluted to appropriate concentrations and allowed to age for two to three weeks before being analyzed with a Contraves LS-30 rheometer.

Results and Discussion

Macromolecular Structure. The compositions of the DAPSAM copolymers were determined using ^{13}C NMR and elemental analysis. Integration of the ^{13}C carbonyl resonances of AM and AMPDAPS gave the mole percent of each monomer in the copolymer. This information agrees favorably with that derived from elemental analysis as shown in Table II.

Table II. Compositions of the DAPSAM Copolymers as Determined by Elemental Analysis and ^{13}C NMR

Sample	%C	%N	%AM	%AMPDAPS[a]	%AMPDAPS[b]
DAPSAM-10	45.5	14.6	88.2	11.9	9.3
DAPSAM-25	43.7	12.0	74.0	26.0	26
DAPSAM-40	44.0	10.7	57.8	42.9	38
DAPSAM-60	44.5	10.0	42.3	57.9	57
DAPSAM-75	44.4	09.4	28.1	71.9	67

[a] Elemental Analysis
[b] ^{13}C NMR

Weight-average molecular weights were determined by low-angle laser light scattering. Table III shows this data obtained at 25°C in 0.512M NaCl. The molecular weights varied from 1.3×10^6g/mol to 1.0×10^7g/mol for the AMPDAPS homopolymer and the DAPSAM-40 copolymer respectively. In general, the more AMPDAPS in the feed, the lower the molecular weight of the resulting polymer. The similar molecular weights of DAPSAM-25, -60, and -75 are important in demonstrating the effect of copolymer composition on their solution behavior.

Table III. Molecular Weight and Degree of Polymerization Data for the DAPSAM Copolymer Series

Sample	dn/dc	MW$_w$(g/mol x 10^{-6})	DP$_w$ (x 10^{-4})
DAPSAM-10	0.116	7.5	7.7
DAPSAM-25	0.155	4.8	3.8
DAPSAM-40	0.134	10.2	6.2
DAPSAM-60	0.148	4.2	2.1
DAPSAM-75	0.159	4.1	1.8
DAPSAM-100	0.163	1.3	0.5

[a] 0.514M NaCl, shear rate = 1.75 sec^{-1}

Reactivity Ratio Studies. Reactivity ratios for the AMPDAPS series were determined from the feed ratios of the monomers and the resultant copolymer compositions obtained by elemental analysis of the low conversion samples. The methods of Fineman-Ross ([14]) and Kelen-Tudos ([15]) were employed to determine the monomer reactivity ratios from the low conversion copolymer samples. The Fineman-Ross method yielded reactivity ratios for AM (M$_1$) and AMPDAPS (M$_2$) of $r_1 = 0.79$ and $r_2 = 0.73$. The Kelen-Tudos method gave reactivity ratios of 0.79 and 0.75 for r_1 and r_2 respectively and $r_1 r_2 = 0.60$, Table IV. The copolymer composition as a function of feed composition for the DAPSAM series is shown in Figure 2. The experimental data indicate random comonomer incorporation with a slight alternating tendency.

Table IV. Reactivity Ratios for the Copolymerization
of AM (M_1) with AMPDAPS (M_2)

Method	r_1	r_2	r_1r_2
K-T[a]	0.79	0.73	0.58
F-R[b]	0.79	0.75	0.60

[a] Kelen and Tüdos
[b] Fineman and Ross

Copolymer Microstructure. To calculate the microstructural features of these
copolymers, the equations of Igarashi (16) and Pyun (17) were employed. The
fraction of AM-AM, AMPDAPS-AMPDAPS, and AM-AMPDAPS units (the mole
% blockiness, the mole % alternation, and the mean sequence length) in the
copolymers were calculated from the reactivity ratios and the copolymer
composition, Table V. The data confirms that the copolymerization was random
with a slight alternating tendency.

Table V. Calculated Structural Data for the DAPSAM
Copolymer Series

Sample	Blockiness Mole % M_1-M_1	M_2-M_2	Alternation Mole% M_1-M_2	Mean Sequence Length M_1	M_2
DAPSAM-10	77.2	00.9	21.9	8.2	1.1
DAPSAM-25	53.0	05.0	42.0	3.4	1.3
DAPSAM-40	30.4	14.8	54.8	2.2	1.5
DAPSAM-60	14.7	30.5	54.8	1.5	2.1
DAPSAM-75	06.0	49.7	44.4	1.3	3.3

Effects of Copolymer Composition. The effects of copolymer composition on the
intrinsic viscosities of the of the DAPSAM copolymers in 0.512-M NaCl are
shown in Figure 3. The decrease in the intrinsic viscosities is due to
intramolecular interactions which constrict the polymer coils. The presence of
the AMPDAPS monomer units is responsible for these interactions. For
DAPSAM-25, -60 and -75, which have similar molecular weights, the increased
amount AMPDAPS decreases the intrinsic viscosities. It is believed that these
interactions are based on electrostatic attractions between the opposite charges
of different AMPDAPS units (11).

Effects of Added Electrolytes. The effects of sodium chloride on the intrinsic
viscosities of the DAPSAM copolymers and the AMPDAPS homopolymer were
determined at a shear rate of 1.25 sec^{-1} as shown in Figure 4. These solutions
of polymers show increasing intrinsic viscosities as the amount of salt in the
solutions is increased. It should be noted that some of the polymers would not
dissolve until salt was present. DAPSAM-60 and -75 required 0.0428-M NaCl,
and DAPSAM-100 needed 0.257-M NaCl for dissolution. Attempts to remove
the salt from the polymers by dialysis resulted in the precipitation of the
polymers from solution. Figure 4 shows polyampholytic behavior at all the
compositions studied. In all cases the intrinsic viscosities increase significantly.
Figure 5 shows the effects of adding the divalent salt calcium chloride to
solutions of the DAPSAM copolymers and the AMPDAPS homopolymer.

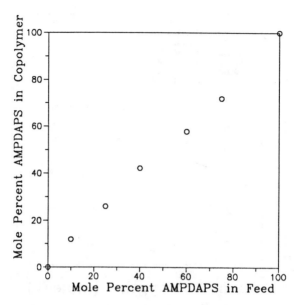

Figure 2. Mole percent AMPDAPS incorporated into the copolymers as a function of comonomer feed ratio.

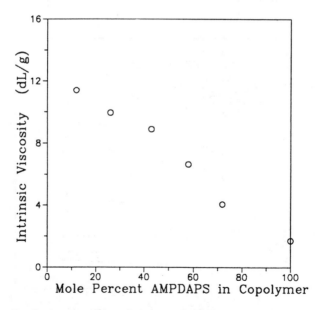

Figure 3. Intrinsic viscosities of the DAPSAM copolymers as a function of mole percent AMPDAPS incorporated. (Determined in 0.514 M NaCl at a shear rate of 1.25 sec[-1].)

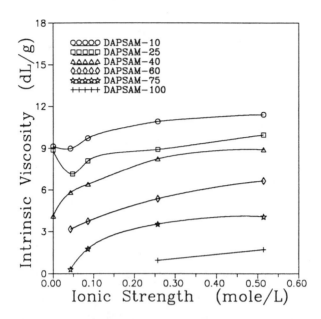

Figure 4. Dependence of the intrinsic viscosities of the DAPSAM copolymers in NaCl solutions of varying ionic strengths. (Determined at 25°C with a shear rate of 1.75 sec[-1].)

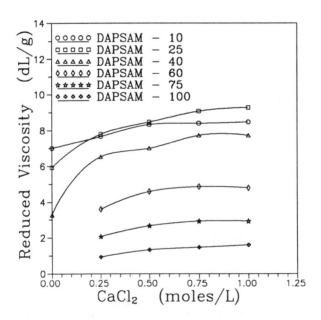

Figure 5. Reduced viscosities of the DAPSAM copolymers as a function of CaCl$_2$ concentration. (Determined at 25°C with a shear rate of 6.0 sec[-1].)

DAPSAM-60, -70, and the AMPDAPS homopolymer are insoluble up to a critical ionic strength.

Effects of pH. Though the use of acrylamide as the comonomer allows these polymers to have very high molecular weights, hydrolysis can be problematic. Figure 6 shows the effects of pH on DAPSAM-25 in deionized water and in 0.512-M NaCl. At high pH values, the acrylamide unit is hydrolyzed to the polyelectrolyte. In the presence of salt, the polyelectrolyte effect is negated and the antipolyelectrolyte effect of the zwitterions dominates. The stability of the AMPDAPS monomer to pH is due to the gem-dimethyl group next to the acrylamido functionality. This combination has been shown in our laboratories to give specially tailored monomers which are stable to hydrolysis and to high temperatures (18).

Effect of Polymer Concentration. At very low polymer concentrations the solutions show behavior consistent with both polyelectrolytes and polyampholytes. In deionized water (Figure 7) the copolymer DAPSAM-10 exhibits the typical polyelectrolyte effect as the concentration is lowered to 0.01-g/dL (this point is off scale). As salt is added this effect disappears. The upswing in η_{red} for increasingly low concentrations of polymer in deionized water is not due to hydrolysis of the acrylamide monomer units along the backbone. The copolymer in question has been analyzed by ^{13}C NMR and FT-IR and no evidence for hydrolysis has been found. Therefore the upswing in deionized water can be attributed to the AMPDAPS units in the copolymer. The bound charges may be responsible for expanding the polymer coil. This mechanism is currently being explored.

Conclusions

A novel sulfobetaine monomer has been synthesized and incorporated into a series of copolymers with acrylamide as the comonomer. These polymers were examined by C-13 NMR, FT-IR, elemental analysis and LALLS. Elemental analysis data from the low conversion copolymers gave the value of 0.60 for r_1r_2 for the AM-AMPDAPS monomer pair. Copolymer microstructures were statistically determined from these reactivity ratios. These data indicate that the copolymers have a random composition with a slight alternating tendency. Weight-average molecular weights in the range of 1.3×10^6 to 1.0×10^7 g/mol have been determined for the polymers. The dilute solution properties were studied as a function of copolymer concentration, composition, pH and added electrolytes. Increasing amounts of AMPDAPS decrease the intrinsic viscosities of the copolymers. The polymer coils expand upon the addition of NaCl and $CaCl_2$. Intramolecular interactions between different AMPDAPS units on the same copolymer coil are responsible for this effect.

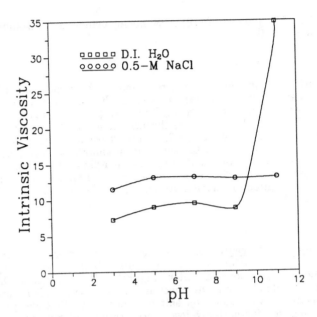

Figure 6. Intrinsic viscosities of DAPSAM-25 as a function of pH in D.I. H_2O and in 0.5 M NaCl. (Determined at 30°C with a shear rate of 1.25 sec^{-1}.)

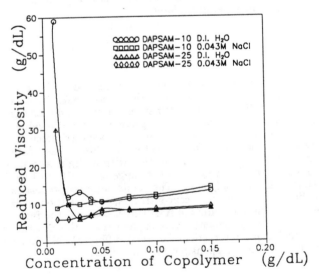

Figure 7. Reduced viscosities for DAPSAM-10 and DAPSAM-25 as a function of polymer concentration in D.I. H_2O and 0.043 M NaCl. (Determined at 30°C with a shear rate of 1.25 sec^{-1}.)

Acknowledgments

Acknowledgement is made to the Department of Energy, the Office of Naval Research, and the Defense Advanced Research Projects Agency for support of portions of this research.

Literature Cited

1. Johnson, C. B.   Ph.D. Thesis, The University of Southern Mississippi, Hattiesburg, MS, 1987.
2. McCormick, C. L.; Salazar, L. C.   Polymer Preprints 1989, 30(2), 344.
3. Tanford, C.   Physical Chemistry of Macromolecules; Wiley & Sons: New York, 1961; p 506.
4. Salamone, J. C.; Tsai, C. C.; Olsen, A. P.; Watterson, A. C.   In Advances in the Chemical Sciences: Ions in Polymers; American Chemical Society, 1980; Vol. 187, p 337.
5. Monroy Soto, V. M.; Galin, J. C.   Polymer 1984, 25, 254.
6. Peiffer, D. G.; Lundberg, R. D.   Polymer 1985, 26, 1058.
7. Salamone, J. C.; Quach, L.; Watterson, A. C.; Krauser, S.; Mahmud, M. U.   J. Macromol. Sci. Chem. 1985, A22(5-7), 653.
8. McCormick, C. L.; Johnson, C. B.   Macromolecules 1988, 21, 686.
9. McCormick, C. L.; Johnson, C. B.   Macromolecules 1988, 21, 694.
10. Monroy Soto, V. M.; Galin, J. C.   Polymer 1984, 25, 121.
11. Schulz, D. N.; Peiffer, D. G.; Agarwal, P. K.; Larabee, J.; Kaladas, J. J.; Soni, L.; Handwerker, B.; Garner, R. T.   Polymer 1987, 27, 1734.
12. McCormick, C. L.; Blackmon, K. P.   Polymer 1986, 27, 1971.
13. McCormick, C. L.; Hutchinson, B. H.   Polymer 1986, 27(4), 623.
14. Fineman, M.; Ross, S.   J. Polym. Sci. 1950, 5(2), 259.
15. Kelen, T.; Tüdos, F.   J. Macromol. Sci. Chem. 1975, A9, 1.
16. Igarashi, S.   J. Polym. Sci., Polym. Lett. Ed. 1963, 1, 359.
17. Pyun, C. W.   J. Polym. Sci. 1970, A2(8), 1111.
18. Blackmon, K. P.   Ph.D. Thesis, The University of Southern Mississippi, Hattiesburg, MS, 1986.

RECEIVED June 4, 1990

Chapter 8

Synthesis and Solution Characterization of Pyrene-Labeled Polyacrylamides

Stephen A. Ezzell and Charles L. McCormick

Department of Polymer Science, The University of Southern Mississippi, Hattiesburg, MS 39406—0076

A new pyrene-containing acrylamido monomer, [1-(β-aminoethyl) sulfonamidopyrene] acrylamide (APS) has been synthesized and carefully purified for copolymerization and subsequent photophysical studies. Pyrenesulfonamide-labeled acrylamide copolymers were prepared in homogeneous and micro-heterogeneous (surfactant) solutions utilizing free radical polymerization techniques. A comparatively blocky microstructure is suggested for the surfactant-polymerized sample relative to the solution-polymerized material as evidenced by enhancement of excimer to monomer intensity (I_E/I_M) for the former. I_E/I_M varies with concentration for both polymer samples in H_2O, reflecting intermolecular hydrophobic interactions of the pyrenesulfonamide labels. Pyrenesulfonamide ground-state aggregates are denoted by excitation spectra. Correlation of polymer dilute solution rheological behavior with the photophysical response of the pyrenesulfonamide label can be related to the degree of intramolecular association.

Our research group has a continuing interest in the synthesis and evaluation of hydrophobically-modified water-soluble polymers. Such materials have potential applications as viscosifiers for enhanced oil recovery, as surface-active polymers for personal care products, as flocculating agents for water treatment, and as frictional drag reducing agents in turbulent flow. One system which has received attention in our laboratory (1) and elsewhere (2) is the acrylamide (AM)-n-decylacrylamide (C10) copolymer (Figure 1) synthesized by a micellar polymerization technique. The technologically interesting behavior of these copolymers at different C10 compositions is illustrated in Figure 2 in which relative viscosity is plotted vs copolymer composition in aqueous solution. Comparison of the behavior of homopolyacrylamide and the AM with 0.75 mol % C10 reveals a much lower value for the critical overlap concentration (C*) for the latter. C* is considered to be the concentration at which the polymer coils overlap; intermolecular hydrophobic intractions of the C10 alkyl sidechains are likely responsible for the low C* value.

An understanding of structure-property relationships in these polymer systems is necessary in order to facilitate the design of materials with desirable

0097–6156/91/0467–0130$06.25/0
© 1991 American Chemical Society

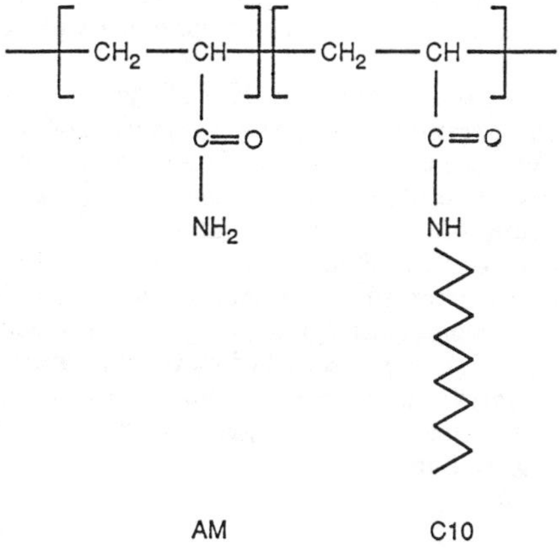

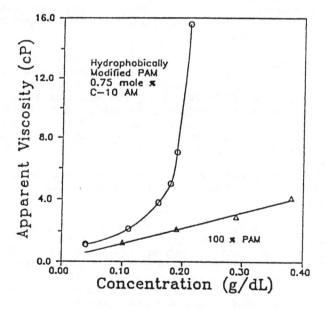

Figure 1. Acrylamide-N-decylacrylamide copolymer system.

Figure 2. Effect of copolymer concentration on the apparent viscosity of a copolymer of acrylamide with decylacrylamide at 25°C and a shear rate of 1.28 sec^{-1}. O, C-10—0.75; △, PAAm.

performance properties. Therefore, we wish to characterize the solution behavior of tailored water-soluble polymers on both molecular and macroscopic scales. Photophysics is one method of choice for the characterization of polymer solution behavior on a molecular level since concentration problems preclude the use of NMR.

Our approach involves the preparation of pyrene-labeled, hydrophobically-modified water-soluble polymers via copolymerization of AM with a new pyrene-containing acrylamido monomer. Two different polymerization techniques are employed to synthesize models for systems such as those of Figure 1. Polymeric incorporation of the pyrene fluorophore allows the photophysical characterization via well-established techniques (3). Polymer solution behavior has been elucidated via photophysical studies of pyrene-labeled water-soluble polymers including polyelectrolytes (4-7), poly(ethylene oxide) (8), hydroxypropylcellulose (9), and poly(N-isopropylacrylamide) (10,11), among others. It is our goal to correlate molecular characteristics derived from photophysics with macroscopic (rheological) properties of hydrophobically-modified water-soluble polymers.

Experimental

Instrumentation. HPLC was performed on a Hewlett-Packard Model 1050 system equipped with a photodiode-array detector. A Waters μ-Bondapak C18 column was employed with methanol as the mobile phase. ^{13}C NMR spectra were measured on a Bruker AC 300 instrument. UV-VIS spectra were recorded with a Perkin-Elmer Lambda 6 spectrophotometer. Steady-state fluorescence spectra were recorded with a Spex Fluorolog 2 Fluorescence Spectrometer equipped with a DM300F data system. Viscosity studies were performed on a Contraves LS-30 rheometer.

Characterization Methods. Copolymer composition was determined via ultra-violet spectroscopy at 350 nm (ε = 24,120 in H_2O). Steady-state fluorescence measurements were performed by excitation at 340 nm and recording the emission intensity from 350 to 600 nm. Spectra were obtained by excitation from 250 to 400 nm, and monitoring of the emission intensity at either 418 nm (monomer emission) or 510 nm (excimer emission). Slits were set at 0.5 nm and the ratio mode was used for both emission and excitation experiments. All spectra were corrected. Solutions were degassed with N_2 for 20 minutes prior to analysis.

Solutions for viscosity studies were prepared gravimetrically. A shear rate of 6 sec^{-1} was employed; sample temperature was maintained at 25°C. All studies were performed in duplicate.

Materials. Acrylamide (AM) was recrystallized from acetone three times and vacuum-dried at room temperature prior to use. Pyrene was purified by flash chromatography (12) (silica gel packing, CH_2Cl_2 eluent). N,N-Dimethyl-formamide (DMF) was allowed to stand overnight over 4 Å molecular sieves, then distilled at reduced pressure. Other starting materials were purchased commercially and used as received. TLC was performed on Merck Kieselgel 60 silica gel plates; developed plates were viewed under 365 nm light.

Monomer Synthesis. Synthesis of the pyrene-labeled monomer [1-(β-aminoethyl)sulfonamidopyrene]acrylamide (APS) (1) is illustrated in Schemes 1-3. A literature method (13) was modified for the preparation of sodium(1-pyrenesulfonate) (2) (Scheme 1). Pyrene (47.60 g, 0.235 M) was dissolved in

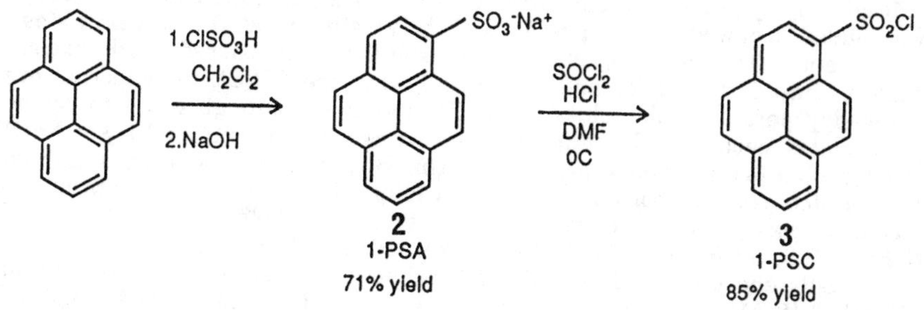

Scheme 1. Synthesis of 1-pyrenesulfonyl chloride.

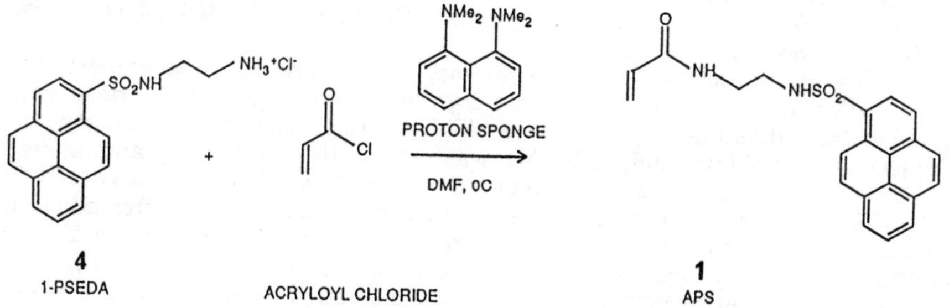

Scheme 2. Condensation of 1-pyrenesulfonyl chloride with ethylene-diamine.

HIGH-DILUTION TECHNIQUE

Scheme 3. APS synthesis.

300 ml CH$_2$Cl$_2$. Chlorosulfonic acid (16 ml, 0.24 M) dissolved in 50 ml CH$_2$Cl$_2$ was added dropwise to the pyrene with brisk stirring at 0°C under a steady N$_2$ stream. The resulting dark green solution was poured (with extreme caution) into 500 cc of ice and stirred, allowing the CH$_2$Cl$_2$ to evaporate over a two-day period. This solution was filtered twice through celite to remove particulates; each time the celite pads were washed with 1x150 ml H$_2$O. NaOH (10.0 g, 0.25 M) was added as a solution in H$_2$O. The yellow sodium salt was precipitated via slow solvent evaporation employing a warm hot plate and N$_2$ bleed. 2 was filtered and vacuum-dried at 65°C. Yield: 51.0 g (71 %). IR (KBr): 3458, 3045, 1194, 1161, 1092, 1060, 1018, 853, 673.

1-Pyrenesulfonylchloride (3) was prepared by addition of thionyl chloride (22 ml, 0.18 M) dropwise to a slurry of 2 (9.1 g, 3x10^{-1} M) in DMF (200 ml) containing 30 ml of an HCl solution (0.10 M) in diethyl ether. Stirring was continued for 3 hrs., then the solution was poured into 400 cc ice. The orange-yellow precipitate was filtered and washed with 500 ml H$_2$O. This material was air-dried overnight then vacuum-dried for 18 hr. at 100°C. Yield: 7.7 g (85 %). MP 172°C. Anal. Calcd. for C$_{16}$H$_9$SO$_2$Cl: C, 63.89; H, 3.00; S, 10.67; Cl, 11.78. Found: C, 63.85; H, 3.09; S, 10.61; Cl, 11.59. IR: 3107, 3046, 1590, 1361, 1172, 850, 644, 603, 521.

N-(1-pyrenesulfonyl)ethylenediamine, hydrochloride (4) was prepared via modification of a literature procedure for the reaction of acid chlorides with symmetrical diamines (14), via a high-dilution technique (Scheme 2). Ethylenediamine (10.0 ml, 0.15 M) was added to 1 l CH$_2$Cl$_2$ and stirred vigorously at 0°C under a N$_2$ blanket. 3 (3.0 g, 1.0x10^{-2} M) was dissolved in 1 l CH$_2$Cl$_2$ and added dropwise to the stirred diamine solution. After addition was completed, the CH$_2$Cl$_2$ layer was extracted with 2 x 3l H$_2$O and 1 x 2l 5 % NaCl. The CH$_2$Cl$_2$ layer was slowly filtered through a pad of MgSO$_4$, then treated with 15 ml of 0.1 N HCl dissolved in diethyl ether. The fine white precipitate was vacuum-dried at room temperature. TLC of this material (silica gel, 3:1 CH$_2$Cl$_2$: acetone) exhibited only one component at R$_f$ = 0; the absence of higher R$_f$ components suggested that no PSC or by-products were present. Anal. Calcd. for C$_{18}$H$_{17}$SO$_2$N$_2$Cl: C, 59.92; H, 4.72; N, 7.77; S, 8.89; Cl, 9.83. Found: C, 59.92; H, 4.59; N, 7.57; S, 8.64; Cl, 9.81. IR (KBr): 3352, 3295, 3048, 2930, 2729, 2656, 2607, 1601, 1315, 1110, 1094. ^{13}C NMR (d-DMSO): 38.61, 39.87 (ethylene resonances); 123.06, 123.28, 124.15, 126.77, 126.96, 129.63, 129.96, 130.40, 131.60, 134.05 (aromatic resonances).

The APS monomer (1) was prepared as follows (Scheme 3): 4(1.0 g, 2.8x10^{-3} M) and [1,8-bis-(dimethylamino)napthalene] (1.19 g, 5.6x10^{-3} M) were stirred with 7 ml DMF under a N$_2$ stream for 15 minutes at 0°C. Acryloyl chloride (2.2 ml, 2.8x10^{-2} M) was added to 7 ml DMF in an addition funnel; this solution was added dropwise to the amine solution under nitrogen. TLC (silica gel, acetone) was used to follow the depletion of starting amine (R$_f$ = 0) and the generation of product (R$_f$ = 0.70). After the addition was complete, the reaction mixture was poured into 150 cc ice. The product precipitated overnight as a yellow solid, which was filtered and vacuum-dried at room temperature. Yield: 0.90 g (88 %). Product recrystallization was performed by dissolution of 0.9 g APS in 300 ml boiling CH$_2$Cl$_2$, decolorization with Norit RB 1 0.6 charcoal pellets, and filtration through a celite pad. A first crop of pale green crystals was recovered in 69 % yield. Purity of this material was determined to be 99.96 % via HPLC. Anal. Calcd. for C$_{21}$H$_{18}$SO$_3$N$_2$: C, 66.67 %; H, 4.76; S, 8.47; N, 7.41. Found: C, 66.83; H, 5.00; S, 8.48; N, 7.49. IR (KBr): 3370, 3334, 3280, 3096, 2931, 2866, 1655, 1624, 1558, 1540, 1160, 854. ^{13}C NMR (d-DMSO): 38.77; 41.95 (ethylene carbons): 127.14, 129.58 (vinylic carbons); 123.09, 123.32, 124.07, 124.27, 125.02, 126.63, 126.79, 126.86, 129.44,

129.73, 130.36, 131.44, 132.23, 133.88 (aromatic carbons); 164.84 (acrylamido ketone carbon).

<u>Model Compound Synthesis</u>. 2,4-dimethylglutaric anhydride (**5**): 2,4-dimethylglutaric acid (mixture of dl and meso isomers, Aldrich) (2.0 g) was added to 5 ml acetic anhydride. Vacuum distillation of this solution at 90°C gave acetic anhydride as the first fraction. The anhydride product then distilled over as a clear liquid which cooled to form a hygroscopic, hard white solid. Although an IR of this product showed the presence of some diacid (O-H stretch 2500-3500 cm^{-1}; C=O stretch due to diacid at 1698 cm^{-1}), this material was successfully used in subsequent reactions without purification. Yield: 1.1 g (62 %). IR (KBr): 3500-2500, 1794, 1752 (asymmetrical and symmetrical anhydride ketone stretching modes); 1698, 1459, 1076, 1016.

Model Compound Synthesis (**6**) (Scheme 4): Compound **4** in the free amine form (0.75 g, 2.31x10^{-3} M) was dissolved in 6 ml DMF. This solution was added dropwise to **5** (0.41 g, 2.54x10^{-3} M) dissolved in 2 ml DMF under N_2 at 0°C. The reactant mixture was stirred for 5 hr. then poured into 50 ml saturated NaCl solution, which was acidified (HCl). A yellow oil immediately formed. The H_2O was decanted and the product dissolved in 30 ml CH_2Cl_2. Extraction of this solution with 50 ml H_2O precipitated the product as a pale green solid. TLC (SiO$_2$, CH$_3$OH) gave a Rf = 0.82 for the product; traces of impurities near the origin were also present. Purification of **6** was performed via dissolution of 0.61 g in DMF followed by flash chromatography on 250 ml silica gel, with CH$_3$OH as the eluent. This procedure was tedious; the product was very slow to elute. Vacuum solvent removal from the pure fractions gave about 0.2 g (33 %) yield of a pale yellow-green product. HPLC purity of this material was determined to be > 99.9 %. ^{13}C NMR (D-DMSO): 20.37, 21.13, 21.56 (aliphatic resonances of the glutaric residue); 42.84, 45.30 (aliphatic resonances of the ethylenediamine residue); 126.94, 127.92, 130.63, 133.23, 135.99, 137.62 (aromatic resonances); 179.01, 180.80 (ketone resonances of the glutoric residue). IR (KBr): 3500-2500 (broad), 3378, 3284, 2966, 1737, 1684, 1655, 1631, 1590, 1549. Anal. Calcd: C, 64.35; H, 5.63; N, 6.01; S, 6.87. Found: C, 64.20; H, 5.69; N, 5.94; S, 6.73.

<u>Polymerizations</u>

<u>AM-APS Copolymerization-Surfactant Technique</u> (Scheme 5). The general method of Turner et al., was employed (<u>15</u>). Monomer feed ratio in this copolymerization was 99.50 mol % AM: 0.50 mol % APS. The polymerization was performed by adding 7.38 g (0.1048 M) AM, 7.92 g (2.74x10^{-2} M) sodium dodecyl sulfate, 0.20 g (5.29x10^{-4} M) APS, and 235 g distilled, deionized H_2O to a 500 ml flask equipped with mechanical stirrer, nitrogen inlet, condenser, bubbler, and heating bath. This mixture was heated to 50°C under a nitrogen purge. All of the APS had dissolved after 15 minutes; polymerization was then initiated via addition of 9.25x10^{-6} M $K_2S_2O_8$ as a deaerated solution in 2 ml of H_2O via syringe. Polymerization continued at 50°C for 12 hours after which time the polymer was recovered via precipitation in acetone. Purification was accomplished by redissolving the polymer in H_2O and dialyzing against H_2O using 12,000-14,000 molecular weight cutoff dialysis tubing. The polymer was recovered by freeze-drying. Conversion was 22 %.

<u>AM-APS Copolymerization-Solution Technique</u> (Scheme 6). Monomer feed ratios, quantities, and equipment in this preparation were the same as in the

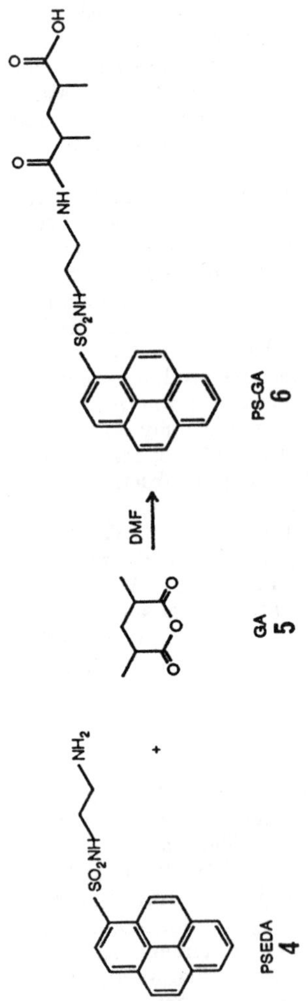

Scheme 4. Model compound synthesis.

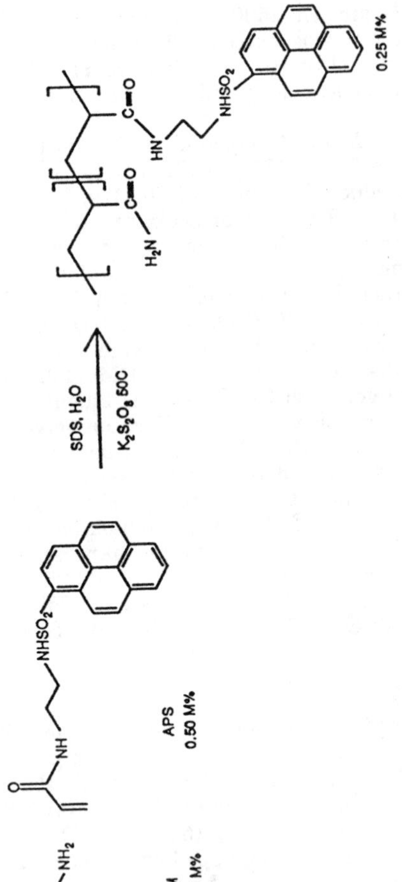

Scheme 5. Surfactant copolymerization technique.

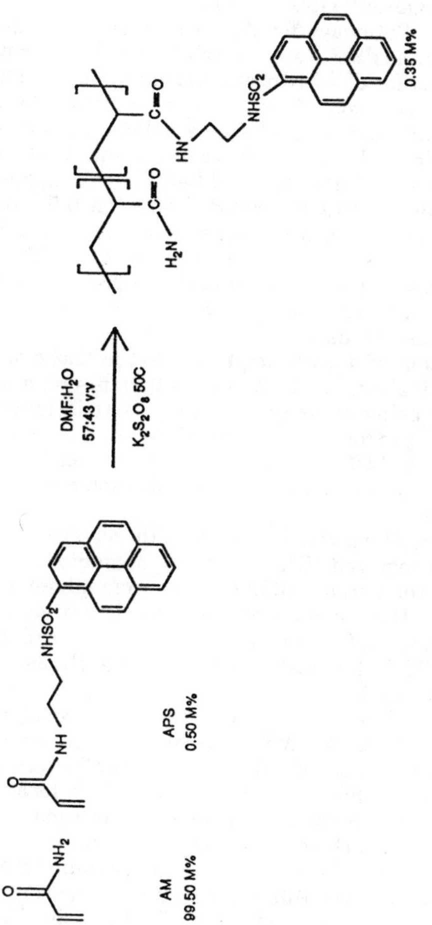

Scheme 6. Solution copolymerization technique.

previous procedure. Comonomers were dissolved in a mixture of 130 ml DMF and 100 ml H_2O. Three freeze-pump-thaw cycles were performed. Subsequent polymerization initiation, recovery and purification procedures were the same as for the surfactant preparation. Conversion was 21 %.

Results

The APS target monomer provides pendent pyrene groups for photophysical studies as well as potential sites for association along the backbone of its resultant copolymers. Pathways for APS synthesis are shown in Schemes 1-3. Synthesis of compounds 2-3 were straightforward; compound 4 proved more problematic. The reaction of 3 with ethylenediamine to generate 4 proved to be a mixing-controlled reaction. The use of seemingly dilute conditions led to production of significant amounts of the ethylene disulfonamide, which was difficult to separate from 4. A synthetic precedent (14) involving diffusion-controlled reactions was successfully modified for this application (Scheme 2). A "high-dilution" technique was employed wherein a 0.5 % (w/w) solution of 3 in CH_2Cl_2 was added dropwise, with vigorous stirring, to a 1 % (w/w) solution of ethylenediamine in CH_2Cl_2 at 0°C. Under these conditions, formation of 4 was facilitated relative to the disubstituted compound. Washing the CH_2Cl_2 solution, followed by neutralization with HCl provided the amine salt in sufficient purity for further use.

The condensation of 4 with acryloyl chloride (Scheme 3) was facilitated by the use of two equivalents of [1,8-bis-(dimethylamino)napthalene] as an acid scavenger. Attempts using other acid scavengers (triethylamine, pyridine) led to undesirable side products. Purification of the reaction mixture by recrystallization from CH_2Cl_2 gave APS in a purity sufficient for free radical copolymerization and subsequent polymer photophysical studies. A 58 % recovery was obtained.

The importance of materials purity with regard to polymerization and photophysical studies required HPLC with UV photodiode array detection. The molar absorptivity of the pyrene sulfonamide chromophore at 350 nm is 24,120; impurities containing this component are therefore readily detectable. The absence of [1,8-bis-(dimethylamino)napthalene] from purified APS was verified by monitoring the HPLC eluent at 220 nm; the napthalene compound absorbs strongly at this wavelength.

Figures 3 and 4 illustrate the two free-radical copolymerization methods employed for AM and APS. A DMF: H_2O cocktail was employed as a solution polymerization medium. Use of this mixture was necessitated since APS is DMF-soluble; AM is H_2O—but not DMF-soluble. This technique was employed in order to provide a microscopically homogeneous copolymerization medium. The surfactant (or micellar) copolymerization technique employs the same monomer feed; in this case, the polymerization medium is an aqueous solution of sodium dodecyl sulfate above its critical micelle concentration (CMC). The surfactant micelles serve to solubilize the hydrophobic APS; this technique therefore provides a microheterogeneous copolymerization medium.

Both techniques were successful in producing pyrene-containing AM copolymers. The solution technique incorporated 0.35 mol % APS, while the surfactant technique introduced 0.25 mol % APS.

Rheological Characterization. Equations 1 (the Huggins equation) and 2 (the "Modified Einstein-Simha" (13) equation) are often utilized to study solution behavior.

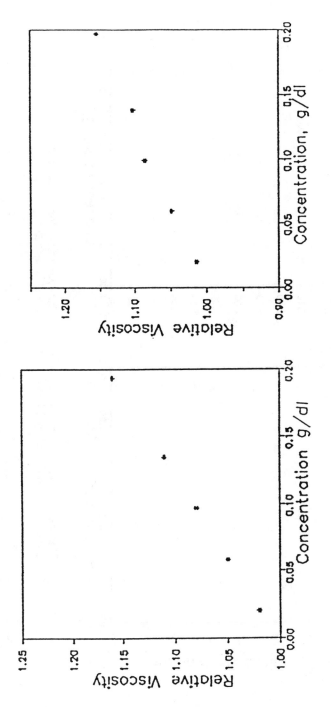

Figure 3. Plot of relative viscosity vs concentration for solution-polymerized AM-APS in DI H_2O (left) and 2 % NaCl (right).

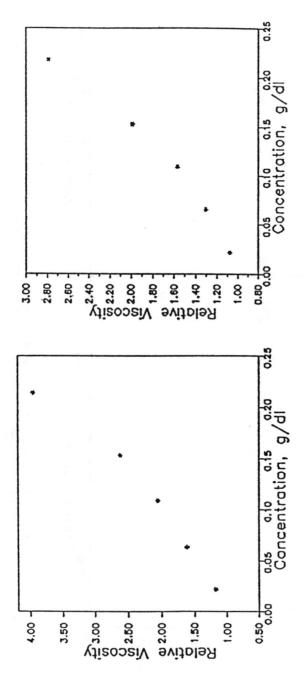

Figure 4. Plot of relative viscosity vs concentration for surfactant-polymerized AM-APS in DI H_2O (left) and 2 % NaCl (right).

$$\eta_{red} = [\eta] + k'[\eta]^2C \qquad (1)$$

$$\eta_{rel} = 1 + [\eta]C \qquad (2)$$

The utility of the Huggins equation is well-known; alternately, Equation 2 has been proposed for the analysis of polymers which behave as suspensions in solution (16). Einstein-Simha plots for the solution-polymerized system (dissolved in H_2O and 2 % NaCl) are represented in Figure 3. The overall linearity of this data and an intercept of 1 suggests that this polymer behaves as a suspension in solution. The higher order concentration terms of the Huggins equation, which account for interpolymer interactions, appear to be unimportant here. Data suggest the presence of intramolecular hydrophobic association of APS moieties. Such associations could result in compaction of the polymer coil, giving the observed suspension-like behavior.

The Einstein-Simha plots of Figure 4 represent data for the surfactant-polymerized sample dissolved in H_2O and 2 % NaCl. A low (< 0.10 g/dl) critical overlap concentration (C*) is exhibited by both of these curves. Increased ionic strength (2 % NaCl) has the effect of lowering the overall viscosity values via contraction of the polymer coil due to enhanced hydrophobic associations. The low C* value likely represents the onset of intermolecular hydrophobic associations of the APS moieties. By comparison, a homopolymer of AM has linear viscosity behavior through this concentration range; C* is not observed.

Photophysical Characterization. Figure 5 shows the absorption spectrum of compound **6** in aqueous solution at a concentration of 5.8×10^{-5} M. Figure 6 is a fluorescence emission spectrum of the same solution. The peaks at 380, 400 and 418 nm represent monomer emission bands of the pyrene sulfonamide chromophore in H_2O. No excimer formation is seen at this concentration as expected of a fluorophore in dilute solution.

Emission spectra of the two model polymers in aqueous solution, at a concentration of 0.2 g/dl, are presented in Figure 7. Accompanying the three monomer bands, an excimer peak is observed in both samples centered around 510 nm. The excimer emission intensity is significantly enhanced for the surfactant-polymerized material relative to the solution-prepared sample. Differences in polymer microstructural characteristics are clearly present. The enhancement of excimer formation in the surfactant-prepared polymer indicates a greater local concentration of pyrenesulfonamide fluorophores, suggestive of a blocky microstructural tendency (Figure 8). The microheterogeneous nature of the surfactant polymerization obviously affects placement of the hydrophobic APS units along the polymer backbone. Blocky microstructural characteristics have been previously suggested for polymers synthesized by various surfactant polymerization techniques (17,18). The homogeneous solution polymerization, on the other hand, apparently produces a more random microstructure (Figure 9). A more random placement of pyrene moieties diminishes the possibility for excimer formation.

Ratios of excimer/monomer intensities as functions of concentration, in H_2O, are tabulated in Tables I and II for the two polymer samples of this study. Data for the surfactant-polymerized sample (Table I) show a diminishing of I_E/I_M from 0.2 to 0.02 g/dl, then relatively little change over the next three orders of magnitude in dilution. This trend is evident whether the I_{III} (400 nm) or I_{IV} (418 nm) monomer peaks are ratioed. I_E/I_M values appear to parallel the viscosity behavior of Figure 5; I_E/I_M increases over the same range in which viscosification effects are noted. Intermolecular polymer

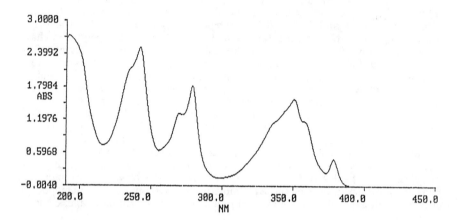

Figure 5. Absorption spectrum of **6**, 5.8x10⁻⁵ $\overline{M}$, in DI H_2O.

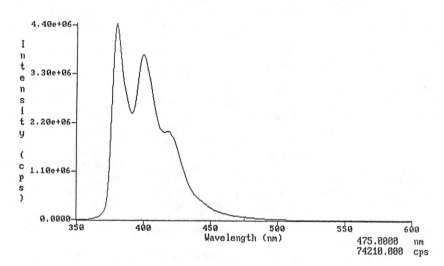

Figure 6. Fluorescence emission spectrum of **6**, 5.8x10⁻⁵ $\overline{M}$, in DI H_2O.

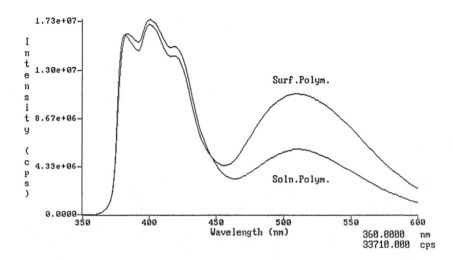

Figure 7. Fluorescence emission spectra of surfactant- and solution-polymerized samples, concentration = 0.2 g/dl.

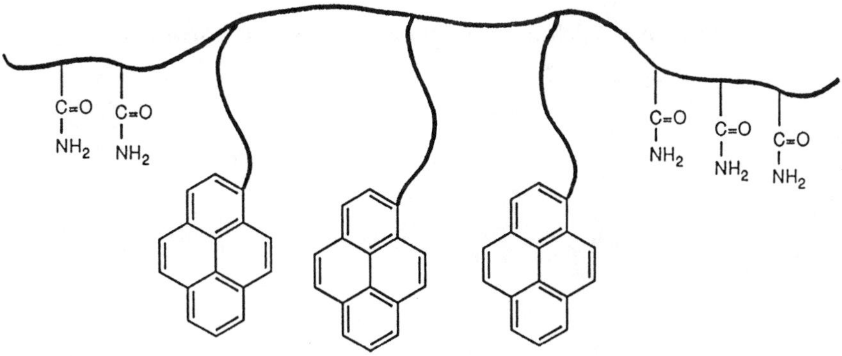

Figure 8. Surfactant-polymerized AM-APS microstructure.

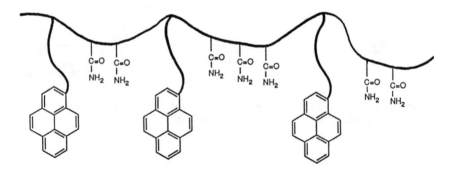

Figure 9. Solution-polymerized AM-APS microstructure.

TABLE I
CONCENTRATION DEPENDENCE OF $I_E:I_M$

Surfactant-Polymerized AM-APS

Polymer C g/dl	APS C $\overline{m}$	$I_E:I_{III}$	$I_E:I_{IV}$
2.18×10^{-1}	7.13×10^{-5}	0.64	0.76
2.21×10^{-2}	7.03×10^{-6}	0.33	0.62
2.18×10^{-3}	7.13×10^{-7}	0.29	0.55
2.18×10^{-4}	7.13×10^{-8}	0.35	0.54

TABLE II
CONCENTRATION DEPENDENCE OF $I_E:I_M$

Solution-Polymerized AM-APS

Polymer C g/dl	APS C $\overline{m}$	$I_E:I_{III}$	$I_E:I_{IV}$
1.93×10^{-1}	9.28×10^{-5}	0.34	0.39
1.94×10^{-2}	9.28×10^{-6}	0.26	0.32
1.93×10^{-3}	9.28×10^{-7}	0.11	0.21
1.93×10^{-4}	9.28×10^{-8}	0.11	0.22

associations are suggested by these trends—hydrophobic associations of the APS residues are reflected on a molecular level by I_E/I_M increases and on a macroscopic level by dramatic viscosity increases with concentration. Macroscopic and microscopic solution properties of this system appear dependent upon hydrophobic interactions of the APS moieties and can be correlated with rheological and photophysical measurements.

I_E/I_M data for the solution-polymerized sample (in H_2O) show the same general trend but the I_E/I_M increase with concentration is less dramatic. As before, an intermolecular association region is observed above about 10^{-2} g/dl, denoted by a significant increase in I_E/I_M. In this instance, however, the correlation of I_E/I_M with viscosity is not so clear. Viscosity measurements (Figure 4) suggest that only flow-field interactions are occurring in this concentration regime—the polymer behaves as a suspension throughout. Work is continuing on this interesting polymer system.

Excitation spectra of the surfactant-polymerized sample (in H_2O, at 0.2 g/dl), Figure 10, are shown wherein monomer emission (λ = 419 nm) and excimer emission (λ = 510 nm) are monitored. An overlay of these spectra reveals a significant 4 nm red shift for the excimer emission relative to the monomer emission. Excitation spectra yield information regarding the fluorophore electronic ground state. The red shifted excimer emission implies that a preformed ground state dimer exists—absorption of the excitation energy is by a dimeric rather than a monomeric species. Absorption of this species is of lower energy due to the electronic interactions of the ground state dimer. Formation of this dimer is driven by hydrophobic associations of the pyrene moieties; such associations appear to be rather static. The same situation is observed for the solution-polymerized sample (Figure 11); excimer emission is red-shifted by 4 nm relative to monomer emission, indicative of the presence of ground-state aggregates.

Additional evidence for the presence of pyrenesulfonamide aggregates is offered by fluorescence lifetime measurements of the excimer emission. Fluorescence decay curves for both the surfactant (Figure 12) and solution (Figure 13) polymerized samples do not exhibit a rise time. It is likely that both possess preformed dimeric associations of the APS residues, a consequence of the hydrophobic character of these species. Similar behavior has been observed for other pyrene-labeled water-soluble polymers; poly(N-isopropylacrylamide) (10) and poly(ethylene oxide) (8) are two examples.

Conclusions

A new pyrene-containing acrylamido monomer, [1-(β-aminoethyl)sulfonamido-pyrene]acrylamide (APS), has been synthesized in sufficient purity for copolymerization and subsequent photophysical studies. Microscopically homogeneous (solution) and heterogeneous (surfactant) free radical polymerization techniques were successfully employed for the preparation of pyrenesulfonamide-labeled acrylamide copolymers. Macroscopic viscosity behavior of the surfactant-polymerized sample suggests the presence of intermolecular associations. The solution-polymerized sample exhibits viscosity behavior demonstrative of an intramolecular hydrophobically associating polymer. A somewhat blocky microstructure is suggested for the surfactant-polymerized polymer relative to the solution-polymerized material due to an observed enhancement of I_E/I_M. Changes in I_E/I_M for both polymers as a function of concentration reflect an onset of intermolecular interactions, correlatable with macroscopic viscosity increases. Excitation spectra indicate

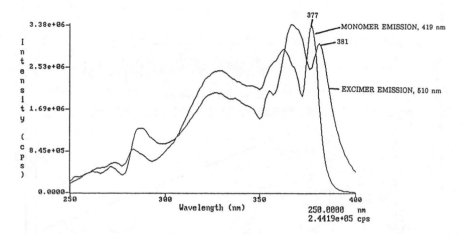

Figure 10. Excitation spectrum of surfactant-polymerized AM-APS.

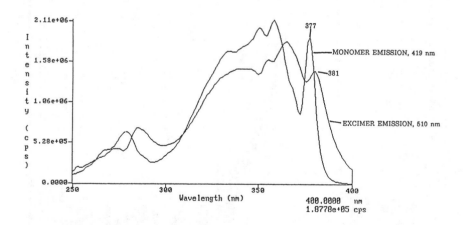

Figure 11. Excitation spectrum of solution-polymerized AM-APS.

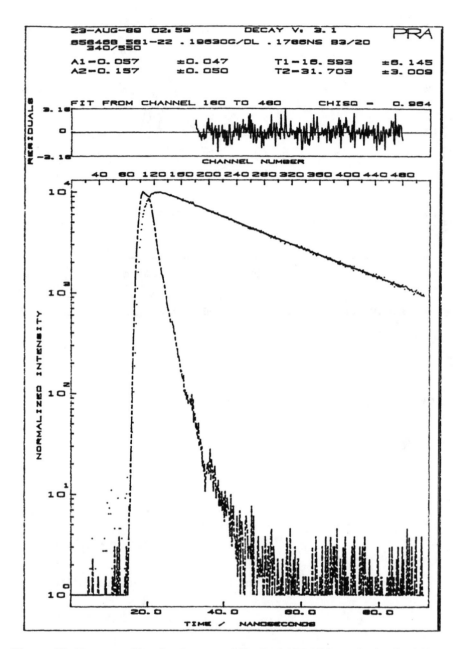

Figure 12. Decay profile of surfactant-polymerized AM-APS, excitation $\lambda = 340$ nm, emission $\lambda = 550$ nm (excimer emission), in H_2O

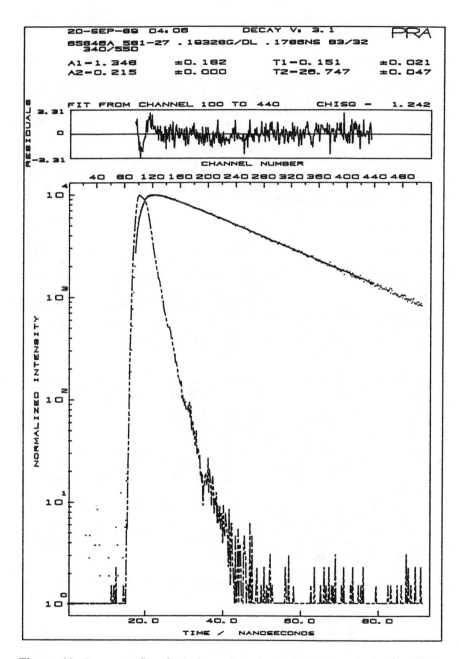

Figure 13. Decay profile of solution-polymerized AM-APS, excitation λ= 340 nm, emission λ = 550 nm (excimer emission), in H_2O

the presence of pyrenesulfonamide ground-state dimers, likely enhanced by the hydrophobic associations in H_2O.

Acknowledgments

Research support from the U.S. Department of Energy, the Defense Advanced Research Projects Agency, the Office of Naval Research, and Unilever is gratefully acknowledged. Original concept of the APS monomer is credited to Dr. Mitchell Winnik's research group of the University of Toronto, which provided us with a small sample and a general synthetic outline. We reworked the synthesis and scaled up APS for our use.

    Dr. Charles Hoyle and Mark Clark (both of The University of Southern Mississippi (USM)) furnished technical advice. Sheila Williamson (Mississippi University for Women) and Kelly Anderson (USM) provided technical assistance with regard to small-molecule studies.

Literature Cited

1. McCormick, C. L.; Nonaka, T.; Johnson, C. B. Polymer 1988, 29, 738.
2. Bock, J.; Siano, D. B.; Valuit, P. L., Jr.; Pace, S. J. In Polymers in Aqueous Media; Glass, J. E., Ed.; Advances in Chemistry Series No. 223; American Chemical Society: Washington, D.C., 1989; p 411.
3. Kalyanasundaram, K. Photochemistry in Microheterogeneous Systems; Academic Press: Orlando, 1987; Chapter 8.
4. Herkstroeter, W. G.; Martic, P. A.; Hartman, S. E.; Williams, J. L. R.; Farid, S. J. Polym Sci. Polym. Chem. 1983, 21, 2473.
5. Chu, D. Y.; Thomas, J. K. Macromolecules 1984, 17, 2142.
6. Turro, N. J.; Arora, K. S. Polymer 1986, 27, 783.
7. Arora, K. S.; Turro, J. J. J. Polym. Sci. Polym. Phys. 1987, 25, 243.
8. Char, K.; Frank, C. W.; Gast, A. P.; Tang, W. T. Macromolecules 1987, 20, 1833.
9. Winnik, F. M. Macromolecules 1987, 20, 2745.
10. Winnik, F. M. Macromolecules 1990, 23, 233.
11. Winnik, F. M. Macromolecules 1990, 23, 1647.
12. Still, W. C.; Kahn, M.; Mitra, A. J. Org. Chem. 1978, 43, 2923.
13. Tietze, E.; Bayer, O. Ann. 1939, 540, 189; Chem. Abstr. 1939, 9318.
14. Jacobson, A. R.; Mahris, A. N.; Sayre, L. M. J. Org. Chem. 1987, 52, 2592.
15. Turner, S. R.; Siano, D. B.; Bock, J. (to Exxon Research and Engineering). U.S. Patent 4 520 182, 1985.
16. Salamone, J. C.; Tsai, C. C.; Olson, A. P.; Watterson, A. C. In Ions in Polymers; Advances in Chemistry Series No. 187; American Chemical Society: Washington, D. C., 1978; p 337.
17. Pear, W. J. In Reference 2, p 381.
18. Dowling, K. C.; Thomas, J. K. Macromolecules 1990, 23, 1059.

RECEIVED June 4, 1990

Chapter 9

Comblike Cyclopolymers of Alkyldiallylamines and Alkyldiallylmethylammonium Chlorides

George B. Butler and Choon H. Do

Department of Chemistry and Center for Macromolecular Science and Engineering, University of Florida, Gainesville, FL 32611-2046

Water-soluble and comb-like cyclopolymers of alkyldiallylamines (alkyl=decyl and hexadecyl, ADAA) and the corresponding alkyldiallylmethylammonium chlorides (ADAMAC) were prepared by radical polymerization at 40°C in water solution. The polymers were characterized by NMR and DSC. The monomers were also cyclocopolymerized with diallyldimethylammonium chloride. Although the protonated ADAA and ADAMAC possess long alkyl chains, they cyclopolymerized like those monomers possessing shorter chains, and formed predominantly pyrrolidinium rings. The ring was largely cis-substituted; however, the ratio of the cis- to the trans-substituted peaks varied among the monomers studied. Because the cyclopolymers possess long alkyl side-chains, they showed a side-chain crystallization. The cyclopolymers also formed polymeric micelles.

Since the discovery that radical polymerization of dialkyldiallylammonium salts yields water-soluble cyclopolymers instead of cross-linked polymers (1-5), homo- and co-cyclopolymers of various diallylamine compounds have been extensively studied and their syntheses, kinetics, ring-sizes, properties, and applications have been reviewed (6-11). Cyclopolymerization of the diallylammonium compounds produces mostly ring structures and the content of double bonds is less than 0.1-3% in the case of poly(diallyldimethylammonium chloride, DADMAC) (9). Free diallylamines do not polymerize but their protonated and quaternary ammonium salts polymerize by radical initiation (10). Although the piperidinium ring is thermodynamically favorable, the kinetically favorable pyrrolidinium ring exists dominantly in the cyclopolymer of DADMAC according to ^{13}C NMR spectroscopic studies (11-13). The size of the ring also depends on the size of N-substituted alkyl groups (11).

　　Complete or partial substitution of one of the methyl groups in the poly(DADMAC) with longer alkyl chains which are hydrophobic would be expected to change the properties of the polymers such as viscosity, solubility, and other applicable properties. Recently, for example, cyclopolymers

0097–6156/91/0467–0151$06.00/0

of diallyldidodecylammonium bromide were studied for vesicle formation (14-15). Introduction of long alkyl chains into the cyclopolymers would produce comb-like polymers which may show side-chain crystallization (16). Furthermore, because the cyclopolymers possess hydrophilic protonated or quaternary ammonium groups and hydrophobic long alkyl chains, the polymers may form polymeric micelles and polymeric aggregates or self-oriented polymeric systems (17-18). Therefore, we became interested in the polymerization of alkyldiallylamines (alkyl=decyl and hexadecyl, ADAA) and the corresponding alkyldiallylmethylammonium chlorides (ADAMAC). Copolymerization of ADAA and ADAMAC with DADMAC, the most common quaternary ammonium monomer, was also studied. We wish to report here the preparation of 1) ADAA and ADAMAC and their homopolymerization (Scheme 1) and copolymerization with DADMAC (Scheme 2) and 2) characterization of the resulting polymers by ^{13}C NMR spectroscopy, DSC, and microscopy.

Scheme 1. Homocyclopolymerization of alkyldiallylamine (ADAA) and alkyldiallylmethylammonium chloride (ADAMAC).

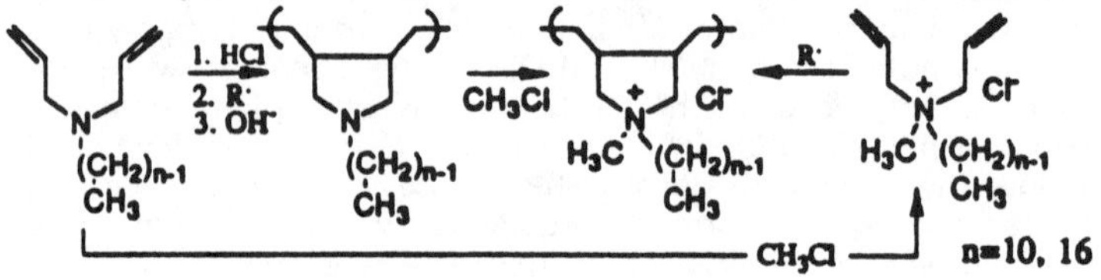

Scheme 2. Cyclocopolymerization of 1) ADAA with DADMAC and 2) ADAMAC with DADMAC.

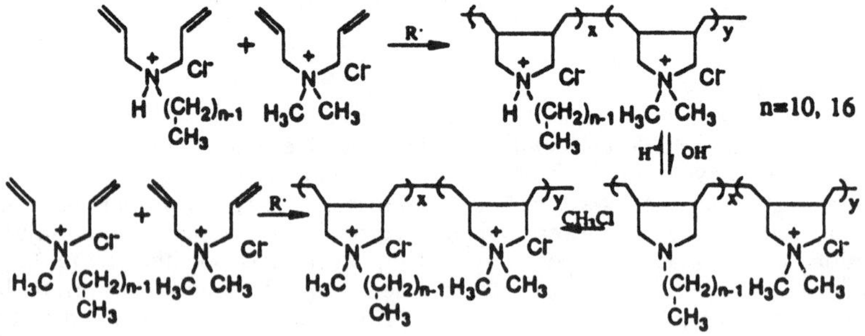

Experimental

Preparation of ADAA and ADAMAC.   ADAA was prepared in high yields (>90%) from diallylamine and 1-bromoalkanes by heating at 110°C for 10 h. ADAMAC was prepared nearly quantitatively by methylating ADAA with CH_3Cl in acetone solution at 80°C under 600 psi for 8 h in a bomb reactor.

Cyclopolymerization of ADAA and ADAMAC. ADAA·HCl was cyclopolymerized or cocyclopolymerized with DADMAC in 40-50 wt % aqueous solution at 40°C for 24 h using 2,2'-azobis(N,N'-dimethyleneisobutyramidine)dihydrochloride (ABDMBA) as radical initiator. After polymerization, the solution was neutralized with 3 N Na_2CO_3. The polymer which separated was washed with water, then dissolved in a small amount of $CHCl_3$ and reprecipitated into acetone. The copolymers were purified by dialysis using a membrane of 8,000 molecular weight cutoff. The yields were higher than 20%.

ADAMAC was homopolymerized or copolymerized with DADMAC using ABDMBA as initiator at 40°C in 40-50 wt% aqueous solution, by following a similar procedure to polymerization of ADAA·HCl. After polymerization, the polymer was purified by dialysis using a membrane of 8,000 cutoff molecular weight. The yields were in the range of 10%.

Characterization. ^{13}C NMR spectra were recorded on a Varian 200 XL NMR Spectrometer at ambient temperature using $CDCl_3$ or D_2O as solvents. Melting points of side-chain crystallization were measured using a Perkin-Elmer DSC-7. The heating rate was 2°C/min. Optical micrographs were obtained by using a Nikon Optiphot microscope.

Results and Discussion

Poly(ADAA) and Its Copolymers. By comparison with the ^{13}C NMR spectra of the monomer and its quaternary ammonium chloride (Figure 1), the peaks of poly(decyldiallylamine, DDAA) were assigned (Figure 2). The spectrum of

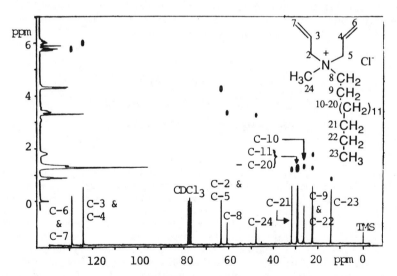

Fig. 1. Heterocorrelated ^1H-^{13}C 2D NMR spectrum of hexadecyldiallylmethylammonium chloride in $CDDI_3$.

poly(hexadecyldiallylamine, HDDAA) was very similar to that of poly-(DDAA). Two ^{13}C NMR peaks around 39-43 ppm represent C-3 and C-4 of the pyrrolidine ring (10,12), and consequently indicate that poly(ADAA) possesses the pyrrolidine ring structure (1 and 2). The ratio of the two

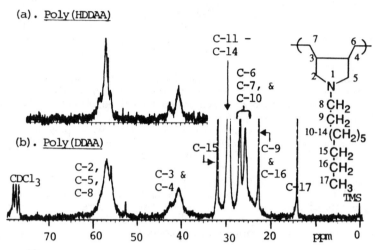

Fig. 2. ^{13}C NMR spectra of (a) poly(HDDAA) and (b) poly(DDAA) in CDCl$_3$.

peaks also indicates cis- and trans-substitutions of the ring (10,12), and the spectra show that the ratios are 3:1 (cis:trans) or less. These ratios are reduced from 5:1, the ratio observed in the cyclopolymer of methyldiallyl-amine. This difference is probably due to steric hindrance of the bulky alkyl groups, but the cis-pyrrolidine ring structure is still dominant.

The positions of C-6, C-7, C-2, and C-5 in the NMR spectra may also support the above conclusions. However, they are obscured and overlapped with the peaks from C-6, C-7, and C-10 which appear between 26 and 27 ppm and those from C-2, C-5, and C-8 which appear around 55-59 ppm. The copolymers prepared from ADAA and DADMAC also possess pyrrolidine rings and are random copolymers rather than block copolymers or mixtures of homopolymers. This conclusion is based on the NMR spectra, the melting points of side-chains, and the formation of micelles. If a mixture of homopolymers or a block copolymer had been obtained, the depression of Tm due to dilution would be linear and small. It also would be observed in a wide range of the composition. Furthermore, such structures would not be expected to form polymeric micelles since the homopolymer showed no tendency to do so.

Poly(ADAMAC) and Its Copolymers. Although decyl- and hexadecyl-diallyl-methylammonium chlorides (DDAMAC and HDDAMAC) possess long alkyl chains, they form polymers possessing pyrrolidinium ring structures accord-ing to ^{13}C NMR spectra (Figure 3). C-3 and C-4 appear around 37-40 ppm

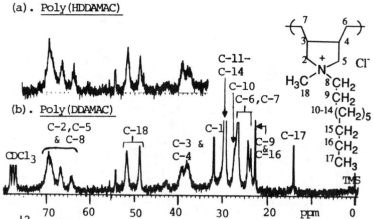

(a). Poly(HDDAMAC)

(b). Poly(DDAMAC)

Fig. 3. ^{13}C NMR spectra of (a) poly(HDDAMAC) and (b) poly(DDAMAC) in CDCl$_3$.

and indicate that it is composed of pyrrolidinium rings (3) (13); the cis- and trans-substitution ratios are 2:1 or less. If piperidinium rings (4) had been formed in addition to the pyrrolidinium ring, the methylene carbon atom (C-4) of the piperidinium ring would appear around 35-41 ppm (13). The methyl carbon atom attached to the nitrogen atom shows two peaks 48.5 and 52 ppm and C-2, C-5, and C-8 show three peaks between 64 and 70 ppm. C-6 and C-7 appear around 24-28 ppm.

Copolymerization of ADAMAC with DADMAC yields random copolymers possessing pyrrolidinium rings. Thus ^{13}C NMR peaks of the methyl groups attached to the quaternary nitrogen atom and C-3 and C-4 regions of poly(DDAMAC-co-DADMAC) made from 1:1 monomer feed are identical with those of poly(DADMAC) (Figure 4) and this result indicates that the copolymer possesses pyrrolidinium ring structure like a poly(DADMAC). The formation of polymeric micelles (see below) suggests that the copolymer is a random copolymer rather than a mixture of homopolymers of each components or a block copolymer.

Side-Chain Crystallization. Poly(HDDAA) and its derivatives and copolymers exhibit endothermic peaks due to the melting of side-chains between 30-50°C (Figure 5). These results indicate that hexadecyl groups crystallize in the side-chains. Apparently, poly(HDDAA) and its HCl salt show more tendency to crystallize than its quaternary ammonium chloride, poly-(HDDAMAC). The limits of the composition of the copolymers and side-chain crystallization will be examined further. Poly(DDAA) and its HCl salts, however, do not show side-chain crystallization clearly.

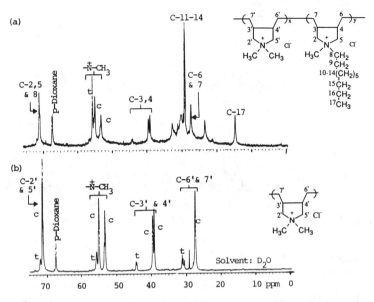

Fig. 4. ^{13}C NMR spectra of (a) poly(DDAMAC-co-DADMAC) and (b) poly-(DADMAC) in D$_2$O.

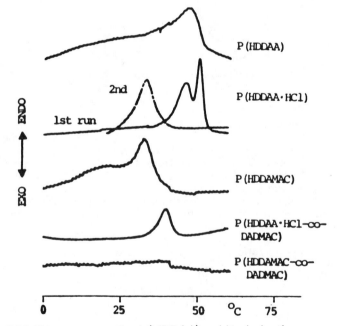

Fig. 5. DSC Thermograms of poly(HDDAA) and its derivatives.

Formation of Polymeric Micelles.   Optical micrograph shows polymeric micelles of 1 μm size (Figure 6a) and polymeric aggregates of < 10 μm size (Figure 6b) formed from an aqueous solution of poly(DDAA-co-DADMAC). Although this shows that the cyclopolymers can form polymeric micelles, their properties and the ranges of the copolymer compositions where the copolymers are able to form micelles have not been fully investigated yet.

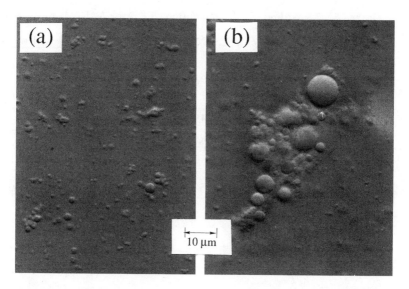

Fig. 6.   Optical micrograph of (a) polymeric micelles and (b) polymeric aggregates formed from aqueous solution of poly(DDAA-co-DADMAC).

Acknowledgments

Financial support of this work was provided by the Department of Energy, Office of Basic Energy Sciences under Grant No. DE-FG05-84ER45104 for which we are grateful.   We are also grateful to Dr. H. C. Aldrich, Dept. of Microbiol., Univ. of Fla., for permission to use his microscope to obtain the micrograph shown in Figure 6.

Literature Cited

1.   Butler, G. B.; Ingley, F. L. J. Am. Chem. Soc. 1951, 73, 895.
2.   Butler, G. B.; Goette, R. L. J. Am. Chem. Soc. 1952, 74, 1939.
3.   Butler, G. B.; Johnson, R. A. J. Am. Chem. Soc. 1954, 76, 713.
4.   Butler, G. B.; Angelo, R. J. J. Am. Chem. Soc. 1956, 78, 4797.
5.   Butler, G. B.; Angelo, R. J. J. Am. Chem. Soc. 1957, 79, 3128.
6.   Butler, G. B. in IUPAC, Polymeric Amines and Ammonium Salts; Goethals, E. J., Ed; Pergamon; New York, 1980; p. 125.
7.   Butler, G. B. Acc. Chem. Res. 1982, 15, 370.
8.   Ottenbrite, R. M.; Ryan, W. S., Jr. Ind. Eng. Chem. Prod. Res. Dev. 1980, 19, 528.

9.   Jaeger, W.; Hong, L. T.; Philipp, B.; Reinisch, G.; Wandrey, Ch. in
     IUPAC, Polymeric Amines and Ammonium Salts; Goethals, E. J., Ed;
     Pergamon; New York, 1980; p. 155.
10. Solomon, D. H.; Hawthorne, D. G. J. Macromol. Sci.-Rev. Macromol.
     Chem. 1976, C15, 143.
11. Ottenbrite, R. M.; Shillady, D. D. in IUPAC, Polymeric Amines and
     Ammonium Salts; Goethals, E. J., Ed; Pergamon; New York, 1980; p.
     143.
12. Johns, S. R.; Willing, R. I.; Middleton, S.; Ong, A. K. J. Macromol.
     Sci.-Chem. 1976, A10, 875.
13. Lancaster, J. E.; Baccei, L.; Panzer, H. P. J. Polym. Sci., Polym.
     Lett. Ed. 1976, 14, 549.
14. Babilis, D.; Dais, P.; Margaritis, L. H.; Paleos, C. M. J. Polym. Sci.,
     Polym. Chem. Ed. 1985, 23, 1089.
15. Babilis, D.; Paleos, C. M.; Dais, P. J. Polym. Sci., Polym. Chem. Ed.
     1988, 26, 2141.
16. Plate, N. A.; Shibaev, V. P. J. Polym. Sci., Macromol. Rev. 1974, 8,
     117.
17. Fendler, J. H.; Tundo, P. Acc. Chem. Res. 1984, 17, 3.
18. Ringsdorf, H.; Schlarb, B.; Venzmer, J. Angew. Chem. Int. Ed. Engl.
     1988, 27, 113.

RECEIVED June 4, 1990

Chapter 10

New Fluorocarbon-Containing Hydrophobically Associating Polyacrylamide Copolymer

Y.-X. Zhang[1], A.-H. Da[1], Thieo E. Hogen-Esch[1,2], and George B. Butler[2]

[1]Department of Chemistry, Loker Hydrocarbon Research Institute, University of Southern California, Los Angeles, CA 90089–1661
[2]Department of Chemistry and Center for Macromolecular Science and Engineering, University of Florida, Gainesville, FL 32611–2046

The synthesis of polyacrylamide copolymers containing fluorine-containing acrylates **1, 2, 4** or **5** or silicone-containing methacrylate **6** was carried out by emulsion polymerization using persulfate as initiators and ionic perfluoro surfactants. The Brookfield viscosities of these copolymers, especially those containing **1** or **2**, were much higher than the laurylacrylate (**3**) containing copolymer. The comonomer content of the **1** or **2** containing copolymers that were most strongly viscosifying was only .07–.28 mole percent, well below that of the corresponding copolymers containing **3**. Viscosity increases of the **1** and **2** copolymers, observed upon increases in temperature or NaCl concentration were consistent with association of these copolymers through intermolecular hydrophobic binding. Reduced viscosity or dynamic light scattering studies are consistent with the existence of polymer assemblies above polymer concentrations of about 100 pm. Addition of surfactants water miscible organic solvents or increasing shear rate was shown to lead to sharp viscosity decreases consistent with the proposed model.

The replacement of hydrocarbon- by perfluoroalkyl groups in hydrophobically associating polymers[1] is of potential interest since the hydrophobic character of perfluoroalkyl groups is more pronounced compared to their hydrocarbon analogs.[2] Thus, the critical micelle concentrations of fluorocarbon surfactants generally are significantly lower than the hydrocarbon analogs of the same chain length and surface tensions of fluorocarbon surfactant solutions are also much lower.[2,3]

Also hydrophobic bonding of fluorocarbon surfactants to β-cyclodextrines is much stronger (about 300 times) compared to similar hydrocarbon containing surfactants.[4] Significantly, the latter process is enthalpy driven (ΔH = -9.0 kcal/mole, ΔS = -12 eu) whereas the bonding of the fluorocarbon

0097–6156/91/0467–0159$06.00/0
© 1991 American Chemical Society

surfactant is mostly favored by entropy (ΔS = +15 eu, ΔH = - 1.7 kcal/mole). These results suggest, at least in this case, that the binding of fluorocarbons is qualitatively different from that of hydrocarbons and is not merely due to the larger surface of the perfluorocarbon chain compared with the hydro- carbon chain of the same carbon number.

As a result of these considerations, we have started a research program aimed at investigating the synthesis of copolymers of polyacryl- amide and other water-soluble polymers containing perfluorocarbon groups (Scheme 1). Included in the group of comonomers investigated was also a silicone containing methacrylate. This monomer was of interest because of the pronounced hydrophobic character of silicones.[5,6]

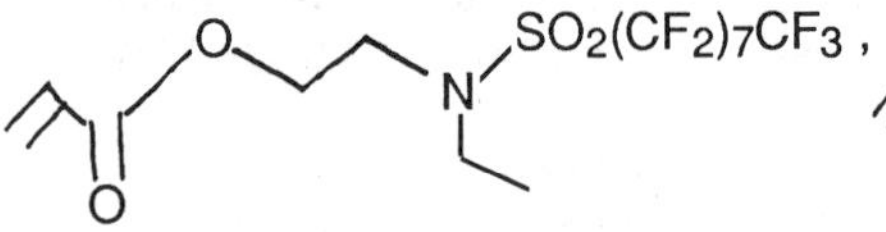

1 (FX - 13) 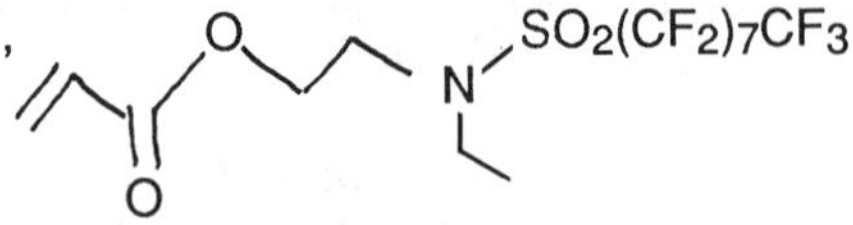   2 (FX - 14)

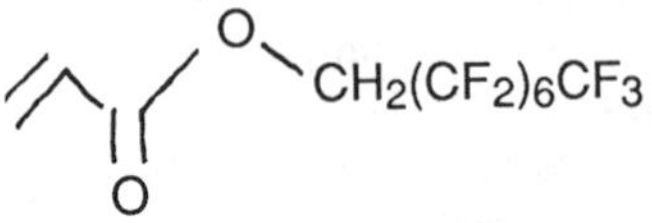

3 (RF - 8) 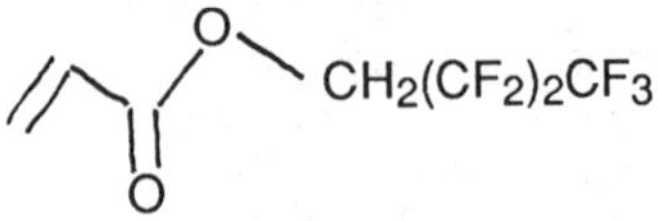   4 (RF - 4)

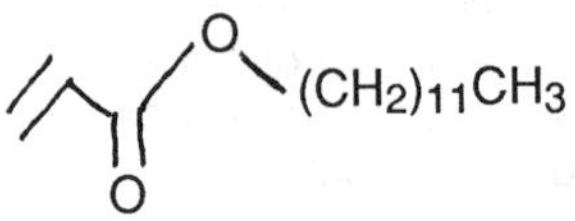

5 (LA) 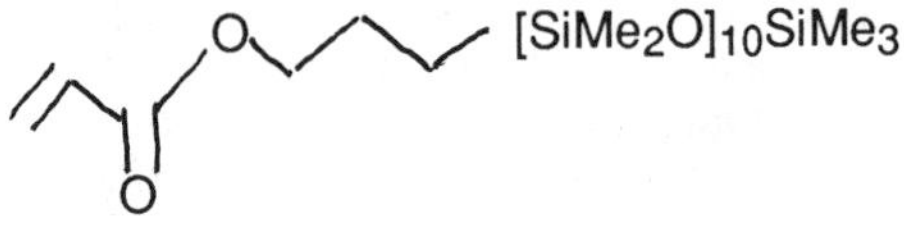   6 (ASi)

Scheme 1   Structures of Hydrophobic Comonomers

Experimental Section

The copolymers were prepared by an emulsion type polymerization under argon in deionized H_2O (20 ml) containing various amounts of acetone (0-5 ml) and a perfluoroalkyl containing surfactant supplied by the 3 M Company (FC-129) at a concentration between .05 - 0.15 %. Acrylamide (2 gr) (Polysciences, "ultrapure") was dissolved in the mixture and the comonomer was usually added as an acetone solution. Comonomers FX-13 (1) and FX-14 (2) were kindly supplied by the 3 M Company. The comonomers 3, 4 and 5 were supplied by Polysciences and comonomer 6 was purchased from Petrarch Systems, Inc. The polymerization was initiated by $(NH_4)_2S_2O_8$ (.0228 gr) and $Na_2S_2O_5$ (.0190 gr) added as aqueous solutions. The mixture was stirred at temperatures varying from 25° to 50° for periods between 6 - 24 hrs. After polymerization, the gel-like product was dissolved in deionized H_2O and dialyzed against deionized water. The dissolution process varied between several hours and several months. The resulting solutions were then diluted to the proper concentrations for rheological studies. In this case, the polymer concentrations were determined by precipitation in excess acetone and weighing of the dried polymer. Alternatively, the concentrated solutions were precipitated in acetone followed by drying in a vacuum oven at 40° for 24 hrs. The solid polymer was then redissolved in a known volume of H_2O or NaCl solutions. The viscosity of the polymer solutions was measured using a Brookfield viscometer (LTV model) with appropriate sample adaptor. Reduced viscosities were determined at 30.0° using a Cannon-Ubbelode viscometer.

Results

A series of copolymers of acrylamide and comonomers 1, 2 and 5 were prepared initially. The Brookfield viscosity results are shown in Tables I - III and in Figures 1 - 2. Table I shows the Brookfield viscosities of 0.5 % solutions of the FX-13 copolymers. It is interesting to see that the viscosities of the 0.5 % solutions are very high and that the comonomer content of the most strongly viscosifying solutions is very low. For instance, the solution of the copolymer containing only .006 mole % (or .05 weight %) of 1 still has a viscosity (at .40 sec $^{-1}$) that is about 10 times larger than that of the homo-polymer prepared under identical conditions. Figure 1 shows the viscosity versus concentration profiles for several of the copolymers as well as the homopolymer prepared under identical conditions. The viscosity differences are especially pronounced at higher concentrations consistent with the pattern expected for associating polymers.[1] A copolymer (AL-3) containing laurylacrylate (5) is shown for comparison. This polymer, as expected, shows a viscosity increase compared to the homopolymer (PAM) at polymer concentrations above 0.5 %. However, the laurylacrylate copolymer seems far less effective as a viscosifier compared to the FX-13 containing samples AF-10 and AF-12. Sample AFT was a terpolymer containing 16 mole % of Na AMPS along with .14 mole % of FX-13. This terpolymer, although less strongly viscosifying than copolymer AF-10, at the higher concentrations gave more viscous solutions at low polymer concentrations (< 0.1 wt %) presumably as a result of polyanion expansion (Fig. 1).

Table I Synthesis of FX-13 containing polyacrylamide [a]

Sample	Mole % Comonomer	Weight % Comonomer	Conversion[d] (%)	Brookfield[b,c] Viscosity (Centerpoise)
PAM	0	0	99.9	50
AF-22	.006	.05	90.3	480
21	.012	.10	90.7	1760
18	.023	.20	88.5	7000
17	.045	.40	95.2	10800
12	.070	.62	95.0	12000
10	.14	1.25	96.3	6200
9	.28	2.40	96.1	300
8	.56	4.80	101.4	50
7	1.12	9.10	100.0	50
AFT-13[e]	.14	1.25	78.9	2500

a. See Experimental Section for details of synthesis. Emulsifier was $CF_3(CF_2)_6COOK$ at .12 wt %. b. Polymer concentration = 0.50gr/dl. c. Shear rate is 0.40 sec^{-1}. d. Refers to conversion of total monomer; determined gravimetrically. e. Terpolymer containing 16% AMPS (see text).

Table II Synthesis of FX-14 containing polyacrylamide [a]

Sample	Mole % Comonomer	Weight % Comonomer	Conversion %	Brookfield[b,c] Viscosity (c_p)
AF-24	.022	.20	101.0	120
23	.044	.40	100.0	2220
15	.069	.62	100.7	6000
16	.14	1.25	83.1	6800
19	.28	2.40	74.3	10,800
20	.55	4.80	75.9	9000
32	1.11	9.10	71.8	200
33	1.66	13.00	67.8	120

Footnotes a-d see Table I.

    Figure 2 shows a plot of the viscosity of 0.5 wt % solutions of a series of FX-13 containing copolymers as a function of comonomer content (mole %). The plot shows a clear maximum at about .07 mole %. Significantly, the decrease in viscosity with increasing comonomer content above .07 mole % is more pronounced than the viscosity decrease observed in the low comonomer content part of the curve (Discussion).
    The pronounced viscosifying tendency of the polyacrylamide copolymers containing FX-13 was also observed for the corresponding copolymers containing the methacrylate (FX-14) (Table II, Figure 2). In this case, the maximum viscosity as a function of comonomer content is observed at slightly higher content of FX-14 (.014%). The asymmetry in the viscosity-comonomer content plot is observed in this case also.

Table III Synthesis and Properties of Laurylacrylate Containing Polyacrylamides[a]

Sample	Comonomer mole %	wt %	Conversion[b] %	Brookfield Viscosity[c] (1.0 wt %)	(0.5 wt %)
Al-1[d]	.74	2.44	100.4	84	10
-2[d]	1.64	4.80	99.5	120	30
-3[d]	2.99	9.13	100.4	170	40
-4[d]	4.52	13.20	92.3	44	10
Al-7[e]	.75	2.40	86.4	400	50
-5[e]	1.50	4.80	86.7	420	100
-6[e]	3.00	9.10	83.4	760	120
-8[e]	4.50	13.00	76.6	250	40
PAM	0	0	99.9	450	50

a. Polymerization Conditions: 2gr AM in 20 ml deionized water. b. Based on weight of precipitated polymer. c. Shear rate is 0.40 sec^{-1}. d. Sodium laurylacrylate (.25 wt %). e. Emulsifier was $CF_3(CF_2)_6COOK$ (obtained from 3M Company) at .12 wt %.

In contrast, the **5** containing copolymers are less strongly viscosifying by at least a two order of magnitude (Table III). Furthermore, the molar content of **5** in these copolymers is almost two orders of magnitude higher than that of the FX-13 series. Thus, the hydrocarbon containing polyacrylamides are both less effective and less efficient than the FX-13 and FX-14 containing copolymers.

Table IV shows the Brookfield viscosities of 0.5 weight percent polyacrylamide copolymers containing comonomers **3** (RF-8) and **4** (RF-4) containing C_7 and C_3 perfluoroalkyl groups respectively. Apparently both comonomers lead to more effectively viscosifying copolymers compared with the laurylacrylate copolymers although the RF-4 and laurylacrylate copolymers are not that different. Again the comonomer content corresponding to the most strongly viscosifying copolymers is lower for the RF-4 and RF-8 series compared to the hydrocarbon analogs both on a mole and weight % basis.

However, the RF-4 and RF-8 containing comonomers are less effectively and efficiently viscosifying compared to the FX-13 and FX-14 series. Although it is possible that the differences between the FX-13/FX-14 type and the RF-8 type copolymers are due to the slightly greater degree of fluorine substitution of the octyl groups in FX-13/FX-4, it appears plausible that the different "spacer" group in the FX-type comonomers plays a role. The polar nature of the sulfonamide group makes it plausible that the hydrophilic spacer in this case may act to decouple the motions of the polymer and the hydrophobic aggregate. Such an effect of spacer groups was recently demonstrated by Schulz et al for hydrocarbon acrylate comonomers containing poly-ethyleneoxide spacers of various lengths.[7]

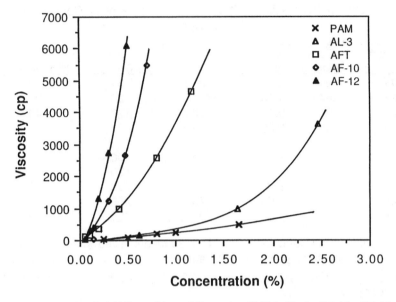

Figure 1. Brookfield Viscosities of Samples PAM, AL-3, AF-10, AF-12 and AFT-13 Measured at 0.40 sec^{-1}.

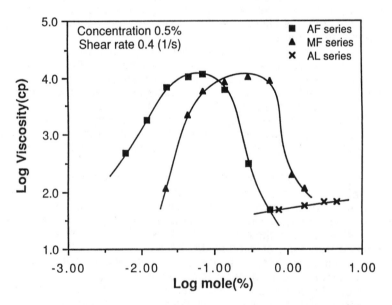

Figure 2. Dependence of Brookfield Viscosities Against Mole % Comonomer for the Copolymers Containing FX-13 (AF Series) and FX-14 (MF Series).

Table IV Synthesis and Properties of RF-4 and RF-8 Containing Polyacrylamides[e]

Comonomer	Sample	Comonomer Mole %	wt %	Conversion[c] %	Brookfield[a,d] Viscosity (cp)
RF$_4$	AF-25	.18	.625	70.7	450
	AF-26	.88	3.03	87.6	1000
	AF-27	3.50	11.11	89.3	400
RF-8	AF-28	.11	.67	88.1	24[b]
	-29	.51	3.03	86.7	3000[b]
	-30	.98	5.88	94.6	2960[b]
	-31	1.96	11.11	90.3	120[b]

a. Polymer concentration is 1.0 wt %. b. Polymer concentration is 0.50 wt %.
c. Based on weight of precipitated polymer. d. Shear rate is 0.40 sec^{-1}.
e. $CF_3(CF_2)COOK$ was emulsifier at .12 wt %.

A comparison of the RF-4 and RF-8 copolymers points up the expected greater viscosifying power of the RF-8 copolymers. Such increases are not unexpected in view of the well documented decreases in CMC of a series of surfactants with increasing hydrocarbon length.[3] Similar trends in fluorocarbon surfactants have also been documented.[3]

Table V Synthesis and Properties of ASi Containing Polyacrylamides[a]

Sample	Comonomer Mole %	Comonomer Wt %	Monomer Conversion	Brookfield[b,c] Viscosity (cp)
ASi-8	.005	.06		250
-7	.009	.12		420
-5	.018	.23	75.8	750
-4	.036	.46	77.3	780
-3	.089	1.13	82.8	600
-2	.446	5.70	85.7	400
-1	.890	11.40	74.7	350

a-c   See footnotes, Table I

Table V shows the Brookfield viscosities of a series of silicone methacrylate **6** containing polyacrylamide copolymers (ASi series). These copolymers are less strongly viscosifying compared with the fluorocarbon containing copolymers but appear to be somewhat more effective than the laurylacrylate copolymers. Thus, the viscosities of 0.5 wt % solutions of sample ASi-4 are from 6.5 to about 20 times more viscous compared with the laurylacrylate copolymers solutions of the same polymer concentration (0.5 wt %). However, since the number average number of -SiMe$_2$O- units is about 10, the hydrophobe, in this case, is essentially almost twice as long as the lauryl group. Interestingly, the optimal molar- and weight-comonomer content for the ASi-series is quite low compared with the laurylacrylate and

RF_4/RF_8 series and comparable with the FX-13/FX-14 series. Thus in comparison, the ASi-comonomers are less effective than the FX-13/FX-14 comonomers but about as efficient.

Characterization

Characterization of these associating polymers was attempted by reduced viscosity measurements and by dynamic light scattering.

The reduced viscosity vs concentration of sample 12-65 in water is shown in Figure 3. The apparent intrinsic viscosity ($[\eta]$) is about 23 dl/gr. However, in H_2O-DMF mixtures of 25/75, 40/60 and 50/50 (V/V), the intrinsic viscosities decrease rapidly to about 18 dl/gr, 10 dl/gr and about 3 dl/gr respectively. In contrast, the $[\eta]$ of homopolyacrylamides typically decrease by about a factor of 1.5-2.0 depending on molecular weight in going from water to 60/40 water DMF. It is therefore plausible that the intrinsic viscosities reported (Table VI) are apparent values and that molecular weights are much lower than predicted from reduced viscosity measurements in water.

The association of these new fluorocarbon containing copolymers is also demonstrated by dynamic light scattering measurements. Figure 4 shows the diffusion constant D as a function of polymer concentration for sample M-23 prepared with .07 mole % of FX-13. Starting at about 100 ppm, D shows a sharp decrease indicating perhaps the onset of aggregation at that concentration. Using scattering techniques of this type aggregate sizes between 1000-5000 nm have been detected.[8] It is not clear yet whether the scattering species below 100 ppm represents the unassociated macromolecules. Studies aimed at this are continuing.

Table VI Intrinsic Viscosities and Huggins Constants of Polyacrylamides Containing Comonomers **1**, **2** and **5**

Copolymer	Comonomer	Mole % Comonomer	$[\eta]^a$ (Dl/gr)	$K_H{}^b$
PAM	none	-	5.22	.81
AL-3	5	3.0	4.37	1.08
AF-17	1	.045	10.13	2.12
AF-12	1	.070	-	-
AF-10	1	.14	9.71	.91
AF-19	2	.28	11.07	2.34
AFT-13	1/AMPS	.14^c	4.81^d	3.47

a. Measured in H_2O at 30° ≠ 0.1°. b. Huggins constant. c. Refers to content of **1**; AMPS content is 16 %. d. Measured in 2.0 % NaCl solution.

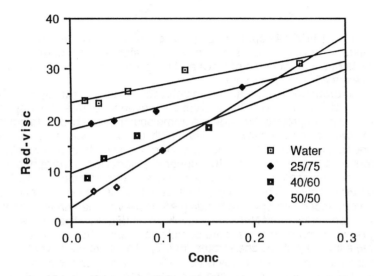

Figure 3. Plots of Reduced Viscosity vs Concentration For Polymer 12-65 in Various H_2O/DMF Mixtures.

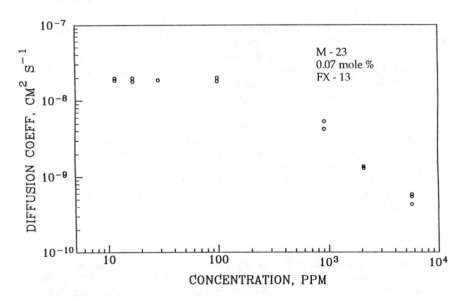

Figure 4. Dependence of Diffusion Coefficient D of Polymer M-23 on Polymer Concentration.

Rheology

Since the FX-13/FX-14 copolymers appeared to be most promising as viscosifiers, they were studied in somewhat greater detail. Figures 5-7 show the effects of shear, addition of NaCl and temperatures on the Brookfield viscosities of AF-12 (FX-13 series).

The pronounced pseudoplastic behavior of this and similar samples (Fig. 5) is not unexpected. The application of even moderate shear apparently efficiently degrades the large macromolecular assemblies into smaller units. However, the viscosity recovers rapidly upon removal of shear. Again, this appears consistent with the occurrence of a "network" of hydrophilic polymers held together by small hydrophobic assemblies. Hydrophobic aggregation of these polymers is also consistent with the effects of NaCl addition and heating. The salting-out effect of NaCl is well known and clearly leads to enhanced hydrophobic bonding.[5,6] However, the enhancement in viscosity upon addition of NaCl is relatively small (Fig. 6). The viscosity changes with temperature are more complex and show both minima and maxima, especially at low shear rates (Fig. 7). The increased hydrophobic association between 40° and 60° is consistent with hydrophobic bonding, an entropy driven process. Above 60° and below 40°, viscosity decreases with increasing temperature. Apparently above 60°, enhanced hydrophobic bonding is outweighed by a general decrease in viscosity at higher temperature (see Discussion).

The effects of addition of ionic surfactants is also consistent with the existence of hydrophobically driven aggregation of macromolecules. Figure 8 shows a decrease in viscosity of a sample AF-90 (.05 - .15 wt %) with increasing concentration of a fluorine containing ionic surfactant FC-149 ($CF_3[CF_2]_7COOK$). Apparently the surfactant competes effectively for the perfluoro chains of the polymer at very low surfactant concentrations. The pattern for non ionic surfactants such as Triton-X-100 and FC-171 ($CF_3(CF_2)_nCH_2[OCH_2CH_2]_mOH$) is quite different (Fig. 9). In the case of FC-171, there is a viscosity maximum as well as a minimum upon addition of surfactant. Increases as well as decreases in viscosity are also observed for Triton-X-100. Increases in viscosity in these cases may be rationalized by micellar bridging in which intermolecular polymer association occurs via a micelle. However, the patterns are complex and further experimentation is needed to better understand these effects.

Further research into this interesting new class of associating water-soluble polymers is continuing.

Discussion

The maxima in the Brookfield viscosity-comonomer content profiles for the FX-13/FX-14 and similar hydrophobic comonomer-containing polymers are of interest (Fig. 2, Tables I-V). The maxima for the fluorocarbon copolymers are especially pronounced. It is plausible that this behavior correlates with the very high efficiency of the fluorocarbon moieties in hydrophobic association. Since hydrophobic association is pronounced for fluorocarbons, the comonomer content in the chain can apparently be decreased without

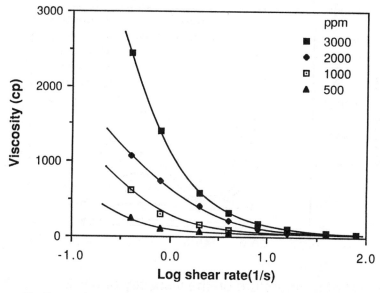

Figure 5. Dependence of Brookfield Viscosity of Polymer AF-12 on Shear Rate at Various Polymer Concentrations.

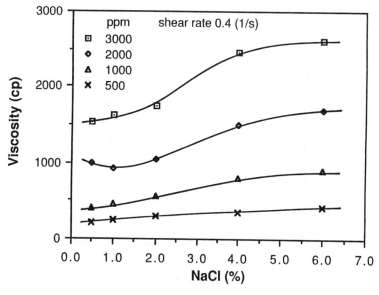

Figure 6. Effect of NaCl Concentration on Brookfield Viscosity of Polymer AF-12 at Various Polymer Concentrations.

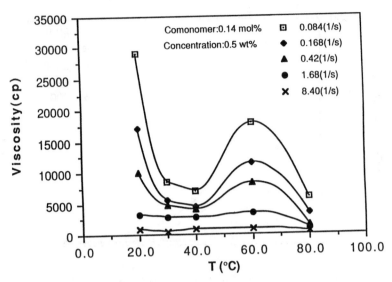

Effect of temperature on the viscosity of MF-25

Figure 7. Effect of Temperature on Brookfield Viscosity of Sample MF-25 at Various Shear Rates.

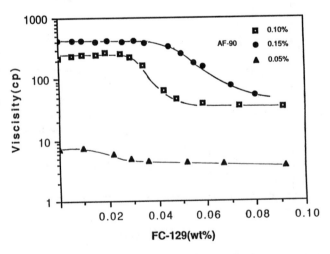

Figure 8. Effect of Concentration (wt %) of Surfactant FC-129 on Brook-field Viscosity of Polymer AF-90 at Various Polymer Concentrations, Shear Rate is 0.40 sec^{-1}.

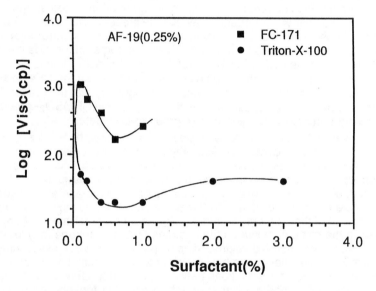

Figure 9. Brookfield Viscosities as a Function of Added Non-Ionic Surfactants FC-171 and Triton-X-100.

decreasing intermolecular association. This decrease in comonomer concentration lowers the degree of intramolecular hydrophobic association and this in turn tends to diminish chain contraction. Effectively associating hydrophobic groups thus would tend to favor intermolecular association even further. Even at .006 mole % of FX-13 intermolecular association appears to be marked (Table I). As the comonomer concentration increases intermolecular association also increases. Eventually increased FX-13 content tends to favor intramolecular association even more and the resulting chain contraction lowers the size and thus the viscosity contribution of the polymer. This viscosity decrease is quite dramatic. A four fold increase in FX-13 content lowers the Brookfield viscosity by a factor of more than a hundred. It is of course, not clear whether the magnitude of the viscosity changes is solely attributable to changes in hydrophobic association patterns. For instance, molecular weight effects may also be present. However, the over-all pattern most likely is due to a balance between inter- and intramolecular association that is affected by the hydrophobic character of the side chain. This tendency of the hydrophobic groups to promote association is strongly influenced by their chain length[1] as a comparison between the RF4 and RF8 copolymers shows (Table IV). The results also indicate that the presence of a water-soluble flexible spacer such as the $-CH_2CH_2N(C_2H_5)SO_2-$ group enhances the ability of the hydrophobic group to associate.

This effect appears to have first been documented by Schulz et al[7] who showed that increasing the oligo ethyleneoxide spacer length in a series of acrylates with the same hydrocarbon group increased the association of the corresponding polyacrylamide copolymers. It may be argued that this effect is due to the decoupling of the motions of the chain and of the hydrophobic moiety in the aggregate. In this sense, the spacer would play a role similar to that of the spacers in liquid crystalline forming polymers. Also, increasing the distance between the chain and the hydrophobe may lessen the importance of excluded volume effects operating between associating polymers in the polymer assembly. Finally, the spacer may lessen unproductive intramolecular associations between the hydrophobe and the hydrophobic carbon chain of the polymer or between the hydrophobic groups of the same chain.

The behavior of the silicone hydrophobe is unusual in that it resembles the hydrocarbon chains in its ability to promote association (Tables III and V). It is not surprising that fluorocarbons are more effective than silicones in promoting association since their hydrophobic character is more pronounced.[3b] However, the silicones are similar to the fluorocarbons in that they are quite effective at very low comonomer levels. Perhaps this is due to the much longer chain-length and perhaps to the greater surface area of the silicone chain. Further studies with comonomers with varying silicone chain lengths should clarify this point.

Although the nature of the polymer association appears reasonably clear, the degree of polymer association is not. From the very high value of the reduced viscosity of about 23 dl/gr of sample 12-65 in water at a concentration of about .015 wt % (~150 ppm) (Fig. 3) it would appear that this still represents an associated polymer assembly. Further evidence for this comes from the much lower value of η_{red} in a 50% mixture of H_2O-DMF. However, the dynamic light scattering data on similar polymers are consistent with association beginning around 100 ppm (Fig. 4). It is of course, possible that there is residual association below 100 ppm but there is no evidence so far. It seems probable that this can be clarified by more extensive light scattering and viscosity studies of copolymers with varying comonomer content and in various water-DMF systems. We have studied the effects of cosolvent addition on the viscosity of aqueous solutions of these copolymers including DMF, DMSO acetone, etc.[9] The addition invariably leads to viscosity decreases presumably as a result of a less hydrophobic environment and thus a lesser driving force toward hydrophobic association. Also, competitive interactions may occur between the hydrophobic groups and the cosolvent similar to the effects caused by the addition of surfactants. Unfortunately, precipitation occurs at 30-50% cosolvent so that the range of solvent compositions is limited.

Although the addition of ionic surfactants results in a simple decrease of viscosity, the addition of non ionic surfactants leads to more complex changes (Fig. 9). These effects are rather puzzling and it is likely that there are several effects operating simultaneously. These effects are expected to include, a) competition of the surfactant molecule for the polymer hydrophobe, b) micellar bridging that would lead to crosslinking by intermolecular association of two or more polymers with a single micelle and c) intramolecular micellar bridging in which several hydrophobic groups of a single polymer associate with a micelle.

An unknown quantity in all of these considerations is the size of the hydrophobic domains in the polymer assemblies specifically the number of hydrophobic groups/ hydrocarbon aggregate. In view of the associating ability of the copolymers with extremely low comonomer content (Tables I-II) it is not probable that this number is large for the FX-13/FX-14 copolymers. A large number would entail unacceptably large excluded volume effects for intermolecular association. Moreover, such a case would be expected to lead to predominant intramolecular micellization at low comonomer content and this is not observed. It is more likely that the number of hydrophobic groups/aggregate is rather small, perhaps as low as two. Therefore, the hydrophobic aggregates themselves would not resemble micelles but much smaller entities. Studies aimed at this are in progress.

Acknowledgments

This research was supported by the Department of Energy, office of Basic Energy Sciences. We wish to thank Professor Eric Amis and Thomas Seery for carrying out the dynamic light scattering measurements.

Literature Cited

1. Evani, S.; Rose, G.D., Polym. Mat. Sci. Eng. Preprs., 1987, 57, 477 ; Landoll, L.M., J. Polym. Sci. Chem., 1982, 20, 443 ; Schulz, D.N.; Kaladas, J.J.; Maurer, T.T.; Bock, J.; Pace, S.J.; Schulz, W.W., Polymer,

1987, 28, 2110; Turner, S.R.; Schulz, D.N.; Siano, D.B.; Bock, J., Polym. Mat. Sci. Eng. Preprs., 1986, 55(2), 355 ; Middleton, J.C.; Cummins, D.; McCormick, C.L., Polym. Preprs., 1989, 30(2), 348.

2. a. Zhang, Y-X; Da, A-H; Hogen-Esch, T.E.; Butler, G.B., Polym. Preprs., 1989, 30(2), 338.
   b. Ibid., J. Polym. Sci. (Polym. Lett.), 1990, 00, 000.
3. a. Schwartz, E.G.; Reid, W.G., Ind. Eng. Chem., 1964, 56, 26.
   b. Myers, D., Surfactant Science and Technology; VCH Publishers, Inc., New York, NY, 1988.
4. Jiang, X-K; Acc. Chem. Res., 1988, ??, 362.
5. Tanford, C., The Hydrophobic Effect; Second Ed., Wiley, New York, 1980.
6. Ben-Naim, A., Hydrophobic Interactions; Plenum Press, New York, 1980.
7. Schulz, D.N. et al, Polymer, 1987, 28, 2110.
8. Amis, E.; Seery, T.J., Unpublished results.
9. Zhang, Y-X; Da, A-H; Hogen-Esch, T.E., Unpublished Results.

RECEIVED July 2, 1990

Chapter 11

Hydrophobically Associating Ionic Copolymers of Methyldiallyl-(1,1-dihydropentadecafluoro-octoxyethyl)ammonium Chloride

Sridhar Gopalkrishnan[1], George B. Butler, Thieo E. Hogen-Esch[2], and Nai Zheng Zhang

Department of Chemistry and Center for Macromolecular Science and Engineering, University of Florida, Gainesville, FL 32611–2046

Water-soluble polymers have gained increasing importance commercially and they continue to be studied extensively in recent years. Many of these polymers find applications in water-treatment, flocculating agents and in enhanced oil recovery (EOR) (1). The polymers widely used in EOR applications are the copolymers of acrylamide. Several copolymers of acrylamide with monomers containing hydrophobic groups (usually long-chain alkyl groups) have been reported in the literature (2a,b,c) and the rheological properties of these copolymers have been studied in detail. However, very few studies have been carried out with hydrophobically associating ionic copolymers. A schematic representation of a typical hydrophobic association is shown in illustration below, where the filled circles represent the hydrophobic groups on the polymer chain:

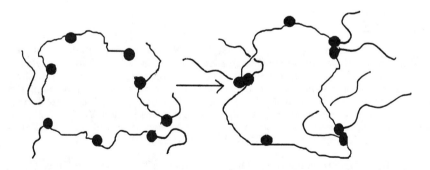

[1]Current address: BASF Corporation, 1419 Biddle Avenue, Functional Fluids Building, Wyandotte, MI 48192
[2]Current address: Department of Chemistry, Loker Hydrocarbon Research Institute, University of Southern California, Los Angeles, CA 90089–1661

0097–6156/91/0467–0175$06.00/0

In aqueous solution, intermolecular association of these hydrophobic groups takes place leading to the formation of very large macromolecular aggregates, and this results in a substantial enhancement of the viscosities. Up to this point the hydrophobic groups present in the main chain or in side groups have been hydrocarbons. Substitution of hydrogen by fluorine has been shown to result in enhanced hydrophobic character as manifested by the much lower critical micelle concentration (CMC) of fluorocarbon surfactants compared to hydrocarbons of the same carbon number.

This enhanced hydrophobic character of the perfluoro carbon chains has prompted us to investigate the association of the corresponding water-soluble polymers. Preliminary investigations show that the viscosities of copolymers of acrylamide and perfluoro alkyl containing acrylates and methacrylates are much greater compared with both the lauryl acrylate-containing copolymers and with the homopoly acrylamides, both prepared under identical conditions. Furthermore, the most strongly associating perfluorocarbon-containing copolymers had a comonomer content far below that (.07-.10 mole %) of the most effectively viscosifying lauryl acrylate containing the co-polyacrylamides (1.0-4.0 mole %).

It therefore became of interest to study the effects of perfluoro carbon groups in other water-soluble copolymers including polyelectrolytes. We have now investigated hydrophobically associating copolymers of diallyldimethylammonium chloride (DADMAC) and suitably modified diallylammonium monomers containing pendant hydrocarbon and fluorocarbon moieties.

Copolymers of diallyldimethylammonium chloride (DADMAC) with diallydecylmethylammonium chloride (DDAMAC) and methyldiallyl-(1,1-dihydropentadecafluorooctoxyethyl)-ammonium chloride (FX-15) were synthesized and the rheological properties of the copolymers were studied. At the same concentration, the viscosity of the DADMAC-DDAMAC and DADMAC-FX-15 copolymers showed significant increases in the viscosities compared to polyDADMAC. The increased hydrophobicity of the perfluoro group is reflected in the dramatically enhanced viscosities above the overlap concentration (at a particular comonomer content), compared to polyDADMAC-DDAMAC. All of the copolymers showed increased viscosities in saline solution. The DADMAC-DDAMAC copolymers show Newtonian behavior whereas the DADMAC-FX-15 copolymer (7) shows non-Newtonian psuedoplastic behavior.

Experimental

Materials. Phosphorus tribromide (Aldrich 99%), hydroxyethyl-1,1-dihydro pentadecafluoro-n-octyl ether (PCR), methyl chloride (Aldrich), and anhydrous ethyl ether were used as received. Diallylamine (Aldrich) was purified by distillation prior to use. Acetone (Fisher) was dried over calcium sulphate. Water soluble initiators (Wako Pure Chemical Industries), VA-044 [2,2'-azo-bis(N,N'-dimethyleneisobutyramidine) dihydrochloride] and V-50 [2,2'-azo-bis(2-amidinopropane) dihydrochloride] were used as received. Diallyldimethylammonium chloride (DADMAC) was used as a 60% solution in water.

Preparation of 2-Bromoethyl-1,1-Dihydropentadecafluoro-n-Octyl Ether (1).

$BrCH_2CH_2OCH_2C_7F_{15}$ (See Scheme 1)

A three-necked flask was equipped with a dropping funnel (containing 0.0316 M 3 ml of phosphorus tribromide) and a reflux condenser. 0.095 M (42.1 g) of 2-hydroxyethyl-1,1-dihydropentadecafluoro-n-octyl ether was added to the flask and the temperature was raised to 70°C. Phosphorus tribromide was gradually added over a period of 1 h. After the addition of phosphorus tribromide was complete, the temperature was raised to 110°C and maintained for 4 h. At the end of 4 h, the product was distilled under reduced pressure. The yield of the product was 75%. B.P. 81°C at 3.7 mm.

Scheme I

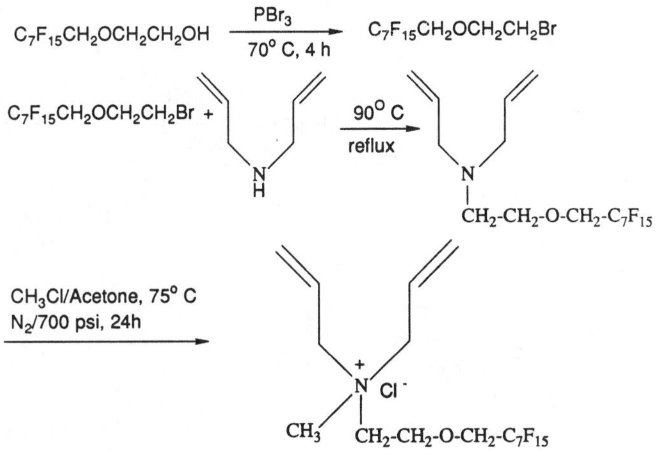

Preparation of Diallyl-(1,1-dihydropentadecafluorooctoxyethyl)-amine (II).

$$[CH_2=CH-CH_2]_2 N-CH_2CH_2OCH_2C_7F_{15}$$

A solution of 0.04 M (22.6 g) of I, 0.13 M (13.0 g) of diallylamine and 100 ml of toluene was refluxed for a period of 3 h at 90°. Temperature control is very important in this reaction. Above 100°C, degradation of II takes place. As the reaction progresses, precipitation of diallylammonium bromide takes place. The salt was removed by suction filtration and the filtrate was distilled under reduced pressure to remove all of the excess diallylamine and toluene. The isolated yield of the product was (22 g) 94%. This material was used without further purification for the next step.

Preparation of Methyldiallyl-(1,1-dihydropentadecafluorooctoxyethyl)-ammon-ium Chloride (FX-15).

$$[CH_2=CH-CH_2]_2 \overset{+}{\underset{CH_3}{N}} -CH_2CH_2OCH_2C_7F_{15} \quad Cl^-$$

A solution of 0.038 M (20 g) of II and 15 ml of dry acetone were placed in a Parr pressure reactor. Methyl chloride, 0.19 M (9.64 g), was transferred to the reactor via a glass tube. The reactor was sealed and then pressurized with nitrogen to 700 psi. The reactor was heated to 75°C and maintained for

24 h. At the end of 24 h, the reactor was depressurized to atmospheric pressure and the excess acetone was removed under reduced pressure. The product was washed several times with anhydrous ethyl ether to remove any unreacted amine and dried in a high vacuum oven. The yield of the product was 15 g (68.4%). Decyldiallylamine was quaternized using methyl chloride, according to the procedure of Do (3) (see Scheme II).

Scheme II

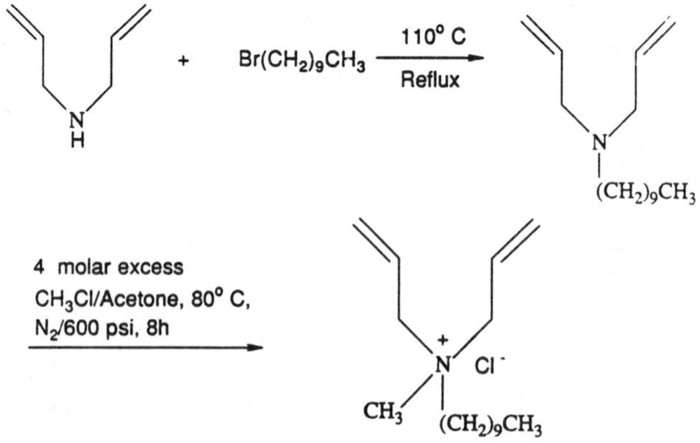

Polymerizations. Homopolymerizations were carried out in Pyrex tubes fitted with serum caps. Initially the required amount of monomer, initiator and water were charged into the tubes. The tubes were fitted with serum stoppers and the contents were degassed in a continuous stream of nitrogen for thirty min. The tubes were then placed in a constant temperature bath maintained at 50°C and shaken for the required period of time. The contents of the tube were then poured slowly into 50 ml of acetone to precipitate the polymers. The homopolymers were recovered by suction filtration, washed with several aliquots of acetone and dried in a high vacuum oven (see Table I). The copolymerizations were carried out in a similar manner with the following modifications. After the initial addition of monomers and initiator, water was added

Table 1   Homopolymerization of the FX-15 monomer

#	Monomer(g)	Initiator(Wt%)	Water(ml)	t °C	time	%Conversion
1	0.46	9.00 (V-50)	1.00	50	24h	34.78
2	1 00	5.00 (V-50)	1.00	50	48h	78.89
3	1.00	5.00 (V-50)	2.50	50	96h	50.10

to adjust the total monomer concentration to 30%. At the end of the polymerizations, the contents were dissolved in 100 ml of water and the copolymers were precipitated by adding the aqueous solution to 1 L of acetone (see Table 2). The copolymers were recovered by suction filtration (see Schemes III and

Table 2 Copolymerization of DADMAC with DDAMAC

DDAMAC mole %	DADMAC Conc.	Initiator V-50, g	Total monomer Conc. (mole)	Time hour	Temp. °C
0.00	60 %	0.1824	0.03715	24	50
1.93	60 %	0.1851	0.03787	24	50
4.54	60%	0.1841	0.0389	24	50
8.22	60 %	0.1845	0.0402	24	50

IV). The DADMAC-DDAMAC copolymers were soxhlet extracted with chloroform to remove any residual homopolymer of DDAMAC. The DADMAC-FX-15 copolymers were soxhlet extracted with acetone to remove any unreacted (FX-15) monomer. These copolymers were then dried in a high vacuum oven. The conversions for all of the copolymers were greater than 95%.

Scheme III

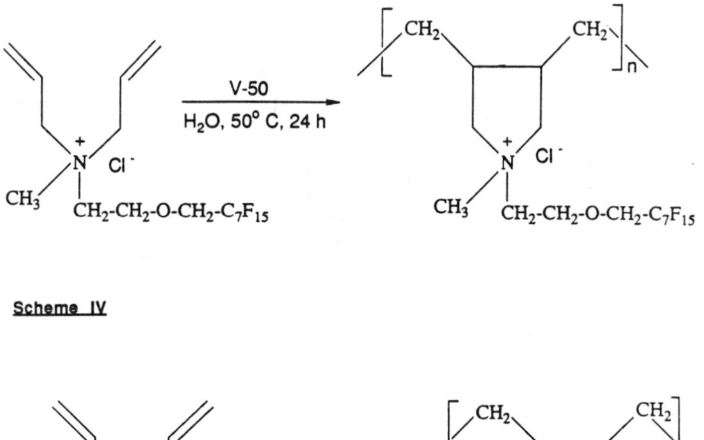

Scheme IV

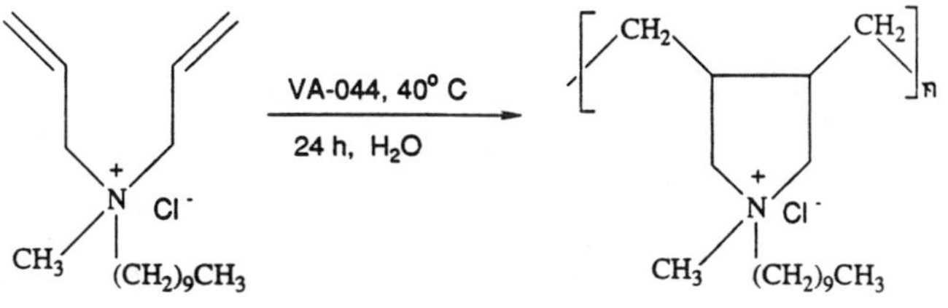

Brookfield Viscosities. Concentrated solution viscosities were measured using a Brookfield dial reading viscometer (LVT model) equipped with a small sample adapter (spindle #SC4-34, chamber size 13R, sample volume 10 ml). The polymers were dissolved in 10 ml of water and placed in a constant temperature bath maintained at 50°C and shaken for a period of 24-72 h, until the solutions were homogeneous. All measurements were carried out at 25°C. Shear rates varied from 0.084 S^{-1}-16.8 S^{-1}.

NMR Measurements. ^{13}C NMR measurements were conducted on a Varian XL-200 instrument with 32K data points and a sweep width of 10,000 Hz. The pulse angle was 90°. Protons were decoupled from the carbon nuclei using a random noise decoupling field. All chemical shifts are reported relative to TMS. A pulse delay of 5s was used for the homopolymer of the FX-15 monomer.

Results and Discussion

Homopolymerizations of FX-15 and DDAMAC. The monomers, FX-15 and DDAMAC, were synthesized as shown in Schemes I and II. Figures I and 2 show the ^{13}C NMR spectra of intermediates I and II prepared during the

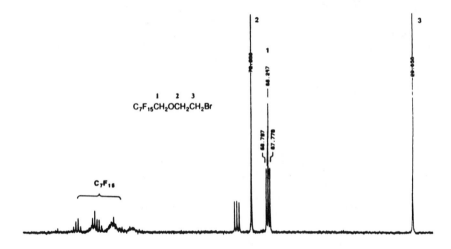

Figure I. ^{13}C NMR spectrum of **I** in CDCl$_3$.

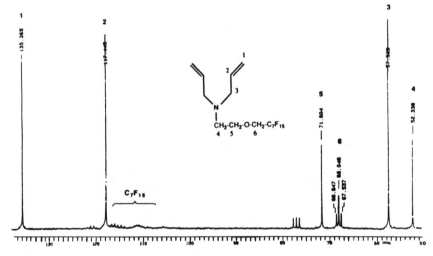

Figure 2. ^{13}C NMR spectrum of **II** in CDCl$_3$.

synthesis of the FX-15 monomer. Figure 3 shows the ^{13}C NMR spectrum of the FX-15 monomer. The assignments for the resonances of the vinylic

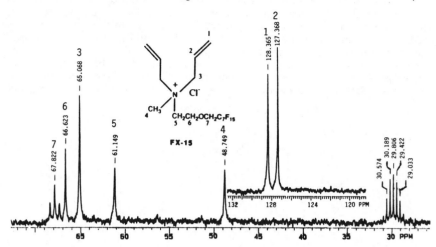

Figure 3. ^{13}C NMR spectrum of the FX-15 monomer in Acetone-d_6.

carbons of the FX-15 monomer were based on the APT spectrum shown in Figure 4. Table I shows the results of the homopolymerization of FX-15 monomer using a water-soluble initiator (V-50) at 50°C (Scheme III). At very high monomer concentrations (#2), the degree of conversion is high, but the resulting homopolymer contains a high percentage of unreacted double

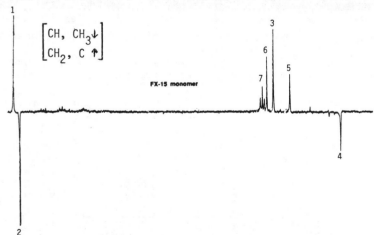

Figure 4. ^{13}C APT spectrum of the FX-15 monomer in D_2O.

bonds. At high monomer concentration open-chain propagation becomes more favorable and the polymer contains linear portions with pendant double bonds which may eventually lead to the formation of a crosslinked polymer. A similar behavior has been reported for homopolymerizations involving N-N dialkyldiallylammonium chlorides (4). At lower monomer concentrations (40%

in H$_2$0) cyclopolymerization is favored and the resulting homopolymer contains a much lower percentage of unreacted double bonds as shown in Figure 5
which shows the ^{13}C NMR spectrum of the homopolymer in d$_4$-acetic acid.
The resonances at ca. 28 ppm, 40 ppm and 70 ppm indicate that the polymer is
predominantly in the cyclic form as shown in Scheme III. The chemical shift
values for the ring carbons are in good agreement with the values obtained for
the ring carbons of cyclic poly(diallyldimethylammonium chloride), a water-
soluble polymer that consists of predominantly five-membered rings (5). The
FX-15 homopolymer is insoluble in water and forms a microemulsion. On

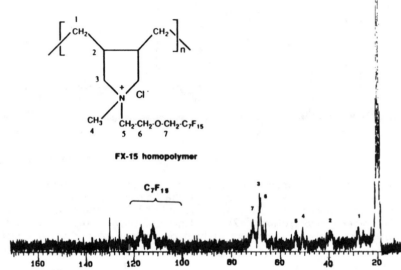

Figure 5. ^{13}C NMR spectrum of the FX-15 homopolymer in d$_4$-acetic acid.

adding the polymer to water, the solution foams, suggesting that the
homopolymer may have surfactant properties. The polymer is soluble in

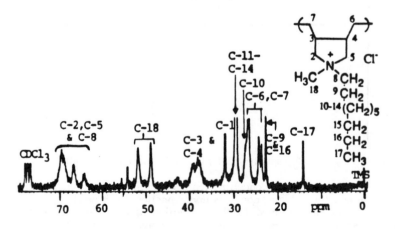

Figure 6. ^{13}C NMR spectrum of poly(DDAMAC) in CDCl$_3$.

trifluoroethanol, ethanol and acetic acid indicating that it is not crosslinked. DDAMAC polymerizes readily in aqueous solutions in the presence of water-soluble initiators to give cyclic poly(decyldiallylmethylammonium chloride) as shown in Scheme IV. This cyclopolymer is also known to consist predominantly of pyrrolidinium ring structures. The ^{13}C NMR spectrum of this cyclopolymer and the respective assignments for the ring carbons as well as the side-chain units are shown in Figure 6.

Copolymerizations of FX-15 and DDAMAC with DADMAC. The copolymers of FX-15 and DDAMAC with DADMAC were synthesized as shown in Schemes V and VI. Tables 2 and 3 show the results of the copolymers of DDAMAC and

Scheme V

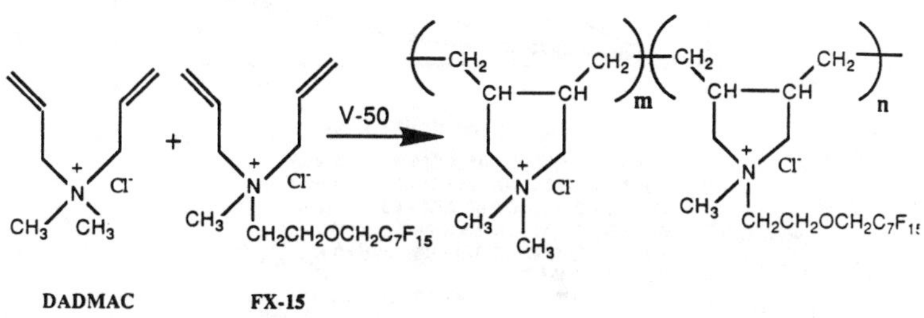

DADMAC   FX-15

Scheme VI

DADMAC   DDAMAC

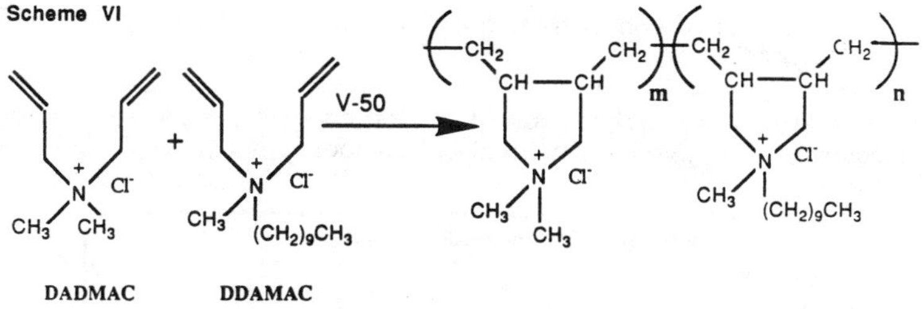

FX-15 with DADMAC. The total monomer concentration was a 30% solution in water and varied only from 0.037-0.040 mole. In all cases, the conversion was over 95%. In order to insure complete removal of the homopolymer of DDAMAC, the copolymer of DDAMAC-DADMAC was soxhlet extracted with

Table 3 Copolymerization of DADMAC with FX-15

FX-15 mole %	DADMAC Conc.	Initiator V-50,g	Total monomer Conc. (mole)	Time hour	Temp °C
1.062	60 %	0.1858	0.0375	24	50
2.496	60 %	0.1862	0.0380	44	50
8.220	60%	0.1858	0.0400	48	50

chloroform. The DADMAC-FX-15 copolymer was completely soluble in water indicating the absence of any homopolymer of FX-15. Also, the copolymer was soxhlet extracted with acetone to insure complete removal of any traces of the FX-15 monomer. Figure 7 shows the viscosity-concentration profile of

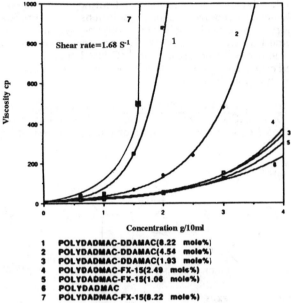

1 POLYDADMAC-DDAMAC(8.22 mole%)
2 POLYDADMAC-DDAMAC(4.54 mole%)
3 POLYDADMAC-DDAMAC(1.93 mole%)
4 POLYDADMAC-FX-15(2.49 mole%)
5 POLYDADMAC-FX-15(1.06 mole%)
6 POLYDADMAC
7 POLYDADMAC-FX-15(8.22 mole%)

Numbers in parentheses refer to the mole% of the comonomer that was added to the feed during the copolymerizations.

Figure 7.   Effect of concentration on the viscosity of copolymers of DAD-
MAC.

DDAMAC and FX-I5 copolymers. All of the copolymers show an increase in
the viscosities at a given concentration compared to polyDADMAC. This

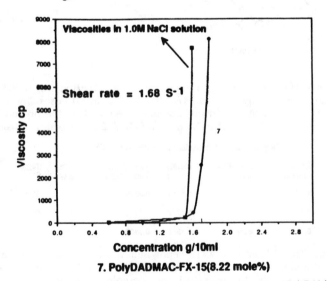

7. PolyDADMAC-FX-15(8.22 mole%)

Figure 8.   Effect of concentration on the viscosity of a DADMAC-FX-I5
copolymer.

increase is more pronounced as the concentration of the comonomer is increased, as shown in samples I and 2. Sample I has a viscosity of 880 cp at a concentration of 20% in water. These increases in viscosity are due to the association between the long-chain alkyl hydrophobic groups, and this association increases as the concentration of the hydrophobic groups in the copolymer is increased. Figure 8 shows the effect of polymer concentration on the viscosity of a DADMAC-FX-15 copolymer (7), whose comonomer content is identical to that of polyDDAMAC-DADMAC (I). Above the overlap concentration range Figure 8 shows a very sharp upsweep in the viscosity curve. This overlap concentration (usually denoted as C*) probably occurs in the concentration range of 16-17% in solution for this copolymer. In this concentration range the viscosity of the DADMAC-FX-15 copolymer (7) is over ten times higher than that of the DADMAC-DDAMAC copolymer (I). The C* concentration for DADMAC-DDAMAC copolymer (I) is in the range of 15-16%. The DADMAC-DDAMAC copolymer (I) flows freely even at 20% concentration in aqueous solution, whereas the DADMAC-FX-15 copolymer (7) behaves as a gel at concentrations greater than 18% in water, and is no longer free-flowing. Figure 9 shows the effect of the comonomer content on the viscosity of

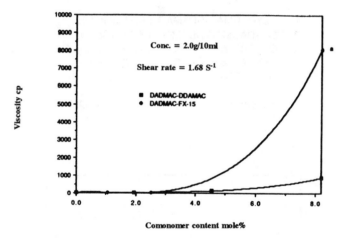

Conc. = 2.0g/10ml

Shear rate = 1.68 S⁻¹

■ DADMAC-DDAMAC
◆ DADMAC-FX-15

Viscosity cp

Comonomer content mole%

a. Since the viscosities of the DADMAC-FX-15(#7) copolymer at concentrations above 1.8g/10ml were beyond the range of the spindle and chamber used, this concentration was chosen for the data point.

Figure 9.   Effect of comonomer content on the viscosity of DADMAC copolymers.

DADMAC copolymers. These results indicate the increased associative ability of the perfluoroalkyl group. A similar effect was observed for the copolymers of acrylamide with monomers containing long chain perfluoro alkyl groups and this was reported in our previous work (2c). Fluorocarbons are potentially more hydrophobic than hydrocarbons and the increased hydrophobicity of the perfluoro groups can be attributed to the low values of the cohesive energy density and the low surface energy of these compounds (2c). Figure 10 shows the effect of shear rate on the viscosity of the copolymers of DADMAC. With the exception of 7, the viscosities of all of the copolymers of DADMAC that were prepared were essentially independent of the shear rate in the range 0.084 S⁻¹ to 16.8 S⁻¹. Figures 11 and 12 show the dependence of the viscosity

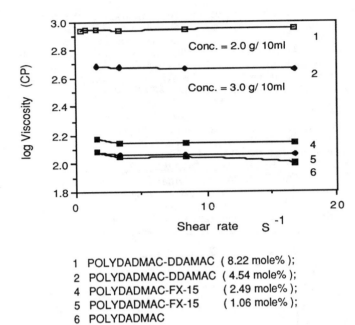

1 POLYDADMAC-DDAMAC (8.22 mole%);
2 POLYDADMAC-DDAMAC (4.54 mole%);
4 POLYDADMAC-FX-15   (2.49 mole%);
5 POLYDADMAC-FX-15   (1.06 mole%);
6 POLYDADMAC

Figure 10.   Effect of shear rate on the viscosity of DADMAC copolymers.

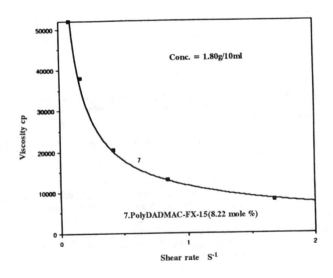

Figure 11.   Effect of shear rate on the viscosity of DADMAC-FX-15 copolymers.

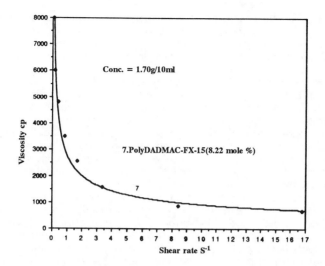

Figure 12.   Effect of shear rate on the viscosity of a DADMAC-FX-15 copolymer.

of a DADMAC-FX-15 copolymer (7) on the shear rate at two different concentrations. This copolymer shows psuedoplastic or shear-thinning behavior, with the viscosities decreasing with increasing shear rate. As the shear rate increases, the intermolecular hydrophobic associations are disrupted and this results in a decrease in the viscosity. Figure 13 shows the effects of the

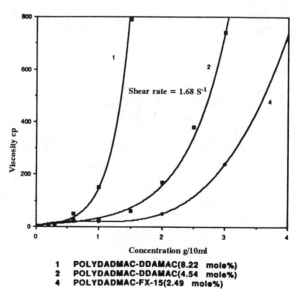

1   **POLYDADMAC-DDAMAC(8.22 mole%)**
2   **POLYDADMAC-DDAMAC(4.54 mole%)**
4   **POLYDADMAC-FX-15(2.49 mole%)**

Figure 13.   Effect of sodium chloride on the viscosity of copolymers of DADMAC.

sodium chloride on the viscosity of the copolymers. The viscosities of these copolymers quickly tend toward zero upon removal of shear. In the presence of salt the viscosity enhancements occur at a much lower concentration. The presence of salt suppresses charge-charge interactions and promotes greater association at a lower concentration.

These hydrophobically associating copolymers may have potential applications in EOR, or serve as flocculating agents and may also be useful in other applications where effective viscosification of water is of interest. However, these hydrophobically associating (ionic) copolymers have a fairly high overlap concentration (C*) compared to the hydrophobically associating (nonionic) copolymers of acrylamide, reported in our previous work. In the case of the copolymers of DADMAC, below the overlap concentration, the repulsive electrostatic forces dominate the polymer solution behavior. Due to these repulsive forces, the polymer chain is extended and very little associa-tion takes place. As the concentration of the polymer in solution is increased (above the overlap concentration C*), the hydrophobic associative forces dominate the polymer solution behavior. Even more effectively viscosifying polymers are possible by judiciously varying the length of the perfluoro carbon group and copolymerizing these monomers with acrylamide. Since acrylamide has an exceptionally high $k_p k_t^{1/2}$ ratio and has a low chain transfer constant (k_{ct}) (6), ultra-high molecular weight polymers of acrylamide with the FX-15 monomer may be prepared. We would like to dedicate our efforts to the further development and testing of these kinds of copolymers. It would also be very important to understand why the DADMAC-FX-15 copolymer (7) shows psuedoplastic behavior whereas the DADMAC-DDAMAC copolymers show Newtonian behavior.

Acknowledgements

Support for this work was provided by the Department of Energy, Office of Basic Energy Sciences, under Grant No. DE-FG05-84ER45104 for which we are grateful.

Literature Cited

1. Glass, J. E. Water Soluble Polymers, Adv. Chem. Ser. 213.
2. a. Ezzell, S. A.; McCormick, C. L. Polym. Preprints, 1989, 30(2), 340.
   b. Middleton, J. C.; Cummins, D.; McCormick, C. L. Polym. Preprints, 1989, 30(2), 348.
   c. Zhang, Y. Y.; Da, A. H.; Hogen-Esch, T. E. Polym. Preprints, 1989, 30(2), 338.
3. Do, C. H.; Butler, G. B. Polym. Preprints, 1989, 30(2), 352.
4. Jaeger, W.; Hang, L. T.; Philipp, B.; Reinisch, G.; Wandrey, C. in Polymer-ic Amines and Ammonium Salts; E. J. Goethals, Ed.; Pergamon Press: Oxford, 1980, pp. 155-161.
5. Lancaster, J. E.; Baccei, L.; Panzer, H. P. Polym. Letters, 1976, 14, 549.
6. Schulz, D. N. Polym. Preprints, 1989, 30(2), 329.

RECEIVED July 17, 1990

Chapter 12

Peptide Graft Copolymers from Soluble Aminodeoxycellulose Acetate

William H. Daly and Soo Lee

Macromolecular Studies Group, Louisiana State University,
Baton Rouge, LA 70803

The synthesis of 6-amino-6-deoxycellulose acetate, 1, by conversion of partially hydrolyzed cellulose triacetate to 6-azido-6-deoxycellulose acetate followed by selective reduction of the azide with 1,3-propane- dithiol is described. Utilizing 1 as a macroinitiator for γ-benzyl-α,L-glutamate N-carboxyanhydride (BLG-NCA) leads to the formation of soluble peptide cellulose conjugates.

We are engaged in the synthesis of a chitin composite analog based upon cellulose derivatives. Completely biodegradable graft copolymers of cellulose and polypeptides could serve as drug delivery systems or as calcium binding matices for prosthetic devices. As we reported earlier,(1) the utilization of insoluble O-(3-aminopropyl)cellulose as a macroinitiator did not produce soluble graft copolymers with predictable properties. The heterogeneous conditions limited the extent of grafting and prevented accurate control of the graft length. The resultant graft copolymers remained insoluble and thus were difficult to characterize. These difficulties reinforced the need for an easily characterizable soluble aminocellulosic, which could be employed as a macroinitiator of graft copolymerization to produce a completely biodegradable polymeric delivery system. In this paper we report the synthesis of soluble 6-amino-6-deoxycellulose acetate, 1, and its utilization as a macroinitiator for the polymerization of γ-benzyl-α,L-glutamate N-carboxyanhydride (BLG-NCA) to produce the desired cellulose polypeptide graft copolymers, 6.

Replacement of one hydroxyl group with an amino function, i.e production of aminodeoxyanhydroglucose repeat units, has received considerable attention. The two most common approaches to introduce the aminodeoxy function have been either a reduction of azidodeoxycellulose, which can be prepared from halodeoxy- or tosylated cellulose, or reduction of the oxime derived from selectively oxidized cellulosics.(2)

Horton and Clode (3) first reported the preparation of 6-azido-6-deoxycellulose acetate with relatively low D.S.(0.25) by tosylating the residual hydroxyls of commercial cellulose acetate, followed by nucleophilic displacement with sodium azide. A regiospecific preparation of 6-azidodeoxycellulose prepared

0097–6156/91/0467–0189$06.00/0

by a blocking-unblocking scheme involving tritylation at C-6, substitution of hydroxyl groups at C-2 and C-3 with acetate, detritylation, tosylation, and nucleophilic substitution of tosylhydroxyl group with sodium azide led to higher DS derivatives, but migration of the acetate groups to the 6-position during the deblocking step prevented absolute regiospecificity.(4) Direct chlorination of cellulose with methanesulfonyl chloride in N,N-dimethylformamide was quite stereospecific; subsequent displacement of the chloride afforded 6-azido-6-deoxycellose with D.S.'s ranging up to 0.67.(5). Photolysis of the azidocelluloses followed by mild hydrolysis converted the derivatives into the corresponding 6-aldehydocellulose. Horton did not convert either of these derivatives into amino celluloses but either derivative is a good precusor for such a conversion.

Reduction of 6-azidodeoxycellulose with LiAlH$_4$ was reported by Usov et al.(6), and Teshirogi et al.(7-10). The azido derivatives were prepared by the same procedure as described above, but to mitigate the problems associated with the acetate migration, phenyl isocyanate was used to protect the 2,3-hydroxyls. Teshirogi reported that the phenylcarbamoyl groups were retained, but in fact we have found, as reported by Usov, that the reduction effects the cleavage of the carbamate linkage as well as the azide. Tritylation of cellulose followed by tosylation yielded 2(3)-O-tosyl-6-O-tritylcellulose which afforded the corresponding 2(3)-aminodeoxy cellulose upon tosyl displacement, reduction with LiAlH$_4$, and detritylation.(11)

6-Amino-6-deoxycellulose was also prepared by the oxidation of 2,3-diphenylcarbamoylcellulose to the corresponding 6-carboxaldehyde derivatives with dicyclohexylcarbodiimde-pyridine trifluoroacetate in dimethyl sulfoxide, oximation, and reduction the oxime with LiAlH$_4$(12). Oxidation of a homogeneous cellulose solution in DMSO-paraformaldehyde with acetic anhydride leads to 3-oxocellulose, in contrast, oxidation of a 6-substituted cellulose, such as 6-O-acetylcellulose, occurred mainly at C-2 under the same conditions.(13) The ketocelluloses were utilized extensively to generate aminodeoxy derivatives; oximation followed by catalytic hydrogenation(14), diborane(14), sodium borohydride(15) or LiAlH$_4$(16) reduction led to the expected products. Both Usov's group(17) and Yalpani and Hall(18,19) have reported a process for 2(-3)-aminodeoxycellulose with various D.S.'s (0.45-0.55) by reaction of oxocellulose with ammonium acetate and sodium cyanoborohydride [NaB(CN)H$_3$] in methanol.

Synthesis of aminodeoxycellulose by the techniques described above led to either insoluble derivatives or derivatives which were soluble in polar solvents only. Further, the procedures tended to be accompanied by extensive chain degradation so the resultant derivatives would not be useful macroinitiators. We elected to use a more direct approach based upon acetate blocking groups, because the resultant aminodeoxycellulose derivatives must remain soluble in non-polar organic solvents to serve as macroinitiators for NCA grafting. 6-Amino-6-deoxycellulose acetate,1, proved to the most interesting macroinitiator.

Experimental

All reagents and solvents used in the synthesis were of reagent grade or purified by standard techniques. Partial hydrolysis of cellulose triacetate,2, under acidic conditions afforded the starting cellulose derivative,3, with a D.S. of acetylation = 2.45.(20) Nuclear magnetic resonance spectra were obtained with a Bruker WP-200 spectrometer at 298°K unless otherwise indicated. Infrared spectra were recorded with a Perkin-Elmer 283 spectrometer; films were cast of polymer samples from suitable solvents.

Synthesis of 6-O-Tosylcellulose Acetate,4. Into a solution of 25.0 g of partially hydrolysed cellulose acetate,3, in 400 mL of anhydrous pyridine were added 40.0 g of p-toluenesulfonyl chloride. The mixture was stirred at 25°C for 24 hr before diluting with 400 mL of acetone. The diluted solution was poured into 5 L of methanol, and the precipitate was recovered by filteration, washed and dried in vacuo at 50°C for 48 h to yield 27.3 g (82.7%) of a light brown derivative,4. ^{13}C Nmr (acetone-d$_6$): 20.7, 20.8, 21.7, (acetate CH$_3$); 63.3 (acetylated C-6); 68.6 (tosylated C-6); 73.4, 76.0 (C-2, C-3, C-5); 77.1 (C-4); 101.2 (acetylyated C-1); 103.8 (C-1); 128.9, 131.4, 134.0, 147.0 (aromatic C's); 170.0, 170.1, 170.9 (acetate C=O).

Synthesis of 6-Azido-6-deoxycellulose Acetate,5. Into a solution of 20.0 g of 4 (D.S. of tosylation = 0.45) in 400 mL of DMF were added 30.0 g of sodium azide. The mixture was stirred for 48 h at 100°C, and then precipitated into 4 L of distilled water. After two reprecipitations (acetone/water), 15.4g (92.6%) of light brown 5 was obtained. ^{13}C Nmr (acetone-d$_6$): 20.6, (acetate CH$_3$); 51.2 (azido-deoxy C-6); 63.2 (acetylated C-6); 72.8, 73.4 (C-2, C-3, C-5); 77.2 (C-4); 101.2 (acetylated C-1); 103.9 (glucose C-1); 169.7, 170.1, 170.9 (acetate C=O). IR (film from THF): 2130 cm^{-1} (N$_3$).

Synthesis of 6-Amino-6-deoxycellulose Acetate, 1. To 5.00 g of 5 in 150 mL of THF and 50 mL of methanol were added 4.1 mL of triethylamine and 7.5 mL of 1,3-propanedithiol. The mixture was stirred for 48 h at 60°C under nitrogen, and then precipitated in 500 mL of ethyl ether. The precipitate was filtered, washed with 2 L of hexane, air dried for 3 h at room temperature and then dried for 48 h at room temperature in vacuo. Partially reduced 6-azido-6-deoxycellulose acetate,1, 4.12 g (0.47 meq/g of amino content) was obtained. This product is insoluble in acetone, THF, and chloroform, but soluble in hot organic solvents, such as pyridine, DMF, DMSO, and THF-ethyl alcohol (5:1 v/v). ^{13}C Nmr (pyridine-d$_5$, 353°K): 20.6, (acetate CH$_3$); 41 (aminodeoxy C-6); 51.2 (azidodeoxy C-6); 63.2 (acetylated C-6); 73.1, 73.7 (C-2, C-3, C-5); 77.1 (C-4); 101.2 (acetylated C-1); 104.4 (C-1); 169.6, 169.9, 170.5 (acetate C=O).

Graft Copolymerization of BLG-NCA with Macroinitiator 1. The designated quantity of 1 was dissolved in hot dried DMF, the solution was cooled to 25°C, and a solution of BLG-NCA in DMF was added. The graft copolymerization remained homogeneous through out the reaction. After stirring for 48 h the copolymer solutions were poured into 100 mL of distilled water. White, fibrous products, 6a-6f, were obtained after freeze drying. The results are summarized in Table I.

Results and Discussion

In order to synthesize soluble aminodeoxycellulose derivatives, commercially available cellulose triacetate, 2, (D.S. = 2.8) was employed as a starting material. As shown in Scheme 1, initial partial hydrolysis of 2 under acidic conditions at room temperature afforded derivative 3 with a low concentration of free hydroxyl groups located primarily at the C-6 position. Derivative 3 (D.S. of acetylation = 2.45) was readily soluble in most organic solvents including acetone, THF, DMF, pyridine, and DMSO. Treatment of 3 with p-toluenesulfonyl chloride in pyridine

Table I. Graft Copolymers of γ-Benzyl-α,L-Glutamate with
6-Amino-6-deoxycellulose Acetate, **1**

Cpd. (g)	ADCA[a] (g)	NCA (mol/mol)	NCA/ADCA Efficiency	Grafting	DP[b]	Conform
6a	0.35[c]	0.15	3.7	38.5%	1.4	Random & α-Helix
6b	0.25	0.25	8.2	64.7	5.3	α-Helix
6c	0.15	0.35	20	92.4	19	α-Helix
6d	0.04	0.46	89	95.0	88	α-Helix
6e	0.35[d]	0.15	3.7	83.8	3.1	Random
6f	0.25	0.25	8.2	86.6	7.1	Random & α-Helix

[a] 6-Amino-6-deoxycellulose acetate, 0.47 meq/g of NH_2

[b] Average D.P. of grafts based upon weight gain.

[c] Reaction conditions: total reactant conc., 0.5 g/10 mL DMF, room temperature, 48 h.

[d] Reaction conditions: total reactant conc., 0.5 g/20 mL DMF, room temperature, 48 h.

effected a quantitative conversion to the 6-tosylcellulose acetate.   Tosylated cellulose acetate, **4**, was easily converted to the corresponding azidodeoxycellulose acetate, **5**, by nucleophilic substitution of the tosyl group with sodium azide in DMF at 100°C.

Scheme 1. Synthesis of ADCA-PBLG graft copolymers, **6**.

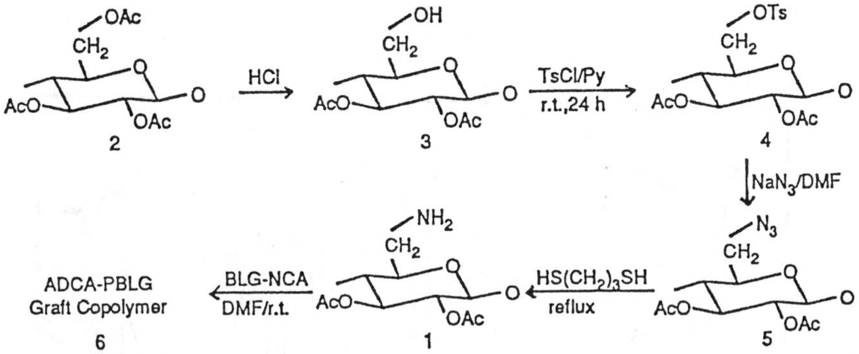

Infrared spectroscopy (Figure 1) can be used effectively to characterize the modified cellulose acetates, **3,4,**and **5**.   Typical absorption bands at 1180 cm^{-1} (SO_3) and 2130 cm^{-1} (N_3) are confirmations for tosylation and nucleophilic substitution with sodium azide, respectively. ^{13}C nmr shows that the nucleophilic substitution occurs predominately on the C-6 position; a quantitative chemical shift of the 68.6 ppm peak for tosylated C-6 of **4** (Figure 2-B) to 51.2 ppm for azidodeoxy C-6 of **5** (Figure 2-C) is observed.   Increasing the reaction time to 72 h failed to effect the displacement of any residual tosyl groups, which may be bound to a secondary carbon.

Several methods for reducing azido groups to the corresponding amino functions are known.   The reagents, $LiAlH_4$ or H_2/catalyst, commonly employed for this purpose are rather unselective.(6)   As expected, only deacetylated aminodeoxycellulose, **7**, was formed upon reduction of **5** with $LiAlH_4$ or a borane complex in refluxing THF; **7** was soluble in 10% aqueous NaOH.   The ^{13}C nmr (10% $NaOH/D_2O$) exhibited a clear peak for aminodeoxy C-6 at 42.1 ppm. However, the product did not meet our objective, i.e., an organic soluble, well characterized macroinitiator.   Catalytic hydrogenation under mild conditions (25°C, 45 psi H_2 in 25 v/v% methanol/THF) did not effect a reduction.   A modified Staudinger reaction (21), which should produce ammonium salts from azido compounds by acid hydrolysis of an intermediate nitrogen-phosphorous complex formed with triethyl phosphite, yielded only **7**.   The vigorous acidic hydrolysis conditions required to cleave the N-P complex effected hydrolysis of the acetate ester.

Selective reductions of various azido compounds containing functionalities such as nitro, ester, amide and unsaturated carbon double and triple bonds with 1,3-propanedithiol and triethylamine have been reported.(22)   This procedure was

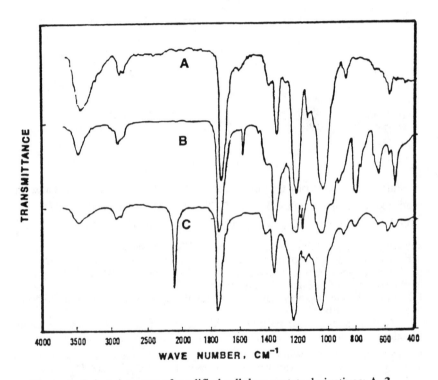

Figure 1. Infrared spectra of modified cellulose acetate derivatives: A, **3**, partially hydrolysed cellulose acetate; B, **4**, tosylated cellulose acetate; C, **5**, azidodeoxycellulose acetate.

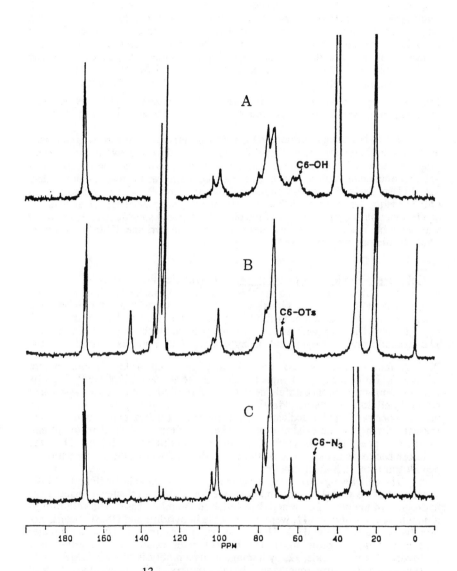

Figure 2. 50.34 MHz ^{13}C NMR spectra of modified cellulose acetate derivatives: A, **3**, partially hydrolysed cellulose acetate (DMSO-d$_6$); B, **4**, tosylated cellulose acetate (acetone-d$_6$); C, **5**, azidodeoxycellulose acetate (acetone-d$_6$).

applied successfully to the reduction of **5**, albeit only partially reduced aminodeoxy-azidodeoxycellulose acetate could be obtained on treatment of **5** with propanedithiol in THF/methanol (3:1 v/v ratio) at 60°C for 48 h. No deacetylation was detected. Extension of the reaction time for up to a week or addition of aliquots of propanedithiol failed to increase the extent of reaction. The partially aminated cellulose acetate, **1**, contained 0.47 meq/g amine; this was quite adequate for a macroinitiator. Polymer **1** formed gels in pyridine, DMF, and DMSO at room temperature, but stable solutions could be formed by heating the gels to > 90°C and recooling to room temperature. The ^{13}C nmr spectra of **1** was not well resolved; only a small broad peak around 41 ppm could be assigned to aminodeoxy C-6.

    2,3-O-bis(phenylcarbamoyl)-6-azido-6-deoxycellulose,**8**, was prepared according to the procedure of Uzov(<u>6</u>), and subjected to selective reduction with propanedithiol and triethylamine. Initially, a homogeneous reaction occurred in a 4:1 THF/methanol mixture, but the product precipitated as the reaction proceeded. 2,3-O-bis(phenylcarbamoyl)-6-amino-6-deoxycellulose, **9**, with an amino content of 0.35 meq/g was isolated; infrared confirmed the complete disappearance of the azido- function and the retention of the phenylcarbamoyl substituents. However, **9** was insoluble in all solvents tried and would swell in pyridine, DMSO and DMF. Thus, we did not attempt to use **9** as a macroinitiator.

    <u>Cellulose-*g*-Peptide Graft Copolymers.</u> Polymerization of BLG-NCA using **1** as a macroinitiator was effected in DMF under homogeneous conditions at room temperature (Scheme 2). Grafting efficiencies were dependent upon the macroinitiator concentration; in diluted systems, i.e., 0.5g of **1** in 20 mL DMF, high grafting efficiency with more controlled peptide block lengths was achieved. The aromatic ring (7.2 ppm) and benzylic CH_2 (5.01 ppm) associated with the BLG grafts on polymers **6a-f** can be clearly identified in the ^{1}H nmr spectra (Figure 3). Unfortunately, attempts to calculate the average DP of the BLG grafts from the nmr were not successful due to the poor resolution of the peaks associated with the cellulose backbone. Identification of the secondary peptide structure by infrared spectra (Figure 4) indicated that in all cases, α-helical conformations were present. Strangely, the existance of a β-sheet structure was not detected in any sample except **6f**, where the grafting was done under diluted conditions. Thus, the cellulose backbone does not appear to be effective in directing the conformation of peptide grafts into a β-sheet structure.

    The cellulose-*g*-PBLG copolymers remained soluble in DMF and DMSO. Treatment of **6b** with 1 N sodium hydroxide for 24 hours effected a quantitative hydrolysis of the benzyl ester and dissolution in the dilute base. The cellulose-*g*-glutamic acid copolymers,**11**, were also soluble in pyridine and DMSO but did not dissolve in water below pH 8. Using 1-hydroxybenzotriazole,**10**, as a promoter (1.2 meq **10**/eq peptide unit), the benzyl ester could be converted to a tris(hydroxymethyl)amide, **12**, by homogeneous aminolysis of **6b** in DMSO with tris(hydroxymethyl)aminomethane. If the conversion is quantitative, **12** will dissolve in water as well as alcohols and polar aprotic solvents. Treatment of **6b** with hydrazine in the presense of **10** in refluxing methanol/THF leads to crosslinked gels, **13** where the crosslink density can be controlled (Scheme 2).

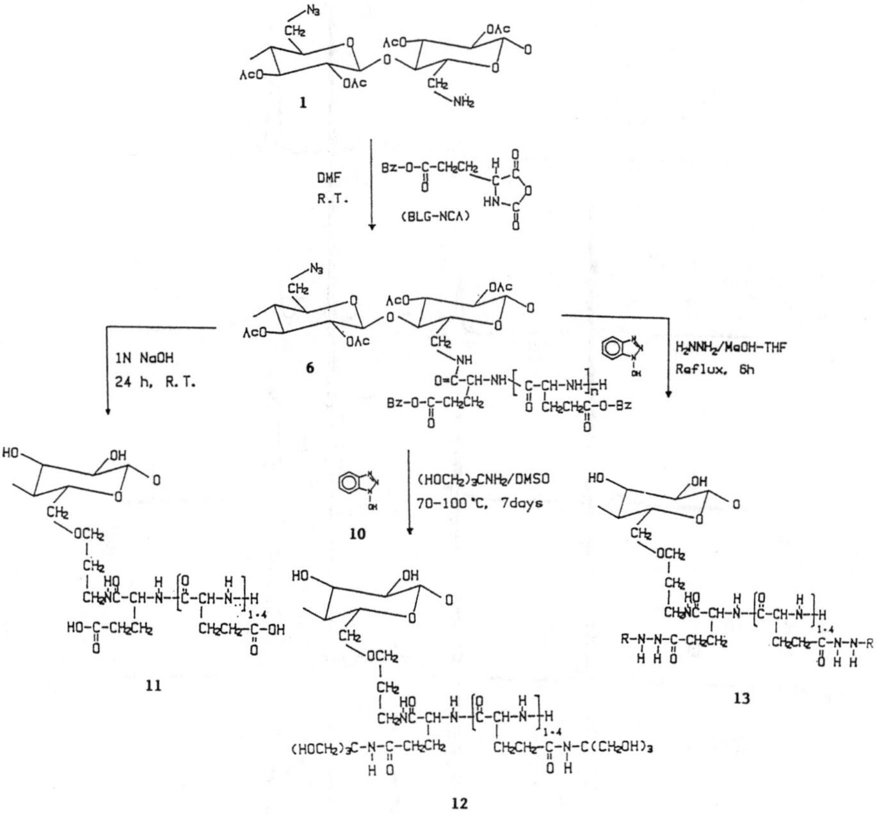

Scheme 2. Synthesis and Modification of
Cellulose-*g*-PBLG Copolymers

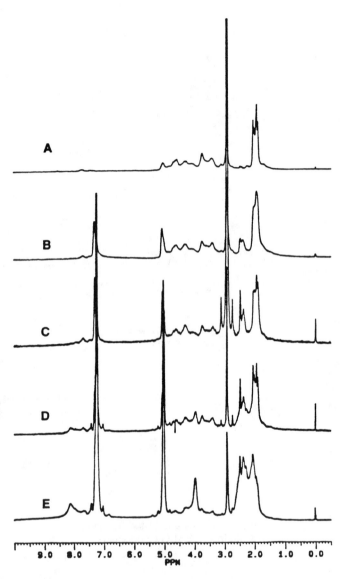

Figure 3. 200.13 MHz ^{1}H NMR spectra in DMSO-d_6 of 6-amino-6-deoxycellulose acetate-g-PBLG copolymers, **6** with different PBLG graft lengths: A, **1**; B, **6a**, D.P.= 1.4; C, **6b**, D.P.= 5.3; D, **6c**, D.P.= 19; E, **6d**, D.P.= 88.

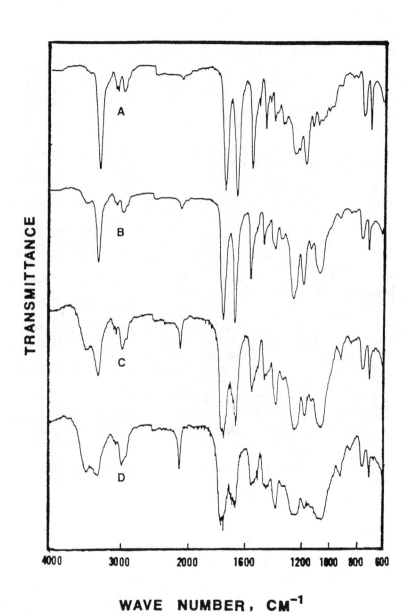

WAVE NUMBER, CM⁻¹

Figure 4. Infrared spectra of 6-amino-6-deoxycellulose acetate-*g*-PBLG copolymers, **6** with different PBLG graft lengths: A, **6d**, D.P.= 88, B, **6c**, D.P.= 19; C, **6e**, D.P.= 7.1; D, **6f**, D.P.= 3.1.

Literature Cited

1. Daly, W.H.; Lee, S. Polym. Mat. Sci. Eng. 1987, 57, 291.
2. Krylova, R. G. Russ. Chem. Rev. 1987, 56(1), 97-105.
3. Clode, D. M.; Horton, D. Carbohyd. Res. 1970, 12(3), 477-479.
4. Clode, D. M.; Horton, D. Carbohyd. Res. 1971, 19, 329-337.
5. Horton, D.; Luetzow, A.E.; Theander, O. Carbohyd. Res. 1973, 26, 1-19.
6. Usov, A. I.; Nosova, N. I.; Firgang, S. I.; Golova, O. P. Vysokomol. soyed. 1973, A15(15), 1150-1153.
7. Cho, H.; Teshirogi, T.; Sakamoto, M.; Tonami, H. Chem. Lett. 1973, 595-597.
8. Teshirogi, T.; Yamamoto, H.; Sakamoto, M.; Tonami, H. Sen'i Gakkaishi 1979, 35(12), T525-T528; Chem. Abst. 1979, 92, 130906k.
9. Teshirogi, T.; Yamamoto, H.; Sakamoto, M.; Tonami, H. Sen'i Gakkaishi 1980, 36(12), T560-T563 ; Chem. Abst. 1981, 94, 67477v.
10. Teshirogi, T.; Kusaoke, H.; Sakamoto, M.; Tonami, H. Sen'i Gakkaishi 1981, 37(12), T528-T532 ; Chem. Abst. 1982, 96, 79450k.
11. Usov, A. I.; Nosova, N. I.; Firgang, S. I.; Golova, O. P. Izv. Akad.Nauk. SSSR. Ser. Khim. 1974, 11, 2575-2581; Chem. Abst. 1975, 82, 112209z.
12. Kuznetsova, Z. I. Izv. Akad. Nauk SSSR. Ser. Khim., 1979, (5), 1101-1103; Chem. Abst. 1979, 91, 75898m.
13. Bosso, C.; Defaye, J.; Gadelle, A.; Wong, C. C.; Pederson, C. J. Chem. Soc., Perkin Trans. I 1982, (7), 1579-1585.
14. Usov, A. I.; Ivanova, V. S. Izv. Akad. Nauk SSSR. Ser. Khim. 1976, (5), 1142-1146; Chem. Abst. 1976, 85, 110284h.
15. Krkoska, P. Czech. Patent 158,310, 1975; Chem. Abst. 1976, 84, 32864k.
16. Teshirogi, T.; Yamamoto, H.; Sakamoto, M.; Tonami, H. Sen'i Gakkaishi 1980, 36(11), T501-T505; Chem. Abst. 1981, 94, 158561j.
17. Usov, A. I.; Ivanova, V. S. Izv. Akad. Nauk SSSR. Ser. Khim. 1986, (9), 2114-2118; Chem. Abst. 1986, 107, 198802g.
18. Yalpani, M.; Hall, L. D. Can. J. Chem. 1984, 62(2), 260-2.
19. Yalpani, M. U.S. Patent 4,531,000, 1985.
20. Tange, L.J.; Genung, L.B.; Mench, J.W. in Methods in Carbohydrate Chemistry, Whistler, R.L. Ed., Academic Press, New York, 1963, p 200.
21. Koziara, A.; Osowska-Pacewicka, K.; Zawaedzki, S.; Zwierzak, A. Synthesis, 202 (1985).
22. Bayley, H.; Standring, D.N.; Knoeles, J.R. Tetrahedron Lett. 1978, 39, 3633.

RECEIVED July 2, 1990

PHYSICOCHEMICAL ASPECTS
OF AQUEOUS SOLUTIONS

Chapter 13

Sodium-Ion Interactions with Polyions in Aqueous Salt-Free Solutions by Diffusion

Paul Ander

Chemistry Department, Seton Hall University, South Orange, NJ 07079

An unambiguous way to study the interactions of counterions with polyelectrolytes in solution is by determining the counterion and coion radioactive tracer diffusion coefficients in the presence of a polyelectrolyte with a common counterion. This has been done principally by Fernandez-Prini, et al.[1,2], Magdelenat, et al.[3,4] and Ander, et al.[5-13]. A capillary diffusion technique was employed. Aqueous solutions have been used in almost all studies. This brief review will focus on the experimental findings of the author. Equations have been developed by Manning for counterion and coion diffusion coefficients in polyelectrolyte solutions in the absence and presence of simple salts for a line charge model for the polyelectrolyte.[14-20]

DIFFUSION MEASUREMENTS

To determine the self-diffusion coefficients of ions, the open-end capillary method originally introduced by Anderson and Saddington[21] was employed without stirring. Due to the high viscosity of polyelectrolyte solutions, a non-stirring technique was found to be more reliable than one employing stirring because of the considerable change of viscosity of the solution as concentration varies. Fawcett and Caton[22] analyzed the source of error in the capillary method of measuring diffusion coefficients. They concluded that the serious source of systematic error results from convective loss of diffusant from the mouth of the capillary, and that accurate results require the use of fine bore capillaries from which convective loss is minimal.

In our laboratory all diffusion measurements were carried out in a constant temperature bath at $25.00 \pm 0.01^{\circ}C$. Precision bore capillaries of 1.00 ± 0.005 mm diameter and 2.82 ± 0.005 cm length were filled with the desired polyelectrolyte solution containing tracer amount of radioactively labeled ion under investigation. The outside of the filled capillaries were then wiped dry and each capillary was placed into an individual test tube (100 mm x 13 mm), which was subsequently filled with a radioactively inert solution of identical composition to the solution inside the capillary. Since the solutions inside and outside the capillary were the same polyelectrolyte concentration and the same simple salt concentration, no concentration

0097–6156/91/0467–0202$06.00/0

gradient existed. The only gradient which existed was that for radioactively-labeled cation, which caused it to diffuse out of the capillary. After sufficient time was allowed for diffusion, usually twenty-four hours, the outer solution was carefully withdrawn by means of an aspirator. Upon removal of the capillary from the test tube, the outside of capillary was dried and place with its open-end down into a vial containing 5 ml of scintillation fluid (Ready-Solve HP, Beckman). Each vial was then centrifuged for five minutes, followed by shaking and mixing for thirty seconds on a Vortex Geni (Scientific Industrial Inc.) to insure through mixing of the capillary contents with the scintillation liquid. The radioactive content of each capillary, C, was determined using a Model LS 7500 Beckman liquid scintillation system. Each sample was counted to a minimum of 20,000 counts, which corresponds to about 0.7% counting error. The initial radioactive content of capillary, C_0, was determined by the same procedure, using a capillary of identical bore and length with no diffusion time allowed. A minimum of six capillaries was used to determine each value of C and C_0.

Diffusion coefficients, D, were calculated using the expression[23,24] which is obtained from integrating of Fick's equation

$$C/C_0 = \sum_{n=0}^{\infty} 8\,\pi^{-2}(2n+1)^{-2}\exp(-\pi^2(2n+1)^2 Dt/4L^2) \qquad (1)$$

where L is the length of the capillary in centimeters and t is the time allowed for diffusion in seconds. The McKay's approximation solution of Fick's equation[25]

$$D = (\pi/4)(1-C/C_0)^2(L^2/t) \qquad (2)$$

could be used. From the approximation used in deriving Equation 2, when the ratio of C/C_0 is equal to 0.45, an error of 0.5% results in the value of D. The higher the ratio, the smaller is the error. However, the higher the ratio, the larger is the error due to experimental manipulation. To avoid these problems, Equation 1 was used to determine the experimental values of D. This expression was inverted and solved for D with the aid of computer.

RESULTS

The most important characteristic of a polyelectrolyte is the reduced charge density or the charge density parameter ξ,

$$\xi = e^2/\,\varepsilon kTb \qquad (3)$$

where $\underline{e}$ is the fundamental charge, ε is the dielectric constant of the medium, $\underline{k}$ is the Boltzmann constant, $\underline{T}$ is the absolute temperature and $\underline{b}$ is the average axial distance between charges. The dominant polyelectrolyte theory is that of Manning,[14,15,16] which developed equations that can be easily correlated with several experimental thermodynamic and transport properties. The polyelectrolyte is modeled as an infinite line charge whose counterions can be totally dissociated from the polyelectrolyte or partially dissociated from (and partially condensed onto) the polyion. The theory has the polyelectrolyte dissociating its counterions completely if $\xi < \xi_C$ where $\xi_C = |Z_C|^{-1}$, with Z_C is the counterion charge, and

$$D_i/D_i^0 = f \ [1 - \frac{z_i^2}{3} A(\xi_c, \ \xi^{-1}X)] \tag{4}$$

where D_i/D_i^0 are the tracer diffusion coefficients of the i th ion in the condensation term

$$f = (\ \xi_c \ \xi^{-1}X + 1)/(X + 1) \tag{5}$$

where $X = N_p/N_s$, the equivalent concentration of polyelectrolyte to that of simple salt. The value of f is unity for coions and for counterions if $\xi < \xi_c$. The value of $A(\xi_c, \ \xi^{-1}X)$ or simply A is 0.87 for salt-free solutions of polyelectrolytes with monovalent counterions, i.e., in the limit of $X = \infty$ or practically at low simple salt concentration. This is an electrostatic interaction term between small ions and the polyion. For polyelectrolytes with $\xi > \xi_c$, some counterion condensation occurs to reduce ξ to its effective value ξ_c and $f \neq 1$. From Equation 5, f is the fraction of total uncondensed counterions. Salt-free diffusion results were usually obtained over the concentration range of polyelectrolyte from $10^{-4} < N_p < 10^{-1}$. If one examines the salt-free counterion diffusion coefficients in aqueous polyelectrolyte solutions in Figures 1 and 2, it is noted generally that a minimum in the curve of D_i/D_i^0 vs. N_p occurs when the tracer ion and the counterion are the same. Sometimes the minimum is gentle, sometimes more pronounced and a few times D_i/D_i^0 is found to be fairly constant as N_p is varied between 10^{-4} and 10^{-1}N. The minimum D_i/D_i^0 value is, of course, a measure of maximum counterion-polyion interaction and usually occurs between 10^{-2} and 10^{-3} N. One can rationalize the minimum by understanding that as the polyion in a dilute solution is further diluted, the concomittant elongation of the polyion and its Debye Hückel atmosphere expansion causes some counterions in the atmosphere to become more free. On the other side of the minimum, increasing concentration causes greater overlap of the polyions and their ionic atmospheres and counterions become more free. A perusal of the salt-free polyelectrolyte solutions literature will show the same concentration trend when the counterion activity coefficients are determined. The D_{Na}/D_{Na}^0 trend for vinylic type polyelectrolytes is illustrated in Figure 1 for fully or almost fully charged polyacrylate (PA) and polystyrenesulfonate (PSS) polyelectrolytes. Table I

Table I. Comparison of the Minimum D_{Na}/D_{Na}^0 Experimental Values
with Those Predicted Theoretically for Several
Polyelectrolytes in Salt-Free Solutions

polyelectrolyte	ionic group on polyelectrolyte	ξ	ξ-1	exptl	eq 4
i-carrageenan	OSO_3^-	1.66	0.60	0.65	0.52
alginate	COO^-	1.43	0.70	0.65	0.61
polyvinylsulfonate	SO_3^-	2.5	0.40	0.63	0.35
polystyrenesulfonate	SO_3^-	2.6	0.38	0.62	0.34
polystyrenecarboxylate	COO^-	2.6	0.38	0.45	0.32
polyacrylate	COO^-	2.7	0.37	0.38	0.32
polyacrylate	COO^-	2.7	0.37	0.40	0.32
dextran sulfate	OSO_3^-	2.85	0.35	0.27	0.30
heparin	COO^-, OSO_3^-, NSO_3^-	3.0	0.33	0.30	0.29

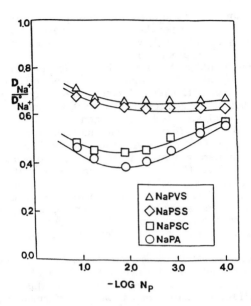

Figure 1. D_{Na}/D_{Na}° dependence on the equivalent concentration and nature of polyelectrolyte in aqueous salt-free solutions.

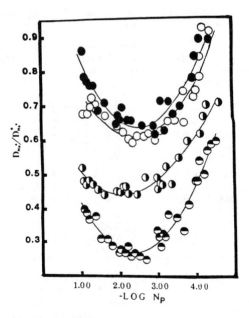

Figure 2.   Na⁺ ion diffusion coefficient dependence on the equivalent concentration for (●)i-carrageenan, (O)alginate, (◑)heparin and (◐)dextran sulfate.

lists the minimum values obtained for these polyelectrolytes, where it is noted that the vinylic carboxylate polymers have values close to their theoretical values of $0.87 \; \xi^{-1}$, while the sulfonates appear to dissociate more counterions that is predicted theoretically.

To test the Manning theory it is best to use polyelectrolytes close to the theoretical model. Thus small ion tracer diffusion studies were performed with long, stiff ionic polysaccharides. Figure 2 shows D_{Na}/D_{Na}^o vs log N_p for sodium dextran sulfate (ξ= 2.85), sodium iota-carrageenan (ξ = 1.66), sodium alginate (ξ= 1.43) and sodium heparin (ξ= 3.0), respectively.[10,11] The first two polysaccharides have pendant sulfate groups, the third pendant carboxyl groups and the fourth pendant carboxyl, sulfate and N-sulfonate groups. Each curve has a pronounced minimum at approximately 10^{-2} to $10^{-3}N$. Table I indicates that the experimental minimum values for ionic polysaccharides are close in accord with the theoretical value of $0.87\xi^{-1}$. Also, the counterion diffusion ratios seem to be independent of the nature of the ionic group, as is predicted from the theory. However, the theory appears to be incorrect in predicting that the counterion diffusion ratio is independent of concentration for salt-free solutions.

An interesting study was the evaluation of the D_{Na}/D_{Na}^o for aqueous salt-free solutions of polyelectrolytes with similar structures whose charge density could be varied. To this end measurements sodium ion diffusion measurements were performed using the sodium salts of acrylate/acrylic acid(NaPA/HPA), acrylate/acrylamide(NaPA/PAM) and acrylate/N,N-dimethacrylamide(NaPA/PDAM) over the ranges 5×10^{-4} $< N_p < 10^{-2}$ and $0.20 < \xi < 2.73$.[8] The results are in Table II, where average

Table II. Average D_{Na}/D_{Na}^o Values of Several Normalities
for Aqueous Salt-Free Solutions at 25°C

ξ	ξ^{-1}	NaPA/HPA	NaPA/PAM	NaPA/PDAM
2.73	.37	0.405 ± 0.010	—	—
2.67	.37	—	—	0.411 ± 0.011
2.61	.38	—	0.429 ± 0.012	—
2.22	.45	0.458 ± 0.011	0.445 ± 0.020	0.431 ± 0.013
1.82	.55	0.532 ± 0.014	0.569 ± 0.015	0.547 ± 0.015
1.54	.65	0.636 ± 0.017	0.669 ± 0.013	0.634 ± 0.014
1.33	.75	0.696 ± 0.014	0.716 ± 0.007	0.699 ± 0.015
1.18	.85	0.792 ± 0.022	0.758 ± 0.017	0.791 ± 0.011
1.05	.95	0.849 ± 0.021	0.793 ± 0.019	0.817 ± 0.009
0.80	1.25	0.649 ± 0.007	0.802 ± 0.009	0.809 ± 0.016
0.67	1.5	0.624 ± 0.016	0.798 ± 0.019	0.813 ± 0.007
0.40	2.5	0.624 ± 0.016	0.795 ± 0.013	0.810 ± 0.015
0.20	5.0	0.638 ± 0.026	0.788 ± 0.007	0.819 ± 0.011

D_{Na}/D_{Na}^o ratios are listed for the concentration range indicated. A glance at Table II shows that for each ξ value for all three polyelectrolytes, the D_{Na}/D_{Na}^o ratios are fairly independent of N_p in the concentration range studied. Very slightly higher diffusion ratio values were found at the lowest concentration, but nevertheless each average D_{Na}/D_{Na}^o value indicates fairly constant ratios at each charge density, independent of the polyelectrolyte concentration. Most striking is the discontinuity found at $\xi_C=1$ for NaPA/HPA when the averages of D_{Na}/D_{Na}^o for the four concentrations are plotted against ξ^{-1}, as is illustrated in Figure 3. Here, for NaPA/HPA, the plot is linear in the range $2.7 > \xi > 0.95$ with a slope of 0.79 ± 0.04 and an intercept of 0.11 ± 0.03. The diffusion ratio falls precipitously (about 22%) at $\xi = 1$ to a constant diffusion ratio of 0.631 ± 0.017 for $\xi < 1$. The few Na^+ ion diffusion coefficients in NaPA/HPA solutions obtained by Wall et al.,[26] shown in Figure 3, agree with those presented here for $\xi > 1$. For NaPA/PAM and NaPA/PDAM, Figure 3 shows that D_{Na}/D_{Na}^o is linear in ξ^{-1} for $2.7 < \xi < 0.95$ with respective slopes of 0.68 ± 0.04 and 0.77 ± 0.05 and respective intercepts of 0.18 ± 0.03 and 0.11 ± 0.04, and D_{Na}/D_{Na}^o remains constant at 0.796 ± 0.013 and 0.813 ± 0.013 respectively, in the range $0.95 < \xi < 0.20$.

The dominant feature in Figure 3 is the abrupt change in slope at $\xi = 1$ for each of the three polyelectrolytes, with a discontinuity occurring for NaPA/HPA. This appears to be strong evidence for counterion condensation since the breaks in the slopes occur just where the Manning theory predicts that it should occur, at $\xi = \xi_C$. That condensation occurs close to $\xi = 1$ for monovalent counterions has been reported by Zana et al.,[27] Leyte et al.,[28] and Gustavsson et al.[29] from magnetic resonance experiments, by Ikagami[30] from refractive index experiments and by Ware et al.[31] from electrophoretic mobility experiments. It should be pointed out these experiments are sensitive to measurements close to the polyelectrolyte chain, while the tracer Na^+ ion diffusion coefficients determined in the present study monitor the long-range, Debye-Huckel interactions.

The Manning theory predicts that D_{Na}/D_{Na}^o in salt-free polyelectrolyte solutions depends only on the nature of the polyelectrolyte through its charge density,

$$D_{Na}/D_{Na}^o = 0.866 \, \xi^{-1} \quad \text{for } \xi > 1 \tag{6}$$

and

$$D_{Na}/D_{Na}^o = 1 - (0.55 \, \xi^2)/(\xi + \pi) \quad \text{for } \xi < 1 \tag{7}$$

It should be noted from Figure 3 that the functional dependency predicted by Equation 6, given by the solid line is obeyed. The D_{Na}/D_{Na}^o ratios at each ξ value for all three polyelectrolytes have close values within experimental error and the line resulting from their average values, which is shown in Figure 3, has an average experimental slope of 0.74 ± 0.04 and an average experimental intercept of 0.13 ± 0.03. Thus good agreement with Equation 6 is achieved. It is interesting that if the zero point ($D_{Na}/D_{Na}^o = 0$; $\xi^{-1} = 0$) is used along with the experimental points for $\xi > 1$, then almost exact accord with the theoretical slope of Equation 6 is achieved, i.e., 0.88, 0.87, and 0.87 for NaPA/HPA, NaPA/PAM, and NaPA/PDAM, respectively. The similar results obtained for the three polyelectrolytes indicate that the polyelectrolytes have their carboxyl groups distributed along the chain in the same manner, most probably random. Other facts support this.

This discontinuity at $\xi = \xi_C$ for NaPA/HPA and the change in slope at $\xi = \xi_C$ for NaPA/PAM and NaPA/PDAM are evident from Figure 3. With counterion condensation occurring at ξ_C, the reduction of charge along the chain caused it to coil, thereby facilitating intramolecular hydrogen bond formation at lower charge densities, i.e., $\xi < 1$, for NaPA/HPA. Cooperative hydrogen bonding by two adjacent carboxylate groups in polycarboxylates has been discussed by Begala and Strauss.[32] This would increase the charge density of the coil and result in lower D_{Na} values, as is observed. Such an amount of coiling is not as pronounced for the other two polyelectrolytes at $\xi_C = 1$ because intramolecular hydrogen bond formation is not probable and hence, only a change in slope occurs at ξ_C. The constant D_{Na}/D_{Na}^0 values for $\xi < \xi_C$ which were obtained for all three polyelectrolytes indicate that the effective charge density of the coils is fairly constant in this charge density region and that rod-like behavior does not occur since Equation 7 is not obeyed. It is interesting that the electrophoretic mobilities of 6-6 ionene bromide in 4.0 mM KBr remained fairly constant for $\xi < 1$, followed by a precipitous drop at $\xi = 1$ to a constant value for $\xi > 1.3$[1] Ware explains these results by noting that near $\xi = 1$, the spacing of the condensed ions predicted by condensation theory is greater than the Debye screening length, resulting in further counterion condensation. Perhaps ξ_C is the effective value for $\xi > \xi_C$ as well as for $\xi > \xi_C$ to give stability to the solution. For $\xi > \xi_C$ this stability is achieved by counterion condensation reducing the charge on the chain. For $\xi < \xi_C$ this stability is achieved by a slight conformation change to increase the charge density of the chain. (It would be interesting to examine this idea by noting the change in the radius of gyration of the chain as ξ decreases below ξ_C.) For each of these polyelectrolytes below ξ_C, the Na^+ ion interacts with a polyion of constant effective charge density no matter what the value of the stoichiometric charge density. The only way this can happen is that the rodlike polyelectrolyte folds when $\xi < \xi_C$ to a constant effective ξ value. Dr. Marie Kowblansky, as a graduate student in my laboratory, first suggested that an explanation of these observations can be that unstable localized "loops" formed when $\xi < \xi_C$ and that the critical charge density parameter ξ_C is a stability point for the conformation of the polyelectrolyte. It is tempting to speculate that this constant effective ξ is ξ_C, i.e., at ξ_C the free energy of the solution is minimized. Recently, Manning used his condensation theory to account for these global polyelectrolyte transitions quantitatively by correlating them with the linear critical charge density for counterion condensation.[33] He showed that a mechanical instability of the locally folded structures can arise at the condensation critical point.

It was then of interest to see if the behavior obtained for sodium polyacrylates of varying charge density would be similar for sodium carboxymethylcellulose (NaCMC). Aqueous salt-free solution of NaCMC with ξ values of 0.53, 0.67, 0.89, 1.00, 1.03, 1.32 and 2.04 were prepared with a concentration range of 1×10^{-4} to 5×10^{-2}N, the same concentration range used for the sodium polyacrylates. The Na^+ ion diffusion results for NaCMC are shown in Figure 4, which have the characteristics of the previous figures for D_{Na}/D_{Na}^0 vs. log N_p, i.e., shallow minima with $(D_{Na}/D_{Na}^0)_{min}$ decreasing as ξ increases. Also, $(D_{Na}/D_{Na}^0)_{min}$ and $(D_{Na}/D_{Na}^0)_{X=10}$ are close in value at each charge density. Table III lists these values, along with the values predicted by Equation 6. A plot for NaCMC according to Equation 6 is in Figure 3 is linear, where the experimental slope 0.32 ± 0.04 and the intercept 0.41 ± 0.03. These values are quite different from those predicted by Equation 6. This might be due to both the hydrophilic nature of the NaCMC surface and the rigid backbone of this polymer as compared to the hydrophobic surface and flexible backbone of the vinylic polymers. In the region of $\xi > 1.32$, for which the NaCMC polymers have more than one carbonyl group per glucose ring (DS > 1), the charge density is more likely high enough to make the NaCMC and

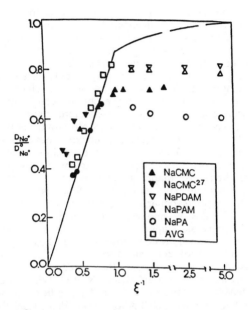

Figure 3. D_{Na}/D_{Na}^o dependence on ξ^{-1} for NaPA/HPA, NaPA/PAM, NaPA/PDAM, and NaPA/HPA[26]. The solid line is the Manning theory prediction.

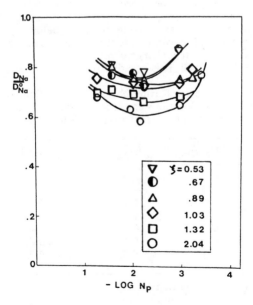

Figure 4. D_{Na}/D_{Na}^o dependence on the normality of NaCMC of varying charge density in aqueous salt-free solutions.

vinylic polymers seem equally rigid. The hydrophilic nature of the NaCMC polymers, due to the hydroxyl groups, becomes more significant in this region. The attraction of water molecules to the NaCMC backbone screens the charges, thereby weakening the long-range polyion-counterion interactions and leads to the observed higher diffusion coefficients for the NaCMC polymers as compared to the vinylic polymers in Figure 3. For intermediate ξ values, $1.32 > \xi > 1$, the weaker electrostatic repulsive forces between the charged groups on the chains permit the polyions to coil to a greater extent and the diffusion coefficient rises. Consequently the sodium ions that interact with the more flexible vinylic polymers, which can coil more easily than the polysaccharides, have greater diffusion coefficients in this region. This results in the smaller slope in the region of $\xi > 1$ for the NaCMC polymers as compared to the vinylic polymers.

The rodlike model of Manning predicts that condensation does not occur for $\xi < 1$ for sodium polyelectrolytes. Since NaCMC was used in this study because ionic polysaccharides are relatively stiffer than vinylic polyelectrolytes, for $\xi < \xi_C$ one would expect the experimental diffusion curve to approach unity as ξ approaches zero. It is obvious from Figure 3 that for NaCMC for $\xi < 1$, a constant value of $D_{Na}/D_{Na}^o = 0.72$ has been reached. When this line of zero slope for $\xi < 1$ is extrapolated to the experimental NaCMC line for $\xi > 1$, they intersect close to $\xi = 1$, the ξ_C value. This again, as for NaPA/PAM and NaPA/PDAM, indicates that the critical charge density parameter is correctly predicted from theory. (The discontinuity observed for NaPA/HPA at $\xi = \xi_C$ also shows this!)

Figure 3 also shows that for $\xi < \xi_C$ the D_{Na}/D_{Na}^o points for each polymer are all below those predicted by the rodlike theories, indicating that these models are inappropriate in this range. Similar results were obtained for sodium ion activity coefficients for aqueous solutions of sodium pectinate,[34] NaCMC,[35] and NaPA/HPA.[36] The zero slope line for NaCMC for $\xi < \xi_C$ lies close to the zero slope lines for NaPA/PAM and NaPA/PDAM. Also, for $\xi < \xi_C$ the D_{Na}/D_{Na}^o values for NaPA/HPA are constant. For each of these polyelectrolytes below ξ_C, the Na^+ ion interacts with a polyion of constant effective charge density, no matter what the value of the stoichiometric charge density.

From Tables I, II and III, it is noted that the quantities $(D_{Na}/D_{Na}^o)_{min}$

Table III. Sodium Ion Diffusion Ratios Obtained from Minimum Values in Salt-
free NaCMC Solutions and from Constant Values in
Solutions of High Polyelectrolyte to NaCl Normalities

ξ	ξ^{-1}	$(D_{Na}/D_{Na}^o)_{min}$	$(D_{Na}/D_{Na}^o)_{X=10}$	$(D_{Na}/D_{Na}^o)^*_{X=\infty}$	$(1-r)$
0.53	1.89	0.73	0.74	0.72	0.72
0.67	1.49	0.72	————	————	————
0.89	1.12	0.72	0.72	0.71	0.71
1.00	1.00	0.72	0.69	0.68	0.68
1.03	0.97	0.70	————	————	————
1.32	0.76	0.65	0.64	0.63	0.63
2.04	0.49	0.56	0.57	0.53	0.53

*Obtained from slope of Equation 10

and of ξ^{-1} are close in value, especially if $\xi > \xi_C$. This suggests that

(D_{Na}/D_{Na}^o) gives the charge fraction of the polyelectrolyte and that the salt-free polyelectrolyte dissociates the same fraction of counterions whether salt is present or not. Counterion additivity rules are based on this idea. For Na^+ ion diffusion the additivity rule could be written as

$$(N_p+N_s)(D_{Na}/D_{Na}^o)_X = N_p(D_{Na}/D_{Na}^o)_{X=\infty} + N_s(D_{Na}^o/D_{Na}^o)X \qquad (8)$$

If Equation 8 is divided by N_s on both sides, it can be written in term of concentration parameter X and can be rearranged as

$$(D_{Na}/D_{Na}^o)_X = (X^{-1} + 1)^{-1}(D_{Na}/D_{Na}^o)_{X=\infty} + (X + 1)^{-1}(D_{Na}^o/D_{Na}^o)_X \qquad (9)$$

where the first term on the right hand side is the polyelectrolyte contribution and the second term on the right hand side is the simple salt contribution. A comparison of the experimental values with those predicted by additivity Equation 9 is made in Figure 5 for copolymers of NaPA/HPA with varying charge densities. The lines in Figure 5 are based on Equation 9, calculated for $(D_{Na}/D_{Na}^o)X$ with knowledge of (D_{Na}/D_{Na}^o) in salt-free solutions, and the points are experimental. Best accord is achieved at X = 0.1, 5.0, and 10.0, where one component is in large excess. A most stringent test would be for X = 1, i.e., where both components contribute equally. Here good accord is achieved for the charge densities in range $1.05 < \xi < 1.82$, and poor accord is obtained for $\xi > 2$. It seems from Figure 5 that above $\xi^{-1} = 1$, an additivity rule for diffusion coefficient does not exist.

It is worthwhile to test whether this additivity rule is valid for NaPA, NaPSC, NaPVS, and NaPSS of constant reduced charge density of 2.6. If Equation 8 is divided by $(X+1)^{-1}$ on both sides, it can be written as

$$(X + 1)(D_{Na}/D_{Na}^o)_X = X(D_{Na}/D_{Na}^o)_{X=\infty} + 1 \qquad (10)$$

Since the diffusion coefficient ratio in salt-free solutions, $(D_{Na}/D_{Na}^o)_{X=\infty}$, is constant for a given ξ value, Equation 9 plots linearly for $(X+1)(D_{Na}/D_{Na}^o)_X$ vs. X with intercept of unity and slope of $(D_{Na}/D_{Na}^o)_{X=\infty}$, if the additivity law is valid for Na^+ ion diffusion in polyelectrolyte solutions. An illustrative plot of Equation 10 is shown in Figure 6. The ordinate intercept of resulting lines, listed in Table IV, are close to unity within the experimental error, giving some validity to the additivity rule. It was of interest to compare the slopes of these lines, $(D_{Na}/D_{Na}^o)_{X=\infty}$, with the minimum values of D_{Na}/D_{Na}^o obtained in salt-free polyelectrolyte solutions from Figure 6. Excellent agreement between these two sets of results, listed in Table IV gives some validity to additivity rule. These

Table IV. Slopes and Intercepts Obtained From Plots of Equations 10 and 11

	EQUATION 10		EQUATION 11		
	Intercept	Slope	Intercept	Slope	(1 - r)
NaPA	1.15 ± 0.03	0.38 ± 0.02	-0.15 ± 0.09	0.62 ± 0.03	0.38 ± 0.02
NaPSC	1.14 ± 0.04	0.43 ± 0.02	-0.14 ± 0.09	0.58 ± 0.03	0.42 ± 0.02
NaPVS	1.06 ± 0.02	0.62 ± 0.03	-0.05 ± 0.04	0.38 ± 0.03	0.62 ± 0.05
NaPSS	1.04 ± 0.02	0.62 ± 0.03	-0.03 ± 0.03	0.39 ± 0.02	0.62 ± 0.03

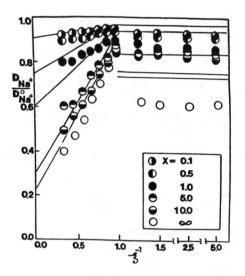

Figure 5. Using NaPA/HPA copolymers, a test of the Na⁺ ion diffusion
additivity rule given by Equation 9 and the lines in the figure.

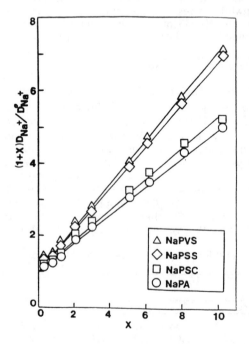

Figure 6. The average $(D_{Na}/D^o_{Na})(X + 1)$ values vs. X for vinylic
polyelectrolytes of approximately the same ξ values in
aqueous NaCl solutions.

experimental results, therefore, suggest that the properties based on long-range coulombic interactions can be approximated for polyelectrolytes in salt-containing solutions, by appropriate measurements in salt-free solutions.

Equation 10 could be arrived at with different reasoning. In aqueous solutions of polyelectrolytes with Na^+ as with counterion, rN_p equivalents of Na^+ ions of the total $(N_p + N_s)$ equivalents are condensed on the polyion and the fraction of condensed Na^+ ions f_{Na}^c in solution is [19,37]

$$f_{Na}^c = rN_p/(N_p + N_s) = rX/(1 + X) \qquad (11)$$

where r is the fraction of condensed or bound Na^+ ions originally on the polyelectrolyte and $(1 - r)$ is the charge fraction of the polyelectrolyte, which is the degree of dissociation of the polyelectrolyte or the charges on the polyelectrolyte uncompensated for by counterions. The fraction of Na^+ ions in the solution that are "free" is $f_{Na} = 1 - f_{Na}^c$ or

$$f_{Na} = ((1 - r)N_p + N_s)/(N_p + N_s) \qquad (12)$$

Counterion condensation onto polyelectrolytes has been operationally defined as association such that the total fraction of polyion sites compensated for with counterion remains invariant over a wide range of X values[18]. If the interaction of Na^+ ions with polyelectrolytes is properly described as "counterion condensation", then a plot of f_{Na}^c (X + 1) vs. X should be linear with slope $\underline{r}$. Note that this would indicate that the fraction of sodium ions dissociated from the polyelectrolyte is constant and independent of the concentration of polyelectrolyte and of simple salt.

To evaluate r from the diffusion measurements presented here, an assumption must be made. It's plausable to assume that f_{Na^+} is given by D_{Na}/D_{Na}^0 and that any small coion-polyion interactions, if at all present, it does not affect the assumption. So, f_{Na}^c is given by

$$f_{Na}^c = 1 - D_{Na}/D_{Na}^0 \qquad (13)$$

and plots were made of $f_{Na}^c(1 + X)$ vs. X for each NaPA/HPA polyelectrolyte with $\xi > 1$. The results from such plots indicated linearity was obtained for $0.1 < X < 10$, as shown in Figure 7 and Table V. The charge fractions $(1 - r)$

Table V. Parameters of Equation 11 Obtained for NaPA/HPA Copolymers[*]

ξ	Slope	Intercept	$(1 - r)$	$(D_{Na}/D_{Na}^0)_{X = \infty}$
2.73	0.57 ± 0.08	-0.13 ± 0.10	0.43 ± 0.06	0.41 ± 0.01
2.22	0.50 ± 0.06	-0.08 ± 0.05	0.50 ± 0.06	0.46 ± 0.01
1.82	0.43 ± 0.06	-0.08 ± 0.06	0.58 ± 0.08	0.53 ± 0.01
1.54	0.34 ± 0.04	-0.06 ± 0.06	0.66 ± 0.08	0.64 ± 0.02
1.33	0.28 ± 0.05	0.08 ± 0.04	0.72 ± 0.13	0.70 ± 0.01
1.18	0.24 ± 0.03	0.06 ± 0.04	0.72 ± 0.10	0.79 ± 0.02
1.05	0.19 ± 0.03	0.00 ± 0.00	0.81 ± 0.13	0.85 ± 0.03

[*] $f_{Na}^c(X + 1)$ vs. X.

is plotted against ξ^{-1} in Figure 8. It should first be noted from Table V and

Figure 8 that within the indicated experimental error, the calculated charge fraction $(1 - r)$ have values close to the theoretical value ($|Z_1| \xi^{-1}$), especially for the range $1.3 < \xi < 2.7$. This gives some validity to calculating a charge fraction from Equation 11, which states that condensation (and not electrostatic) is the dominant interaction. In fact, Equation 12 can be obtained from Equation 4 by letting the interaction term vanish. Also, for the higher charge density values, the contribution of condensation term to D_{Na}/D_{Na}^o in Equation 4 is much greater than the contribution of the interaction term, while for lower charge densities both terms contribute about equally. This explains why the lower charge density points in Figure 8 show the greatest deviation from the theoretical line. Because of agreement Wall obtained in calculating $(1 - r)$ from electrical transference experiment for salt-free solutions,[26] his values for salt-free $(D_{Na+}/D_{Na}^o)_{X = \infty}$ are included in Figure 8. The agreement between these values and the $(1 - r)$ values for each charge density gives further creditability to the evaluation of a charge fraction for each polyelectrolyte. Also, it seems that Manning is correct in stating that the charge fraction depends only on ξ.

It is worthwhile to evaluate charge fraction for NaPA, NaPSC, NaPVS, and NaPSS of constant charge density of 2.6 to note if the charge fraction depends on the nature of the polyelectrolyte as well as on the charge density. From the D_{Na}/D_{Na}^o values, and using Equations 11 and 13, $f_{Na}^c(1+X)$ was plotted against X for each polyelectrolyte. These plots were always found to be linear in several salt concentrations. (This is essentially Figure 6.) It was assumed, again, that all coion-polyion interactions, if at all present, to be negligible, i.e., $D_{Cl^-}/D_{Cl^-}^o = 1$. From average D_{Na}/D_{Na}^o values and using Equations 11 and 13, the same plots were made for each polyelectrolyte in Figure 8. The simple linear relation between $f_{Na}^c(1 + X)$ and X was always obtained which shows that r is constant over the whole X range and therefore, the polyelectrolyte charge fraction $(1 - r)$ can be calculated from the Na^+ ion diffusion measurements. Furthermore, the linearity is consistent with the result expected from the condensation model that the polyelectrolyte charge fraction is a constant value and independent of the salt concentration. Similar results have been obtained for other polyelectrolytes[37] and from ^{23}Na NMR measurements in salt-containing DNA solutions[38]. In accord with Equation 11, the intercept of lines obtained, listed in Table IV, are close to the origin within experimental error and the slopes are 0.62 ± 0.03 for NaPA; 0.58 ± 0.03 for NaPSC; 0.38 ± 0.03 for NaPVS; and 0.39 ± 0.02 for NaPSS. For carboxylate polyelectrolytes, the resulting charge fraction, $(1-r)$, of 0.38 and 0.42 for NaPA and NaPSC, respectively, compares excellently with the theoretical value of ($|Z_1| \xi)^{-1}$ for the polyelectrolyte. This gives some validity to the calculation of a charge fraction from Equation 11, which states that the condensation is the dominant interaction. For the sulfonate polyelectrolytes, however, the calculated charge fractions of 0.62 and 0.61, for NaPVS and NaPSS, respectively, deviate from the predicted theoretical value of ($|Z_1|\xi$)$^{-1}$. It appears that the fraction of condensed or bound Na^+ ions originally on the polyelectrolyte, and therefore the polyelectrolyte charge fraction, depends on the nature of the charge groups on the polyion.

It is interesting that the additivity rule (Equation 9) and the charge fraction (Equation 11) are of the same form, with $(D_{Na+}/D_{Na+}^o)_{X = \infty}$ identified with $(1 - r)$. When plotted as $(D_{Na+}/D_{Na+}^o)_x(1 + X)$ vs. X, both Equations 9 and 11 gave straight lines for $\xi > 1$, with the phenomenological parameters $(D_{Na}/D_{Na}^o)_{X = \infty}$ and $(1 - r)$ independent of N_s and both parameters are close in value for each polyelectrolyte studied here.

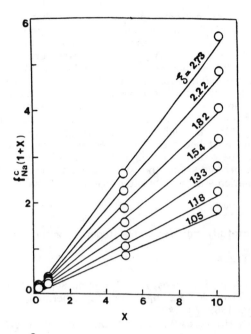

Figure 7. $f^C_{Na}(X + 1)$ vs. X for NaPA/HPA with varying ξ.

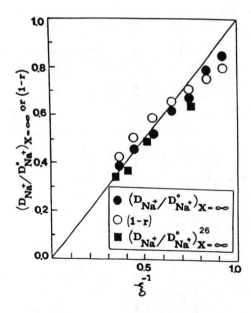

Figure 8. Experimental values of $(1 - r)$ and D_{Na}/D^o_{Na} dependence on ξ^{-1} for NaPA/HPA. The line is from the Manning theory.

Equations 9 and 11 stem from the same concept that Na^+ ions are either condensed onto the polyion and do not contribute to the measured D_{Na^+} or that the Na^+ ions are free so as to "not" interact with the polyion and contribute to the measured D_{Na^+}. The interpretation of Wall[26] for $(D_{Na}/D_{Na}^o)_{X} = \infty$ is the degree of dissociation of the polyelectrolyte is consistent with the value obtained for $(1 - r)$ from Equation 11. The theoretical equation of Manning, Equation 4, can be cast in the same form as Equations 17 and 19 if the Debye-Hückel interaction term vanishes. Then, the theoretical value for the fraction of Na^+ ions dissociated from the polyion $(|Z_1|\xi)^{-1}$ is identified with $(D_{Na}/D_{Na}^o)_{X} = \infty$ of Eq 10 and $(1-r)$ of Equation 9. It becomes clear then as to why Equations 9 and 11 should appear to be valid. For high charge density polyelectrolytes, i.e., those used in the present study, the contribution of the condensation term to D_{Na}/D_{Na}^o in Equation 4 is much greater than that of the interaction term A.

ACKNOWLEDGMENTS

The author is gratefully indebted to his students for their contributions.

LITERATURE CITED

(1) Fernandez-Prini, R.; Lagos, A.E. J. Polym. Sci. 1964, 2, 2917.
(2) Fernandez-Prini, R.; Baumgartner, E.; Liberman, S.; Lagos, A.E. J. Phys. Chem. 1969, 73, 1420.
(3) Magdelenat, H.; Turq, P.; Chemla, M.; Para, B. Biopolymers 1976, 15, 175.
(4) Magdelenat, H.; Turq, P.; Chemla, M.; Para, B. Biopolymers 1974, 13 1535.
(5) Kowblansky, M.; Ander, P. J. Phys. Chem 1967, 80, 297.
(6) Kowblansky, A.; Sasso, R.; Spagnuola, V.; Ander, P. Macromolecules 1977, 10, 78.
(7) Ander, P.; Gangi, G.; Kowblansky, A. Macromolecules 1978, 11, 904.
(8) Ander, P.; Kardon, M., Macromolecules 1984, 17, 2431, 2436.
(9) Dixler, D.; Ander, P. J. Phys. Chem. 1973, 77, 2684.
(10) Ander, P.; Lubas, W. Macromolecules 1981, 14, 1058.
(11) Lubas, W.; Ander, P. Macromolecules 1980, 13, 318.
(12) Trifiletti, R.; Ander, P. Macromolecules 1979, 12, 1197.
(13) Henningson, C.T.; Karluk, D.; Ander, P. Macromolecules 1987, 20, 1986.
(14) Manning, G.S. J. Chem. Phys. 1969, 51, 924.
(15) Manning, G.S. J. Chem. Phys. 1969, 51, 934.
(16) Manning, G.S. J. Chem. Phys. 1969, 51, 3249.
(17) Manning, G.S. Annu. Rev. Phys. Chem 1972, 23, 117.
(18) Manning, G.S. Q. Rev. Biophys. 1978, 11, 179.
(19) Manning, G.S. Acc. Chem. Res. 1979, 12, 443.
(20) Manning, G.S. J. Phys. Chem. 1981, 85, 870.
(21) Anderson, J.S.; Saddington, K. J. Chem. Soc. 1949, S381.
(22) Fawcett, N.C.; Caton, R.D. Anal. Chem. 1976, 48, 229.
(23) Sanborn, R.H.; Orlemon, E.F. J. Amer. Chem. Soc. 1955, 77, 3726.
(24) Mills, R. J. Amer. Chem. Soc. 1955, 77, 6116.
(25) McKay, R. Proc. Phys. Soc. 1930, 42, 547.
(26) Huizenga, J.R.; Grieger, P.F.; Wall, F.T. J. Am. Chem. Soc. 1950, 72, 4228.
(27) Zana, R.; Tondre, C.; Rinaudo, M.; Milas, M. J. Chim. Phys. Phys.-Chim. Biol. 1971, 68 1258.

Chim. Biol. 1971, 68 1258.
(28) Van der Klink, J.J.; Zuiderweg, L.H.; Leyte, J.C. J. Chem. Phys. 1974, 60, 2391.
(29) Gustavsson, H.; Lindman, B.; Bull.T. J. Am. Chem. Soc. 1978, 100 4655.
(30) Ikagami, A. J. Polym. Sci., Part A 1964, A2, 907.
(31) Klein, J.W.; Ware, B.R. J. Chem. Phys. 1984, 80, 1334.
(32) Begala, A.J.; Strauss, U.P. J. Phys. Chem. 1972, 76, 254.
(33) Manning, G.S. J. Chem. Phys. 1988, 89, 3772.
(34) Joshi, Y.M.; Kwak, J.C.T. J. Phys. Chem 1979, 83, 1978.
(35) Rinaudo, M.; Milas, M. J. Polym. Sci. 1974, 12, 2073.
(36) Kowblansky, M.; Zema, P. Macromolecules 1981, 14, 166, 1448.
(37) Ander, P. "The Charge Fraction of Ionic Polysaccharides", in Solution Properties of Polysaccharides; Brant, D., Ed.; American Chemical Society: Washington, DC, 1981; ACS Symp. Ser. No. 150, Chapter 28.
(38) Anderson, C.; Record, M.; Hart, P. Biophys. Chem. 1978, 7, 301.

RECEIVED June 4, 1990

Chapter 14

Aqueous-Solution Behavior of Hydrophobically Modified Poly(acrylic Acid)

T. K. Wang[1], I. Iliopoulos, and R. Audebert

Laboratoire de Physico-Chimie Macromoléculaire, Université Pierre et Marie Curie, URA Centre Nationale de Recherche Scientifique No. 278, ESPCI-10, rue Vauquelin, 75231 Paris Cedex 05, France

Hydrophobically modified poly(acrylic acid) was obtained by reaction of a small amount of an alkylamine on the carboxyl groups of the polyacid. Alkylamines with 8, 14 or 18 carbon atoms were used and the molar content of substitution ranged from 0 to 10%. Above a critical polymer concentration the viscosity in pure water of the modified samples (in the sodium salt form) drastically increases and can be several orders of magnitude higher than the one of the corresponding precursor polymer. Further enhancement of viscosity is achieved by increasing the ionic strength of the solution or by addition of a surfactant, Solutions exibiting high viscosities also present a marked solubilization of pyrene. These results were interpreted in terms of conformational changes and interchain aggregation due to hydrophobic interactions.

It is well known that thickening of solutions can be achieved by polymers of very high molecular weight. A more efficient viscosification or gelation can be obtained by the so-called "associating polymers". The most popular example of associating polymers in low-polarity solvents are the ionomers (1-4) : i.e. nonpolar macromolecules containing a low molar content of ionic groups (up to 5 %). In low-polarity solvents, these polymers aggregate because of the attraction between the ion pairs and at sufficiently high concentrations the intermolecular association leads to gel formation. Similarly, viscosification or gelation of aqueous solutions can be achieved by using amphiphilic polymers (5-13). That is water soluble polymers bearing a few of very hydrophobic groups, typically up to 5 mol%. In aqueous solution, aggregation of the hydrophobic parts may occur resulting in an increase in the apparent molecular weight. Such thickeners may

[1]Current address : Beijing Institute of Chemical Technology, Heping Street, Beijing 100013, China

0097–6156/91/0467–0218$06.00/0

control very efficiently the flow properties of aqueous-based fluids in many industrial applications/formulations, e.g. latex paints, drilling muds, hydraulic fracturing fluids, foods, cosmetics and drag reduction.

The last few years there has been substantial interest in these amphiphilic associating polymers. Most of the studies deal with nonionic polymers obtained by micellar copolymerization of a hydrophilic and a hydrophobic monomer (7,12). Because of the large difference in the water solubility of the two types of monomer, both molecular weight and local composition of the corresponding copolymers are difficult to control. In a recent work Ezzell and McCormick (14) shown that copolymers obtained by micellar copolymerization or by solution copolymerization exibit differences in their solution properties. Amphiphilic polymers have also been synthesized by modification of a precursor water soluble polymer. In this way a rather random distribution is expected. Still, most of the studied associating polymers are non-ionic (5,9). Very recently efforts were made to design associating polyelectrolytes (15).

In a previous paper, we have reported a very simple reaction of poly(acrylic acid) (PAA) with alkylamines (13). By this way a series of samples with the same molecular weight but controlled extent of modification were obtained.

In the present work we give typical viscometric results related to the aqueous solution behavior of these amphiphilic poly(acrylic acids). The influence of parameters such as extent of modification, alkyl chain length, ionic strength, pH, presence of a surfactant, is discussed. In order to clearly point out the formation of hydrophobic microaggregates, the solubilization of a very hydrophobic probe, pyrene, is also studied.

Experimental

Materials. Poly(acrylic acid) (PAA) in concentrated aqueous solution was purchased from Polysciences. The average molecular weight given by the supplier was 150 000. Solid PAA in the acid form was obtained by ultrafiltration of the commercial solution at first with an aqueous HCl solution (0.01M), then with a large excess of pure water and finally by freeze-drying. Technical grade alkylamines, octyl-(Genamin 8R 100D) and octadecyl-(Genamin 18R 100D) were kindly supplied by Société Française Hoechst, and tetradecylamine was supplied by Aldrich. All alkylamines were used without further purification. 99 % purity 1-methyl-2-pyrrolidone (MPD) and N,N'-dicyclohexylcarbodiimide (CDI) from Janssen and dodecyltrimethylammonium bromide (DTAB) from Aldrich were used. Pyrene (refence standard) was purchased by Polysciences. All other reagents were of analytical grade and water was purified by a Milli-Q system (Millipore).

Modificiation of poly(acrylic acid). The classical reaction of amines with carboxylic acids in an aprotic solvent (MPD) and in the presence of CDI was used for the modification of poly(acrylic acid) :

$$-COOH + H_2NR \xrightarrow[\text{in MPD}]{60°C, \, CDI} -CONHR + H_2O$$

A typical example of this reaction is given in reference (13). After purification the polymers were isolated in the sodium salt form. The ^{1}H NMR spectra of the modified polymers indicate that the yield of modification reaction reaches 100% (16). Therefore the degree of modification (or hydrophobic group content) was adjusted by the molar ratio of alkylamine to acrylic acid units. Precursor and modified polymers are listed in table 1. The sample designation is the following : e.g. PAA-150 is the precursor polymer 150 000 molecular weight ; PAA-150-3-C18 is derived from the above polymer containing 3 mol% of N-octadecylacrylamide groups. Since the modification reaction was performed in homogeneous solution, a random distribution of the alkyl groups along the PAA chain can be expected. Furthermore we verified, by means of intrinsic viscosity measurements (16), that the modified polymers have the same polymerization degree as the original PAA.

Table 1

PAA-150	PAA-150-1-C18	PAA-150-3-C18	PAA-150-10-C18
		PAA-150-3-C14	
		PAA-150-3-C8	

Apparatus. Ultrafiltration of PAA commercial samples was carried out in a Pellicon Cassette system (Millipore) using IRIS-3026 (Rhône Poulenc) ultrafiltration membranes of 10 000 nominal molecular weight cut-off.

Both a Contraves LS-30 Couette viscometer and a Carri-Med controlled stress rheometer with a cone and plate geometry were used to obtain viscosity values. Except for some studies of the shear rate effect, all our viscosity measurements were performed at low shear rates (1.28 s^{-1} and 0.06 s^{-1}) corresponding to the newtonian viscosity of the system. For samples with viscosities higher than about 1 000 cp the two types of apparatus give very similar results. Measurements of pH were performed with a Tacussel TAT-5 pH-meter using a glass-calomel unitubular electrode. UV-visible spectra for solutions containing pyrene, were recorded between 200 and 450 nm on a UV-vis spectrophotometer 552 (Perkin Elmer). The peak at 334 nm were used as a measurement of the pyrene concentration.

Conditions. Firstly, concentrated stock solutions were prepared under magnetic stirring at least 24 hours before use. Then, solutions of desired concentration (by weight %) were obtained by dilution of the appropriate stock solution with water and, if necessary, addition of solid NaCl (or of a concentrated surfactant solution). Solutions at various pH were obtained by adjustement of the stock solution pH and then by further dilution with water.

For solubilization experiments the following procedure was adopted : in 3 ml of a solution with the desired polymer concentration, a very small amount of an ethanolic solution of pyrene was injected in order to obtain a well known final concentration in pyrene. The final concentration of ethanol in the solution never exceeds 5 ml/l. We had verified that the presence of such a low concentration of ethanol does not alter the solubility behavior of pyrene in water. For solutions with high concentrations in pyrene a slight diffusion was observed. In this case base-line correction was made to obtain contribution due to the pyrene peak only (maximum at 334 nm). All solutions were equilibrated for about 24 hours before measurements have been made.

Viscosity measurements were conducted at 30°C ($\pm$ 0.1°C), while solubilization experiments were performed at room temperature ($\approx$ 20°C). Air bubbles in the highly viscous samples were readily eliminated by brief centrifugation.

Results

Effects of Alkyl Chain Content and Length on Aqueous Solution Properties. Typical viscosity results of modified and precursor poly(acrylic acid) in pure water are given in figure 1 as a function of the polymer concentration. A semilogarithmic scale is suitable for adequate representation of the observed viscosity variations. A classical polyelectrolyte behavior is found for the precursor polymer, PAA-150 (figure 1). By increasing polymer concentration, the viscosity firstly rises sharply (C_p< 1%) due to the electrostatic repulsions between charged groups along the polymer chains. Further increase in polymer concentration (C_p> 1%) leads to the smoothness of the slope of the viscosity curve due to the progressive self-screening of the electrostatic interactions in semi-dilute solutions.

Introduction of small amounts of alkyl chains into the PAA molecule completely changes its viscometric behavior. At concentrations lower than a critical value, C_p^c, depending on the degree of modification and alkyl chain length, the modified polymers behave similarly to the precursor chain. When polymer concentration exceeds C_p^c, the viscosity of the solution increases sharply and, for some samples, geletion may occur at sufficiently high polymer concentrations. The critical polymer concentration and the sharpness of the viscosity curve beyond C_p^c depend on the degree of modification and the alkyl chain length (figure 1). The change of the polymer molecular weight does not modify the general behavior of the systems. However, for a given modification degree and alkyl chain length, the larger the polymer molecular weight, the stronger the viscosification efficiency (16,17).

Effects of Ionic Strength and Shear Rate. Figure 2 displays typical viscometric behavior of original and hydrophobically modified polymers in the presence of NaCl and at constant polymer concentration C_p= 2%. Unmodified PAA as well as PAA-150-3-C8 show a continuous decrease in viscosity upon addition of NaCl. By increasing alkyl chain length or alkyl chain content the enhancement of viscosity with ionic strength becomes all the more pronounced and a maximum in the curve appears for the sample PAA-150-3-C18 at NaCl concentration of 1%. At this ionic strength, the viscosity of the modified polymer (PAA-150-3-C18) is four orders of magnitude higher than the viscosity of the precursor polymer (PAA-150).The very hydrophobic PAA-150-10-C18, although it is a very efficient thickener in pure water (completely gelled at C_p= 2%), loses its

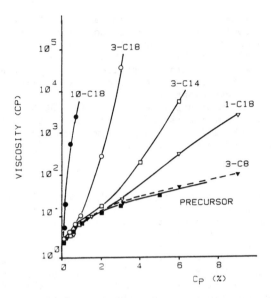

Figure 1 : Viscosity versus polymer concentration for PAA-150 series in pure water. Shear rate = 1.28 s^{-1}. (■) PAA-150, (▼) PAA-150-3-C8, (▽) PAA-150-1-C18, (□) PAA-150-3-C14, (○) PAA-150-3-C18, (●) PAA-150-10-C18.

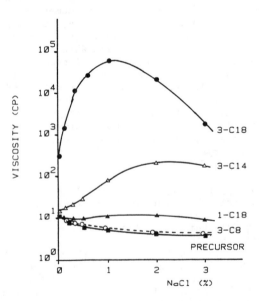

Figure 2 : Viscosity versus NaCl concentration for PAA-150 series. Polymer concentration = 2%. Shear rate = 0.06 s^{-1}. (■) PAA-150, (○) PAA-150-3-C8, (▲) PAA-150-1-C18, (△) PAA-150-3-C14, (●) PAA-150-3-C18.

viscosifying properties as soon as traces of NaCl are added in the solution. When the salt concentration exceeds 0.1%, the above polymer phase separates, giving a concentrated gel-like phase in equilibrium with a dilute supernatant.

Some examples of the influence of shear rate on the viscosity of modified polymers are given in figure 3. A typical pseudoplastic behavior is observed with most of these polymers. Almost in all the cases a newtonian plateau is found at the low shear rates, except for the very modified PAA-150-10-C18.

Influence of pH. Since polymer samples are obtained in their sodium salt form, their aqueous solutions are basic (pH > 9). In order to study the influence of pH on the viscometric behavior of these polymers the pH was adjasted by addition of the required amount of a strong acid solution (HCl). However, the addition of HCl results in the formation of an equivalent amount of NaCl in the polymer solution and consequently the viscosity behavior will be affected by both decrease in pH and increase in ionic strength (NaCl). In order to minimize the importance of the salt effect on the viscosity, we have chosen a polymer sample (PAA-150-1-C18) which present practically no viscosity change with addition of NaCl.

A typical variation of the viscosity versus pH, at fixed polymer concentration (C_p= 2%), is given in figure 4. The viscosity remains practically constant when the pH decreases from 9.7 to 6.5 and then increases sharply and reaches a maximum at about pH ≈ 5. Further decrease in pH results in a dramatic decrease in viscosity and finally phase separation occurs for pH values lower than 4.

Effect of Surfactants. It has been reported that addition of surfactants to solutions of associative acrylamide copolymers (10,12) or to solutions of hydrophobically modified hydroxyethylcellulose (HMHEC) (9) results in a decrease in viscosity. Other authors have claimed that the viscosity of HMHEC (5) or hydrophobically modified ethoxylate urethanes (8) can be enhanced in the presence of surfactants.

The viscosity of our polymers increases by addition of anionic, cationic or nonionic surfactants. Even an anionic surfactant, which presents unfavourable electrostatic repulsions with the anionically charged polymer chains, leads to a noticeable increase in viscosity. For instance, the viscosity of an aqueous solution of PAA-150-1-C18 (C_p= 3%) rises from 23 cp in pure water to 300 cp in the presence of sodium dodecylsulfate (4 10^{-3} mol.1^{-1}).

The most striking behavior is observed when a cationic surfactant is used, dodecyltrimethylammonium bromide (DTAB). Typical results are given in figure 5 where the viscosity of the precursor (PAA-150) and of a modified polymer (PAA-150-1-C18) is plotted as a function of the polymer concentration in pure water and in 4 10^{-3} mol.1^{-1} DTAB.

Addition of DTAB to the precursor polymer does not imply any viscosification of the system, at least in the range of concentrations used. For polymer concentrations higher than 1% the viscosity is the same in the absence or in the presence of DTAB (lower curve in figure 5). When the polymer concentration is lower than 1%, corresponding to a ratio R = [acrylate residue] / [surfactant] < 26, addition of DTAB results in a precipitation of

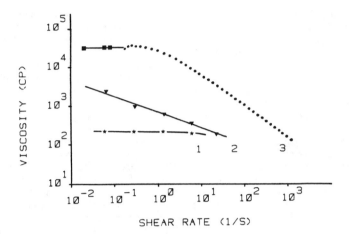

Figure 3 : Dependence of viscosity on the shear rate for three systems : 1 ($\star$) PAA-150-3-C18, C_p = 2% in pure water ; 2 ($\blacktriangledown$) PAA-150-10-C18, C_p = 0.35% in pure water ; 3 ($\blacksquare$ and $\bullet$) PAA-150-3-C18, C_p= 2% in 0.6% NaCl. $\star$, $\blacktriangledown$, $\blacksquare$, were obtained with a Contraves LS-30 viscometer and $\bullet$ with a Carri-Med rheometer.

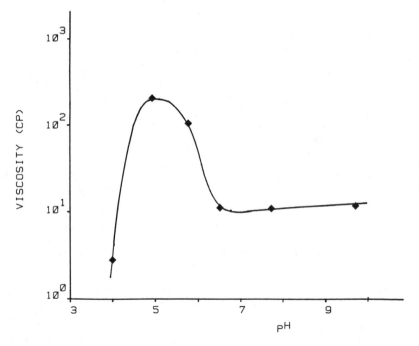

Figure 4 : Variation of viscosity versus pH for the PAA-150-1-C18 modified polymer. Polymer concentration = 2%. Shear rate = $1.28s^{-1}$.

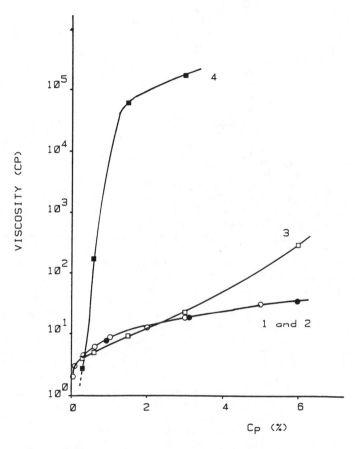

Figure 5 : Effect of a cationic surfactant (DTAB) on the viscosity of the precursor and a modified polymer. Viscosity is plotted as a function of polymer concentration : curve 1 (O) and 2 (●) for PAA-150 ; curves 3 (□) and 4 (■) for PAA-150-1-C18. 1 (O) and 3 (□) were obtained in pure water while 2 (●) and 4 (■) in an aqueous solution of DTAB (4 10^{-3} mol.l^{-1}). Shear rate = 1.28 s^{-1}. All measurements were performed with a Contraves LS-30 viscometer except for the two upper points of the curve 4 obtained with a Carri-Med rheometer.

the complex DTAB/PAA-150. Such a behavior is well known for mixtures
of a polyelectrolyte and a oppositely charged surfactant (18-20).
Undoubtedly, at a very low polymer concentrations, the system
becomes homogeneous again. Although the less modified PAA (PAA-150-
1-C18) does not exhibit a very pronounced thickening behavior in
pure water as compared with the precursor polymer, in the presence
of DTAB ($4\ 10^{-3}$ mol.1^{-1}) it shows a surprisingly important increase
in viscosity. For instance, when the polymer concentration is 1.5%,
the viscosity in the presence of DTAB is four order of magnitude
higher than in pure water. Furthermore, precipitation of the complex
DTAB/PAA-150-1-C18 occurs only at polymer concentrations lower
than 0.3% (R = [acrylate residue]/[surfactant] $\lesssim$ 7).

Solubilization of Hydrophobic Additives. It is well known that
solubility of aliphatic and aromatic hydrocarbons in water can be
enhanced in the presence of surfactants or hydrophobically
associating polymers (21,22). Pyrene is an aromatic hydrocarbon
exhibiting a very low solubility in pure water
($[Py]_{S,water} \simeq 7\ 10^{-7}$ M) (23,24) and it is largely used as a probe
for the study of micelles and other hydrophobic aggregates in water.
We report here some results concerning the solubilization of pyrene
by the hydrophobically modified PAA. In figure 6, the absorbance of
pyrene at the maximum of the peak (334 nm) is plotted against
pyrene concentration. Typical saturation curves are obtained : at
first, the absorbance linearly increases with pyrene concentration
and finally levels off. The slope of the linear part of the curves
is the same for all systems indicating, that the molar extinction
coefficient does not change in the presence of polymers.
     We will consider that the saturation concentration of pyrene,
$[Py]_S$, is given by the upper limit of the linear part of the curve.
As a consequence, in pure water (figure 6a) the saturation value
found ($\simeq 8\ 10^{-7}$ M) is close to the literature data ($\simeq 7\ 10^{-7}$ M).
Addition of unmodified PAA (at C_p= 0.5%) induces a very slight
change in the saturation value : $[Py]_S \simeq 1.1\ 10^{-6}$ M. On the other
hand, if 0.5% of a modified polymer (PAA-150-3-C18) is added, a
noticeable increase in $[Py]_S$ is found (1.9 10^{-6} M). An increase in
modified polymer concentration results in a higher solubility of
pyrene : $[Py]_S$= 2.8 10^{-6} M for PAA-150-3-C18 at C_p= 1% (figure 6a).
     The most interesting result is shown in the figure 6b. The
lower curve is already given in the figure 6a and corresponds to the
system PAA-150-3-C18, C_p= 0.5%, in water. The upper curve was
obtained with the same polymer and at the same concentration but
1% NaCl solution is used as solvent. It is clear that in the
presence of NaCl, the solubility of pyrene is very increased. It
must be noted that addition of NaCl brings about an enhancement of
pyrene solubility only if a modified PAA is present in the solution.

Discussion

The drastic viscosity increase in the aqueous solutions of modified
poly(acrylic acid) beyond a critical polymer concentration, C_p^c, can
be ascribed to the interchain association through formation of
hydrophobic microdomains. In fact, it has been suggested that water
soluble polymers with a low content of very hydrophobic groups
aggregate in a similar way as surfactant molecules micellize (9,11).

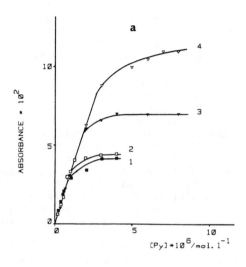

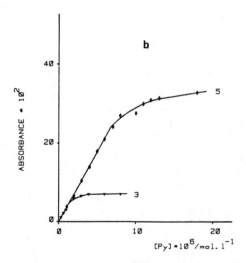

Figure 6 : Solubility curves for pyrene expressed as the absorbance at 334 nm versus pyrene concentration. 1 in pure water ; 2 in aqueous solution of PAA-150, C_p= 0.5% ; 3 in aqueous solution of PAA-150-3-C18, C_p= 0.5% ; 4 in aqueous solution of PAA-150-3-C18, C_p= 1% ; 5 in 1% NaCl solution of PAA-150-3-C18, C_p= 0.5%.

Our viscometric results plotted in figure 1, can be compared
qualitatively to the micellization of ionic surfactants. The
viscosification of our samples in pure water and consequently the
formation of hydrophobic aggregates, is all the more pronounced as
the alkyl chain length or the alkyl chain content increases. In a
similar way, the micellization of ionic surfactants becomes more
effective by increasing the surfactant concentration or the alkyl
chain length (22).
On the other hand, it is very interesting to compare the
behavior of our polymers with that of hydrophobically modified
nonionic polymers (HMNIP) (9,10,12). Despite the differences in the
molecular weight, ionic and nininonic hydrophobically modified
polymers show similar trends in their behavior in water solution
(effect of alkyl chain content and length) (9,12). However,
viscosification of a HMNIP solution occurs at polymer concentrations
and alkyl chain contents of one or two orders of magnitude lower
than in hydrophobically modified PAA (HMPAA) (10,12). These
differences in the behavior of HMNIP and HMPAA can be compared to
the differences in the c.m.c. value of nonionic and ionic
surfactants. In general, the c.m.c. is lower for nonionic than for
ionic surfactants (22,25).
The thickening of aqueous solutions of hydrophobically modified
water soluble polymers can be attributed to the formation of highly
branched multichain aggregates or of a physical gel in which crosslinking is assured by hydrophobic interchain aggregation. As a
consequence, viscosification occurs only at polymer concentrations
above the critical overlap concentration, C^*. In pure water, the
higher the number of interchain hydrophobic aggregates, the higher
the viscosity of the system.
Addition of a salt to an aqueous solution of HMPAA results in
the screening of the electrostatic repulsions between charges along
the polymer chain and brings about, at the same time, a retraction
of the polyelectrolyte chain and a more effective aggregation of the
alkyl chains. As a consequence, addition of NaCl to associating
polyelectrolytes solutions leads either to a more pronounced
decrease in viscosity or to a very effective thickening of the
solution. For relatively low polymer concentrations ($C_p < 0.5\%$), the
enhancement of hydrophobic aggregation upon addition of NaCl results
rather in an intrachain association which causes a more pronounced
viscosity shift (not shown in figure 2). When the polymer
concentration is high enough, in general higher than C_p^c, addition of
NaCl brings about increasing interchain hydrophobic aggregation and
finally leads to a substantial enhancement in the viscosity (curves
3-C14 and 3-C18 in figure 2). However, chain retraction is in
competition with interchain aggregation regarding the effect on the
viscosity of the system. Above a given level of NaCl concentration,
depending on the hydrophobically modified PAA used, the chain
retraction effect prevails over the aggregation effect and the
viscosity of the system starts to decrease (see curve 3-C18 in
figure 2). At higher salt concentrations (> 3%) phase separation
may occur. A similar maximum in viscosity upon addition of NaCl was
reported recently for ionic terpolymers of acrylamide, N-n-
decylacrylamide and sodium acrylate or sodium 3-acrylamido-3-
methylbutanoate (15). The HMNIP exhibit only small viscosity
enhancement with addition of salts (12,26). On the other hand, the

HMPAA sample, which does not exhibit associating behavior in pure water, in the presence of NaCl behaves like the precursor polymer (see curves 3-C8 and PRECURSOR in figure 2).

The influence of pH on the viscosity of HMPAA (figure 4) can be explained in the same terms as the influence of salts. By decreasing pH the actual charge density of the PAA chain decreases bringing about a retraction of the chain and a more effective hydrophobic aggregation. Even the less modified polymer (PAA-150-1-C18) which does not exhibit any noticeable thickening behavior upon addition of NaCl, presents an increase in viscosity of one order of magnitude when the pH shifts from ≈ 9.7 to ≈ 5. Presumably, addition of NaCl or decrease in pH do not have exactly the same influence on the association of the hydrophobic groups and on the polymer chain retraction. The details of this phenomenon are to be elucidated.

Whatever the hydrophobically modified PAA, the most striking thickening behavior was observed in the presence of surfactants. It is well established that surfactants interact with nonionic polymers mainly through hydrophobic interactions and with ionic polymers through both hydrophobic and electrostatic interactions (27). In general, the more hydrophobic the polymer is, the stronger the interaction with surfactants. In a sense, our polymers (HMPAA) behave as polymeric surfactants and may form a kind of mixed aggregates with low molecular weight surfactants. This seems to be actually the case as emerged from the viscosity behavior of modified and precursor polymer in the presence of DTAB shown in figure 5. In pure water, because of the strong electrostatic repulsions between the charged groups of the chains, only a small fraction of polymer-alkyl groups is involved in hydrophobic aggregates. Addition of a surfactant leads to the formation of mixed aggregates and thus to a more effective cross-linking and to the viscosification of the system : mixed aggregates containing at least two polymer-alkyl groups carried on distinct polymer chains act as effective cross-links. If a large excess of surfactant is added, mixed aggregates containing only one polymer-alkyl group can be formed and consequently the effective cross-linking and the viscosity of the system decrease. Such a behavior for associating polymers in the presence of surfactants was reported by Lundberg, Glass and Eley (8) and by Sau (5) and observed recently in our laboratory (T.K. WANG, unpublished results).

The idea of mixed aggregates formation is also supported by the differences in the value of the ratio R = [acrylate residue]/ [surfactant] at which phase separation occurs. This ratio is much lower for the modified PAA-150-3-C18 (≈ 7) than for the precursor PAA (≈ 26), indicating that the nature of association between anionic polymers and cationic surfactants is changed when hydrophobic groups are attached on the polymer chain.

On the other hand, the hydrophobically modified nonionic polymers aggregate very efficiently in pure water (probably most of the alkyl groups are involved in aggregates) and thus addition of surfactants and formation of mixed aggregates lead to the destruction of some cross-links and, consequently to a decrease of the viscosity.

The pseudoplastic behavior of our polymers (figure 3) is similar to that observed with other associating systems (11,12) and confirm the picture of a physical reversible highly branched

structure or gel formed by interchain hydrophobic aggregation.
Practically, the same plots of viscosity were obtained by increasing
or decreasing the shear rate.

Finally, enhanced solubilization of pyrene in aqueous solutions
of HMPAA, clearly bear out the statement of hydrophobic aggregation.
By increasing polymer concentration as well as by addition of NaCl
enhanced solubility of pyrene was observed (figure 6). This is in
agreement with the viscosity results of figures 1 and 2. In pure
water, near to the viscosification threshold ($C_p = 0.5\%$) a relatively
small increase in pyrene solubility is observed (figure 6a)
indicating some extent of hydrophobic aggregation. Addition of NaCl
(1%) promotes hydrophobic aggregation, mainly intrachain in the
range of concentration studied, and consequently it leads to lower
viscosity and increased pyrene solubility (figure 6b).

Conclusions

The classical viscosity reduction of polyelectrolyte solutions upon
addition of salts can be prevented if a small amount of highly
hydrophobic groups are covalently bounded on the polyelectrolyte
chain. Under suitable conditions the hydrophobic moieties aggregate
and the solution viscosity is found increased. Addition of a salt or
a surfactant or decrease in pH promotes hydrophobic groups
aggregation and may result in a very important enhancement of
viscosity and in an increased solubilization power towards aromatic
hydrocarbons.

Acknowledgments

This study was supported by the "Société Française HOECHST". We wish
to acknowledge Miss M.N. CHAUSSET for performance of solubilization
experiments.

Literature Cited

1. Lundberg, R.D. ; Phillips, R.R. J. Polym. Sci. Polym. Phys.
    1982, 20, 1143.
2. Peiffer, D.G. ; Kim, M.W. ; Schulz, D.N. J. Polym. Sci. Polym.
    Phys. 1987, 25, 1615.
3. Hara, M. ; Wu, J.L. ; Lee, A.H. Macromolecules 1988, 21, 2214
4. Granville, M. ; Jerome, R.J. ; Teyssie, P. ; De Schryver, F.C.
    Macromolecules 1988, 21, 2894.
5. Sau, A.C. Proceedings ACS Div. Polym. Mater. Sci. Engin. 1987,
    57, 497.
6. King, M.T. ; Constien V.G. Proceedings ACS Div. Polym. Mater.
    Sci. Engin. 1986, 55, 869.
7. Bock, J. ; Siano, D.B. ; Schulz, D.N. ; Turner, S.R. ; Valint,
    P.L. ; Pace, S.J. Proceedings ACS Div. Polym. Mater. Sci. Engin.
    1986, 55, 355.
8. Lundberg, D.J. ; Glass, J.E. ; Eley, R.R. Proceedings ACS Div.
    Polym. Mater. Sci. Engin. 1989, 61, 533.
9. Landoll, L.M. J. Polym. Sci. Polym. Chem. 1982, 20, 443.
10. Schulz, D.N. ; Kaladas, J.J. ; Maurer, J.J. ; Bock, J. ; Pace,
    S.J. ; Schulz, W.W. Polymer 1987, 28, 2110.
11. Valint, P.L. ; Bock, J. Macromolecules 1988, 21, 175.

12. McCormick, C.L. ; Nonaka, T. ; Johnson, C.B. Polymer 1988, 29, 731.
13. Wang, T.K. ; Iliopoulos, I. ; Audebert, R. Polym. Bull. 1988, 20, 577.
14. Ezzell, S.A. ; McCormick, C.L. Polym. Preprints ACS 1989, 30(2), 340.
15. Middleton, J.C. ; Cummins D. ; McCormick, C.L. Polym. Preprints ACS 1989, 30(2), 348.
16. Wang, T.K. ; Iliopoulos, I. ; Audebert, R. Manuscript in preparation.
17. Wang, T.K. ; Iliopoulos, I. ; Audebert R. Polym. Preprints ACS 1989, 30(2), 377.
18. Goddard, E.D. ; Leung, P.S. in Microdomains in Polymer Solutions Dubin, P., Ed ; Plenum Press : New York, 1985 ; p 407.
19. Dubin, P.L. ; Rigsbee, D.R. ; Gan, L.M. ; Fallon, M.A. Macromolecules 1988, 21, 2555.
20. Chandar, P. ; Somasundaram, P. ; Turro, N.J. Macromolecules 1988, 21, 950.
21. Strauss, U.P. ; Gershfeld, N.L. J. Phys. Chem. 1954, 58, 747.
22. Tanford, C. The Hydrophobic Effect : Formation of Micelles and Biological Membranes ; John Wiley and Sons : New York, 1980.
23. Binana-Limbele, W. ; Zana, R. Macromolecules 1987, 20, 1331.
24. Schnarz, F.P. J. Chem. Eng. Data 1977, 22, 273.
25. Lindman, B. ; Wennerström, H. in Topics in Current Chemistry ; Springer-Verlag : Berlin, 1980 ; Vol 87, p 1.
26. Zhang, Y.-X. ; Da, A.-H. ; Hogen-Esch, T.E. ; Butler, G.B. Polym. Preprints ACS 1989, 30(2), 338.
27. Robb, I.D. in Chemistry and Technology of Water-Soluble Polymers Finch, C.A. , Ed. ; Plenum Press : New York, 1983 ; p 193.

RECEIVED June 4, 1990

Chapter 15

Role of Labile Cross-Links in the Behavior of Water-Soluble Polymers

Hydrophobically Associating Copolymers

Philip Molyneux

Macrophile Associates, 53 Crestway, Roehampton, London SW15 5DB, United Kingdom

The labile crosslinks (herein called "liaisons"), involving noncovalent and/or reversible covalent interactions, that arise between polymer chains when they
come into contact, play an important part in the solution behavior of water-soluble polymers. Three cases
are discussed in this paper: (1) a simple homopolymer
alone in solution, where intrachain liaisons are the
so-called "long-range interactions" which influence
the conformational size of the molecule; (2) copolymers with a minor content of a comonomer unit which associates by hydrophobic or other interactions, where
the application of liaison theory to two types of hyd-
rophobically-associating copolymers - poly(vinyl acet-
ate-co-vinyl alcohol), and alkyl-substituted hydroxyethylcellulose - leads to the establishment of a scale
of free energies of hydrophobic-bonding between alkyl
chains which is linear with chain length for the range
$n = 1$ to 16; (3) cosolute-binding systems, where extra
liaisons may arise from associations between the bound
cosolute molecules.

The aim of this paper is to show how the behavior of water-soluble
polymers - especially their conformational size, their precipitabil-
ity and phase-separation behavior, and their rheology - is affected
by the labile crosslinks that arise between their chains when they
come into contact. For convenience and brevity, these labile crosslinks will be called liaisons throughout the rest of the paper.

1. Characteristics of Liaisons

A *liaison* is taken to be formed wherever there is a contact either
between two distantly-connected parts of the same polymer chain
("intrachain") or between parts of two different polymer chains

0097–6156/91/0467–0232$06.00/0

("interchain"). Such a contact must at least involve noncovalent interactions, which individually are generally weak so that the "crosslinking" that occurs is reversible; in certain cases it may involve weak covalent bonds, as with the poly(vinyl alcohol)/borax system discussed in Section 7. In the present paper the discussion of these liaisons has been restricted to aqueous systems, but it should be evident that many of the conclusions also apply to non-aqueous systems. Furthermore, in the context of molecular biology, the present effects are also involved in the folding and "molecular recognition" aspects of biopolymers (proteins, nucleic acids, poly-saccharides, etc.) in their native, aqueous solution environment.

The specific term "liaison" is introduced since the behavior of the two parts of polymer chain concerned lies somewhere between that of a simple "contact" (a term that does sufficiently reflect the well-defined forces of attraction and interaction energies that are involved) and that of a true "crosslink" (a term which is associated with a fixed, strong covalent bond between the polymer chains). In fairly concentrated polymer solutions (i.e., above the overlap con-centration), these liaisons are also "entanglements", although this latter term focuses more on the repulsions between the two parts of the polymer chain when they are forced into closer contact.

In the following section the different types of noncovalent force are presented. In Section 3, the role of liaisons in the behavior of a simple homopolymer alone in solution is discussed. In Section 4 the effects of the extra specific liaisons in polymer systems are considered, and in Section 5 the theory of their conformational effects is outlined. Two specific cases are then considered: that of a copolymer where the units of the minor comonomer can associate to give the extra liaisons (Section 6), and that of a polymer in the presence of a bindable cosolute (small-molecule solute), where extra liaisons may be formed through the bound cosolute molecules (Section 7).

2. Types of Interaction Forces in Liaisons

If we consider the case of two molecular species (i.e., molecules, or groups of atoms on a molecule) X and Y that are brought into contact under given conditions, then between these species there will arise a particular combination of noncovalent (or "secondary") forces, which can be represented by "X∼Y". In the present context, bringing such species that are in *aqueous solution* into contact requires the removal of a neighbouring water molecule (or molecules) W, with the concomitant formation of noncovalent interactions (i.e., mainly, hydrogen bonds) between the released water molecules, so that the process is actually an *exchange equilibrium*:

$$X{\sim}W \; + \; Y{\sim}W \; \Longleftrightarrow \; X{\sim}Y \; + \; W{\sim}W \tag{1}$$

In the case of liaison formation, X and Y would be the parts of the two chains that come into contact. The various types of such force that may be involved in liaison formation with water-soluble polymers are listed below; we have recently discussed the specific effects of these forces on the behavior of aqueous polymer/cosolute binding systems (1, 2).

Van der Waals forces (3,4). These are classified according to the polarity of the two interacting groups into: (a) van der Waals-Keesom (dipole-dipole) forces; (b) van der Waals-Debye (dipole/induced dipole) forces; and (c) van der Waals/London (dispersion) forces.

Hydrogen bonds (4). These are evidently very important in aqueous systems. As Equation 1 indicates, even if there is the possibility of such bonding between the monomer units in the liaison formation, there will also be a disfavorable effect from the need to break such bonds between these units and the water molecules hydrating them, as well as a favorable effect from the formation of hydrogen bonds between the water molecules thus released.

Hydrophobic interactions(5). This type of interaction is, of course, specific to aqueous systems. In this case, X and Y are alkyl or other nonpolar groups, while the "W" in Equation 1 is that part of the "structured water" about such groups which is released when X and Y are brought into contact.

Charge transfer interactions (6). Although this is a frequently discussed type of noncovalent interaction for nonaqueous systems, there seems to be little information on the strengths of charge transfer interactions in aqueous systems.

Coordination-complex forces. This is the type of interaction which occurs between, for example, metal ions and anionic ligand groups. The strength of the interaction depends markedly on the natures of the two interacting species, and the strongest of them approach in character to true covalent forces.

Coulombic (electrostatic) forces. Although these are greatly reduced in their strength in water because of its high dielectric constant, they are still significant in aqueous systems, as witness the special behavior of polyelectrolytes. They are truly long range in their nature, in contrast with the other noncovalent forces listed; however, in the present context, it is their behavior at short range, that is, for species essentially in contact, that is involved in liaison formation. The theory of liaison formation developed below applies strictly speaking only to nonelectrolyte polymers, and it requires modification to be applicable to polyelectrolytes.

3. Liaisons in Simple Homopolymer Solutions

The starting point for discussing liaisons must be the consideration of a single molecule of a simple polymer, such as a homopolymer, alone in solution; this is the "extreme dilution" state to which such experimentally obtainable quantities as the limiting viscosity number (LVN) apply. We consider its behavior firstly in an "indifferent" solvent, which a convenient term for a solvent where the strength of the interaction forces between a chain unit and the solvent is essentially the same as the average of those between each of the two species; this could, for example, be a solvent with a very similar molecular structure to that of the chain unit, so that all three types of interaction are equally strong. Even in such an

"energetically neutral" situation, there must be a definite probability that a chain unit finds itself in close proximity to another
distantly-connected unit, so that within the domain of this polymer
molecule there must be a definite density of contacts (each involving noncovalent forces) between distantly-connected monomer units.
These contact-species (i.e., *intrachain liaisons*) will remain essentially constant in number (although their locations will be continu-
ally changing), as controlled by the exchange equilibrium:-

$$M{\sim}S \ + \ M{\sim}S \quad <=> \quad M{\sim}M \ + \ S{\sim}S \qquad (2)$$

where M is the monomer unit and S is the solvent molecule, and "$\sim$"
represents (as in Equation 1) all of the noncovalent interactions
between each pair of species.
    The species M$\sim$M are the *liaisons* for this system; when they
occur within the polymer coil they are more conventionally referred
to as the "long-range interactions". Although for simplicity they
are represented here as involving only one monomer unit from each
part of the chain, in fact it is possible that a number of monomer
units would be involved - although there seems to be absolutely no
information available on such aspects of this basic process (Equation 2) for any polymer/solvent system. Correspondingly, S must then
represent the number of solvent molecules that occupy essentially
the same volume (or area) as the M-species.
    It is useful to make at least an estimate of the *average number*
of these liaisons in the molecule of a simple homopolymer in solution. For this purpose, we take the isolated polymer molecule to be
a droplet of polymer solution, with the average volume fraction of
polymer in its domain related to the LVN by the Einstein equation;
we will also use a simple cubic lattice model (i.e., coordination
number six) for the solution, so that each monomer unit will have
four sites (normally occupied by solvent molecules) for forming a
liaison with another distantly-connected section of the polymer
chain (the other two sites being occupied by the neighboring monomer
units of the chain). Thus, for the case of poly(vinylpyrrolidone)
grade K-90 in aqueous solution, the LVN is about 200 cm^3 g^{-1} (2),
and with the partial specific volume as about 0.8 cm^3 g^{-1}, using
the Einstein viscosity equation gives the average volume fraction as
2.5 x 0.8/200, that is, 0.01. This grade has a molecular weight of
about 1 million, corresponding to about 10,000 units in the chain,
so that if we go along the chain counting the liaisons by taking the
above volume fraction to represent the probability that any one of
the sites is in fact occupied by a distantly-connected chain unit,
the total number of liaisons will be given by 4 x 0.01 x 10,000/2
(where the division by 2 is needed because in this process each
liaison is counted twice, i.e., firstly as "M$\sim$M'" and later as
"M'$\sim$M"); this gives a total of about 200 liaisons, so that about 4%
of the monomer units are at any time in contact with another distantly-connected unit of the same chain.
    In an "indifferent" solvent as already defined, the value in
mole fraction units of the equilibrium constant, K_E, for this process (Equation 2) will be unity by definition. If the solvent is
made poorer, then K_E will be larger, leading to more liaisons, and
hence to a reduction in conformational size and in the value of the

LVN. Conversely, with a better solvent, K_E will decrease, and there will then be fewer liaisons in the molecule and an increase in LVN.

For the more practically important situation of polymer molecules in solution at *finite concentration*, the interactions between different chains when they come into contact (whose effects are characterized by the second virial coefficient, and the Huggins viscosity parameter) will also involve the same type of liaison formation as represented by Equation 2, so that the thermodynamics of this process will also profoundly influence the rheology of the system.

In a similar fashion, from the viewpoint of *solubility*, any effect which leads to an increase in the number of liaisons (e.g., a change in temperature) will favour precipitation (phase-separation) of the polymer, and vice versa. This is reflected in the fact that such precipitation is commonly presaged by a decrease in the LVN of the polymer, since both effects are associated with the formation of more liaisons.

The value of the equilibrium constant K_E is closely related to the conventional *Flory-Huggins interaction parameter* χ, since this is derived by considering the reverse of the liaison formation (Equation 2) as a quasi-chemical mixing of monomer units and solvent molecules, which in the Flory treatment (7) is put into the form

$$\tfrac{1}{2}M\!\sim\!M + \tfrac{1}{2}S\!\sim\!S \quad \Rightarrow \quad M\!\sim\!S \qquad\qquad (3)$$

Taking the system to have a lattice structure with (high) coordination number z, on the Flory theory the process (3) is formalised as the transfer of the chain unit M from the pure polymer (where it is interacting with z adjacent chain units) to the pure solvent (where it is interacting with z solvent molecules); the interaction parameter is defined as the standard free energy change for one mole of this process divided by RT (where R is the Gas Constant and T is the thermodynamic temperature)(7). Comparing equations (2) and (3), the equilibrium constant for (2), K_E, is then $\exp(2\chi/z)$. Thus, for an indifferent solvent, with $K_E = 1$, the parameter χ will be zero; the positive values for χ obtained with most systems reflect the fact that the solvent can be poorer ($K_E > 1$, with liaison formation favoured) and yet the polymer is still soluble, although the practical upper limit is the "ideal solvent" with $\chi = 0.5$. The ubiquity of the interaction parameter χ in the thermodynamics of polymer solutions (7) is a clear indication of the primary role of liaison formation and disruption in polymer conformation and polymer solubility, although we believe that few of those who use these mathematical relations understand that this is the underlying physical model.

The present liaison picture also seems to be equivalent to the alternative conventional approach using cluster integrals (8).

The two main advantages of the present picture over the more conventional ones, are that it provides a more concrete picture of the interactions occurring within and between the polymer molecules (not forgetting the solvent molecules which mediate these interactions), and that it also should allow correlations with the data becoming available from direct measurements of the strengths of noncovalent interactions between small-molecule solutes in aqueous solution (9-12).

4. Effects of Extra Liaisons on Polymer Behavior

Having started with a certain number of liaisons in the system,
these can be increased by modifying the system (see Sections 6 and
7, below). These extra liaisons affect water-soluble polymers in a
variety of ways, such as their spectroscopic and their solubility
behavior (considered next), and especially their conformational (and
hence, rheological) behavior (Section 5).

Spectroscopic Behavior. From this viewpoint, it is the fluorescence
properties of the polymer that may be most directly affected by
liaison formation, particularly where there are fluorogenic groups
on the polymer which can associate, leading to "excimers" in the
fluorescence process. This type of effect has been studied recently
with water-soluble polymers by a number of research groups, most
commonly with pyrene as the fluorogen (13-15). The fluorescence
effects are useful for studying liaison formation in systems where
fluorogenic groups have been deliberately introduced, but they are
not so important from the technical viewpoint because few commercially important polymers contain such groups. Nevertheless, spectroscopy provides the most promising technique for the direct observation of liaisons.

Solubility Behavior. Since the formation of liaisons between the
monomer units increases the number of such monomer/monomer contacts
in the system, then this in turn favours the transfer of the polymer
to a more concentrated phase, and hence reduces the solubility of
the polymer - in practical terms, it makes the polymer precipitate
"earlier" when the conditions (temperature, polymer concentration,
cosolute concentration,...) are changed in the direction that eventually leads to precipitation. Liaison formation can therefore explain, for example, the ability of phenols to precipitate many water-soluble polymers, through the reversible cross-linking effect of
the bound phenol molecules (see Section 7, below). Cosolute-induced
liaison formation can alternatively result in gelation, which may be
viewed as a kind of *homogeneous precipitation*.

5. Effects of Extra Liaisons on Conformational Behavior

Even a few extra liaisons can have a marked effect on the conformational size of a flexible chain polymer molecule in solution, as reflected in the limiting viscosity number (LVN), [η], which is proportional to the molecular hydrodynamic volume. This can be illustrated by the typical example of poly(vinylpyrrolidone) (PVP), a fairly
flexible-chain polymer, of molecular weight (say) about a million,
in water (a fairly good solvent), where by using a relation for the
effect of crosslinks on the LVN derived by Kuhn and coworkers
(16,17), we have shown (2) that each additional liaison will reduce
the LVN by about 5%, while it requires only about 10 such liaisons
to halve the value of the LVN. Since in this example the chain
contains about 10,000 monomer units, it is evident that it needs
only a very small fraction (<0.1%) of these to be involved in extra
liaisons for the conformational size to be markedly affected.

Quantitatively, the effects of such liaisons will depend upon the manner in which they are formed, but there are two models outlined below which are both simple and also apply in practice to a number of actual systems (see Sections 6 and 7, below).

Unimolecular Liaison Formation. Here the liaisons are formed between the main units in the chain A and a small proportion of a different monomer unit B, where for example B may be a comonomer (Section 6), or a unit of A with a noncovalently-attached cosolute molecule (Section 7):

$$A + B \iff A \sim B \qquad (4)$$

where for simplicity the neighbouring solvent (water) molecules which mediate this behavior are omitted. In this case, so long as the liaisons involve only a small fraction of the A units, it turns out ($\underline{2}$) that (as might be expected intuitively) their effect on the LVN has the "first order" form:

$$V \equiv [\eta]_r/[\eta]_0 = 1 - K_1 r/Q \qquad (5)$$

where V is the <u>relative limiting viscosity ratio</u>, defined as the ratio of the LVN of the polymer in the presence of the extra liaisons ($[\eta]_r$) to that in their absence ($[\eta]_0$),
K_1 is the equilibrium constant for liaison-formation (Equation 4),
r is the mole-fraction amount of B units in the copolymer (i.e. expressed as a fraction of the total number of chain units),
and Q is a coefficient characteristic of the system which is defined ($\underline{2}$) by:

$$Q \equiv 2^{1/2} abK_v^{1/3} M_0 [\eta]_0^{(2a-1)/3} \Phi^{1/3}/2.5 \qquad (6)$$

where a and K_v are the parameters in the Mark-Houwink-Sakurada equation relating $[\eta]$ with the polymer molar mass M;
b is the length of the bond in the polymer chain (e.g., the C-C bond length for vinyl polymers, or the span of the glucose ring for polysaccharides);
M_0 is the molar mass of the monomer unit;
$[\eta]_0$ is the LVN of the polymer in the absence of extra liaisons;
and Φ is the "Flory constant", taken here as 2.8×10^{23} mol^{-1} ($\underline{18}$).
Thus, for example, for polyvinylpyrrolidone (PVP) grade K-90, with a molecular weight of about one million, in aqueous solution at 25 °C, Q is 20×10^{-3} M^{-1} ($\underline{2}$).
It is this unimolecular behavior that is shown, for example, in the cosolute-binding liaisons of some phenols and related aromatic compounds with PVP (see Section 7). It should also occur with copolymers where the units of the minor comonomer can associate with those of the major one in a specific way, as with hydrogen bond donor/acceptor systems (see Section 6).

Bimolecular Liaison Formation. Here the extra liaisons are formed between pairs of identical species, such as hydrophobic comonomer units on a mainly hydrophilic chain (Section 6), or associating bound cosolute molecules on the polymer chain (Section 7):

$$B + B \iff B{\sim}B \tag{7}$$

In this case, as might be expected intuitively, the reduction in LVN is linear in the *square* of the fraction, r, of these species B relative to the main (inactive) units A (2):

$$V = 1 - K_2 r^2 / Q \tag{8}$$

where K_2 is the equilibrium constant for the bimolecular association (Equation 7), and Q is the same characteristic coefficient given by Equation 6.

6. Liaisons in Associative Copolymer Systems

As already noted (Section 5), one way to increase the number of liaisons in a polymer solution is to introduce into the chain some comonomer units which associate more strongly than the main units. For example, these could be hydrogen-bond donor/acceptor pairs (e.g., acrylic acid/vinylpyrrolidone), or they could be anion/cation pairs, since in each of these cases the aqueous solution behavior of mixtures of the corresponding homopolymers shows that these lead to a net attraction even in aqueous solution (19). Alternatively, this may be done using hydrophobic groups; two examples of technically important water-soluble copolymers that associate in this way, via a minor content of a hydrophobic monomer, are considered next.

Poly(vinyl acetate-co-vinyl alcohol) (PVAC-VAL). These copolymers, which for brevity will be referred to as PVA, are one form of the commercial so-called "polyvinyl alcohol" obtained by the partial saponification of PVAC (20). It is well known that these residual acetate groups greatly modify the properties of the polymer with respect to water, so that for example with more than about 30 mole % (0.30 mole fraction) the polymer is insoluble.

An estimate of the equilibrium constant for the association process (Equation 7) in this system can be obtained by using the results of some classic work on PVA carried out 30 years ago by Beresniewicz (21); having prepared five series of fractions of PVA with molar masses between 80,000 and 400,000, and with mole fraction residual acetate contents, r, from 0.24 (i.e., approaching the solubility limit) down to 0.002 (i.e., essentially pure PVAL), he determined the LVN for each of these fractions, and for the unfractionated materials. Inspection of his data shows that the relative LVN, V (as defined by Equation 5), is a falling linear function of the square of the acetate content (Figure 1):-

$$V = 1 - k r^2 \tag{9}$$

with slope-coefficient $k = 5.5 \pm 0.5$; note that in fitting the line to the plotted points in Figure 1, because the "origin" (at 0, 1.0) is a multiple experimental point, since it represents the reference value of the LVN for each of the main data points, the best line is taken to go through this point.

On the basis of the discussion in Section 5 above, this corresponds to the bimolecular hydrophobic association between the acetate groups (compare Equations 8 and 9), with the equilibrium constant K_H (equivalent to that for Equation 7) given by:-

$$K_H = Qk \qquad (10)$$

Substituting the appropriate values ($\underline{20},\underline{21}$) into Equation 6 gives $Q = 20 \ (\pm 1) \times 10^{-3}$ M^{-1} for these PVA samples in water. Using this value gives the equilibrium constant for hydrophobic association between the acetate groups as $K_H = 0.11 \pm 0.01$ M^{-1}. This equilibrium constant can be converted into the standard free energy change for the association, ΔG^0_H, using the thermodynamic relation:

$$\Delta G^0_H = - RT \ln K_H \qquad (11)$$

which gives $\Delta G^0_H = +5.5$ kJ mol^{-1}.

It should be borne in mind that with PVA (i.e., PVAC-VAL), the distribution of the acetate groups depends very much on the route that is used for its production, and that the alkaline methanolysis route used in the above work leads to a rather blocky distribution. However, since this is the commonest commercial method, then this value of the association constant should be applicable to a wide range of commercial products of this type ($\underline{20}$).

Hydrophobically-Modified Hydroxyethylcellulose (HM-HEC). This type of copolymer has been produced commercially, for example, by the Hercules Corporation ($\underline{22}$). Gelman and Barth at Hercules have published rheology data on HM-HEC samples with octyl and with hexadecyl (cetyl) chains as the hydrophobic groups ($\underline{23}$). For these materials, it turns out that the LVN is less than that of the parent HEC (as with PVA), but the practical solution viscosity is greater, e.g., an HM-HEC with 1.2 w/w % cetyl groups has an LVN which is one-third that of the parent HEC, but its 2% solution has a viscosity which is 100 times that of the parent polymer at this concentration.

Figure 2 shows that, as with PVA, both for the octyl and for the cetyl series the LVN is a falling linear function of the square of the alkyl group content (Equation 8, above), showing again the bimolecular character of the hydrophobic association between the alkyl chains; note that in these plots, as with Figure 1, the "origin" (0, 1.0) is a multiple reference point, so that the straight lines are defined by three (octyl) and four (cetyl) points, with one other data point in each case evidently "wild" and hence disregarded. In these cases the effects are much greater than with PVA (as would be expected from the much longer alkyl chains), with values of the slope-coefficient k (Equation 9) of 110 for the octyl series and 9.0 x 10^3 for the cetyl, i.e., an 80-fold increase from a doubling in alkyl chain length. For these polymers the available data ($\underline{23}$) gives the value of the coefficient Q (Equation 6) as 62 x 10^{-3} M^{-1}, which in turn gives the values for the hydrophobic association constant K_H as 6.8 M^{-1} and 558 M^{-1}, respectively. The corresponding values of the standard free energy change, ΔG^0_H, are -4.7 and -15.7 kJ mol^{-1}, respectively.

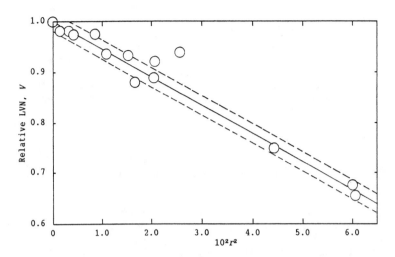

Figure 1. Relative limiting viscosity number, V, versus the square of the mole fraction acetate group content, r, for partially hydrolyzed poly(vinyl acetate)s: 25 °C. Data adapted from Ref. 21.

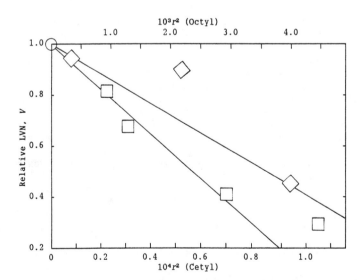

Figure 2. Relative limiting viscosity number, V, versus the square of the mole fraction content, r, of alkyl group R, for hydrophobically modified hydroxyethyl-celluloses from the same parent HEC. Diamonds and upper abscissa scale, R = octyl; squares and lower abscissa scale, R = cetyl (hexadecyl): 25 °C. Data adapted from Ref. 23.

Other Hydrophobically Associating Systems. The data presented here
for PVAC-VAL and for HM-HEC seem to be the only ones available to
which the present theory can be applied directly; we have not been
able to find any new data during the scanning we ourselves carry out
in collecting the references for our "Updates" (24,25) which cover
the most recent work on the physical chemistry of water-soluble synthetic polymers. Most syntheses of hydrophobically-modified polymers
has been carried out by copolymerization of a hydrophilic monomer
with varying minor proportions of the hydrophobic comonomer, where
it is most unlikely that the degree of polymerization is constant
as required by the present treatment. However, our treatment should
be applicable once the LVN value for each such copolymer can be related to that of the LVN expected for the equivalent homopolymer of
the same degree of polymerization.

From the more general viewpoint, these hydrophobic interactions
must also be involved in such recently reported work as the precipitation temperatures of N-alkylacrylamide copolymers determined by
Priest and coworkers (26), and the precipitation and gelation behavior of methylcelluloses studied by Takahashi and coworkers (27).

These associative effects are also important in *nonaqueous* systems, such as kerosene solutions of the ICI antimisting additive FM9
(which is a butylstyrene/methacrylic acid copolymer with about 12
mole % of the carboxylic acid monomer) studied by Ballard and coworkers (28), who present a theoretical treatment of these effects.

Free Energies of Hydrophobic Bonding between Alkyl Chains. It is revealing to correlate the data obtained above for hydrophobic bonding
between alkyl chains in these two different copolymers. Taking the
data for PVA to correspond to an alkyl chain length $n = 1$, i.e. that
for the methyl group on the acetate, then Figure 3 shows that the
standard free energy change of the interaction is a linear function
of the alkyl chain length from $n = 1$ up to $n = 16$. This indicates
that each methylene group of the alkyl chain makes the same contribution, i.e., close to -1.4 kJ mol^{-1}, to ΔG°_H.

This seems to be the first estimate of this methylene-group
contribution to true hydrophobic bonding for such a wide range of
alkyl chain lengths, and indeed it seems to be the first indication
that it is constant over such a range. That this is a reasonable value is supported by the fact that it is comparable to the value of
-1.0 kJ mol^{-1} we obtained for the equivalent contribution to the
free energy of the binding of alkyl 4-hydroxybenzoates by PVP in
aqueous solution (29), where each additional methylene chain in the
cosolute ester group is taken to interact with a methylene group on
the PVP chain.

Furthermore, viewing this hydrophobic association as the
transfer of each methylene group from a wholly aqueous environment
to the hydrophobic-association environment (i.e., adjacent to the
methylene group on the other alkyl chain), this gives the free energy of transfer for a single methylene group as -0.7 kJ mol^{-1}. This
value is compared in Table I with data from the literature for methylene-group transfers to another environment "x":

$$(-CH_2-)(aq) \Rightarrow (-CH_2-)(x) \tag{12}$$

Table I. Thermodynamics of hydrophobic bonding: comparison of the
standard free energies ΔG°_{TR} for the transfer of an alkyl-chain
methylene group from aqueous solution into another environment
(Equation 12)

State "x"	$-\Delta G^{\circ}_{TR}$ /kJ mol^{-1}	Reference
Free alkyl chain in aqueous solution[a]	0.0[a]	---
Hydrophobic bond[b]	0.5	29
Gas (vapor) state[c]	0.6	30
Hydrophobic bond[d]	0.7	This work
Hydrophobic bond[e]	0.7	31
Air/water interface[f]	2.6	32
Alkyl-amphiphile micelle[g]	2.7	32
Octanol[h]	3.1	29
Alkane/water interface[f]	3.5	32
Heptane[h]	3.5	33
Cyclohexane[h]	3.6	29
Liquid alkanes[i]	3.6	32

[a] By definition - reference state
[b] From the thermodynamics of the binding of C_1 to C_4 alkyl 4-hydroxy-
benzoates to polyvinylpyrrolidone, assuming the the increment in
binding strength to correspond to the pairwise interaction between a
methylene group on the cosolute and one on the polymer chain (29)
[c] From the vapor pressure behavior of aqueous C_1–C_8 alkanols (30)
[d] Half the value of ΔG°_H, since this corresponds to the transfer of
two methylene groups into each others' vicinity
[e] Theoretical estimate from the Nemethy-Scheraga theory (31)
[f] Dilute "gaseous" film, i.e., isolated adsorbate molecules
[g] All types of head-group (anionic, cationic, zwitterionic and nonionic) on the amphiphile
[h] From partition coefficient data, i.e. solvent is water-saturated
[i] From the solubilities of the alkanes

The values for the transfer into a liquid alkane (-3.6 kJ mol^{-1}),
and to an alkane/water interface (-3.5 kJ mol^{-1}) (31) are about five
times that of the hydrophobic bonding effect, which is reasonable
since in these fully alkane environments there should be something
like five or six methylene groups from other alkane chains surrounding (and hence interacting with) the methylene group in question,
compared with one such methylene group in hydrophobic association.
   Two points of detail arise from the present interpretation of
the behavior of these hydrophobically-modified polymers. The first
point is that in the case of PVA, it may be questionable to take the
methyl group of the acetate as corresponding to alkyl chain length
n = 1, because the adjacent hydrophilic ester group might be
expected to interfere with the hydrophobic hydration around the
methyl part. However, although indeed the methyl group does not

always follow the behavior expected for it as the first member of the alkyl series, it is significant that we ourselves found, in our studies on the binding of alkyl (i.e., methyl to butyl) 4-hydroxy-benzoates to PVP (29), that the free energy of binding (represented by the logarithm of the binding constant) was a linear function of the alkyl chain length for all four chain lengths; this is a parallel case to the present systems both because of the clear involvement of hydrophobic bonding, and also because with these co-solutes the alkyl chain is in a similar environment to that in PVA, albeit on the other side of the ester group.

The second point is that the free energy might not be expected to be so accurately a linear function of chain length (Figure 3), because the longer alkyl chains should fold up in water, leading to a shorter effective chain length. However, in the parallel case of the partition of aliphatic carboxylic acids between heptane (where clearly the chains do not fold up) and water, the free energy of the transfer is a linear function of chain length from (at least) $n = 7$ up to $n = 21$, supporting the absence of any effect from such folding on the behavior of these chains in aqueous solution (33).

The significance of these results for the behavior of water-soluble polymers in general, is that the hydrophobic effect of the alkyl group should be very much the same whatever the nature of the parent polymer, so long as its chain is fairly flexible. Further-more, the free energy contributions of other alkyl chains may be obtained using the pair of equivalent relations representing the best fit to the data in Figure 3:

$$\log (K_H/M^{-1}) \quad = \quad -1.18 \quad + \quad 0.245n \qquad (13)$$

$$\Delta G^0{}_H/kJ \ mol^{-1} \quad = \quad 6.78 \quad - \quad 1.41n \qquad (14)$$

Thus, the present results should be applicable to predicting and interpreting the LVN values for almost any hydrophobically-modified water-soluble polymer. Although the theory presented only applies directly to the "extreme dilution" state, it is likely that it could be extended to the rheological behavior of finite (i.e. practical) concentration systems. For example, so long as the alkyl-chain association remains bimolecular, the quotient $K_H r^2$ should be a scaling factor for correlating the rheological behavior of copoly-mers with different alkyl chain lengths and/or different degrees of substitution derived from the same parent polymer. However, at polymer concentrations well above the overlap point, and particu-larly also with the longer alkyl chains, the association may become multimolecular with a cooperative character approaching that of micellization, whereby for example the methylene-group free energy contribution would increase towards the value of about -2.7 kJ mol^{-1} seen in the true micellization of small-molecule amphiphiles (32).

7. Liaisons in Cosolute-Binding Systems

The third case of extra liaison formation to be considered is that of cosolute-binding systems, where the extra liaisons will arise if the bound cosolute molecule associates either with a section of free polymer chain ("unimolecularly") or with another bound cosolute mol-

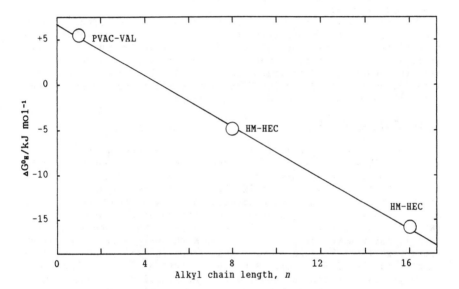

Figure 3.   Standard free energy change, ΔG^{0}_{H}, for the hydrophobic association of alkyl groups attached to the chains of water-soluble polymers, versus the alkyl chain length, n: 25 °C.

ecule on the chain ("bimolecularly") (see Section 5). In fact, the
present concepts and the theory of liaisons outlined in Section 5
stem from our previous interpretation of the binding, viscosity and
precipitation results we obtained for phenols and *O*-substituted
phenols as cosolutes with PVP, where each type of behavior was seen,
and where the theory was applied to evaluate the equilibrium constants for the associations (1, 2).
    Other examples are the gelation of PVAL induced by cosolutes
such as azo dyes or borax. In this last case, the equilibria involved in the cosolute binding have been studied extensively (34-38),
while the gelation observed under suitable conditions (39) has been
used as the basis of a striking teaching and lecture-demonstration
experiment (40, 41).
    Another type of this liaison effect which is technically impor-
tant (e.g., in petroleum extraction processes) is the gelation or
precipitation of polyelectrolytes such as partially hydrolysed poly-
(acrylamide) by transition metals such as Cr(III). However, in this
case the presence of the ionic groups on the chain and their neutralization by the bound cosolute ions are complicating factors in interpreting even the extreme dilution behavior.

8. Concluding Remarks

    In the present paper it is shown that the concept of noncoval-
ent reversible crosslinks or "liaisons" can be used quantitatively,
to interpret and explain a number of aspects of the behavior of
water-soluble polymers, that is, for simple homopolymers alone in
solution, for copolymers with association between the monomer units,
and for cosolute-binding systems where the bound cosolute molecules
can induce liaison formation. In this respect, the concept of liaisons is not a novel one, but represents the extension of a concept
already well-established (i.e., "long-range interactions") and fundamental to the Flory theory of polymer solutions, to these more complex systems. In case of copolymers, application of liaison theory
has led to a hydrophobic bonding scale for alkyl groups which should
be useful in interpreting the behavior of these technically important materials, as well as being applicable more widely in molecular
biology and other systems involving aqueous solutions.

Literature Cited

1. Molyneux, P.; Vekavakayanondha, S. J. Chem. Soc. Faraday Trans.
    I 1986, 82, 291-317.
2. Molyneux, P.; Vekavakayanondha, S. J. Chem. Soc. Faraday Trans.
    I 1986, 82, 635-661 and 3287 (correction).
3. Hildebrand, J. H.; Scott, R. L. The Solubility of Nonelectroly-
    tes; Dover: New York, 1964.
4. Speakman, J. C. The Hydrogen Bond and Other Intermolecular
    Forces; The Chemical Society; London, 1975.
5. Tanford, C. The Hydrophobic Effect: Formation of Micelles and
    Biological Membranes; Wiley-Interscience: New York, 2nd edi-
    tion, 1980.
6. Bender, C. Chem. Soc. Rev. 1986, 15, 475-502.
7. Flory, P. J. (a) Principles of Polymer Chemistry; Cornell Univ-

ersity Press: Ithaca, NY, 1953; Chapter XII (b) Discuss. Faraday Soc. 1970, 49, 7-29.

8. Morawetz, H. Macromolecules in Solution; Wiley-Interscience: New York, NY; 2nd edition, 1975, Chapter IV.
9. Stahl, N.; Jencks, W. P. J. Amer. Chem. Soc. 1986, 108, 4196-4205.
10. Bernal, P. J.; Christian, S. D.; Tucker, E. E. J. Solut. Chem. 1986, 15, 1031-1039 and earlier papers.
11. Franks, F.; Desnoyers, J. E. Water Sci. Rev. 1985, 1, 171-190.
12. Oakenfull, D.; Fenwick, D. E. J. Chem. Soc. Faraday Trans. I 1979, 75, 636-645.
13. Char, K.; Frank, C. W.; Gast, A. P.; Tang, W. T. Macromolecules 1987, 20, 1833-1838.
14. Oyama, H. T.; Tang, W. T.; Frank, C. W. Macromolecules 1987, 20, 1839-1847.
15. Yamazaki, I.; Winnik, F. M.; Winnik, M. A.; Tazuke, S. J. Phys. Chem. 1987, 91, 4213-4216.
16. Kuhn, W.; Majer, H. Makromol. Chem. 1955, 18/19, 239-253.
17. Kuhn, W.; Balmer, G. J. Polym. Sci. 1962, 57, 311-319.
18. Elias, H. G. Macromolecules. 1. Structure and Properties; Wiley: London, 1977; p 360.
19. Molyneux, P. Water-Soluble Synthetic Polymers: Properties and Behavior; CRC Press: Boca Raton, FL, 1984; Vol. II, Chapter 3.
20. Molyneux, P. Water-Soluble Synthetic Polymers: Properties and Behavior; CRC Press: Boca Raton, FL, 1983; Vol. I, Chapter 4.
21. Beresniewicz, A. J. Polym. Sci. 1959, 34, 63-79.
22. Natrosol Plus Hydrophobically Modified Hydroxyethylcellulose, Technical Bulletin, Hercules Inc: Wilmington, DE, 1986.
23. Gelman, R. A.; Barth, H. G. In Water-Soluble Polymers - Beauty with Performance; Glass, J. E., Ed.; ACS Symposium Series No. 213; American Chemical Society: Washington, DC, 1986; Chapter 6.
24. Molyneux, P. Water-Soluble Synthetic Polymers: Update I; Macrorophile Associates: London, 1987.
25. Molyneux, P. Water-Soluble Synthetic Polymers: Update II; Macrorophile Associates: London, 1989.
26. Priest, J. H.; Murray, S. L.; Nelson, R. J.; Hoffman, A. S. In Reversible Polymeric Gels and Related Systems; Russo, P. S., Ed.; ACS Symposium Series 350; American Chemical Society: Washington, DC, 1987, Chapter 18.
27. Takahashi, S.-I.; Fujimoto, T.; Miyamoto, T.; Inagaki, H. J. Polym. Sci. A - Polym. Chem, 1987, 25, 987-994.
28. Ballard, M. J.; Buscall, R.; Waite, F. A. Polymer, 1988, 29, 1287-1293.
29. Molyneux, P.; Cornarakis-Lentzos, M. Colloid Polym. Sci. 1979, 257, 855-873.
30. Butler, J. A. V. Trans. Faraday Soc. 1937, 33, 229-236.
31. Nemethy, G.; Scheraga, H. A. J.Phys.Chem. 1962, 66, 1773-1789.
32. Molyneux, P.; Rhodes, C. T.; Swarbrick, J. Trans. Faraday Soc. 1965, 61, 1043-1052.
33. Smith, R.; Tanford, C. Proc. Natl. Acad. Sci. USA 1973, 70, 289-293.
34. Ochiai, H.; Kurita, Y.; Murakami, I. Makromol. Chem. 1984, 185, 167-172.

35. Matsuzawa, S.; Yamaura, K.; Tanigami, T.; Somura, T.; Nakata, M.
     Polym. Commun. 1987, 28, 105-6.
36. Ochiai, H.; Kohno, R.; Murakami, I. Polym. Commun. 1986, 27,
     366-368.
37. Shibayama, M.; Sato, M.; Kimura, Y.; Fujiwara, H.; Nomura, S.
     Polymer 1988, 29, 336-408.
38. Sinton, S. W. Macromolecules 1987, 20, 2430-2441.
39. Shibayama, M.; Yoshizawa, H.; Kurokawa, H.; Fujiwara, H.;
     Nomura, S. Polymer 1988, 29, 2066-2071.
40. Cassasa, E. Z.; Sarquis, A. M.; Van Dyke, C. H. J. Chem. Educ.
     1986, 63, 57-60.
41. Sarquis, A. M. J. Chem. Educ. 1986, 63, 60-61.

RECEIVED June 4, 1990

Chapter 16

Probes of the Lower Critical Solution Temperature of Poly(*N*-isopropylacrylamide)

Howard G. Schild

Polymer Science and Engineering Department, University
of Massachusetts, Amherst, MA 01003

Cloud point, microcalorimetric, and fluorescent probe methods were in good agreement in determining the lower critical solution temperature (LCST) of an aqueous solution of a high molecular weight ($\bar{M}_n$ = 160,000), polydisperse ($\bar{M}_w/\bar{M}_n$ = 2.8) sample of poly(N-isopropylacrylamide) (PNIPAAM). The cloud point (32.2°C) concurred with the temperatures of abrupt changes observed in the emission spectra of micromolar quantities of solubilized fluorophores: the I_1/I_3 ratio of pyrene and the emission maxima of 1-pyrenecarboxaldehyde (pycho), sodium 2-(N-dodecylamino)naphthalene-6-sulfonate ($C_{12}NS$), and ammonium 8-anilinonaphthalene-1-sulfonate (ANS) successfully monitored the formation of the precipitated polymer phase. The observed microcalorimetric endotherm had a peak maximum of 32.4 ± 0.1°C and an enthalpy on the order of the strength of hydrogen bonds.

Poly(N-isopropylacrylamide) (PNIPAAM, **1**) exhibits a lower critical solution temperature (LCST) in aqueous solution; our work has focused on characterizing perturbations of the accompanying change in polymer solubility by the addition of cosolutes and by copolymerization.

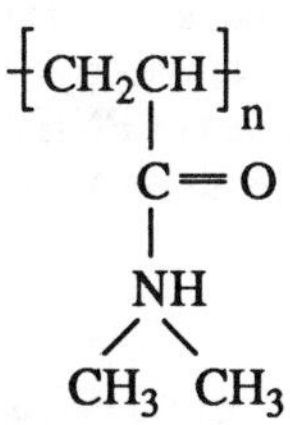

$$\begin{array}{c} \left. \begin{array}{c} CH_2CH \\ | \end{array} \right\}_n \\ C = O \\ | \\ NH \\ \diagup \ \diagdown \\ CH_3 \ \ CH_3 \end{array} \qquad (1)$$

The LCST is a consequence of the hydrogen bonding present in the system (1,2). Although exothermic hydrogen bond formation between polymer and solvent molecules lowers the free energy of mixing, the specific orientations required by these bonds lead to a negative entropy change and thus to a positive contribution to the free energy.

0097–6156/91/0467–0249$06.00/0
© 1991 American Chemical Society

Hydrophobic effects that are also operative as a consequence of the presence of the N-isopropyl group in this polyacrylamide contribute an additional negative entropic change on dissolution (3). This leads to domination over the enthalpic contribution to the free energy as the temperature of the solution increases. Experimentally, as this occurs below the boiling point, one observes precipitation (or a coil-to-globule transition in very dilute solutions (4-6)) of PNIPAAM as polymer-polymer and water-water contacts become preferred to those between unlike species.

Previous investigations of aqueous PNIPAAM indeed demonstrated that theta conditions exist near room temperature, although an expanded polymer conformation relative to those observed in organic solvents was evident (7,8). Recent light scattering experiments have shown that the LCST transition does involve a collapse in coil dimensions prior to phase separation (5,6). The thermal response of PNIPAAM in aqueous media has been exploited for diverse practical applications. Single chains conjugated with monoclonal antibodies have been used as an immunoassay system for the corresponding antigens (9). Crosslinked hydrogels of PNIPAAM exhibit a macroscopic swelling transition (10,11) that has been applied in the selective removal and delivery of species in aqueous solution (12-14). PNIPAAM membranes are novel as they do not possess an Arrhenius response in their permeability (15); colloids incorporating PNIPPAM show both temperature-sensitive stability (16) and electrophoretic mobility (17).

In all of these instances, not only are water and PNIPAAM present, but also variations in composition and structure in the system certain to influence both hydrogen bonding and hydrophobic associations. Therefore, the LCST of PNIPAAM will be perturbed if the interactions between the polymer and water are sufficiently modified. One of our goals is to increase the understanding and design of PNIPAAM-based devices; one cannot deliver a substance that prefers to bind to PNIPAAM. In addition, "superfluous" cosolutes not directly involved with the intended use of the PNIPAAM, such as sodium n-dodecyl sulfate, have been reported to interfere with various applications of PNIPAAM (9,16). We approach this complex system by investigations that manipulate only a single variable such as through study of ternary solutions of PNIPAAM, water, and a cosolute. This also allows exploration of distortions in the water structure present, to obtain data that may be applied to increase understanding of this ubiquitous yet mysterious medium. However, we must first demonstrate the precision and validity of our methods to probe the LCST in binary solutions of single chains of PNIPAAM. This report concerns such an initial exploration of techniques to monitor the LCST of PNIPAAM solutions, and focuses on classical cloud point measurements, microcalorimetry and fluorescence spectroscopy. As we will intensively apply fluorescence methods to the study of mixtures of PNIPAAM and surfactants, we must first elucidate the nature of the interactions with binary polymer solutions alone.

Experimental

Materials. N-isopropylacrylamide was obtained from Eastman Kodak Co. and recrystallized (mp 64-66°C) from a 65/35 mixture of hexane and benzene (Fisher Scientific Co.). Acetone (HPLC grade), sodium carbonate, and hydrochloric acid (HCl) were also obtained from Fisher. Azobisisobutyronitrile (AIBN) from Alfa Chemical Co. was recrystallized from methanol (Aldrich Chemical Co.), avoiding decomposition by maintaining the temperature below 40°C. Pyrene, 1-pyrenecarboxaldehyde, 9-fluorenemethanol (99%), tetrahydrofuran (THF, HPLC grade), triethylamine (Gold Label), acryloyl chloride (98%), chloroform (HPLC grade), p-methoxyphenol and methanol

(HPLC) were used as received from Aldrich Chemical Co. The sodium salt of 2-(N-dodecylamino)naphthalene-6-sulfonic acid ($C_{12}NS$) was used as received from Molecular Probes, Inc. 8-Anilino-naphthalene-1-sulfonic acid, ammonium salt (ANS) was obtained as puriss. biochem. (> 99%) grade from Fluka Chemical Corporation. Distilled water was analyzed (Barnstead Co., Newton, MA) to contain 0.66 ppm total ionized solids (as NaCl) and 0.17 ppm total organic carbon (as C). Sodium azide (Fisher) was employed as a bactericide (0.10 w/v %) in stock polymer solutions. Magnesium sulfate and ethyl ether were obtained from Mallinkrodt.

Synthesis of PNIPAAM. N-isopropylacrylamide (5 g), dissolved in 40 ml benzene with 1 mole % recrystallized AIBN, was degassed through three cycles of freezing and thawing. After polymerization by stirring in an oil bath at 49°C for 22 hr under a positive nitrogen pressure, the solvent was evaporated. The crude solid was vacuum-dried, crushed, dissolved (acetone, 47 ml), and precipitated by dropwise addition to hexane (600 ml). Upon filtering and drying, 3.62 g (76% yield) of polymer was obtained. Anal. calcd. for $C_6H_{11}NO$: C, 63.7%; H, 9.8%; N, 12.4%. Found: C, 63.5%; H, 9.9%; N, 12.2%. ^{1}H NMR (200 MHz, D_2O) δ: 1.0 (CH_3, 6H), 1.2-2.1 (–CH_2CH–, 3H), 3.7 (CH, 1H). No vinyl protons were detected. IR ($CHCl_3$ cast film) cm^{-1}: 3300, 2960, 2925, 2860, 1635, 1530, 1455, 1375, 1390, 1170, 1130, 750. Absent were the 1620 cm^{-1} (C=C), 1410 cm^{-1} (CH_2=), and C–H vinyl out of plane bending vibrations observed in the spectrum of the monomer. GPC: $\overline{M}_w$ = 440,000, $\overline{M}_n$ = 160,000, $\overline{M}_w/\overline{M}_n$ = 2.8.

Synthesis of Fluorene-Labelled PNIPAAM. Nitrogen was bubbled through a solution of 9-fluorenemethanol (9.81 g, 50 mmol) in THF (350 ml) at 0°C with stirring. Triethylamine (9.8 ml, 70 mmol) was then added by syringe. Afterwards, acryloyl chloride (6.0 ml, 74 mmol) in THF (35 ml) was added dropwise over 10 min with magnetic stirring. Stirring followed for 31 hr under nitrogen, allowing the system to gradually reach room temperature. Triethylamine hydrochloride was filtered off and the solvent was removed on the rotary evaporator. The oil was vacuum dried, and then dissolved in a mixture of water (100 ml), ether (200 ml), and HCl (ca. 2.4 N, 100 ml). The organic layer was washed with 10 w/v % sodium carbonate solution (100 ml) and then with water until the pH of the washings was neutral. The ether layer was dried with magnesium sulfate, filtered, and the solvent removed on the rotary evaporator to leave a crude solid. The product repeatedly polymerized to insoluble particles upon attempting recrystallizations even in the dark. Therefore, p-methoxyphenol (92 mg) was added as an inhibitor and the remaining crystals were dissolved in methanol and filtered with no polymerization occurring. The filtrate was ultimately vacuum dried to yield ca. 4 g (ca. 33% yield) of crystalline 9-fluorenemethylol acrylate. Anal. calcd. for $C_{17}H_{14}O_2$: C, 81.6%; H, 5.6%. Found: C, 81.4%; H, 5.7%. ^{1}H NMR (300 MHz, $CDCl_3$) δ: 7.8 (2H, d), 7.6 (2H, d), 7.4 (4H, m), 6.5 (1H, m), 6.3 (1H, m), 5.9 (1H, m), 4.5 (2H, d), 4.3 (1H, t). Signals were also observed for p-methoxyphenol. IR ($CHCl_3$ cast film) cm^{-1}: 3430 (br), 3070, 3040, 2950, 2900, 1950, 1930, 1730, 1640, 1620, 1520, 1480, 1460, 1420, 1380, 1300, 1280, 1190, 1110, 1060, 990, 920, 820, 740.

Nitrogen was bubbled for 15 min through a solution of recrystallized PNIPAAM (ca. 5 g), 9-fluorenemethylol acrylate (monomer feed of 105 mg (0.93 mol %)), and recrystallized AIBN (ca. 90 mg, ca. 1.2 mol %) in benzene (100 ml). The polymerization was then done under nitrogen while stirring the flask in an oil bath at ca. 50°C for 24 hr. The benzene was then removed on the rotary evaporator and the resulting solid was dissolved in chloroform (100 ml). Upon precipitation in hexane (ca. 1000 ml), the polymer was filtered and vacuum dried (ca. 85% yields). Copolymer composition was determined by UV spectroscopy using Beer's Law with a solution of the copolymer in methanol. The extinction coefficient of the 9-fluorenemethylol acrylate units was assumed to be that of 9-fluorene-methanol (ε_{300} = 6300) (18). The copolymer contained 0.42 ± 0.04 mol % of the fluorene-labelled repeating unit. Molecular weight distributions of the homopolymer and copolymer were indistinguishable by GPC.

Sample Preparation. The PNIPAAM concentration in all measurements was 0.40 mg/ml. Stock solutions of concentration 4.00 mg/ml were prepared through dissolution of the polymer (PNIPAAM or fluorene-labelled PNIPAAM) in distilled water with 0.1 w/v % sodium azide for several days. This polymer solution (0.20 ml) was diluted to 2.00 ml with distilled water. Stock solutions for fluorescence measurements were prepared as described above, with slight modification for probe incorporation. Where pyrene or 1-pyrenecarboxaldehyde (pycho) were used, microliters of a millimolar acetone solution were added to sample vials with subsequent evaporation of the acetone prior to the addition of the polymer solution. Enough probe was incorporated for ca. 1 μM aqueous solution. For $C_{12}NS$, samples were prepared by diluting the PNIPAAM stock solution with replacement of 0.10 ml distilled water with 0.10 ml of a 12.6 μM $C_{12}NS$ stock solution (final concentration 0.63 μM). An aqueous stock solution (1.80 ml) of the ammonium salt of 8-anilino-naphthalene-1-sulfonic acid (ANS) was added to 0.20 ml polymer stock solution to yield 290 μM ANS. All samples were vortexed prior to spectral measurements.

Measurements. Infrared spectra were obtained on films cast from chloroform on NaCl plates with a Perkin-Elmer 1320 infrared spectrophotometer. NMR spectra were obtained on a Varian XL-300 spectrometer. Gel permeation chromatography (GPC) was performed with a Waters M45 solvent pump coupled to a R410 differential refractometer and a Hewlett-Packard 3380A digital integrator. Degassed tetrahydrofuran (THF, Aldrich, HPLC grade) was eluted at 1.1 ml/min through four Waters μstyragel columns (10^6, 10^5, 10^4, 10^3 Å). N,N-dimethylformamide (DMF, Aldrich, HPLC grade) was eluted at 1.0 ml/min through three Waters μbondagel columns (E-1000, E-500, E-125). Polystyrene standards (Polysciences) were used for calibration; molecular weights are thus estimated as those of polystyrenes of equivalent elution volume.

Optical density (OD) measurements were done at 500 nm on a Beckman DU-7 spectrophotometer with a water-jacketed cell holder coupled with a Lauda RM-6 circulating bath. Temperatures were manually ramped at rates of ca. 0.5°C/min and monitored by an Omega 450-ATH themistor thermometer. Cloud points were taken as the initial break points in the resulting optical density versus temperature curves and were independent (to within ± 0.5°C) of slight fluctuations in the heating rate.

Calorimetric (DSC) scans were obtained on a Microcal, Inc. MC-1 scanning microcalorimeter at a heating rate of 15°C/hr unless indicated. Samples were degassed and transferred to the sample cell with a calibrated syringe. Polymer-free solutions of the same solvent composition were similarly placed in the reference cell. Calibration was achieved by supplying a precisely known current to the reference cell of the calorimeter. LCSTs and transition widths are accurate to within ± 0.1°C; enthalpies (ΔH) and relative peak heights (ΔC_p) are reproducible to within ± 2 units (cal/g of polymer for ΔH and cal/°C-g of polymer for ΔC_p).

Pyrene, $C_{12}NS$, pycho, and ANS emission spectra were obtained with a Perkin-Elmer MPF-66 fluorescence spectrophotometer exciting at 337, 303, 365.5 and 377 nm, respectively. Slit widths were 3 nm for pyrene, 5 nm for pycho and $C_{12}NS$, and 10 nm for ANS. Emission (using an excitation wavelength of 289.7 nm) spectra for NRET experiments were obtained using 5 nm slit widths unless otherwise indicated. The temperature control system for fluorescence measurements was identical to that used for cloud point measurements.

Results and Discussion

Polymer Synthesis. Free-radical polymerization of N-isopropylacrylamide in benzene with AIBN as initiator results in precipitation of PNIPAAM. The conditions used in this work yielded a high molecular weight, polydisperse sample ($\overline{M}_w$ = 440,000; $\overline{M}_n$ = 160,000; based on polystyrene standards). We used this sample throughout these studies at a fixed concentration of 0.40 mg/ml for experimental control.

The fluorene comonomer was synthesized from the reaction between 9-fluorene-methanol and acryloyl chloride in THF. Elemental, infrared, and ^{1}H NMR analyses were consistent with this previously unreported monomer. The purification of the fluorene-based monomer was complicated by its apparent facile polymerization. We did not extensively analyze our side product; however, its insolubility (in water and a wide range of organic solvents), and the absence of vinyl protons in NMR of suspensions (signals only at δ < 3 ppm) suggest an undesired homopolymerization. We were able to inhibit this reaction with added p-methoxyphenol and with the stringent absence of light. The polymerization was done similar to that used to prepare the homopolymer with the initiator concentration adjusted to yield a copolymer of essentially the same molecular weight distribution. The copolymer contains 0.42 mol % of the fluorene moiety, corresponding to ca. 1 fluorene unit every 238 repeat units.

Cloud Point Measurements. Figure 1 illustrates a typical cloud point curve for the PNIPAAM sample examined in this work; the optical density of the solution abruptly increases as the polymer precipitates. Previous studies of PNIPAAM solutions usually applied such a measurement (6,19,20); however, the thermodynamic interpretation of this observation is not straightforward. Taylor and coworkers (21) have reported that dilute PNIPAAM solutions heated above their LCST do not show two distinct layers. This was unlike the expected consequence of phase equilibria yielding one polymer-rich and one solvent-rich layer as was observed with poly(2-ethyl-2-oxazoline) solutions, which also possess an LCST in water. Moreover, the photon correlation spectroscopy of Yamamoto, Iwasaki and Hirotsu (5) has indicated that collapse in the hydrodynamic radius occurs several degrees lower than that of phase separation. Finally, the research group of Hoffman has observed that modification of PNIPAAM can result in an inability to visually detect aggregation above the LCST at all due to the formation of submicron-size particles (22). Therefore, cloud point measurements for determination of the LCST suffer as a result of dependence on instrumental sensitivity to scattering.

Microcalorimetry. Although we could measure a precise cloud point (32.2 ± 0.5°C), we desired comparison with other methods to determine the LCST, as variation in the settling of PNIPAAM precipitate was also evident. Heskins and Guillet reported a calorimetric endotherm in the first major study performed with PNIPAAM (19); by applying a much more sensitive scanning microcalorimeter developed to study structural transitions in proteins, lipids, and nucleic acids, (23-26), we hoped to measure transition parameters with greater precision.

Figure 2 illustrates the calorimetric endotherm we could detect upon heating our PNIPAAM sample in aqueous solution at 15°C/hr. Defining the LCST as the position of the peak maximum (32.4 ± 0.1°C) leads to excellent independent verification of the cloud point results. The heat of the transition (ΔH) is ca. 13 cal/g of polymer which corresponds to ca. 1.5 kcal/mol of repeating units, consistent with the loss of ca. 1 hydrogen bond per repeating unit upon phase separation (27). Studies of the endotherm associated with the LCST of this sample (at 0.4 mg/ml) show that the peak shape (transition width ($\Delta T_{1/2}$) = 0.4 ± 0.1°C, relative height (ΔC_p) = 13 ± 1 cal/°C-g of polymer) and maximum (LCST) are independent of heating rate over a range from 3.3 to 30°C/hr. Transition kinetics are thus unimportant on the timescale of our experiments, and thermodynamic parameters may be obtained from our calorimetric measurements. Similar results were obtained at polymer concentrations of 0.40-4.00 mg/ml, consistent with the rather "flat" phase diagrams typical of aqueous polymer solutions in such narrow concentration regions (6,15). This is to be expected if the LCST phenomenon involves a coil-globule transition with subsequent aggregation as proposed (4).

The calorimetric endotherms associated with the LCST can be analyzed by a two-state model to provide a measure of the cooperativity of the transition (25,26). The size of the cooperative unit is defined as the ratio of the van't Hoff enthalpy to the calorimetric enthalpy calculated on the basis of the area under the observed peak. The van't Hoff enthalpy can be obtained either from ΔC_p (25) or $\Delta T_{1/2}$ (26). By either method, we obtain a value of ca. 430 repeating units for the size of the cooperative unit associated with the LCST of this sample. Given the uncertainty of our estimated molecular weight and the polydispersity of our sample, a cooperative unit of this magnitude can be interpreted only to mean that the process reported by microcalorimetry involves long segments, and perhaps the entire length, of the PNIPAAM chain.

Fujishige and coworkers (6) reported a similar enthalpy (ΔH = 10 cal/g of polymer) for the LCST transition of aqueous solutions of PNIPAAM, although their experimental methods were unreported. They also found the LCST to be independent of concentration over a similar range as our investigation. However, they claim the cloud point to be independent of chain length (molecular weights 5 - 840 x 10^4). This contrasts with our observations (Schild, H. G.; Tirrell, D. A. *J. Phys. Chem.*, in press) which demonstrate that not only does the LCST (determined by both microcalorimetry and cloud points) depend upon molecular weight, but so does the observed shape of the calorimetric endotherm. We discovered the LCST of PNIPAAM generally increasing with decreasing $\overline{M}_n$ in the range from $\overline{M}_n$ = 1.6 x 10^5 to $\overline{M}_n$ = 5.4 x 10^3 as expected theoretically (4).

Fluorescent Probes. Study of aqueous solutions has been enhanced by the addition of minute quantities of fluorescent species whose emission spectra are dependent upon the micropolarity of the environment within which they are solubilized (28,29). These are usually sparingly soluble hydrocarbons which when "free" in solution "wander" to the most nonpolar site present in the system. Although, they have been typically applied to elucidate the structure of microheterogeneous ordered assemblies such as micelles and vesicles or the polarity of solvent mixtures (28,29), we apply them to detect the LCST through their preferential solubilization into the less polar precipitated polymer phase. What we require is that they do not perturb the LCST; therefore, we desire species that do not bind to PNIPAAM below the LCST. These species would thus exhibit the same spectra below the LCST in PNIPAAM solution as they would in polymer-free water. The abrupt transition in their emission spectra as they partition into the less polar, precipitated phase at the LCST would thus serve as an indicator of the transition. We

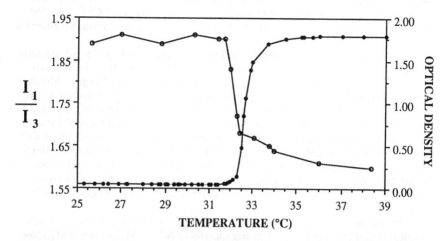

Figure 1. I_1/I_3 of pyrene (O, 1 µM) solubilized in aqueous PNIPAAM (0.40 mg/ml). $I_1/I_3 = 1.86 \pm 0.05$ in water alone, independent of temperature in this range. Also plotted is the optical density (●, measured at 500 nm) of aqueous PNIPAAM (0.40 mg/ml).

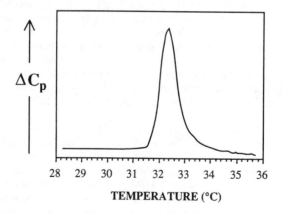

Figure 2.   Microcalorimetric endotherm for aqueous PNIPAAM (0.40 mg/ml) heated at 15°C/hr.

selected several fluorescent probes; none interacts with PNIPAAM below the LCST as indicated by the fact that cloud point and microcalorimetric results are unperturbed upon the addition of the cited concentration of probe. Moreover, the emission yield of all of these probes under the conditions used are high enough such that the spectral parameters we employ are unaffected by the increased scattering above the LCST. Spectral subtraction of a blank (no probe present) yields the same spectrum within experimental error.

Pyrene is perhaps the most widely used micropolarity indicator (28-30); the ratio (I_1/I_3) of the intensities of the 0,0 (373 nm) and 0,2 (385 nm) bands in the emission spectrum decreases as pyrene is solubilized in more nonpolar environments. Absolute results vary among different research groups; Street and Acree (30) have done a systematic study of experimental artifacts and recommend a concentration of 1 μM, use of reference calibrants, and a report of slit settings (cf. Experimental section) for a standard assay. Following such guidelines, we obtained an I_1/I_3 value of 1.86 ± 0.05 for 1 μM pyrene solubilized in water over the temperature range of interest (26-38°C). Winnik reports a continuous decrease of I_1/I_3 with increasing temperature (31); our experiments cover a much smaller range in temperature and we observe no decrease outside of the scatter of our data. We also obtained I_1/I_3 values of methanol (1.47), and ethanol (1.37) for reference.

In the presence of PNIPAAM (0.40 mg/ml) below the LCST, the value of I_1/I_3 is identical to that in water alone. This indicates that PNIPAAM does not bind pyrene in a less polar environment in contrast to early reports of Winnik (18,32) regarding hydroxypropylcellulose (HPC). These suggest that HPC does bind pyrene in aqueous solution below its LCST. On heating the aqueous PNIPAAM solution, we observe in Figure 1 a sharp drop in I_1/I_3 at the LCST determined by the cloud point method and microcalorimetry. The plateau value obtained ($I_1/I_3 = 1.60$) reflects a slightly more polar environment for the precipitated polymer phase than that of methanol. We can assume that the vast majority of the pyrene partitions into the polymer-rich phase; nonetheless, the emission intensity of pyrene also sharply increases as its micropolarity is lowered. Therefore, the emission we detect is probably only from the population of pyrene in the precipitate. Similar success in using free pyrene to detect the LCST of an aqueous polymer solution has been achieved for HPC by Winnik (31). This type of experiment is analogous to the use of pyrene to detect the pH-dependent conformational transition of hydrophobic poly(carboxylic acids) (33). In this case, microscopic hydrophobic sites form and solubilize pyrene as opposed to the macroscopic phase Winnik and we observe.

Figures 3 and 4 indicate that pycho, ANS, and $C_{12}NS$ can also detect the LCST of PNIPAAM in aqueous solution. Each exhibits an emission maximum below the LCST which concurs with the value for polymer-free solution, and a blue shift indicative (28,29) of solubilization in a less polar phase above the LCST. Stoichiometrically, 0.40 mg/ml of polymer corresponds to 3.5 μM polymer chains (based on an assumed degree of polymerization of 1000). With pyrene, pycho and $C_{12}NS$, we clearly have more chains than probes, but with ANS (290 μM) this is not so: ca. 80 molecules exist per polymer chain of 1000 monomer repeating units. Thus rephrased, this does not appear overwhelming, but there is probably an excess of ANS in free solution. However, these molecules contribute little to the observed emission as again the emission intensity sharply increases as ANS is solubilized in nonpolar solvents. Indeed, the emission of 1 μM ANS in water alone is too low for our instrumental detection. In the millimolar range, there is enough ANS to perturb the LCST. The LCST is elevated as bound negatively charged species oppose PNIPAAM collapse and aggregation.

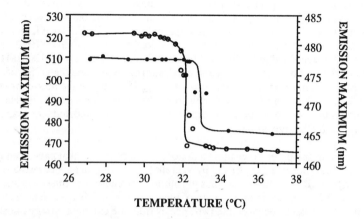

Figure 3.   Emission maxima for aqueous PNIPAAM (0.40 mg/ml) solutions of pycho (●, 1 μM) or ANS (○, 290 μM). The respective emission maxima in water alone are 477.6 ± 0.2 nm and 521 ± 1nm, independent of temperature.

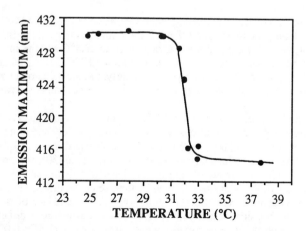

Figure 4.   Emission maxima for aqueous PNIPAAM (0.40 mg/ml) solutions of $C_{12}NS$ (0.63 μM). The emission maximum in water alone is 429 ± 1nm, independent of temperature.

Thus the concentration of ANS we apply is optimal for this system, but results cannot be accepted as unambiguously as with pycho and pyrene. The plateau emission maximum in $C_{12}NS$ (0.63 μM) above the LCST is 415 nm (Figure 4). Reference spectra obtained methanol (412 nm) and ethanol (409 nm) again demonstrate that the precipitated PNIPAAM has a polarity slightly above that of methanol. As $C_{12}NS$ is amphiphilic and interactions between nonionic polymers and surfactants are well known (34), we have explored (Schild, H. G.; Tirrell, D. A., to be submitted.) the interaction of higher concentrations of $C_{12}NS$ with PNIPAAM. We find that $C_{12}NS$ does bind to PNIPAAM (over the range $[C_{12}NS] = 10$-100 μM) with mutual perturbations of both the emission maximum of $C_{12}NS$ and the LCST of PNIPAAM. Thus in experiments where we use $C_{12}NS$ or any of the above probes to explore the interactions of PNIPAAM with species such as micelles and vesicles, we must ascertain that we are not seeing interactions promoted or hindered by modification of the polymer by the free probe.

NRET. Fluorene and pyrene constitute a donor and acceptor pair that have been used by Watanabe and Matsuda (35,36) and by Winnik (18) to investigate nonradiative energy transfer (NRET) between species in solution. Because the emission spectrum of fluorene overlaps the excitation spectra of pyrene, it is possible to excite the former and estimate the distance between the two species from the extent of energy transfer, as reflected by an increase in emission intensity of the acceptor. Using HPC with co-valently bound fluorene, Winnik (18) was able to show enhanced emission from added free pyrene. This was interpreted to indicate that on the average in polymer solutions pyrene resides in a polymer-rich environment. However, the HPC contained 1 fluor-ene unit per 33 monomer repeating units, a loading that perturbs the solution behavior of the homopolymer.

Indeed, we synthesized a fluorene-labelled PNIPAAM (PNIPAAM-Fl) that contained only 1 fluorene repeat per 238 units and found a change in the LCST relative to the homopolymer. The cloud point was depressed to 29.4°C and the shape of the microcalorimetric endotherm was altered (maximum = 30.7°C, 31.7°C (shoulder), $\Delta T_{1/2} = 2.6$°C, $\Delta C_p = 5$ cal/°C-g of polymer). Clearly, the inclusion of a small amount (0.42 mol %) of a more hydrophobic monomer can yield a noticable decrease in solubility.

We nonetheless proceeded with the NRET experiment. Aqueous PNIPAAM-Fl (4.0 mg/ml) and 1 μM pyrene solutions were prepared separately and as a mixture. Excitation of the mixture and both experimental controls was done at 289.7 nm at 24.5 ± 0.1°C. The emission intensity associated with fluorene (305 nm) and pyrene (377 nm) measured in the mixture of the fluorene-labelled polymer and pyrene was found to be the sum of the intensities emitted by their separate, directly excited solu-tions. A lack of a decrease in the fluorene emission intensity and lack of an enhance-ment in the pyrene emission indicates no NRET occurs. Furthermore, direct excitation (at 337 nm) of pyrene alone in the mixture yields an I_1/I_3 value (1.86 ± 0.03) equiva-lent to that quoted above for pyrene in water or aqueous PNIPAAM (A). We conclude that pyrene is not preferentially bound to the PNIPAAM in any of our aqueous solutions at 24.5°C; as noted above, this differs from the studies of HPC by Winnik (18) where enhancement of the pyrene intensity in the presence of labelled polymer is observed. This not only serves as a source of comparison between the solution behavior of the two polymers, but also gives us further evidence that pyrene is a valid probe of the LCST of PNIPAAM; it does not even bind to a slightly more hydrophobic PNIPAAM copolymer. Our attempts to do NRET experiments at temperatures above the LCST show evidence of NRET as can be expected but were inconclusive since in this case the contribution from scattering from these turbid solutions was overwhelming compared to the observed fluorescence.

Conclusions

We have found three general methods to monitor the LCST of PNIPAAM in aqueous solution. Classical cloud point measurements that are sometimes ambiguous can be supplemented by microcalorimetry and fluorescence. Use of a differential scanning microcalorimeter permits the precise measurement of thermodynamic parameters associated with the transition; micropolarity-sensitive fluorescence probes report the LCST through solubilization in the precipitated polymer phase. These basic experiments are necessary for exploration of modified aqueous PNIPAAM systems that contain hydrophobic comomomers (37) or surfactants (38) that also lead to the formation of less polar sites (even below the LCST) that can competitively solubilize the fluorescent probes. For the binary mixture of PNIPAAM and water, we have demonstrated agreement among the three techniques; this creates a sound foundation for study of more complex systems.

Acknowledgments

I would like to thank Professor David A. Tirrell for his guidance during my stay at the University of Massachusetts. This work was supported by a National Science Foundation Graduate Fellowship and by a grant from the U.S. Army Research Office (DAAL03-88-K-0038).

Literature Cited

1. Walker, J. A.; Vause, C. A. Scientific American 1987, 253, 98.
2. Tager, A. Physical Chemistry of Polymers; Mir Publishers: Moscow, 1972.
3. Tanford, C. The Hydrophobic Effect , 2nd Edition; J. Wiley and Sons: New York, 1973.
4. Williams, C.; Brochard, F.; Frisch, H. L. Ann. Rev. Phys. Chem. 1981, 32, 433.
5. Yamamoto, I.; Iwasaki, K.; Hirotsu, S. J. Phys. Soc. Jap. 1989, 58, 210.
6. Fujishige, S.; Kubota, K.; Ando, I. J. Phys. Chem. 1989, 93, 3311.
7. Chiantore, O.; Guaita , M.; Trossarelli, L. Makromol. Chem. 1979, 180, 969.
8. Fujishige, S. Polymer J. 1987, 19(3), 297.
9. Monji, N.; Hoffman, A.S. Appl. Bioch. & Biotech. 1987, 14, 107.
10. Hirotsu, S. J. Phys. Soc. Jap. 1987, 56, 233.
11. Hirotsu, S. J. Chem. Phys. 1988, 88, 427.
12. Hoffman, A. S. J. Control. Release 1987, 6, 297.
13. Bae, Y. H.; Okano, T.; Hsu, R.; Kim, S. W. Makromol. Chem., Rapid Commun. 1987, 8, 481.
14. Freitas, R. F. S.; Cussler, E. L. Sep. Sci. & Tech. 1987, 22, 911.
15. Taylor, L. D.; Cerankowski, L. D. J. Polym. Sci., Pt. A: Polym. Chem. 1975, 13, 2551.
16. Pelton, R. H. J. Polym. Sci., Pt. A: Polym. Chem. 1988, 26, 9.
17. Pelton, R. H.; Pelton, H. M.; Morphesis, A.; Rowell, R. L. Langmuir 1989, 5, 816.

18. Winnik, F. M. Macromolecules 1989, 22, 734.
19. Heskins, M.; Guillet, J. E. J. Macromol. Sci.-Chem. 1968, A2(8), 1441.
20. Priest, J. H.; Murray, S. L.; Nelson, R. J.; Hoffman, A. S. ACS Symposium Series 1987, 350, 255.
21. Chen, F. P.; Ames, A.; Taylor, L. D. Proceedings IUPAC MACRO 82, 1982, p. 532.
22. Cole, C.; Schreiner, S. M.; Priest, J. H.; Monji, N.; Hoffman, A. S. ACS Symposium Series 1987, 350, 245.
23. Hinz, H. J. Methods in Enzymology 1986, 130, 59.
24. Jackson, W.; Brandts, J. Biochemistry 1970, 9, 2294.
25. Privalov, P. Pure & Appl. Chem. 1976, 52, 479.
26. Mabrey, S.; Sturtevant, J. M. Methods in Membrane Biology 1978, 9, 237.
27. Israelachvili, J. N. Intermolecular and Surface Forces; Academic Press: London, 1985.
28. Kalyansundaram, K. Photochemistry in Microheterogeneous Systems; Academic Press: New York, 1987.
29. Thomas, J. K. The Chemistry of Excitation at Interfaces; American Chemical Society: Washington, D.C., 1984.
30. Street, K. W.; Acree, W. E. Analyst 1986, 111, 1197.
31. Winnik, F. M. J. Phys. Chem. 1989, 93, 7452.
32. Winnik, F. M.; Winnik, M. A.; Tazuke, S. J. Phys. Chem. 1987, 91, 594.
33. Borden, K. A.; Eum, K. M.; Langley, K. H.; Tirrell, D. A. Macromolecules 1987, 20, 454.
34. Goddard, E. D. Colloids & Surfaces 1986, 19, 255.
35. Watanabe, A.; Matsuda, M. Macromolecules 1986, 19, 2253.
36. Watanabe, A.; Matsuda, M. Macromolecules 1985, 18, 273.
37. Schild, H. G.; Tirrell, D. A. Polymer Preprints 1989, 30(2), 342.
38. Schild, H. G.; Tirrell, D. A. Polymer Preprints 1989, 30(2), 350.

RECEIVED June 4, 1990

Chapter 17

Complexation Between Poly(N-isopropylacrylamide) and Sodium n-Dodecyl Sulfate

Comparison of Free and Covalently Bound Fluorescent Probes

Howard G. Schild

Polymer Science and Engineering Department, University
of Massachusetts, Amherst, MA 01003

Cloud point, calorimetric, and fluorescence probe methods were combined to examine aqueous mixtures of poly(N-isopropylacrylamide)
(PNIPAAM) and sodium n-dodecyl sulfate (SDS). Addition of the
surfactant to PNIPAAM (0.40 mg/ml) leads to a continuous elevation of
the LCST above an SDS concentration of 0.75 mm. This concentration
was essentially identical to the critical aggregation concentration (CAC)
of SDS in the presence of the polymer as detected by both free and
covalently bound fluorescent probes. Thus the binding of SDS micelles
to PNIPAAM is the driving force for the enhanced polymer solubility.
The CAC was ten-fold lower than the critical micelle concentration (CMC)
and could be successfully predicted by the theory of Nagarajan and
Ruckenstein.

On the basis of neutron scattering studies, Cabane and Duplessix ($\underline{1}$) have
proposed that single poly(ethylene oxide) (PEO) chains cooperatively bind a number of
miniature sodium n-dodecyl sulfate (SDS) micelles along the chain contour in aqueous
solution. Numerous investigations indicate that nonionic polymers in general complex
best with anionic surfactants ($\underline{2}$). Other techniques have been employed to determine
the critical aggregate concentration (CAC) at which these attached micelles form ($\underline{2}$).
    Our investigations have been exploring aqueous mixtures of the nonionic polymer
poly(N-isopropylacrylamide) (PNIPAAM, **1**) and diverse cosolutes.

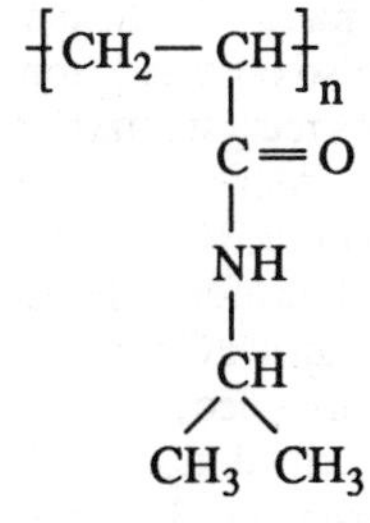

$$\left\{ CH_2 - \underset{\substack{| \\ C=O \\ | \\ NH \\ | \\ CH \\ \diagup \diagdown \\ CH_3 \ \ CH_3}}{CH} \right\}_n$$

(1)

0097–6156/91/0467–0261$06.00/0
© 1991 American Chemical Society

Complexation between PNIPAAM and surfactants such as SDS would have important implications for the emerging applications of the polymer (3-7). Moreover, though many studies of polymer-surfactant complexes measure the change in critical micellar concentration of the surfactant (2), fewer studies have probed the corresponding change in polymer conformation. Because the binding of anionic surfactant micelles converts nonionic polymers into charged species, polyelectrolyte behavior is anticipated. Evidence of characteristic polyelectrolyte viscosity behavior (2,8) has indeed been found in such systems, although the quantitative treatment of such findings is not straightforward (8). PNIPAAM exhibits a lower critical solution temperature (LCST) in water above which it precipitates upon heating (9,10). Since attached surfactant aggregates create an electrostatic barrier that should oppose polymer collapse and aggregation, one would also expect enhanced polymer solubility in surfactant solutions as detected by an elevated LCST. This phenomenon can not be observed in solutions of PEO, since its critical temperature lies above the boiling point of the solution (11). Comblike polymers (12) of PEO do display LCSTs, as does poly(vinyl acetate) (70% hydrolyzed) (13); both classes of polymers show elevated cloud point temperatures upon surfactant addition. However, neither study probed for changes in the state of surfactant association.

We report herein a systematic investigation of the interaction of PNIPAAM with SDS. Solution microcalorimetry (DSC) and cloud point methods are used to study the LCST of PNIPAAM. At the same time we apply fluorescence techniques to monitor changes in the critical micelle concentrations (CMCs) of each surfactant, in order to relate the association behavior of the surfactants to changes in the LCST. Fluorescence experiments typically employ free probes or, in some cases, amphiphilic probes (14). This investigation will use both of these approaches as well as a polymer-bound fluorophore to examine complexation.

The most closely related work has involved the interactions of another nonionic polymer, hydroxypropylcellulose (HPC) (15,16) with SDS. Other studies have been concerned with fluorescent, charged polymers (17-20) in the presence of amphiphiles. Using covalently-bound probes removes doubts concerning solubilization sites (18); however, such probes must be used cautiously as higher degrees of substitution can lead to additional contributions to hydrophobic driving forces for surfactant binding (17) or otherwise modify the original solution behavior of the polymer (20-24). Finally, the theory of Nagarajan and Ruckenstein (25-28) will be applied to predict the observed CAC.

Experimental

Materials. N-isopropylacrylamide (NIPAAM)was obtained from Eastman Kodak Co. and recrystallized (mp 64-66°C) from a 65/35 mixture of hexane and benzene (Fisher Scientific Co.). Acetone (HPLC grade) was also obtained from Fisher. Azobisisobutyronitrile (AIBN) from Alfa Chemical Co. was recrystallized from methanol (Aldrich Chemical Co.), avoiding decomposition by maintaining the temperature below 40°C. Tetrahydrofuran (THF, HPLC grade), triethylamine (Gold Label), acryloyl chloride (98%), benzene (Spectrophotometric grade), chloroform (HPLC grade), pyrene and 1-pyrenecarboxaldehyde were obtained from Aldrich Chemical Co.

Elemental analysis was done on sodium n-dodecyl sulfate obtained from J.T. Baker to verify its purity. The sodium salt of 2-(N-dodecylamino)naphthalene-6-sulfonic acid ($C_{12}NS$) and 1-pyrenemethanol were used as received from Molecular Probes, Inc. Hydrocarbons for surface tensiometry were pentane (HPLC grade,

Aldrich), hexane (OmniSolv, EM Science), heptane (HPLC grade, Fisher), octane (99+%, Aldrich), and decane (99+%, Aldrich). Distilled water was analyzed (Barnstead Co., Newton, MA) to contain 0.66 ppm total ionized solids (as NaCl) and 0.17 ppm total organic carbon (as C). Sodium azide (Fisher) was employed as a bactericide (0.10 w/v %) in stock polymer solutions.

Synthesis. The synthesis of the PNIPAAM homopolymer is described in detail in the previous chapter (14). For the synthesis of the pyrene-labelled PNIPAAM, we first made a pyrene-functionalized comonomer. 1-Pyrenemethanol (0.919 g, 3.96 mmol) was dissolved in THF (150 ml) and cooled to ca. 0°C while bubbling nitrogen through the solution. Triethylamine (0.69 ml, 4.96 mmol) was added by syringe. Afterwards, acryloyl chloride (0.41 ml, 4.96 mmol) in THF (15 ml) was added dropwise over 20 min with magnetic stirring. Stirring followed for 28 hr under nitrogen, allowing the system to gradually reach room temperature. The triethylamine hydrochloride was filtered off and the solvent was removed on the rotary evaporator. The crude solid was vacuum dried, triturated with methanol, filtered, and vacuum dried again. The reaction and work-up were done in the absence of light. The off-white crystals (0.18 g, 16% yield) did not melt but polymerized upon heating. ^{1}H NMR (300 MHz, CDCl$_3$) δ: 8.2 (9H, m), 6.48 (1H, m), 6.20 (1H, m), 5.94 (2H, s), 5.85 (1H, m). IR (CHCl$_3$ cast film) cm^{-1}: 3050, 2960, 2930, 2860, 1750, 1660, 1610, 1470, 1415, 1380, 1300, 1270, 1190, 1070, 1050, 970, 850, 820.

Nitrogen was bubbled for 15 min through a benzene solution (100 ml) of recrystallized NIPAAM (ca. 5 g), 1-pyrenemethylol acrylate (monomer feeds of 22 and 100 mg, respectively, 0.17 and 0.79 mol %), and recrystallized AIBN (ca. 73 mg, ca. 1 mol %). The polymerization then proceeded under nitrogen with stirring in an oil bath at 50°C for 22.5 hr. The benzene was removed on the rotary evaporator and the solids were dissolved in chloroform (100 ml). Upon precipitation in hexane (ca. 1000 ml), the polymer was filtered and vacuum dried (ca. 80% yields). Copolymer compositions were determined by UV spectroscopy using Beer's Law with solutions of the copolymers in methanol. The extinction coefficient of the 1-pyrenemethylol acrylate units was assumed to be that of 1-ethylpyrene (log e_{343} = 4.6) (22). The two copolymers possessed respectively 0.06 ± 0.01 and 0.36 ± 0.02 mol % of the pyrene-labelled repeating units. GPC (DMF): M_p = 580,000 ± 50,000 g/mol; molecular weight distributions of both copolymers and the homopolymer were indistinguishable. Table I summarizes characterization of the two copolymers and the homopolymer used in this work.

Sample Preparation. The PNIPAAM concentration in all measurements was 0.40 mg/ml. Stock solutions of concentration 4.00 mg/ml were prepared through dissolution of the polymer in distilled water with 0.1 w/v % sodium azide for several days. Refrigeration was required to obtain optically clear solutions of the more highly labelled sample. Aliquots (0.20 ml) of this polymer solution were diluted to 2.00 ml with distilled water or surfactant stock solutions.

Stock solutions for fluorescence measurements were prepared as described above, with slight modification for probe incorporation. Where pyrene or 1-pyrene-carboxaldehyde were used, microliters of a millimolar acetone solution were added to sample vials with subsequent evaporation of the acetone prior to the addition of the polymer solution. Enough probe was incorporated for ca. 1 mM aqueous solution. For C$_{12}$NS, samples were prepared by diluting the PNIPAAM stock solution with replacement of 0.10 ml distilled water with 0.10 ml of a 12.6 mM C$_{12}$NS stock solution (final concentration 0.63 mM). All samples were vortexed prior to spectral measurements.

Measurements. Infrared spectra were obtained on films cast from chloroform on NaCl plates with a Perkin-Elmer 1320 infrared spectrophotometer. NMR spectra were obtained on a Varian XL-300 spectrometer. Gel permeation chromatography (GPC) was performed with a Waters M45 solvent pump coupled to a R410 differential refractometer and a Hewlett-Packard 3380A digital integrator. Degassed tetrahydrofuran (Aldrich, HPLC grade) was eluted at 1.1 ml/min through four Waters μstyragel columns (10^6, 10^5, 10^4, 10^3 Å). N,N-dimethylformamide (DMF, Aldrich, HPLC grade) was eluted at 1.0 ml/min through three Waters μbondagel columns (E-1000, E-500, E-125). Polystyrene standards (Polysciences) were used for calibration; molecular weights are thus estimated as those of polystyrenes of equivalent elution volume. Copolymer composition (343 nm) and optical density (OD) measurements (500 nm) were acquired on a Beckman DU-7 spectrophotometer with a water-jacketed cell holder coupled with a Lauda RM-6 circulating bath, monitored by an Omega 450-ATH thermistor thermometer.

Our standard approach for obtaining cloud points is described in the previous chapter (14). The same temperature control system was used for 90° scattering studies, which employed a Perkin-Elmer MPF-66 fluorescence spectrophotometer. These measurements utilized 500 nm as both the excitation and emission wavelength and monitored the change in scattering intensity observed at 90°. Calorimetric transition temperatures were obtained to within ± 0.1°C with a Microcal, Inc. MC-1 scanning microcalorimeter (DSC) at a scanning rate of 15°C/hr as in the previous chapter (14).

Pyrene, $C_{12}NS$, 1-pyrenecarboxaldehyde, and polymer-bound pyrene emission spectra were obtained with a Perkin-Elmer MPF-66 fluorescence spectrophotometer exciting at 337, 303, 365.5 and 344.2 nm, respectively. Slit widths were 3 nm for pyrene and 5 nm for the other probes. All measurements were done at 24.5 ± 0.5°C unless indicated, by coupling the spectrophotometer with a Lauda RM-6 circulating bath. Calculations were done using the TK Solver program on a Macintosh SE computer. Interfacial tensions were measured by the Wilhelmy plate method with a Rosano surface tensiometer (Biolar Corp.) and by the du Nouy ring method with a Fisher Model 21 Tensiomat following ASTM methods (29) at room temperature (ca. 24.5°C).

Results and Discussion

Polymer synthesis. Pyrene was selected as the polymer-bound fluorophore since it has been most widely applied in investigations of polymers in aqueous solution with covalently bound fluorophores (15-17,19-24,30,31). Monomeric label constructed by the reaction between 1-pyrenemethanol and acryloyl chloride in THF had IR and NMR spectra consistent with the expected structure. Using AIBN as initiator, the free-radical polymerization of N-isopropylacrylamide in benzene resulted in precipitation of high molecular weight, polydisperse PNIPAAM. Figure 1 illustrates these polymers; Table I summarizes their characterization. Polymerizations carried out under identical conditions for the homopolymer and the two copolymers yielded identical molecular weight distributions to within experimental error. This effectively eliminates one variable in our investigations. Further control was achieved by using a fixed polymer concentration of 0.40 mg/ml.

Table I. Characterization of PNIPAAM Polymers

Sample	Pyrene (mol %)[a]	Fluorophores per chain[b]	Cloud Point (°C)[c]	Microcalorimetry[d]			
				LCST(°C)	$\Delta T_{1/2}$ (°C)[e]	ΔH^f	ΔC_p^g
A	0	0	32.2	32.4	0.9	13	13
B	0.06	0.6	31.5	32.1 / 32.6	1.5	13	7
C	0.36	3.6	29.4	31.1/ 31.8	3.1	13	4

[a] Measured by UV spectroscopy.
[b] Number average value based on an assumed number average dp of 1000 from GPC.
[c] Obtained from the temperature of the onset of the increment in optical density.
[d] See previous chapter (14) for detailed explanations.
[e] Width of the endotherm at half the relative peak height.
[f] Calorimetric enthalpy of endotherm (cal/g of polymer).
[g] Relative peak height (cal/°C - g of polymer).

Aqueous solutions of PNIPAAM. The previous chapter (14) established that the coupling of microcalorimetry with cloud point measurements provides a reproducible and information-rich approach to exploring the LCST of PNIPAAM in aqueous solution. Less than one fluorophore per polymer chain is present in sample B (Table I); such a criterion has been suggested (21,22) for minimizing any changes in the aqueous solution behavior of the polymer. We checked this hypothesis for our system through solution microcalorimetry and cloud point measurements of the LCST associated with PNIPAAM. The summary in Table I reveals that slight depression of the LCST and a broadening in the transitions results from copolymerization. These observations can be attributed to the increased hydrophobicity associated with the introduction of pendant pyrene as the polymers have equivalent molecular weights. Interestingly, our copolymers (B and C, respectively) have 1 pyrene unit every 278 and 1667 monomer repeat units; Winnik (16,22-24) typically labels HPC at a level similar to that of sample C or even an order of magnitude more heavily; this resulted in a "double-LCST" for one such sample (24). Turro notes (17) that environmental response can not be explained according to the nature of the polymer alone at this level of labelling. Indeed, using fluorescence, Thomas (21) has observed differences in the conformational transition of poly(methacrylic acid) with positioning the pyrene unit at polymer ends versus randomly along the chain. The true test of whether pyrene-labelled PNIPAAM can be used to study the properties of PNIPAAM will be whether the three polymers exhibit equivalent responses to the addition of SDS.

Effects of SDS on the LCST of PNIPAAM. Eliassaf (32) did not observe any precipitation upon boiling a PNIPAAM solution containing 1% (ca. 35 mM) SDS. Indeed, our standard cloud point method (cf. Experimental Section) measures a change in optical density upon heating through the LCST only less than 10 µM SDS was in solution, i.e., ca. 6 SDS molecules per polymer chain (based on an assumed degree of polymerization of 1000). However, microcalorimetric results (Figure 2) support the persistence of the LCST up to considerably higher surfactant concentrations. The calorimetric enthalpies and peak heights are seen to decrease with increasing [SDS], while the temperature and width of the demixing transition increase rapidly. More

1-y
y
C=O
NH
C=O
O
CH2
A : y = 0
B : y = 0.0006
C : y = 0.0036

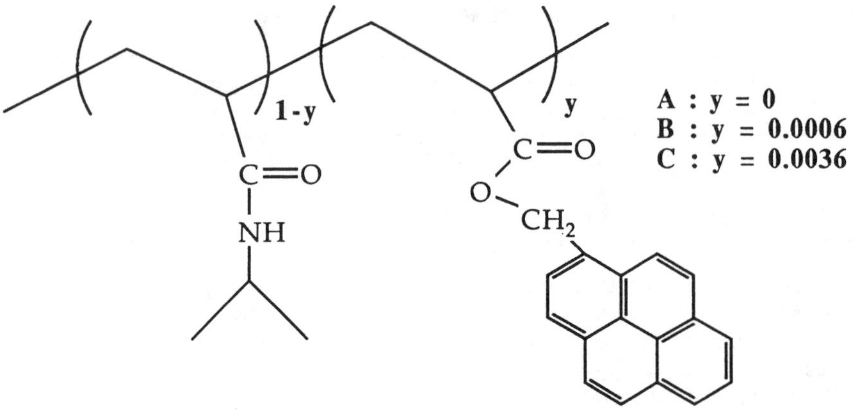

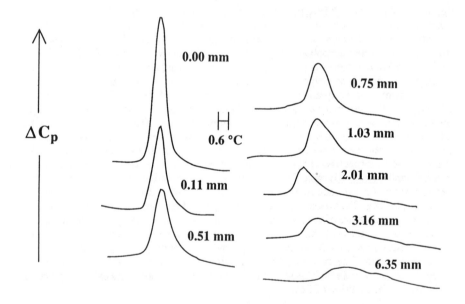

Figure 1. Structures of PNIPAAM polymers.

ΔCp
0.00 mm
0.6 °C
0.11 mm
0.51 mm
0.75 mm
1.03 mm
2.01 mm
3.16 mm
6.35 mm
TEMPERATURE (°C)

Figure 2. Microcalorimetric endotherms for aqueous PNIPAAM (0.40 mg/ml) with added SDS. Temperatures of peak maxima are plotted in Figure 3.

sensitive instrumentation resolves the discrepancy; measurement of 90° scattering (cf. Experimental Section) yields well-defined cloud points at least up to 5 mM SDS. The precipitated particles are apparently too small to cause visible turbidity at high surfactant concentrations (33). We can rationalize the retarded intermolecular aggregation as a consequence of the polymer-bound charged micelles leading to interparticle repulsion.

These cloud points agree very well with the calorimetric LCSTs (Figure 3). The rapid rise in the LCST with increasing SDS concentration is almost certainly associated with surfactant binding and intermicellar repulsion; above ca. 17 mM SDS, the LCST exceeds the boiling point. Moreover, we plot several points of the data set for the LCST curves of the pyrene-labelled PNIPAAM copolymers. The excellent overlap of the general shape of the curves shows that the same phenomenological response is present without any drastic quantitative differences; the degree of labelling is not high enough to perturb the characteristic behavior of PNIPAAM.

Free probes of the CAC. The ratio (I_1/I_3) of the intensities of the 0,0 (373 nm) and 0,2 (385 nm) bands in the emission spectrum abruptly decreases as pyrene is solubilized in the more hydrophobic environment of the micellar surfactant (2,16,34,35). The presence of pyrene at a concentration of 1 μM does not affect the LCST of PNIPAAM as determined by cloud point, microcalorimetric, and fluorescence methods (14). Turro and coworkers have shown that pyrene does not perturb the CMCs of anionic surfactants as determined by surface tension measurements (36). Thus use of this probe should not perturb the system. Data for micelle formation in polymer-free solution are illustrated in Figure 4a. As the surfactant concentration increases in solution, I_1/I_3 remains at its surfactant-free value until the CMC, where it abruptly decreases to a lower plateau value. In solutions of 0.40 mg/ml PNIPAAM, the inflection point of the abrupt drop in I_1/I_3 is shifted from 7.1 to 0.79 mm. We attribute the decrease in microenvironmental polarity in this case to the formation of polymer-bound micelles at a CAC some tenfold lower than the CMC. Pyrene has been solubilized at a nonpolar site under conditions where either PNIPAAM or SDS alone do not form such environments. The lower I_1/I_3 value seen in the concentration regime where only polymer-attached micelles exist (0.79-7.1 mm), compared with that of SDS micelles (> 7.1 mm in water alone), suggests the former species are less polar. This contrasts with reports on PEO and poly(vinyl pyrrolidone) (PVP) (35) solutions where I_1/I_3, in the presence of polymer, never drops below the value characteristic of SDS micelles. The minimum observed in PNIPAAM solutions must be associated with the greater hydrophobic character of the polymer (PNIPAAM possesses a lower LCST in aqueous solution than either PEO or PVP (2)). Above the CMC, I_1/I_3 averages to the micellar value since pyrene is partitioned in large part into free micelles. Goddard (2) has questioned the validity of the pyrene technique on the basis of a PVP/SDS study (35) that did not show this convergence in I_1/I_3.

The blue shift in the emission maximum of 1-pyrenecarboxaldehyde (1 μM) in nonpolar media can be used to monitor surfactant aggregation. In fact, Turro and coworkers (36) have suggested this probe reports CMCs in closer agreement with those obtained by surface tension measurements than pyrene does. As shown in Figure 4a, 1-pyrenecarboxaldehyde reports a CMC of 8.0 mm for SDS in water. In the PNIPAAM/SDS mixture, the initial break point in the curve (1.20 mm) lies closer to the pyrene-associated CAC (0.79 mm) than the inflection point (2.00 mm). Nonetheless, the complex still appears to be less polar than free micelles between the CAC and CMC, and we also find the averaging effect at higher concentrations. Given the likely

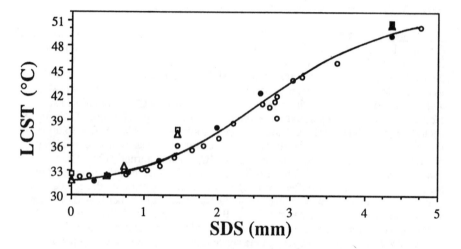

Figure 3.   LCSTs of PNIPAAM samples (0.40 mg/ml) in aqueous solutions
with added SDS. Filled symbols refer to cloud points, open symbols to microcalorimetric peak maxima. (O, ●) polymer A; (▢, ■) polymer B; (Δ, ▲) polymer C.

differences in the nature of the solubilization of the two probes (36), small differences of this order are to be expected.

Bound probes of the CAC. To more closely follow the SDS in solution, we selected the sodium salt of $C_{12}NS$ since it is an amphiphilic fluorophore. Waggoner and Stryer employed the octadecyl analog (ONS) as a polarity probe (37); in a similar manner, the emission spectrum of $C_{12}NS$ also blue shifts with decreasing polarity: micromolar solutions in water, 50/50 water/ methanol, methanol, and ethanol emit at 430, 421, 415 and 411 nm, respectively. Figure 4b shows the transition curves obtained in SDS solutions and in PNIPAAM/ SDS mixtures. The curves are similar to those obtained using pyrene, although lower critical surfactant concentrations (CAC of 0.67 mM and a CMC of 7.1 mM) are observed. At the concentration (0.63 µM) used in these experiments, $C_{12}NS$ itself has no effect on the LCST of PNIPAAM (14). Fluorescence spectra for pyrene-labelled PNIPAAM (0.40 mg/ml) are similar to those reported in the literature (20,21) for other types of polymers with bound pyrene; we observe similar spectra for samples B and C although higher intensities are measured for the latter at equivalent concentrations. However, most noteworthy at this pyrene concentration (ca. 2 mM for sample B) is the absence of excimer. Although excimers are not ordinarily seen in such dilute solu-tions for free pyrene, the local concentration can be much higher with polymer-bound pyrene. Indeed, Winnik (16) and Turro (17) apply the change in the excimer to monomer ratio of intensities in the emission spectra to probe polymer-surfactant complexation; both researchers used more highly labelled polymers than we have synthesized. Instead, we must rely on environmentally sensitive changes in the absolute intensity of the monomer emission (377 nm) as Thomas (21) and Tirrell (19) have; their polymers were labelled at levels similar to those of samples B and C. A lack of excimers does avoid secondary LCSTs resulting from their dissociation (24). No hydrolysis of bound pyrene was found in our investigations; no changes in the fluorescence, microcalorimetry or cloud point measurements were observed over the course of the experiments.

Figure 4b shows that abrupt increases in fluorescence intensity for both pyrene-tagged PNIPAAM (B and C) solutions result from the addition of SDS. These occur at approximately the same concentration (based on the inflection point (16)) as the free probe CAC transitions (Table II). Since the fluorescence intensity of pyrene increases in less polar environments, such increased emission is logical if the covalently-bound pyrene were solubilized inside attached SDS micelles. At higher concentrations of SDS, the intensity goes through a slight maximum prior to levelling off. As the poly-mer chain is expanded upon conversion into a polyelectrolyte (2,8), we might expect such conformations to result in greater exposure of the pendant pyrene to water. This would explain the slight decrease in pyrene intensity observed. Such a fine balance detected by locally sensitive probes has been discussed previously (17). We can conclude that bound-pyrene did not promote SDS aggregation since similar CACs are obtained for all three polymers.

Modelling polymer-surfactant complexation. Nagarajan has applied a multiple equilib-rium model (25-27) to the description of polymer-surfactant complexation. The model assumes competition between the formation of free micelles of aggregation number m with an equilibrium constant K_m and binding of micelles of aggregation number g with equilibrium constant K_b to z available polymer sites. Despite simplifying assumptions, this model serves as a guide to the cooperativity and extent of binding under defined conditions.

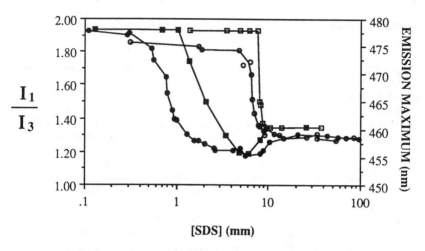

Figure 4a. I_1/I_3 of pyrene (○, ●) and emission maxima of 1-pyrenecarbox-
aldehyde (□, ■) with added SDS. Measurements were made with 1 μM probe at
24.5°C. Open symbols refer to water, filled symbols to aqueous PNIPAAM (0.40
mg/ml).

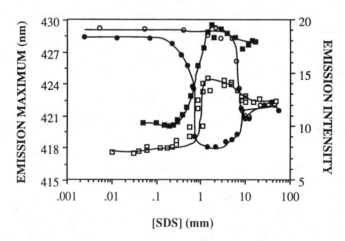

Figure 4b.   Emission maxima of $C_{12}NS$ (0.63 μM) in water (○) and in aqueous
PNIPAAM sample A (●). Emission intensities (377 nm) of PNIPAAM sample B
(□) and sample C (■). Measurements were made at 24.5°C with 0.40 mg/ml
polymer with added SDS.

Nagarajan and Ruckenstein endow this simple model with physical meaning by calculating the equilibrium constants directly from the free energy of aggregation (25-28):

$$K_i = \exp(-\Delta\mu_i/kT) \tag{1}$$

The free energy change to form micelles of size i in the absence of polymer is the sum of several contributions:

$$\Delta\mu_i = \Delta\mu_{HC/w} + \Delta\mu_C + \sigma(a - a_S) - kT\ln(1 - a_p/a) + \Delta\mu_{electr} \tag{2}$$

where $\Delta\mu_{HC/w} = -(2.05 + 1.49n)kT$, the free energy advantage conferred when the tail of the surfactant of chain length n is transferred from water to a liquid hydrocarbon medium. The second term, $\Delta\mu_C = -(0.50 - 0.24n)kT$, corrects for the constraining of these tails by attachment to polar headgroups "tied" to the micelle-water interface. The third term accounts for residual contact between the hydrocarbon core and water; σ represents (28) the hydrocarbon-water interfacial tension of 50 dyne/cm, a is the (unknown) optimal area per surfactant molecule at the micellar interface, and a_S is the area per surfactant molecule shielded from water by the polar head group (which for sulfates (28) is estimated as $a_p = 17$ Å^2, the polar headgroup area, since it is smaller than the cross-sectional area of the hydrocarbon chain ($a_h = 21$ Å^2)). The fourth term is due to surface exclusion of the head groups caused by their finite size. The final term is a Debye-Huckel approximation (28) to the electrostatic repulsion between the ionic head groups; it involves a number of molecular parameters, the ionic strength, and the degree of dissociation of the head groups, β.

Following Ruckenstein's approach (28), we set the derivative of the free energy expression (Equation 2) with respect to the average aggregation number, m, equal to zero:

$$\partial(\Delta\mu_i)/\partial i = 0 \qquad \text{for } i = m \tag{3}$$

The resulting equation is then iteratively solved through use of our experimental CMCs and thermodynamic relationships (28), to obtain β.

Ruckenstein (28) suggests that the polymer primarily contributes by shielding the hydrocarbon cores of the micelles from water. The third term of Equation 2 (for the case where $a_p < a_h$) then becomes

$$(\sigma - \Delta\sigma)(a - a_p) + a_p\Delta\sigma_p$$

where $\Delta\sigma$ and $\Delta\sigma_p$ are the changes in the interfacial tension between the hydrocarbon core and water and between the headgroups and water, respectively, caused by polymer. Ruckenstein takes these two parameters to be equal, and simplfies this term to

$$\sigma(a - a_p) - a\Delta\sigma + 2a_p\Delta\sigma$$

One then estimates $\Delta\sigma$ as the difference of the interfacial tension between water and hydrocarbon, and the interfacial tension between aqueous polymer solution and hydrocarbon. We measured $\Delta\sigma$ to be 33 ± 2 dyn/cm using 0.40 mg/ml PNIPAAM (as was used in all our experiments) and hydrocarbons ranging from pentane to decane. This value was independent of the hydrocarbon selected and of PNIPAAM concentration over two orders of magnitude. For PEO (28), $\Delta\sigma$ was found to be only 20 dyn/cm. Equation 2 can be solved for the CAC, subject to Equation 3, using the β calculated from the experimental CMC data (28) and the modified term that takes into account the interfacial properties of the polymer. We obtain (Table II) a calculated CAC of 0.73 ± 0.18 mm that exhibits good agreement with the experimental CACs also listed.

Relationship between the CAC and the LCST. Table II lists all of the above determined critical concentrations. The range of the CMC is in accord with literature reports (38). Excellent agreement is found between the CACs and the concentration at which the LCST rises above its surfactant-free value.

Table II. Summary of Critical Concentrations in Aqueous PNIPAAM/SDS Mixtures

	CMC	CAC
pyrene (1 μM)	7.1	0.79
1-pyrenecarboxaldehyde (1 μM)	8.0	1.20
$C_{12}NS$ (0.63 μM)	7.1	0.67
PNIPAAM-co-pyrene (0.06 mol %)	–	0.83
PNIPAAM-co-pyrene (0.36 mol %)	–	0.71
Nagarajan-Ruckenstein Model	8.2	0.73 ± 0.18
Concentration at which the LCST rises above the surfactant-free value	–	0.75

Superposition of plots of the concentration dependences of I_1/I_3 of free pyrene in PNIPAAM/SDS mixtures and the LCST (Figure 5) affords the best support that the transitions are coincident for added SDS. This clearly suggests that it is the *binding of micelles that causes elevation of the LCST*. We propose that elevation of the LCST is a result of electrostatic repulsion between charged, polymer-bound micelles, which oppose polymer collapse and aggregation.

Conclusions

Codissolution of PNIPAAM and SDS in aqueous media is accompanied by marked changes in the solubility properties of both polymer and surfactant. In general, PNIPAAM enjoys enhanced water solubility (as reflected in elevation of the LCST) as a result of the binding of surfactant micelles. The polymer promotes surfactant aggregation; the CAC detected by both bound and free fluorescent probes and predicted by theory is an order of magnitude lower than the CMC. We have observed such an agreement between the CAC and the concentration at which the LCST rises in aqueous PNIPAAM solutions containing other sodium n-alkyl sulfates with hydrocarbon chain lengths longer than butyl (39).

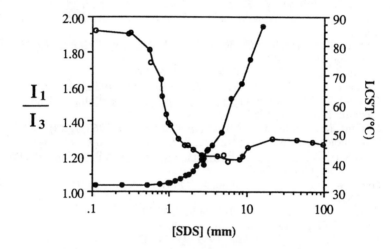

Figure 5.   The LCSTs of aqueous PNIPAAM (●, 0.40 mg/ml) with added SDS and I_1/I_3 of pyrene (○) in such solutions at 24.5°C.

Acknowledgments

This work was supported by a National Science Foundation Predoctoral Fellowship to Howard G. Schild and by a grant from the U.S. Army Research Office (DAAL03-88-K-0038). We thank Professor D. A. Hoagland for helpful suggestions and Professor R. Nagarajan for instructive correspondence.

Literature Cited

1. Cabane, B.; Duplessix, R. *J. Physique* 1982, 43, 1529.
2. Goddard, E. D. *Colloids and Surfaces* 1986, 19, 255.
3. Monji, N.; Hoffman, A. S. *Appl. Bioch. & Biotech.* 1987, 14, 107.
4. Taylor, L. D.; Cerankowski, L. D. *J. Polym. Sci., Pt. A: Polym. Chem.* 1975, 13, 2551.
5. Bae, Y. H.; Okano, T.; Hsu, R.; Kim, S. W. *Makromol. Chem., Rapid Commun.* 1987, 8, 481.
6. Hoffman, A. S. *J. Control. Release* 1987, 6, 297.
7. Okahata, Y.; Noguchi, H.; Seki, T. *Macromolecules* 1986, 19, 493.
8. Gilyami, T.; Wolfram, E. in Microdomains in Polymer Solution, Dubin, P., Ed.; Plenum: New York, 1985.
9. Heskins, M.; Guillet, J. E. *J. Macromol. Sci.-Chem.* 1968, A2(8), 1441.
10. Yamamoto, I.; Iwasaki, K.; Hirotsu, S. *J. Phys. Soc. Jap.* 1989, 58, 210.
11. Molyneux, P. Water Soluble Polymers: Properties and Behavior, CRC Press: Boca Raton, Florida, 1983.
12. Nwankwo, I.; Xia, D. W.; Smid, J. *J. Polym. Sci., Pt. B: Polym. Phys.* 1988, 26, 581.
13. Saito, S. *J. Polym. Sci., Pt. A-1* 1969, 7, 1789.
14. Schild, H. G. previous chapter in this volume.
15. Winnik, F. M. *J. Phys. Chem.* 1989, 93, 7452.
16. Winnik, F. M.; Winnik, M. A.; Tazuke, S. *J. Phys. Chem.* 1987, 91, 594.
17. Chandar, P.; Somasundaran, P.; Turro, N. J. *Macromolecules* 1988, 21, 950.
18. McGlade, M. J.; Olufs, J. L. *Macromolecules* 1988, 21, 2346.
19. Borden, K. A.; Eum, K. M.; Langley, K. H.; Tirrell, D. A. *Macromolecules* 1987, 20, 454.
20. Herkstroeter, W. G.; Martic, P. A.; Hartman, S. E.; Williams, J. L. R.; Farid, S. *J. Polym. Sci., Pt. A: Polym. Chem.* 1983, 21, 2473.
21. Chu, D.-Y.; Thomas, J. K. *Macromolecules* 1984, 17, 2142.
22. Winnik, F. M. *Macromolecules* 1989, 22, 734.
23. Winnik, F. M.; Winnik, M. A.; Tazuke, S.; Ober, C. K. *Macromolecules* 1987, 20, 38.
24. Winnik, F. M. *Macromolecules* 1987, 20, 2745.
25. Nagarajan, R.; Ruckenstein, E. *J. Coll. Inter. Sci.* 1979, 71, 580.
26. Nagarajan, R. *Polymer Preprints* 1981, 22(2), 33.
27. Nagarajan, R. *J. Chem. Phys.* 1989, 90, 1980.
28. Ruckenstein, E.; Huber, G.; Hoffmann, H. *Langmuir* 1987, 3, 382.
29. Annual Book of ASTM Standards, Volume 8.02, D 971 and D1331, American Society for Testing and Materials: Philadelphia, 1986.
30. Turro, N. J.; Arora, K. S. *Polymer* 1986, 27, 783.
31. Chandar, P.; Somasundaram, P.; Turro, N. J.; Waterman, K. C. *Langmuir* 1987, 3, 298.

32. Eliassaf, J. <u>J. Appl. Polym. Sci.</u> 1978, <u>22</u>, 873.
33. Cole, C.; Schreiner, S. M.; Priest, J. H.; Monji, N.; Hoffman, A. S. <u>ACS Symposium Series</u> 1987, <u>350</u>, 245.
34. Kalyanasundaran, K.; Thomas, J. K. <u>J. Am. Chem. Soc.</u> 1977, <u>99</u>, 2039.
35. Turro, N. J.; Baretz, B. H.; Kuo, P. L. <u>Macromolecules</u> 1984, <u>17</u>, 1321.
36. Ananthapadmanabhan, K. P.; Goddard, E. D.; Turro, N. J.; Kuo, P. L. <u>Langmuir</u> 1985, <u>1</u>, 352.
37. Waggoner, A. S.; Stryer, L. <u>Proc. Nat. Acad. Sci.</u> 1970, <u>67</u>(2), 579.
38. Mukerjee, P.; Mysels, K. J. <u>Critical Micelle Concentrations of Aqueous Surfactant Systems: NSRDA-NBS-36</u>, U.S. Government Printing Office: Washington, D.C., 1971.
39. Schild, H. G.; Tirrell, D. A. <u>Polymer Preprints</u> 1989, 30(2), 350.

RECEIVED June 4, 1990

Chapter 18

Determination of Molecular-Weight Distribution of Water-Soluble Macromolecules by Dynamic Light Scattering

Michael J. Mettille and Roger D. Hester

Department of Polymer Science, The University of Southern Mississippi, Hattiesburg, MS 39406–0076

An algorithm for the extraction of molecular weight distributions of typical high molecular weight, water-soluble polymers from dynamic light scattering has been developed. A function fit method was used to determine the best distribution parameters of a generalized exponential function. The algorithm employs a Marquardt-Levenburg constrained nonlinear least squares regression. A series approximation of the integral equation which is generated was attempted. This approach was abandoned because of difficulties with series convergence. A modified Rhomberg numerical integration was utilized with limited success. Two water-soluble polymer molecular weight distributions were obtained. The distributions of a polyethyleneoxide and a polyacrylamide were determined and compared with distributions obtained using gel permeation chromatography. The two methods were in fair agreement.

Molecular weight characterization of macromolecular systems is extremely important. Many solution properties are functions of the macromolecular size (and hence the molecular weight). An example of one property of major importance is viscosity. High viscosity solutions are useful as polymer flooding agents in enhanced oil recovery applications. A high molecular weight system exhibits higher viscosity than one of lower molecular weight for a single polymer at a specified concentration. In general, high molecular weight water-soluble polymer systems are polydisperse. In other words, the macromolecules are not of a single molecular weight but rather have a distribution of molecular weights. The fraction of polymer molecules having lower molecular weight is much less effective in producing high viscosity solutions. Thus, knowledge of this distribution is very important in designing an economical, effective EOR system. At the present there is no adequate tool for the determination of molecular weight distributions of the large water-soluble macromolecular systems.

Various methods have been employed to obtain molecular weight information. Vapor phase osmometry, membrane osmometry, and end group analysis have all been used to obtain number average molecular weights. These techniques are unable to measure molecular weights in the high

0097–6156/91/0467–0276$06.00/0

molecular weight regime and also suffer from an inability to determine information on the molecular weight distribution (MWD). Classical light scattering has also been used to acquire weight average molecular weights. Although this technique can determine molecular weights of high molecular weight systems, it does not provide information on the distribution of molecular weights. Size exclusion chromatography (SEC) is the classical method of obtaining MWDs. SEC is very difficult with aqueous solutions of very high molecular weight macromolecules because packing materials are not efficient at separating very large polymers. SEC requires molecular weight standards that span the appropriate molecular weight range. Water-soluble standards do not exist of the required high molecular weight. Thus, the utility of this method is limited. One method for determining the molecular weight distribution which does not rely on the existence of high molecular weight water-soluble standards is dynamic light scattering (DLS).

Background and Theory

DLS is also known as photon correlation spectroscopy, quasi-elastic light scattering, intensity fluctuation spectroscopy, and several other names. Thorough discussions of the theory and mathematics of DLS may be found elsewhere (1). Briefly, the phenomenon that is observed in DLS experiments is intensity fluctuations in the light scattered from a dilute solution of macromolecules. As the solute molecules which are scattering light undergo brownian diffusion, the relative phase of the scattered light changes. These changes in phase cause fluctuations in the detected scattering intensity. The rate at which these fluctuations occur can be, in turn, related to the rate of diffusion of the scattering molecules through a time autocorrelation of the scattered intensity (2). The intensity time autocorrelation function can be defined as shown in Equation 1.

$$G(\tau) = \lim_{T \to \infty} \frac{1}{2T} \int_{-T}^{+T} I(t)\, I(t+\tau)\, dt \qquad (1)$$

In this equation I(t) and I(t+τ) are the intensity of scattered light at time t and time t+τ. The integral equation shown is normally approximated as a discrete sum in most instrumentation. For most applications, the measured autocorrelation function G(τ) is converted to what is termed the first order normalized autocorrelation function. The transformation is shown in Equation 2.

$$g(\tau) = \left[\frac{G(\tau) - K_B}{K_0} \right]^{\frac{1}{2}} \qquad (2)$$

Where : $G(\tau)$ = Measured autocorrelation function

$g(\tau)$ = First order normalize autocorrelation function

K_0 = Baseline value, equal to $< I>^2$ and $G(\infty)$

K_B = Measured constant, theoretically equal to $\dfrac{< I^2 >}{< I>^2}$

$< I>^2$ = Square of the average value of the intensity

$< I^2 >$ = Average value of the intensity squared

K_B is measured during the experiment and K_o is determined during the data analysis.

In systems that consist of only a single size diffusing species, the relationship between the translational diffusion coefficient of the species in dilute solution and the first order normalized autocorrelation function is simply a single decaying exponential as shown in Equation 3.

$$g(\tau) = \exp(-q^2 D \tau) \qquad (3)$$

Where : q = Scattering vector = $\dfrac{4 \pi n \sin(\theta/2)}{\lambda_0}$

D = Diffusion coefficient
n = Refractive index of solvent
λ_0 = Radiation wavelength in vacuum

However, polymers have a distribution of molecular weights and this must be accounted for in the relationship. The above equation can be modified to become a summation over the various sizes of polymer molecules and is shown below.

$$g(\tau) = \sum_{i=0}^{\infty} F(D_i) \exp(q^2 D_i \tau) \qquad (4)$$

Where : $F(D_i)$ = The discrete distribution function of diffusion coefficients

D_i = The diffusion coefficient of the ith size molecule

If one assumes that the discrete distribution can be approximated as a continuous distribution, the summation equation may be transformed into an integral equation. It is assumed that the diffusion coefficient may be related to molecular weight as shown in Equation 5.

$$D_i = \alpha M_i^{-\beta} \qquad (5)$$

The α and β parameters are tabulated for many polymer-solvent systems. Those values not appearing directly in the literature may be estimated from the Mark-Houwink constants.

One may then transform the integral equation from decay constant space to molecular weight space (3). This transform is shown in Equation 6.

$$g(\tau) = \dfrac{\displaystyle\int_0^{\infty} C_I \, F(M) \, M^{7\beta-1} \exp(-\alpha q^2 M^{-\beta}\tau) \, dM}{\displaystyle\int_0^{\infty} C_I \, F(M) M^{7\beta-1} \, dM} \qquad (6)$$

Where : $F(M)$ = The number distribution of molecular weights

C_I is an intraparticle interference factor. This factor arises from the interference of light that is scattered from a single particle. As the size of scattering particles approach the wavelength of the incident radiation, the light scattered from one portion of a particle may interfere with the light scattered from another portion of the same particle. For monodisperse systems the result is simply a change in overall intensity without changing the shape of the correlation function. For polydisperse systems the result is markedly different. Particles of a size such that C_I is high (or the intensity of scatter is low) contribute little to the integral equation and particles where C_I is small contribute greatly. C_I acts to appropriately weight the scattering intensity from each molecular size. If this factor were not applied the MWD may show anomalous results. In its most general form for spherical particles, C_I takes a form known as the Mie scattering equation and is dependent upon the size of the scattering particle, the refractive indices of both the particle and solvent, the angle of scattered light, and the wavelength of the scattered radiation (4). This form, however is very complicated and cumbersome. Fortunately, this very general form is necessary only in cases of extremely large particles. For many macromolecular systems the Rayleigh Debye approximation to the Mie equation is adequate. This approximation of the intraparticle interference factor is shown in Equation 7 (5).

$$C_I = \frac{9[\sin(qR) - qR\cos(qR)]^2}{(qR)^6} \tag{7a}$$

As shown, the Rayleigh Debye approximation is dependent upon the scattering vector, q, and the particle radius, R. The hydrodynamic radius, R, is related to the diffusion coefficient through the Stokes-Einstein relationship (6). Equation 5 can be used to represent the diffusion coefficients in terms of molecular weight.

$$R_i = \frac{k_bT}{6\pi\eta D_i} = \frac{k_bTM_i^\beta}{6\pi\eta\alpha} \tag{7b}$$

Where : T = Absolute temperature
k_b = Boltzman constant
η = Solvent viscosity

Theoretically with knowledge of $g(\tau)$, α, β, and the scattering vector, q, the MWD may be extracted by inverting Integral Equation 6. The inversion of this equation to find the molecular weight distribution is not a simple task. Equations of this type appear frequently in science and are known as Fredholm Integral Equations. The mathematical properties of this class of equations are ill-conditioned. Thus, without making any priori assumptions about the properties of the distribution there is no guarantee that a solution for the distribution, F(M), exists. In addition, if there does exist a solution, the uniqueness of that solution is not assured.

Algorithm Development

Various methods have been used in the past to extract distribution information (7-11). The approach we have chosen is a type of function fit method. This approach assumes a distribution type for the MWD and then attempts to find the best set of distribution parameters consistant with the scattering data. These "best" set of parameters consistant with th scattering data define the "best" F(M). The distribution type chosen was the generalized exponential (GEX) distribution. The GEX distribution function is shown in Equation 8.

$$h = h_m \, L^{B-1} \exp\left[\left(1 - L^A\right)\frac{(B-1)}{A}\right] \qquad (8)$$

$$\text{Where}: L = \frac{M - M_0}{M_M - M_0}$$

In its most general form it has five adjustable parameters. Three of these parameters have easily observed physical meaning, the maximum peak height (h_m), the initial point at which the value of the function is defined to be zero (M_0), and the location of the peak maximum (M_M). These are diagrammatically illustrated in Figure 1. The other two parameters are shape parameters (A,B) and are functions of the inflection points of the distribution (12). The GEX distribution function was chosen for two reasons. First, it is an extremely versatile distribution function and can approximate a wide variety of distribution types (13). Also the GEX function has been shown to approximate MWDs fround from chromatagraphic analysis (14).

The ill-conditioned nature of the equations which govern the relationship between the MWD and the scattered intensity correlation function can make the inversion of the integral equations very difficult if not impossible. By placing appropriate restrictions on the search space many of the pitfalls associated with the inversion may be possibly avoided. The choice of the GEX may be viewed as just such a limitation of search space. In this case the choice of a GEX distribution function constrains the polymer MWD to be both positive and unimodal. The positivity constraint is completely valid because it is physically meaningless to discuss a negative number of molecules. The restriction that the distribution be unimodal is more limiting. However, most unfractionated water-soluble polymer systems are unimodal or near unimodal. Two other constraints may also be applied. The first is that M_0 is zero. This means that any real polymer MWD has some molecules of very low molecular weight and at this low molecular weight limit the number of molecules approaches zero. This assumption simplifies many of the calculations involved and reduces the number of parameters required to be determined for the MWD function.

The weight average molecular weight, M_w, is easily attainable through the use of low angle laser light scattering. With the M_w information and the assumption that M_0 is zero, it is possible to define M_M in terms only the weight average molecular weight and the two shape parameters, A and B. This eliminates the need to solve for another parameter in the MWD function. These constraints have reduced a four parameter problem to a two parameter problem. This lessens the complexity and time requirements for any regression technique which is utilized to find the remaining two parameters, A and B.

The GEX function was also initially chosen because it has moments which are analytically integrable (15). The advantage of integrability is that it diminishes computer computation time. Numerical integration is a time consuming procedure. Initial attempts to integrate both the numerator and denominator of Equation 6 did not produce a solution in closed form. Prior to the inclusion of the intrapartical interference factor the denominator was simply the second moment about zero of the GEX function and as such was analytically integrable without having to resort to numerical integration. When one includes C_I, the intraparticle interference factor, the resulting integral equation is no longer analytically integrable. Thus a larger amount of computer time is required to calculate the value of $g(\tau)$.

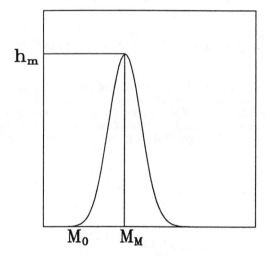

Figure 1. GEX distribution.

Series Use for Integration. Attempts were made to solve integral Equation 6 for the shape parameters, A and B, in a manner that did not require lengthy numerical integration routines. One very powerful technique that is often applied is to approximate the kernel of the integral by a power series. The order of summation and integration may be exchanged. Thus, the integral may now be represented by a summation of integrals. It is also possible that each separate integral in the series can be expressed in closed form and that the series is convergent.

First let us examine the numerator of the Equation 6. C_I is dependent upon the radius of the particle. The radius of the particle can be related to the diffusion equation by the Einstein-Stokes relationship (6). As previously explained, from a Mark-Houwink-like relationship and Equation 5, one may relate the particle radius to the molecular weight. By expanding the square into its subsequent terms, the following relationship is obtained.

$$C_I = \frac{9}{(qR)^6}\left[\sin^2(qR) - qR\sin(2qR) + (qR)^2\cos^2(qR)\right] \tag{9}$$

Substituting the expanded C_I into the integral and exchanging the order of integration and summation provides a series of three integrals. Each is comprised of the product of a trigonometric function, an exponential term and the GEX distribution function. These trigonometric functions and the exponential function can be expanded in power series form as shown below.

$$\sin^2(x) = \sum_{i=1}^{\infty}\frac{(-1)^i\,2^{2i-1}\,x^{2i}}{(2i)!} \tag{10}$$

$$\sin(2x) = \sum_{i=0}^{\infty}\frac{(-1)^i\,2^{2i+1}\,x^{2i+1}}{(2i+1)!}$$

$$\cos^2(x) = 1 - \sin^2(x) = 1 - \sum_{i=1}^{\infty}\frac{(-1)^{i+1}\,2^{2i-1}\,x^{2i}}{(2i)!}$$

Substituting these summations for the appropriate functions in Equation 9 is the next step in determining a power series expansion for the numerator of Equation 6. The integral formed by the product of the cosine squared and exponential terms can subsequently be separated into two integrals. The original integral has now been separated into four integrals. Three of these integrals have kernels which are comprised of the product of two power series, the GEX function and the molecular weight raised to a power. The product of two power series may be transformed into a single power series by performing the multiplication on a term-by-term basis. Gathering terms of like power provides the power series coefficient for each term in the series. The kernel of the remaining integral is composed of the product of only a single power series, the GEX function and the molecular weight raised to a power. When the order of integration and summation is exchanged the resulting equation is obtained for the numerator of Equation 6, I_N.

$$I_N = \sum_{N=-\infty}^{\infty} C_{N,1} \int_0^{\infty} M^{\beta(N+1)-1}F(M)\,dM \ - \ K_1 \sum_{N=-\infty}^{\infty} C_{N,2} \int_0^{\infty} M^{\beta(N+2)-1}F(M)\,dM$$

$$\text{(11)}$$

$$+ K_1^2 \left[\sum_{j=0}^{\infty} B_j \int_0^{\infty} M^{(3-j)\beta-1}F(M)\,dM \ - \ \sum_{N=-\infty}^{\infty} C_{N,1} \int_0^{\infty} M^{(N+3)\beta-1} F(M)\,dM \right]$$

Where α, β = Parameters in Diffusion Equation, $D = \alpha M^{-\beta}$

$$K_1 = \frac{q k_b T}{6\eta\pi\alpha}$$

k_b = Boltzman Constant
q = Scattering Vector
T = Temperature
η = Viscosity of solvent

The power series coefficients in Equation 11 are defined to be :

$$C_{N,1} = \sum_{k=\xi}^{\infty} A_{k,1} B_{2k-N} \qquad \text{(12)}$$

$$C_{N,2} = \sum_{k=\zeta}^{\infty} A_{k,2} B_{2k+1-N}$$

$$\text{With}: A_{k,1} = \frac{(-1)^{k+1} 2^{2k-1} (K_1)^{2k}}{(2k)!}$$

$$A_{k,2} = \frac{(-1)^k 2^{2k+1} (K_1)^{2k+2}}{(2k+1)!}$$

$$B_j = (-\alpha q^2 \tau)^j$$

$$\xi = 1 \text{ for } N \le 1$$

$$\xi = \frac{N}{2} \text{ for } N \ge 2 \text{ and } N \text{ is even}$$

$$\xi = \frac{N+1}{2} \text{ for } N \ge 3 \text{ and } N \text{ is odd}$$

$$\zeta = 1 \text{ for } N \le 1$$

$$\zeta = \frac{N}{2} \text{ for } N \ge 2 \text{ and } N \text{ is even}$$

$$\zeta = \frac{N-1}{2} \text{ for } N \ge 3 \text{ and } N \text{ is odd}$$

Note then the integrals in Equation 11 are simply molecular weight raised to a power times the GEX function. This integral defines a moment about zero of the GEX molecular weight distribution. This function is analytically integrable if one assumes M_0 is zero (16), an assumption which was made and has been previously adressed. In a similar manner the denominator of Equation 6 may also be formulated in terms of a power series. The resulting equation is shown below.

$$I_D = \sum_{k=1}^{\infty} A_{k,1} \int_0^{\infty} M^{\beta(2k+1)-1} F(M)\, dM \; - \; K_1 \sum_{k=0}^{\infty} A_{k,2} \int_0^{\infty} M^{\beta(2k+3)-1} F(M)\, dM$$

$$+ K_1^2 \left[\int_0^{\infty} M^{3\beta-1} F(M)\, dM - \sum_{k=1}^{\infty} A_{k,1} \int_0^{\infty} M^{(2k+3)\beta-1} F(M)\, dM \right]$$

(13)

The coefficients in this equation are completely analogous the ones defined for the series approximation of the numerator of Equation 6. It is now possible to formulate the autocorrelation function in terms of a series which requires no numerical integration. It does, however, require calculation of several complex sums.

Regression Algorithm. The regression algorithm presently employed to solve for the shape parameters, A and B, is a Levenburg-Marquardt nonlinear least squares technique (LM) (17). The original regression algorithm has been modified slightly to accommodate the use of weighting factors.

$$S = \sum_{i=1}^{N} W_i \left[(g_i(\tau,\theta,c))_{meas} - (g_i(\tau,\theta,c))_{calc} \right]^2$$

(14)

In Equation 14, W_i are weighting factors. The determination of the values for the weighting factors will be discussed later. In brief, the LM method attempts to find the shape parameters A and B such that the sum of squared residuals times the weighting factor (S as defined in Equation 14) is minimized. The advantages and disadvantages of the LM have been discussed previously (18).

Experimental

All experiments were performed using a light scattering apparatus that was constructed in our laboratories. This apparatus has been discussed in detail in a previous publication (19). Briefly the scattering apparatus consists of a detector system capable of observing scattered radiation from a polymer solution at various angles. The dynamic scattering signal obtained is fed into a Brookhaven BI2030AT 136 channel correlator. The resulting measured autocorrelation functions, $G(\tau)$, are saved on disk for later data analysis.

Sample Preparation. Dust is a major problem in DLS measurements of water-based polymer systems because dust particles are strongly attracted to water. Dust particles are powerful scatterers. Unfortunately, even distilled and filtered water usually contains some dust. The detector treats this scattering as it would scattering from a polymer particle. Normally, dust particles are much larger than the polymer molecules in the sample and this high intensity scattering can be digitally filtered from the data. However, when the polymer size is large, the scattering produced by dust is very similar to that produced by the polymer. Thus, the data obtained from a system contaminated with dust would show distributions having a very large size component. Consequently, if noise from dust is not minimized, the measurements obtained for the molecular weight distributions can have significant high molecular weight error.

The dust in a sample must be removed by filtration. Choice of filter pore size is very important. The pore size should be chosen such that all the polymer molecules can pass while retaining the maximum number of dust particles. Choosing a filter that has a pore size of about 4 times the average hydrodynamic diameter of the polymer is usually optimum. The average polymer hydrodynamic diameter is calculated from Equation 15,

$$d_{ave} = 4.8(kM^{a+1})^{1/3} \qquad (15)$$

where 'M_w' is the weight average molecular weight as measured by classical light scattering. We have used a Chromatix KMX-6 low angle spectrophotometer (20). 'k' and 'a' are the constants in the Mark-Houwink equation which relate molecular weight to intrinsic viscosity. The two constants are dependent on polymer and solvent type, however values can be found in the literature for many polymer/solvent systems.

Two high molecular weight polymer samples were investigated, a polyethyleneoxide (PEO) and a polyacrylamide (PAM). In each system, a series of five concentrations were prepared for light scattering analysis. Each concentration was placed in a sample vial. The five concentrations prepared were 3.4E-4, 6.6E-4, 1.00E-3, 1.24E-3, 1.54E-3 grams per milliliter for the PEO and 3.4E-4, 6.5E-4, 1.05E-3, 1.32E-3, and 1.66E-3 grams per milliliter for the PAM. Sample cleanup was accomplished by passing each sample through a filter loop with a 0.45 micron average pore size, one inch diameter disk, in line filter (manufactured by Gelman Inc.) for 24 to 48 hours. The flow rate through the filter was extremely low to eliminate any mechanical degredation of macromolecules. This extensive filtration was necessary to limit the amount of dust in a sample.

Data Collection.   After each of the five concentrations had undergone the filtration procedure, DLS measurements were taken. Scattering data were taken at angles of 30, 45, 60, 75, 90, 105, 120 degrees. Analysis using multiple angles helps to ensure the correctness of the calculated molecular weight distribution. The time delay increment, $\Delta\tau$, was chosen for the 90 degree angle-concentration pair such that the correlation function decayed to within one to two percent of the baseline. The exponential function in the numerator of Equation 6 is dependant upon the product of the scattering vector squared and the delay time. For the calculations at different angles to be directly comparable $q^2\Delta\tau$ must vary over the same range. Thus, the delay times at other angles were chosen such that this product, $q^2\Delta\tau$, was a constant. The total sample time used to generate the autocorrelation function was chosen such that a smooth correlation function was usually obtained. Measurements at every angle-concentration pair were repeated five times holding the delay increment and sample time constant. The resulting correlation functions were stored as a series of data files for subsequent analysis. Upon completion of the DLS measurements for a single polymer experiment at all angles, there were 175 data files each consisting of 136 data channels and 8 baseline channels. This large raw data array (24,500 data points) was preprocessed to reduce size prior to being used to estimate a MWD.

Data Preprocessing.   All raw data preprocessing was performed on a IBM-AT compatible computer having a math coprocessor. As discussed previously, one must convert the measured autocorrelation function, $G(\tau)$, to the first-order normalized correlation function, $g(\tau)$. A Pascal program, using Turbo Pascal 4.0 (Borland, Inc.), was written to perform this conversion. The program first calculated a baseline value from the 8 delay channels obtained by the correlator. Thereafter, this baseline was subtracted from the correlation function. Next the program extrapolated the resulting baseline corrected function to zero delay time. A linear extrapolation of the logarithm of the correlation function at small delay times was used to estimate the zero delay time correlation function value. Dividing the baseline corrected correlation function data by this extrapolated value and subsequently taking the square root of the result provided the first-order normalized correlation function. This operation was performed on each of the data files. During the course of these calculations several of the data files showed negative values in the baseline corrected correlation function. This is physically meaningless and makes determination of a first order normalized correlation function impossible (because of the difficulty in calculating the square root of a negative number). These anomalous data files were discarded.

As was noted, the amount of preprocessed data was very large. Prior to attempting a nonlinear regression, the data set size must be reduced in order to complete the regression in a reasonable amount of time. Each angle-concentration pair had been repeated five times with identical delay time increments. The data at each value of τ was averaged to generate a single value of $g(\tau)$. This not only reduced the size of the regression data set by a factor of five, but also provided a lower signal to noise ratio in the resulting averaged correlation functions. In addition, with these repeats the variance of the measurements may be estimated as shown in Equation 16.

$$\sigma^2(\tau,\theta,c) = \sum_{i=1}^{N} \frac{[\ g_i(\tau,\theta,c) - \overline{g_i(\tau,\theta,c)}\ ]^2}{N-1} \tag{16}$$

Where : $\overline{g(\tau,\theta,c)}$ = average value of $g_i(\tau,\theta,c)$

The resulting correlation data for a single angle had a series of five correlation functions, one for each concentration. At each τ value, the five $g(\tau)$-concentration pairs were assumed to have a linear relationship. Linear regression was used to extrapolate $g(\tau)$ to zero concentration. Extrapolation to zero concentration is necessary to eliminate any polymer coil to coil interactions. After completing the extrapolation to zero concentration, the data set size was limited to just seven correlation functions each at a different scattering angle and was sufficiently small to perform the nonlinear regression. Linear regression can allow an estimation of the variance of the zero concentration intercept. This is accomplished through the use of Equation 17 (21).

$$\sigma^2_{intercept} = \frac{1}{S_\sigma}\left(1 + \frac{S_c^2}{S_\sigma S_{tt}}\right) \tag{17}$$

where : $S_\sigma = \sum_{i=1}^{N} \frac{1}{\sigma^2(\tau,\theta,c_i)}$

$S_c = \sum_{i=1}^{N} \frac{c_i}{\sigma^2(\tau,\theta,c_i)}$

$$S_{tt} = \sum_{i=1}^{N} \frac{1}{\sigma^2(\tau,\theta,c_i)} \left[c_i - \frac{S_x}{S_\sigma} \right]^2$$

N = the number of concentrations

This intercept is the zero concentration extrapolated correlation function data point. The reciprocal of the calculated variance may now be used as the weighting factor required in Equation 14.

The regression algorithm requires knowledge of alpha and beta values. For POE these values are not in the literature. Thus, one must calculate these two parameters from information that is available. It is possible to determine the alpha and beta values from the Mark-Houwink 'k' and 'a' values. The transformation is shown below.

$$\alpha = \frac{k_b T \times 10^8}{14.4 \pi \, \eta \, k^{\frac{1}{3}}}$$

$$\beta = \frac{a+1}{3} \tag{18}$$

Where : k,a = Mark–Houwink parameters
k_b = Boltzman Constant
q = Scattering Vector
T = Temperature in Kelvin
η = Viscosity of solvent in Poise

The values of 'k' and 'a' used were 12.5E-5 and 0.78, respectively. For the PAM, the diffusion equations parameters, α and β, are available (22). Both sets of parameters are listed below.

$$\alpha_{PEO} = 2.04 \times 10^{-4} \qquad \beta_{PEO} = 0.593$$

$$\alpha_{PAM} = 8.46 \times 10^{-4} \qquad \beta_{PEO} = 0.69$$

In addition, the algorithm requires the wavelength of the incident radiation (the red line of a HeNe laser) and the refractive index of the scattering solvent (water at 298K) which are 6.328E-5 cm and 1.3324 respectively.

Results

DLS measurements on both PEO and PAM produced correlation functions which were free enough of noise to attempt extraction of the GEX function parameters. For both samples the preprocessed data consisted of 945 data points. Each point contains the τ, $g(\tau)$, the scattering vector, q, and weighting factor values. In order to reduce the time required for computation a reduced size data set was initially used. The reduced data set was obtained by retaining one data point out of every fifteen in the full data set. This provides a data set of only sixty-three data points. Upon convergence to a solution using the reduced data set, the calculated parameters were then used as starting approximations for the analysis upon the full data set.

The series approximations for Equation 6, discussed previously, was implemented. However, use of the series approximation was abandoned after several attempts to perform the nonlinear regression. The major problem with the series approximation is its slow convergence. It was found that the leading terms of the series cancel each other. Thus, the numerical value of the remaining terms define the value of the series. These terms, however, can be as much as 10 to 20 orders of magnitude less than the leading terms. Thus, precision of some of the numerical calculations involved in calculating the kernel of the integrals may become limiting. One example is the calculations of the gamma function. The numerical method used to calculate the value of the gamma function is a method developed by Lanczos (23). Initially the absolute error produced by this algorithm was 2E-10 using a five term series. Extension of the series using the same method decreased this absolute error to a value of 2E-16. Even this low error level was not sufficient. The series approximation simply breaks down under the normal conditions using the Levenburg-Marquardt algorithm. Thus, we were compelled to implement numeric integration for solving Equation 6.

The analysis of the PEO sample was computer time intensive. The analysis took 12 to 24 hours of CPU time. The weight average molecular weight (M_w) as determined by low angle laser light scattering was 2E6 daltons. The shape parameters, A and B, determined for the GEX function were 3.38E-2 and 33.544, respectively. The resulting generalized exponential distribution function gives a value of 8.5E5 daltons for the number average molecular weight and 4.6E6 daltons for the z-average molecular weight. The complete distribution function is shown in Figure 2.

The analysis of the PAM sample was also very time consuming. The regression usually took about 24 hours of computer time to converge to a solution. The M_w was determined to be 2.4 million daltons. The values of the shape parameters, A and B, were 2.5E-1 and 8.05 respectfully. This resulted in number average and Z-average molecular weights of 1.5 and 3.6 million daltons. Again the complete distribution is shown in Figure 3.

Figures 2 and 3 show not only the distribution calculated from light scattering (the solid line), but also the distribution (show as data points) found using a prep scale gel permiation chromatography (GPC) apparatus. The details concerning both the apparatus and the data analysis associated with the determination of the MWD can be found elsewhere (Mettille, M. J. and Hester, R. D., in Polymers for Mobility Control in Enhanced Oil Recovery—Final Report, U.S. Dept. of Energy, contract DE-AC19-85BC10844, in press).

Discussion

As can be seen in the Figures 2 and 3, the MWD calculated by light scattering is in fair agreement with the distributions obtained from the prep scale GPC. The high molecular weight region shows some variation between the two methods. This variation may be due to several causes. The GPC system was never constructed with analytical work in mind. It was designed to facilitate the fractionation of large water-soluble macromolecule. The data collected during this fractionation, fortuitously, provided a method with which we could compare the results from light scattering MWD determination. The molecular weight averages calculated from the GPC chromatogram for the PEO were in agreement. The MWD obtained for the PAM appears to be in fair agreement with the GPC.

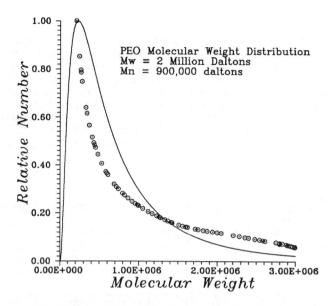

Figure 2. PEO molecular weight distribution.

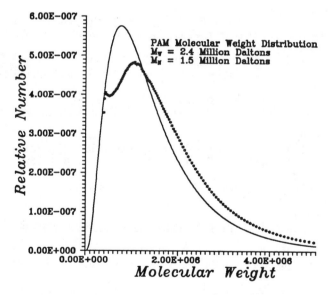

Figure 3. PAM molecular weight distribution.

Conclusions

The PEO and PAM distributions calculated from the dynamic light scattering data were in fairly good agreement with the distributions obtained through the use of gel permeation chromatography. The computation time required by the regression algorithm is high, however. Although there is more work to be done, the regression algorithm does show utility in the extraction of MWDs from dynamic light scattering of water-soluble macromolecules. Further study on other water-soluble polymer systems should provide additional confirmation of the utility of this method for MWD analysis from DLS data. Several other high molecular weight water-soluble polymer systems are being prepared for DLS and subsequent MWD analysis.

Literature Cited

1. Berne, B. J.; Pecora, R. Dynamic Light Scattering; John Wiley and Sons: New York, 1976; pp 1-38).
2. Zwanzig, R. Ann. Rev. Phys. Chem. 1965, 16, 65.
3. Mettille, M. J.; Hester, R. D. In Polymers for Mobility Control in Enhanced Oil Recovery—Third Annual Report; Contract DE-AC19-85BC10844; U.S. Department of Energy: Bartlesville, OK, 1989; p 73.
4. Bott, S. In Measurement of Suspended Particles by Quasi-Elastic Light Scattering; Dahneke, B. E., Ed.; John Wiley and Sons: New York, 1983; p 132.
5. Bohren, C. F.; Huffman, D. R. Absorption of Light by Small Particles; John Wiley and Sons: New York, 1983; p 163.
6. Reference 1, p 60.
7. McWhirter, J. G.; Pike, E. R. J. Phys. A: Math. Gen. 1978, 11, 1729.
8. Gardener, D. G.; Gardener, J. C.; Laush, G.; Meinke, W. W. J. Chem. Phys. 1959, 31, 978.
9. Koppel, D. E. J. Chem. Phys. 1972, 57, 4814-20.
10. Chu, B.; Gulari, Es.; Gulari, Er.; Tsunashima, Y. J. Chem. Phys. 1979, 70, 3965.
11. Provencher, S. W. Makromol. Chem. 1979, 180, 201.
12. Hester, R. D.; Vaidya, R. A.; Dickerson, J. P. J. Chromatogr. 1989, 462, 5.
13. Ibid., 8-10.
14. Vaidya, R. A.; Hester, R. D. J. Chromatogr. 1984, 287, 231.
15. Vaidya, R. A.; Hester, R. D. J. Chromatogr. 1985, 333, 152.
16. Reference 3, p 79.
17. Marquardt,D. W. J. Soc. Ind. Appl. Math. 1963, 11, 431.
18. Reference 3, pp 80-81.
19. Ibid., 88-89.
20. KMX-6DC Low Angle Spectrophotometer Instruction Manual; Chromatix, Inc., 1977.
21. Press, W. H.; Flannery, B. P.; Teukolsky, S. A.; Vetterling, W. T. Numerical Recipes; Cambridge University Press: New York, 1986; p 507-508.
22. Klarner, P. E. O.; Ende, H. A. In Polymer Handbook; Brandrup, J.; Immergut, E. H., Eds.; John Wiley & Sons: New York, 1975; IV-69.
23. Lanczos, C. Journal S.I.A.M. Numerical Analysis 1964, Series B, 1, 86.

RECEIVED June 4, 1990

Chapter 19

Photophysical and Rheological Studies of the Aqueous-Solution Properties of Naphthalene-Pendent Acrylic Copolymers

Mark D. Clark, Charles L. McCormick, and Charles E. Hoyle

Department of Polymer Science, The University of Southern Mississippi, Hattiesburg, MS 39406–0076

The effects of pH and the addition of various additives on the photophysical solution properties of the amphiphilic polyelectrolyte poly(2-(1-naphtylyacetyl)ethylacrylate-*co*-methacrylic acid) (NAEA-MAA) have been investigated. Unlike previously reported systems, an almost four-fold *increase* in excimer emission relative to monomer emission (I_E/I_M) is observed upon increasing pH and, consequently, the degree of ionization. An increase in the dilute solution viscosity of NAEA-MAA solutions with increasing degree of ionization indicates a somewhat expected transition from a compact coil to a more expanded coil at pH 7.5. The hydrophobic character of both the naphthyl groups of NAEA and the pendent methyl groups of MAA and/or the effective decoupling of the naphthyl chromophores from the polymer backbone via a spacer linkage are believed to be responsible, at least in part, for this behavior. Experiments utilizing urea, a water-structure breaker, indicate that at high pH hydrophobic interactions between naphthyl chromophores stabilize the copolymer in an intramolecular "hypercoil". Finally, synthesis of a hydrolytically stable analog of NAEA, necessary for further investigations, is outlined, and initial photophysical studies of its copolymers with methacrylic and acrylic acid are presented.

Amphiphilic polyelectrolytes containing aromatic chromophores have been the focus of a number of studies. These polymers can form heterogeneous, often hydrophobic, microenvironments in dilute aqueous solution. These "pseudo-micellar" microenvironments or "hypercoils" are formed as the hydrophobic aromatic groups are clustered toward the center of the coil, while the solubilizing hydrophilic ionic groups are located on the exterior (1,2). Not only does this pseudo-micellar conformation enhance the migration of energy within the polymer coil (3), it is also able to solubilize large hydrophobic compounds (4,5) which may in turn act as energy traps. Since the initial work of Guillet and coworkers with copolymers of acrylic acid and naphthylmethyl methacrylate (1), recognition of the potential importance of these types of synthetic polymers in "photon-harvesting" or artificial photosynthetic applications has fueled efforts to create a number of modified water-soluble antennae systems (6-10).

0097–6156/91/0467–0291$06.00/0
© 1991 American Chemical Society

However, most investigations to date have been conducted on solutions of polymers in the electrolyte form. Few comprehensive studies directed toward understanding the effect of pH on the conformation, and consequently the photophysical and rheological properties of the pseudo-micellar polymers have been conducted. Similarly the effect of "decoupling" the chromphore from the polymer backbone via a spacer group has only recently received attention.

In this chapter, we report the results of both fluorescence emission and viscosity studies of a polymer of methacrylic acid containing 20 mol% of 2-(1-naphtylacetyl)ethylacrylate (NAEA). Data suggest that the observed behavior relates, at least in part, to hydrophobic naphthyl group associations within the polymer coil. However, further investigations of this system dictate that a more hydrolytically stable label than NAEA (11) be used. Therefore, we also outline the preparation of the amide analog to NAEA and present initial photophysical data from those respective copolymers with methacrylic acid.

Experimental

Materials.   The synthesis of the 2-(1-naphtylyacetyl)ethylacrylate (NAEA) monomer and copolymers with methacrylic acid have been outlined elsewhere (12). Scheme 1 illustrates the synthesis of the hydrolytically stable amide analog of NAEA, 2-(1-naphtylyacetamido)ethylacrylamide (NAEAm). The activated succinimido ester of naphtylacetic acid (NAA-NHS) was prepared by reacting the acid with

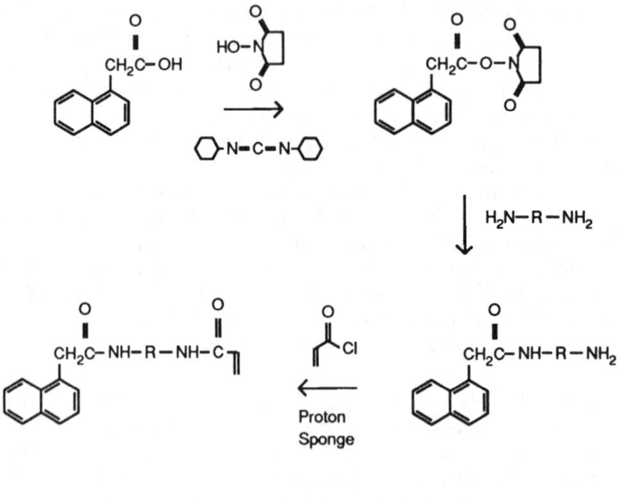

Scheme I.   NAEAm Monomer Synthesis.

N-hydroxysuccinimide in $CHCl_3$ utilizing dicyclohexylcarbodiimide (DCC) as a coupling agent. 2-(1-Naphthylacetamido)ethylamine (NAA-EDA) was prepared via modification of a high-dilution procedure for the reaction of acid derivatives with symmetrical diamines (13). The NAEAm monomer was prepared via addition of acryloyl chloride to NAA-EDA at 0°C using "Proton Sponge" (1,8-dimethylamino)naphthylene (Aldrich) in $CHCl_3$. Purification of the material was accomplished via multiple recrystallizations from a mixture   of $CHCl_3$ and CH_3OH.

Urea (Aldrich, 99+%) was recrystallized three times from methanol. Copper nitrate (Aldrich 99.99%) was used as received.

All polymer solutions were prepared in deionized water. Due to the inherent hydrolytic instability of NAEA (11), solutions of NAEA-MAA were discarded after 48 hrs., and fresh solutions prepared for subsequent studies.

<u>Methods</u>. *Polymer Characterization.* UV absorbance measurements (Perkin-Elmer Lambda 6B) and elemental analysis (MHW Laboratories, Phoenix, AZ) were used to determine copolymer compositions. The number average molecular weight, M_n, of the NAEA-MAA copolymer was estimated to be 140,000, based on osmotic pressure measurements (Knauer Osmometer with 600W membrane, Arro Laboratories, Inc.) in N,N-dimethylacetamide. Residual NAEA monomer in the purified polymers was found to be <0.1 mol% via a dialysis/liquid chromatography method.

Fluorescence Spectroscopy. The concentration of the NAEA-MAA copolymer in solution was ca. 10 mg/dL such that the concentration of naphthyl moieties in solution was always $<10^{-4}$ mol/L. Similarly, concentrations of NAEAm(10)MAA and NAEAm(1)MAA copolymers were 9.50 and 86.2 mg/dL, respectively, such that [NAEAm] = 10^{-4} mol/L. Sample solutions were degassed with N_2 for 15 min. prior to emission measurements. Steady-state emission spectra of NAEA-MAA were measured with a Perkin-Elmer 650-10B spectrophotometer; spectra were not corrected for photomultiplier response. Emission spectra of the NAEAm copolymers were recorded with a SPEX Fluorolog-2 fluorescence spectrometer. NAEAm emission spectra were corrected for the wavelength dependence of the detector response using an internal correction function provided by the manufacturer. All samples were excited at 280nm, and monomer and excimer intensities were measured at 330 and 420nm, respectively.

Viscosity Measurements. An NAEA-MAA polymer concentration of 100 mg/dL were used for rheological studies. Solution viscosities were measured at 25°C with a Contraves LS-30 low shear rheometer at a shear rate of 6.0 s^{-1}.

<u>Results and Discussion</u>

<u>NAEA-MAA Copolymer</u>. The copolymer of 2-(1-naphthylacetyl)ethylacrylate (NAEA) and methacrylic acid (MAA) was prepared by free-radical polymerization in N,N-dimethylformamide to yield a 20:80 mol% copolymer of NAEA:MAA. Copolymer microstructural data as determined by monomer Q and e values indicate that NAEA units are essentially isolated from one another by relatively long intervening sequences of MAA mers.

Structure of NAEA-MAA copolymer.

The normalized fluorescence emission spectra of the NAEA-MAA copolymer for a wide range of pH values are shown in Figure 1. The emission spectra consist of two bands: a structured band at shorter wavelength (λ_{max} = 340 nm) assigned to monomer emission and a broad, structureless emission (λ_{max} = 392 nm) ascribed to excimer fluorescence.

The intensity of excimer emission relative to monomer emission for the NAEA-MAA copolymer exhibits a considerable dependence on the pH of the aqueous solution. The maximum value of the excimer (420 nm) to monomer (330 nm) intensity ratio (I_E/I_M) is observed under alkaline conditions (pH 9.0) while the minimum occurs at a pH of 5.0 (Figure 2). A sharp transition occurs at pH 7.5. In N,N-dimethylformamide (DMF), however, a good solvent for the copolymer and more specifically the hydrophobic NAEA units, the degree of excimer formed is negligble (Figure 1), indicating a distinctly different conformation of the copolymer in DMF. We feel that this observation indicates that the formation of excimer in aqueous solutions is largely due to non-nearest neighbor interactions. The increase in I_E/I_M is accompanied by a concordant pH-induced increase in the reduced viscosity (η_{sp}/c) of the polymer solution, a general indicator of the polymer hydrodynamic size (Figure 2). This behavior is rather unusual for a labeled polyelectrolyte as an increase in hydrodynamic size is more often associated with a decrease in excimer formation as chromophore density within the polymer coil is decreased and will be discussed further in light of experimental evidence presented in the following sections.

In order to determine the role of hydrophobic association of pendent NAEA units at high pH, we have employed the water-structure breaker urea to disrupt such interactions (14,15). At pH 8.0, a significant decrease in I_E/I_M is observed with the addition of urea to the polymer solution (Figure 3). Since urea was found not to quench the fluorescence of the NAEA model compound, this decrease in I_E/I_M is attributed to the effective solubilization of the hydrophobic naphthyl groups by the urea. As these hydrophobic interactions are disrupted, the chromophores can no longer maintain the minimum separation necessary for excimer formation to occur, such that excimer intensity, and therefore I_E/I_M, is decreased. Unfortunately, these observations do not definitively indicate whether the hydrophobic associations are inter- or intramolecular.

One might expect that upon disruption of specific NAEA interactions, characteristic changes in viscosity should indicate the nature of these associations. Intuitively, disruption of intermolecular associations should provide a decrease in the reduced viscosity as the effective hydrodynamic size of the polymer coils would be reduced. Conversely, if the interactions were largely intramolecular, the overall hydrodynamic size of the polymer would be expected to increase as the constricting associations were disrupted and electrostatic repulsions cause the coil to expand. Figure 3 indicates that for the NAEA-MAA copolymer the latter of these is indeed the case. An increase in η_{sp}/c of the polymer solution is observed with an increase in urea concentration, indicating that the NAEA associations are predominantly intramolecular in nature.

Further clarification of the nature of the associations among naphthyl groups at high pH dictated that the effect of polymer concentration on the solution properties of be investigated. Intermolecularly associating polymers would be expected to yield a steady, though not necessarily linear, increase in both I_E/I_M and the reduced viscosity of the solution with increasing polymer concentration. However, an intramolecularly associating polymer should yield only small, if any, changes in these two solution properties with increasing polymer concentration. At pH 8.0, there is no change within experimental

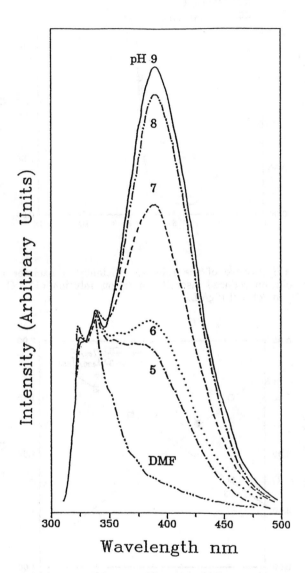

Figure 1. Steady-state fluorescence spectra (λ_{ex} = 280 nm) of NAEA-MAA (C_p = 0.01g/dl) in aqueous solution at different values of pH and in N,N-dimethylformamide.

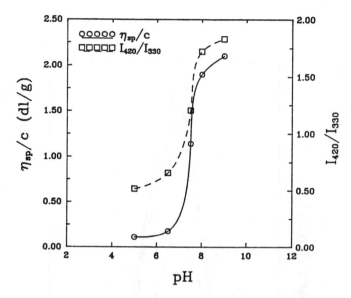

Figure 2. Dependence of the relative efficiency of excimer formation (I_E/I_M) and solution reduced viscosity (η_{sp}/c) on solution pH. C_p (I_E/I_M) = 0.01g/dL. C_p (η_{sp}/c) = 0.10g/dL.

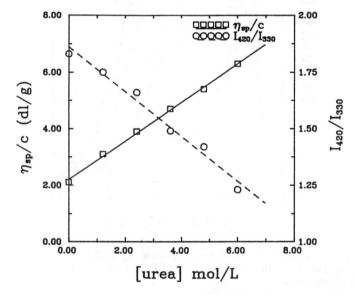

Figure 3. Dependence of the relative efficiency of excimer formation (I_E/I_M) and solution reduced viscosity (η_{sp}/c) at pH 8.0 on the concentration of added urea in the solution. C_p (I_E/I_M) = 0.01g/dL. C_p (η_{sp}/c) = 0.10g/dL.

error of either the I_E/I_M ratio or η_{sp}/c above a limiting polymer concentration (7 - 10 mg/100 mL) up to 150 mg polymer/100 mL (Figure 4). Below this limiting concentration, I_E/I_M decreases and η_{sp}/c increases almost asymptotically. This can be attributed to an increase in Coulombic repulsions often associated with polyelectrolytes in extremely dilute solution, i.e. the "polyelectrolyte effect" (16). Therefore, it can be concluded from both steady state fluorescence and viscosity measurements that, above a limiting polymer concentration, the NAEA-MAA copolymer forms "pseudomicellar", intrapolymer associations among naphthyl groups at high pH.

NAEAm Copolymers.   As mentioned earlier, the desire for further study of this polymer system and the lack of long-term hydrolytic stability of the NAEA monomer necessitated the synthesis of a hydrolytically stable analog of this material. Using a modified technique for the monosubstitution of symmetrical diamines, the amide analog of NAEA, 2-(1-naphthylacetamido)ethylacrylamide (NAEAm) was synthesized.

An interesting feature of this synthesis is that it is possible to make use of diamines of varying lengths to investigate the effect of the spacer group on the photophysical and rheological properties of the polymer in solution. Furthermore, model compound studies indicate that substitution of the amide bond for the ester bond does not affect the photophysical properties of the chromophore (steady-state or transient), presumably due to an insulating effect provided by the methylene group between the amide bond and the naphthyl group.   Finally, this synthetic technique may permit the conversion of a number of traditional probe molecules into free-radically polymerizable labels.

R=H, CH₃
x=0.01, 0.10

General Structure of NAEAm Copolymers

Initial investigations with the NAEAm monomer have been targeted at determining the effect of comonomer on the observed photophysical properties of the label. Polymers of methacrylic acid undergo a non-uniform coil transition with changes in pH due to hydrophobic methyl group interactions (17-19). In order to determine what effect these interactions have on the fluorescence properties of the NAEAm chromophore, a copolymer of MAA with a small amount of NAEAm (approximately 1 mol%) was synthesized. Furthermore, a polymer of MAA containing approximately 10 mol% NAEAm was prepared to elucidate the effect that higher degrees of incorporation of the chromophore have on the polymer conformation.

Figures 5 and 6 depict steady-state fluorescence spectra of NAEAm(10)MAA and NAEAm(1)MAA, respectively. It is important to point out that for both systems, the difference in excimer emission (I_E/I_M) between the polymers in DMF and in aqueous solutions is believed to be due to hydrophobic

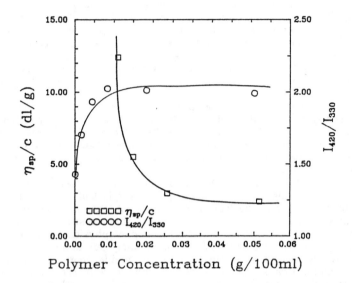

Polymer Concentration (g/100ml)

Figure 4. Dependence of the relative efficiency of excimer formation (I_E/I_M) and solution reduced viscosity (η_{sp}/c) at pH 8.0 on polymer concentration.

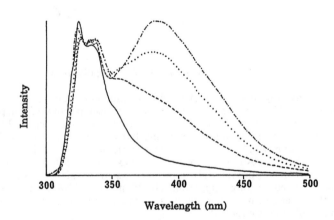

Wavelength (nm)

Figure 5. Steady-state fluorescence spectra (λ_{ex} = 280 nm) of NAEAm(10)MAA (C_{NAEAm} = 10^{-4} mol/L) in aqueous solution at different values of pH and in N,N-dimethylformamide. (–·–·–·–) pH 8.0; (·········) pH 6.5; (– – – –) pH 5.0; (————) DMF.

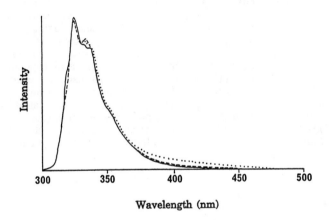

Figure 6. Steady-state fluorescence spectra (λ_{ex} = 280 nm) of NAEAm(1)-MAA (C_{NAEAm} = 10^{-4} mol/L) in aqueous solution at different values of pH and in N,N-dimethylformamide. (·········) pH 8.0; (− − − − −) pH 5.0; (————) DMF.

interactions between chromophores, which have been shown to occur even at low degrees of incorporation of the chromophore in the polymer (20). Interestingly, for both polymers, an increase in I_E/I_M is observed with increasing pH (Figure 7). The fact that these transitions take place at approximately the same pH for two polymers of different compositions, indicates that the hydrophobic methyl group interactions often associated with methacrylic acid polymers, along with the "decoupling" of the chromphore from the polymer backbone, are responsible for this behavior.

In order to confirm the role of hydrophobic methyl group interactions, analogous copolymers of NAEAm with acrylic acid, which exhibits no such interactions, were investigated. Figure 8 depicts the effect of pH on I_E/I_M for both NAEAm(1)AA and NAEAm(10)AA. The absence of any appreciable change in excimer intensity for the NAEAm(1)AA copolymer suggests that hydrophobic methyl group interactions do indeed play a large role in the transitions observed for the methacrylic acid copolymers. After a slight initial increase in excimer intensity, believed to be due to the decreasing degree of ionization, NAEAm(10)AA also undergoes a decrease in I_E/I_M with decreasing pH. However, it is important to point out that this transition occurs some 2 pH units below those observed for the three methacrylic acid copolymers. We speculate that for NAEAm(10)AA the decrease in excimer intensity with decreasing pH is due, not to increased hydrophobic methyl group interactions as with the methacrylic acid copolymers, but to the decreased ionization associated with a pH below the pK of poly(acrylic acid) (4.8) (21). Once NAEAm(10)AA falls below this critical degree of ionization, the hydrophobic naphthyl groups may cause the coil to collapse in much the same fashion as do the methyl groups in the methacrylic acid copolymers.

Conclusions

Photophysical data presented above indicate that the conformations adopted by the NAEA and NAEAm copolymers in aqueous solution are dependent on the pH of the solution as well as hydrophobic interactions inherent within the polymer. At high pH, the NAEA-MAA and NAEAm(10)-MAA copolymers apparently assume a pseudomicellar conformation with the naphthyl groups forming a hydrophobic core which is surrounded by a charged shell of methacrylic acid units. The behavior, both photophysical and rheological, of the NAEA-MAA copolymer observed upon the addition of the water-structure breaker urea indicates that these hydrophobic interactions are largely intramolecular in nature. As the pH is lowered, we speculate that a "dilution" of NAEA units within the hydrophobic portion of the coil and/or nonfavorable chromophore orientation due to decreased mobility are responsible for the observed decrease in excimer formation. Similar changes in I_E/I_M with decreasing pH for the NAEAm(1)MAA copolymer and the lack of such changes for the analogous acrylic acid copolymers indicate that hydrophobic interactions between the pendent methyl groups of MAA are in fact at least partially responsible for this behavior.

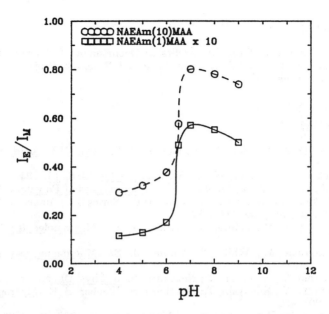

Figure 7. Dependence of the relative efficiency of excimer formation (I_E/I_M) on solution pH for NAEAm(10)MAA and NAEAm(1)MAA copolymers.

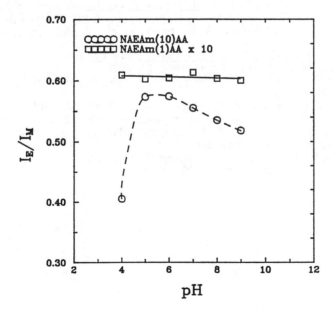

Figure 8. Dependence of the relative efficiency of excimer formation (I_E/I_M) on solution pH for NAEAm(10)AA and NAEAm(1)AA copolymers.

Acknowledgments

The financial support of the United States Department of Energy, the Office of Naval Research, and the Defense Advanced Research Projects Agency is gratefully acknowledged.

Literature Cited

1. Holden, D. A.; Rendall, W. A.; Guillet, J. E. Ann. N.Y. Acad. Sci. 1981, 366, 11.
2. Guillet, J. E.; Rendall, W. A. Macromolecules 1986, 19, 224.
3. Bai, F.; Chang, C.-H.; Webber, S. E. Photophysics of Polymers; Hoyle, C. E.; Torkelson, J. M., Eds.; ACS Symposium Series 358; American Chemical Society: Washington, D.C., 1987.
4. Nowakowska, M.; White, B.; Guillet, J. E. Macromolecules 1989, 22, 2317.
5. Nowakowska, M.; White, B.; Guillet, J. E. Macromolecules 1988, 21, 3430.
6. Morishima, Y.; Kobayshi, T.; Nozakura, S. J. Phys. Chem. 1985, 89, 4081.
7. Morishima, Y.; Kobayshi, T.; Nozakura, S.; Webber, S. E. Macromolecules 1987, 20, 807.
8. Guillet, J. E.; Wang, J.; Gu, L. Macromolecules 1986, 19, 2793.
9. Morishima, Y.; Kobayshi, T.; Nozakura, S. Polymer J. 1989, 21, 267.
10. Morishima, Y.; Kobayshi, T.; Nozakura, S.; Furui, T. Macromolecules 1987, 20, 1707.
11. McCormick, C. L.; Kim, K. J. Macromol. Sci. Chem. 1988, A25, 285.
12. McCormick, C. L.; Kim, K. J. Macromol. Sci. Chem. 1988, A25, 307.
13. Jacobson, A. R.; Mahris, A. N.; Sayre, L. M. J. Org. Chem. 1978, 43, 2923.
14. Mukerjee, P.; Ray, A. J. Phys. Chem. 1963, 67, 190.
15. Schick, M. J. J. Phys. Chem. 1964, 68, 3585.
16. Tanford, C. Physical Chemistry of Macromolecules; John Wiley and Sons: New York, 1961.
17. Chu, D.-Y.; Thomas, J. K. Macromolecules 1984, 17, 2142.
18. Olea, A. F.; Thomas, J. K. Macromolecules 1989, 22, 1165.
19. Ghiggino, K. P.; Tan, K. L. Polymer Photophysics; Phillips, D., Ed.; Chapman and Hall: New York, 1985.
20. Turro, N. J.; Arora, K. S. Polymer 1986, 27, 783.
21. Mandel, M. Eur. Polym. J. 1970, 6, 807.

RECEIVED June 4, 1990

Chapter 20

Hydrophobic Effects on Complexation and Aggregation in Water-Soluble Polymers

Fluorescence, pH, and Dynamic Light-Scattering Measurements

Curtis W. Frank, David J. Hemker, and Hideko T. Oyama

Department of Chemical Engineering, Stanford University, Stanford, CA 94305–5025

Complexation and aggregation between poly(ethylene glycol) (PEG) and either poly (acrylic acid) (PAA) or poly (methacrylic acid) (PMAA) are studied in aqueous solution using fluorescence and dynamic light scattering. Excimer fluorescence from pyrene groups terminally attached to the PEG allow study of intermolecular and intramolecular interactions. Hydrophobic attraction is shown to be important in understanding the photophysical behavior. This influences the chain configuration of PMAA, the formation of ground state interactions of pyrenes in PEG*, the sequestering of pyrenes in hydrophobic regions in PMAA:PEG* complexes and the aggregation of the complexes at low pH.

A sizeable body of literature has resulted from investigations of complexation reactions between synthetic polymers (1-6). Objectives have been to describe the interaction forces (hydrogen bonding, ionic, hydrophobic), to determine structural effects (molecular weight, stoichiometry, chemical composition) and to study the effect of reaction conditions (temperature, pH, solvent). Many of these parameters have been addressed in our previous studies on complexation of poly(ethylene glycol) (PEG) with either poly(acrylic acid) (PAA) or poly(methacrylic acid) (PMAA) in water (7-9). Incorporation of probe molecules consisting of PEG containing terminal pyrene labels (denoted PEG*) has allowed intramolecular and intermolecular excimer fluorescence measurements to be used to monitor local concentration of the pyrene labels in the complexes. From this we have inferred some of the structural details occurring on the scale of individual chains and small clusters.

At the same time, however, we are also interested in the existence of large scale aggregates. Tschida and co-workers observed that as the pH of a solution containing a water soluble polymer complex was lowered, the solution became more turbid as a result of aggregation of the polymer complexes (10,11). Using total intensity light scattering and turbidity measurements, they showed

0097–6156/91/0467–0303$06.00/0
© 1991 American Chemical Society

that aggregation increased with time for PMAA:PEG solutions. Aggregation was also enhanced by increasing the polymer concentration or increasing the temperature. A critical pH was found above which the complexes did not aggregate over a time scale of tens of minutes but below which aggregation was greatly enhanced.

Our objective in this chapter is to provide some unity to the combined topics of complexation and aggregation. We begin by reviewing previous work from other laboratories on complexation using classical as well as fluorescence techniques. We then summarize our current thinking on the molecular interactions giving rise to the photophysical behavior in PAA:PEG* and PMAA:PEG*. Finally, we present a very simple model to account for the size distribution of the aggregates as well as the time dependence of their growth. A common thread that runs through much of the discussion involves hydrophobic interactions of several forms. We will see that the structures of the isolated PMAA chain, of the isolated pyrene labeled PEG* chain, of the PMAA:PEG* complex, and of the large scale aggregates are all influenced by hydrophobic interactions that may dominate the hydrogen bond interactions. We begin, however, by providing a brief review of hydrogen bonding and its role in complexation.

BACKGROUND ON HYDROGEN BONDING AND POLYMER COMPLEXATION

The prerequisites for a hydrogen bond of significant strength are twofold: i) a hydrogen atom covalently bound to an electron withdrawing atom and ii) an acceptor with donatable electrons oriented at about $180°$ with respect to the first bond. The geometric consideration is quite important with the energy falling off rapidly with angle. The potential energy of the hydrogen bond may be explained by the Stockmayer equation (12) based on the electrostatic potential, the Lippincott-Schroder equation (13) based on chemical bonds, and the Scheraga equation (14) based on van der Waals and Coulombic interactions. The hydrogen bond energy is comparatively low, between 10 and 40 kJ/mol, which makes it stronger than a typical van der Waals bond (~1 kJ/mol) but still much weaker than covalent bonds (~500 kJ/mol). It is now accepted that the hydrogen bond is predominantly an electrostatic interaction (15,16). With few exceptions, the H atom is not shared but remains closer to and covalently bound to its parent atom.

A complex can be formed when a pair of water-soluble proton-donating and accepting polymers are mixed. The poly(carboxylic acid):PEG system has been studied by a variety of conventional methods including potentiometry (3,17-21), viscometry (3,17-21), turbidimetry (11,20), sedimentation (22), scanning electron microscopy (20), conductometry (20) and elemental analysis (4). Recent analytical work has been directed at providing a semiquantitative framework for complexation (23). Fluorescence methods have also been employed over the same time period. In most cases, a fluorescent label has been attached to either the poly(carboxylic acid) or the PEG. Anthracene (1,24-26), dansyl (5,27-29), 8-anilino-1-naphthalene sulfonic acid (ANS) (30) and pyrene (6,29) have been utilized. With anthracene and ANS labels, the most important technique is fluorescence depolarization from which relaxation times characterizing intramolecular chain mobility are determined. The intensity of emission from the dansyl label is very sensitive to the environment with an increase in intensity occurring upon complexation. Finally, pyrene excimer fluorescence and the modification of the monomer vibronic band structure provide an indication of local probe concentration and dielectric constant.

It is surprising that complexes are able to form so readily in aqueous solutions where there is such strong competition for hydrogen bonds from the water (31). In fact, for wholy small molecules it appears that the complexation is only marginally favored. However, when the interacting components are polymers, the summation for successive units along the two chains gives a sufficiently favorable enthalpy change to outweigh the unfavorable entropy contribution from aligning the two chains in forming the complex (3,32). Obviously, complex formation is caused by the cooperative interactions of long continuous sequences of functional groups on the polymer chain. Antipina concluded that in order for a complex to be formed between PMAA and PEG, the PEG molecular weight should not be less than 2000, while any significant interaction between PAA and PEG started only when the molecular weight was around 6000 (3). In addition, Ikawa did not observe any change in viscometric data when the PEG molecular weight was less than 1760 for the PMAA:PEG complex or 8800 for the PAA:PEG complex (32).

Complexation is also strongly affected by the degree of dissociation of the acid. The existence of a certain number of undissociated carboxy groups is necessary for PMAA and PEO to form a stable complex through hydrogen bonds. This dissociation is suppressed in the presence of poly(ethylene oxide) (PEO). For example, the dissociation constant (pK_a) 7.5 for PMAA:PEO (MW of PEO = 1300) and 7.9 for PMAA:PEO (MW of PEO = 25,000), whereas it is 7.3 for PMAA (2). At high pH where the number of active sites is insufficient, it is assumed that the enthalpy afforded by hydrogen bonds does not compensate for the decrease in entropy; this critical pH is about 5.7 for PMAA and about 4.8 for PAA.

EXPERIMENTAL

The source and purification of the PMAA and PAA are described elsewhere. (7,8) The viscosity average molecular weights were 9500 for the PMAA and 1850 for the PAA. Narrow distribution (polydispersity < 1.10) PEG of molecular weight 9200 was obtained from Polysciences. Solutions of PEG were made with glass-distilled deionized water. The pH was measured upon addition of either PMAA or PAA using a Beckman Phi 44 pH Meter calibrated to within 0.02 pH units. For blank pH measurements in which no complexation can occur, distilled, deionized water was used in place of the PEG solution. All fluorescence measurements were performed on a SPEX Fluorolog 212. To date we have examined emission and excitation spectra, lifetimes and absorption spectra. Only selected measurements will be described in this chapter, however. In the dynamic light scattering apparatus, the incident radiation is supplied by a Lexel two watt argon-ion laser operating at 514.5 nm. All measurements were done using 12 mm cylindrical cuvettes and a scattering angle of 90°. The autocorrelator receives a signal from the PMT proportional to the light scattering intensity and computes the second-order temporal correlation function (33-35). After normalization by the measured baseline, the second-order function may be related to the first-order correlation function to yield

$$g^{(1)}(\tau) = \int_0^\infty F(R) \exp\left\{-\left[\left(\frac{q^2 kT}{6\pi\eta}\right)R^{-1}\tau\right]\right\} dR \qquad (1)$$

where q is the scattering vector, k is Boltzmann's constant, T is absolute temperature, eta is the solvent viscosity, n is the refractive index of the solution and F(R) is the amount of light scattered by particles of size R.

The primary goal of the DLS data analysis is to obtain a reasonable estimate of F(R) given a series of measured second-order temporal correlation functions. We used the program CONTIN developed by Provencher (36,37) to perform an inverse Laplace transform of Equation (1) yielding the smoothest nonnegative solution for F(R) that is consistent with the signal to noise ratio for the data (38). The ability of CONTIN to determine the relaxation times of bimodal systems accurately has been demonstrated (35) along with its ability to determine F(R) correctly for polydisperse monomodal experimental systems with known particle size distributions (39).

RESULTS AND DISCUSSION

COMPLEXATION

FLUORESCENCE MEASUREMENTS Figure 1 shows earlier results for the normalized intramolecular and intermolecular I_D/I_M ratios for pyrene end-tagged PEG upon the addition of PMAA or PAA (7,8) In general, the intramolecular I_D/I_M decreases and the intermolecular I_D/I_M increases as polyacid is added. The PMAA data show a much stronger initial dependence on molar ratio, [PMAA]/[PEG]. During the initial stages of complexation, the intramolecular excimer in the PMAA:PEG* system is over twenty times more sensitive to additional polyacid than in the PAA:PEG* system. Similarly, the intermolecular excimer is initially over nine times more sensitive to the addition of PMAA, as compared to the PAA.

To gain a better understanding of the complexation in PMAA:PEG and PAA:PEG systems, we recently have examined their pH behavior (40). Such measurements allow calculation of the complexation equilibrium constant and the degree of complexation. In order to make comparisons of the pH measurements with the earlier spectroscopic work, we followed the identical experimental procedure that was used earlier. To be sure, this protocol was quite complex, with simultaneous variations in stoichiometery and in degree of dissociation occurring as the poly(carboxylic acid) was added to the PEG solution. (Buffered solutions could not be used because of insolubility of the pyrene labeled PEG). Nevertheless, the addition of pH results permit a consistent picture to emerge that unifies all of the fluorescence data.

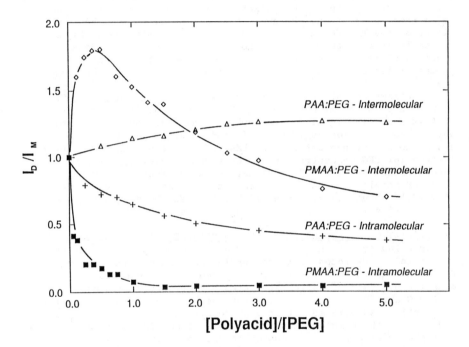

Figure 1. Normalized intermolecular and intramolecular contributions to I_D/I_M for complexes between PEG and either PMAA or PAA as a function of the stoichiometry. (Reprinted from Ref. 40. Copyright 1990 American Chemical Society.)

<u>pH MEASUREMENTS</u>   Figure 2 shows the change in pH of a 10 ml solution of $2x10^{-3}$ M or $5x10^{-2}$ M PEG when $2x10^{-1}$ M PAA is added. The monotonic drop in pH is solely due to the intrinsic acidity of the added polyacid. Most of the complexation in this system is expected to occur between molar ratios of zero and one because PAA and PEG form a 1:1 complex (2). PAA was also added to 10 ml of distilled water and the pH was monitored. If complexation were to occur between PAA and PEG, the pH of the solution containing PAA:PEG would be higher than that of the PAA:H_2O solution at a given molar ratio. This is because some of the acidic hydrogen atoms in the PAA will be participating in complexation hydrogen bonds, rather than being free in solution. However, there is no difference between either of the PAA:PEG mixtures and the PAA:H_2O solution in Figure 2, indicating that either PAA and PEG do not complex under these conditions, or that complexation is not detectable for PAA:PEG using pH measurements. We feel that the latter conclusion is the correct one.

The same set of experiments were also done on PMAA:PEG. The change in pH of 10 ml of $1x10^{-3}$ M PEG upon the addition of $1x10^{-1}$ M PMAA is shown in Figure 3. A blank run was also done to determine whether complexation could be monitored under these conditions. Again, no difference between the PMAA:PEG and the PMAA:H_2O results for the low PEG concentration was observed. However, for PEG concentration of $5x10^{-2}$ M complexation was detectable with pH. These pH results may be used to show that at molar ratios above above 0.1, about 80% of the available hydrogen bonding sites are participating in complexation hydrogen bonds. (40)

<u>INTERPRETATION OF FLUORESCENCE AND pH MEASUREMENTS</u>   Fluorescence data previously obtained for PMAA:PEG* are cross plotted against pH in Figure 4. We focus on these results because of their stronger dependence on pH than the PAA:PEG* system and because DLS measurements (41) of large aggregates have recently been made for PMAA:PEG and PMAA:PEG*. Our current thinking on the molecular level interactions associated with the photophysical measurements and, indirectly, the complexation process places considerable emphasis on hydrophobic effects, as discussed in the following.

We first consider the isolated PMAA chains. The hydrophobic nature of PMAA due to the methyl side group has been observed experimentally in many different systems (42-44) and has been successfully modelled (45-47). One manifestation of the hydrophobic nature of PMAA is that it undergoes a rather sharp coil contraction as the pH is lowered, leading to the formation of regions of greater hydrophobicity than the surroundings. For example, Chu and Thomas (47) studied PMAA using pyrene probes covalently attached at random points along the chain. At high pH the pyrenes produced a small amount of fluorescence, indicating that the pyrenes were in a water-rich environment. Conversely, at low pH there was a strong pyrene emission indicating that the pyrenes were in a hydrophobic environment. This abrupt change in chain configuration, occurring between pH 6 and 4, was attributed to the fact that at high pH a significant fraction of the carboxy groups are ionized. The tendency to maximize the separation between the ionic charges thus

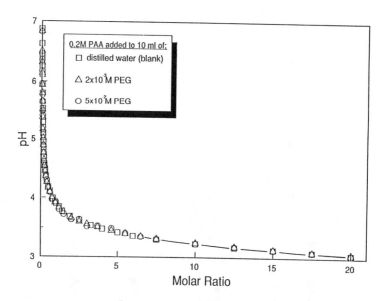

Figure 2. Dependence of pH on stoichiometry for addition of PAA to distilled water or PEG solutions. (Reprinted from Ref. 40. Copyright 1990 American Chemical Society.)

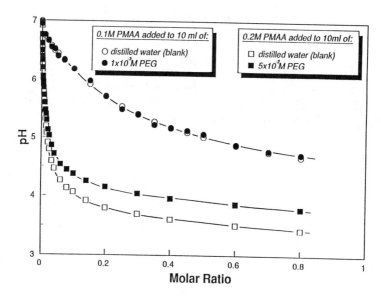

Figure 3. Dependence of pH on stoichiometry for addition of PMAA to distilled water or PEG solutions. (Reprinted from Ref. 40. Copyright 1990 American Chemical Society.)

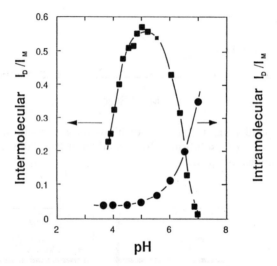

Figure 4.   Dependence of intramolecular and intermolecular I_D/I_M on pH for addition of PMAA to 1 X 10^{-3} M PEG.   (Reprinted from Ref.   40. Copyright 1990 American Chemical Society.)

produces an expanded chain. As the pH is lowered, the carboxylate groups become protonated and the chain relaxes into a more compact structure. They also showed that as the pH was lowered, it became more difficult for fluorescence quenchers to reach the pyrenes. A sharp transition in quenching ability with pH was found, further confirming the collapse of PMAA with decreasing pH.

Next, we consider the interactions possible in aqueous PEG* solutions containing no PMAA. We have shown earlier (8) that at low concentration excimer formation was considerably enhanced in water compared to the expected value resulting from intramolecuar end-to-end cyclization in an organic solvent. We attributed this observation to hydrophobic attraction between the pyrene groups and modeled the attraction in terms of a "hydrophobic capture radius". (47). This radius was about 20 Angstroms in pure water, but could be reduced to about 6 Angstroms upon addition of 30% methanol. Knowledge of the tendency for the pyrene groups to seek a hydrophobic environment is important for understanding complexation and aggregation in these labelled polymers, as we will see.

We are now in a position to discuss complex formation. When the first PMAA is added to the PEG solution, two events occur simultaneously. Not only do hydrogen bonds form between the acid and PEG, but some pyrenes on the end-tagged PEG are expected to be attracted to the hydrophobic regions of the PMAA. Both phenomena will lead to a reduced end-to-end cyclization rate and a much more rapid decrease in intramolecular I_D/I_M for the PMAA:PEG* complex than for the PAA:PEG* complex, where only complexation hydrogen bonds occur.

Differences in initial intermolecular I_D/I_M increases for PMAA:PEG* and PAA:PEG* also may be related to the hydrophobic effects. For PMAA:PEG* not only is there an increase in local pyrene concentration due to complexation, but there is also a contribution to that increase from the hydrophobic attraction of pyrene groups to the hydrophobic regions in the PMAA complex. It is interesting to note that the increase in intermolecular I_D/I_M occurs appreciably past the point at which PMAA coil collapse is expected, as shown in Figure 3. This implies that the pyrenes which are "hydrophobically aggregating" are still mobile enough, or are located correct distances apart, to form intermolecular excimers. As more PMAA is added, however, these hydrophobic regions will most likely increase in local density and thus the steric constraint on the intermolecular excimers resulting from loss of pyrene mobility will continue to increase. If this trend continues, eventually the excimers could become so destabilized that the excimer configuration is destroyed causing the intermolecular I_D/I_M to decrease.

Additional photophysical parameters can be used to support the general picture of complexation that we have described. For example, red shifts in the monomer excitation spectra in the complexes have been interpreted in terms of ground state interactions of the pyrene groups in the hydrophobic clusters (48). When the carboxy groups of the PMAA were 30% ionized by the addition of NaOH, no monomer excitation shift was observed in either the 100% or 1% tagged systems. The neutralization of some of the complexation bonds effectively breaks up the PMAA hydrophobic regions by forcing the PMAA chains into a more extended configuration. Without these consolidated hydrophobic domains, there is no hydrophobically enhanced pyrene aggregation and no ground-state interactions are observed.

Similarly, excimer excitation spectral blue shifts are explained by this model. Earlier work showed that as PMAA was added, the maximum in the excimer excitation spectrum shifted by 10 nm to shorter wavelengths (8). This is evidence of increasing excimer destabilization with increasing complexation with PMAA. No such shift in excimer peak was found for PAA:PEG. When the PMAA was 30% neutralized, no excimer peak shift was observed, further supporting the more extended configuration for PMAA.

These phenomena are also reflected in the intermolecular I_D/I_M data (48). At 30% neutralization, the PMAA:PEG* intermolecular I_D/I_M increases monotonically and does not pass through a maximum, qualitatively very similar to that for PAA shown in Figure 1. Comparable results were obtained for addition of methanol, which also causes dispersal of the hydrophobic regions in PMAA:PEG*. (48).

AGGREGATION

DYNAMIC LIGHT SCATTERING MEASUREMENTS The initial system chosen for examination with DLS (41) consisted of a solution of 2×10^{-3} M PMAA and 2×10^{-3} M unlabeled PEG in water at 25° C. The pH was adjusted with concentrated HCl between values of 3.0 and 1.7 and the fraction of light scattered by particles of a given radius in those solutions immediately after mixing was determined by CONTIN. The size distribution is rather broad; for pH = 2.75 the particles range in size from ~20 nm to ~126 nm. In order to compare this result with other aggregation work, we have used Flamberg and Pecora's definition of the average radius of the distribution (49). Based on this definition, the solution giving rise to the scattered light intensity distribution for pH 2.75 has an average radius of 48 nm, which did not significantly change over the course of 100 minutes. Figure 5 shows the initial average aggregate radius as a function of pH. This figure exhibits the same general behavior as the total scattered light intensity vs pH curve of Tsuchida, which showed a critical pH of 3. In the case of Figure 5, however, the critical pH appears to have been shifted to the lower value of about 1.9 because of the lower molecular weight of our PMAA.

There are many studies that show that complexation occurs rapidly, on the order of milliseconds (11). In order to investigate this aggregation process further, we attempted to find a set of conditions for which the kinetics would be slow enough to allow us to monitor the aggregation with DLS. When the pH of the PMAA:PEG system was lowered to the critical value of 1.9, a definite time dependence of the average radius with time was observed. This increase is plotted in Figure 6 and is mathematically described by a power law relationship $R = R't^b$ where t is time, $R' = 68$ nm and b = 0.11. A value for b less than one indicates that the rate of size increase decreases with time, whereas a value for b greater than one indicates an ever accelerating reaction where the rate of increase in size increases with time. Note, R' should be thought of as a measure of aggregate size at an arbitrary time and not as the initial aggregate size.

The power-law relationship accurately describes the time dependence of the average radius of this system only over a

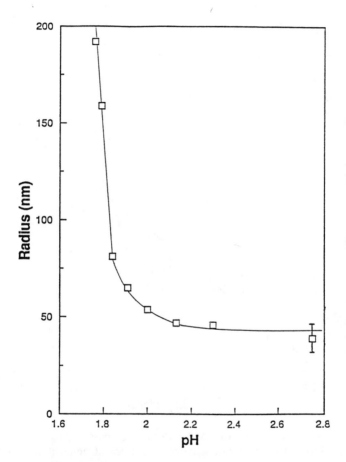

Figure 5. Average aggregate radius as a function of pH for PMAA:PEG immediately after mixing. (Reprinted from Ref. 41. Copyright 1990 American Chemical Society.)

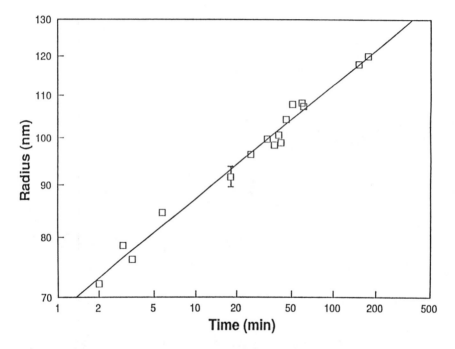

Figure 6. Rate of growth of PMAA:PEG aggregate radius for pH= 1.9. (Reprinted from Ref. 41. Copyright 1990 American Chemical Society.)

restricted time range. Nevertheless, the constants can be used to compare rate data for different systems. For example, from our earlier discussion on hydrophobic effects in the complexes we are very interested in the influence of the pyrene labels on the overall aggregation process. To examine this, an identical solution used to obtain Figure 6 was made with the exception of using labeled PEG rather than unlabeled. However, when the first DLS measurement was made on the solution, approximately one minute after mixing, the solution had already aggregated so rapidly that the average radius had reached ~1000 nm. In order to slow down the aggregation process, a new solution with pH of 3.0 was prepared. The kinetic data for this solution also follow a power-law with constants R' = 123 nm and b = 0.31.

DIFFUSION LIMITED CLUSTER-CLUSTER AGGREGATION MODEL. In considering the basic problem of aggregation, there are two questions that must be answered. The first concerns the limiting step in the aggregation. The general process of aggregation involves two particles diffusing toward each other. When they are some characteristic distance apart, there is a certain probability that the interaction forces between the two particles will be such that the particles stick together via e.g. a hydrogen bond. If the sticking probability is high in comparison to the diffusion rate of the particles, the process is known as diffusion limited aggregation (DLA). If, on the other hand, the diffusion rate is high in comparison to the sticking reaction rate, the process is known as reaction limited aggregation (RLA).

The second question concerns the relative mobility of aggregates. Again, two limiting cases exist. In the first, only the individual sub-particles that make up all larger aggregates are allowed to diffuse. This model was first investigated in a computer simulation by Witten and Sander (50). Aggregates grow via the addition of individual sub-particles. This process is known as particle-cluster aggregation. The second limiting case allows both individual particles and all larger aggregates to diffuse freely. In this model, aggregates grow by the combination of two aggregates of any size. This process is known as cluster-cluster aggregation.

As a starting point for the development of our model, we have assumed that our aggregation system is described by the diffusion limited aggregation, and that cluster-cluster aggregation is the dominant mode for aggregate growth. This implies that aggregates are noninteracting until they stick together irreversibly on contact. The effects on aggregation of long range interactions, such as dipole or screened Coulomb, have been investigated by Hurd (51) and are assumed to be negligible.

The time dependence of aggregate mass, assuming cluster-cluster aggregation, for a solution initially consisting of N_o particles of radius R_o can be described by the generalized Smoluchowski equations (52,53). For DLA the diffusion kernel is equal to the sum of the capture radii for the two clusters times the sum of their diffusion coefficients (54). If we assume that the capture radius scales with the cluster radius and that the clusters undergo Stokes-Einstein diffusion, the rate of aggregation of an i-mer with a j-mer will be independent of cluster size.

Using the constant kernel assumption, the Smoluchowski

equations can be integrated to give the time dependence of an
aggregation number distribution for cluster-cluster, diffusion
limited aggregation.

$$\frac{N_i}{N_0} = \frac{A^{(i-1)}}{(1+A)^{(i+1)}} \tag{2}$$

where $A = k_s N_0 t$. The constant k_s is the rate constant for the
aggregation of any two clusters.

The next step is to write Equation (2) in terms of aggregate
radius, R_i, rather than number of sub-particles. A general scaling
to accomplish this is provided by

$$i = \left(\frac{R}{R_0}\right)^D \tag{3}$$

where D is the scaling dimensionality of the system. By allowing D
to vary in the fitting procedure, a best-fit estimate of the system
dimensionality can be obtained.

Equation (3) can be inserted into Equation (2) to give an
expression for the number distribution of particles in terms of
radius. This number distribution cannot, however, be directly
compared with the light scattering intensity distribution calculated
by CONTIN. In order to relate the two distributions, we follow the
method of Flamberg and Pecora (49) to obtain

$$F(R) = R^D \left[\frac{A^{\left(\frac{R}{R_0}\right)^D}}{(1+A)^{\left(\frac{R}{R_0}\right)^D}} \right] \tag{4}$$

for the normalized intensity distribution. When Equation (4) was
fit to the CONTIN output, convergence was obtained. This least-
squares fitting process was repeated for all distributions and an
average dimensionality of 1.7 +/- 0.25 was found. This exponent was
found in both the PMAA:PEG and PMAA:PEG* systems and was constant in
time.

The non-integral exponent implies a system with dilational
symmetry, i.e., the aggregates can be described as having "fractal"
dimensionality (55). Several computer simulations (56-58) have
shown that cluster-cluster aggregation in the DLA regime should
produce a dimensionality of 1.8. Microscopy experiments on
diffusion limited colloidal gold aggregates done by Weitz (54)
produced a fractal dimension of 1.7. Likewise, neutron and light
scattering done on the same particles gave rise to a fractal

exponent of 1.8. However, under reaction limited aggregation conditions, Weitz (59) found a fractal dimension of 2.0. The close agreement between these studies and our data supports our proposal that the PMAA:PEG aggregates are fractal in nature and grow via DLA and cluster-cluster aggregation.

SUMMARY

As either PMAA or PAA is added to an aqueous PEG solution, complexation occurs and is stabilized by the hydrogen bonds between the polyacid and the PEG chains. Additionally, in the PMAA system, some pyrenes covalently attached to the ends of PEG chains are attracted to hydrophobic regions. The pyrenes interact in the ground state and participate in preformed excimers. This phenomenon manifests itself in red-shifted monomer excitation and absorption spectra, blue shifted excimer excitation spectra, increases in monomer lifetimes, the absence of excimer transient rise times (not discussed here), a rapid initial decrease in intramolecular I_D/I_M, and a rapid initial increase in intermolecular I_D/I_M. Since there are no hydrophobic regions in the analogous PAA:PEG systems, none of the non-hydrogen bonding phenomena are observed. In fact, all of these observations can be eliminated by destroying the consolidated hydrophobic regions in PMAA complexes. This has been experimentally accomplished by the ionization of the carboxy groups in PMAA or by adding methanol. At sufficiently low pH, macroscopic flocculation occurs with a power law dependence of the average particle size on time. Interestingly, the pyrene labeled PEG exhibits much faster aggregation with PMAA than does the unlabeled PMAA. A diffusion limited cluster-cluster aggregation model with fractal dimension of 1.7 has been shown to be adequate to interpret the data.

ACKNOWLEDGEMENT

This work was supported by the Polymers Program of the National Science Foundation Division of Materials Research under DMR 84-07847.

REFERENCES

1. Bekturov, E.A.; Bimendina, L.A. Adv. Polym. Sci. 1981, 41, 99.
2. Tsuchida, E.; Abe, K. Adv. Polym. Sci. 1982, 45.
3. Antipina, A.D.; Baranovskii, V. Yu.; Papisov, I.M.; Kabanov, V.A. Vysokomol. Soedin, Ser. A 1972, A14, 941; Polym Sci. USSR (Engl. Transl.) 1972, 14, 1047.
4. Abe, K.; Koide, M.; Tsuchida, E. Macromolecules 1977, 10, 1259.
5. Chen, H.L.; Morawetz, H. Macromolecules 1982, 15 1445.
6. Turro, N.J.; Arora, K.S. Polymer 1986, 27, 783.
7. Oyama, H.T.; Tang, W.T.; Frank, C.W. Macromolecules 1987, 20. 474.

8.   Oyama, H.T.; Tang, W.T.; Frank, C.W. Macromolecules 1987,
     20, 1839.
9.   Frank, C.W.; Oyama, H.T.; Hemker, D.J. in "Frontiers of
     Macromolecular Science," T. Saegusa, T. Higashimura, A.
     Abe, eds., 1989, pp 337-342.
10. Tsuchida, E.; Osada, Y.; Ohno, H. J. Macromol. Sci. Phys.
     1980, B17(4), 683.
11. Ohno, H.; Matsuda, H,; Tsuchida, E. Makromol. Chem. 1981,
     182, 2267.
12. Stockmayer, W.H. J. Phys. Chem., 1941, 9, 398.
13. Lippincott, E.R. J. Chem. Phys., 1955, 23, 1099.
14. Scheraga, H.A. Biochemistry 1967, 6, 3719.
15. Coulson. C.A; "Valence", 2nd ed., Ch, XIII, Oxford
     University Press, London and New York.
16. Umeyama, M.; Morokuma, K. J. Am. Chem. Soc., 1977, 99,
     1316.
17. Bailey, Jr., F.E.; Lundberg, R.D.; Callard, R.W. J. Polym.
     Sci. 1964, A2, 845.
18. Papisov, I.M.; Baranovskii, V. Yu.; Sergieva, Ye. I.;
     Antipina, A.D.; Kabanov, V.A. Vysokomol. soyed. 1974, A16,
     1133.
19. Osada, Y.; Sato, M. J. Poly. Sci., Polym. Lett. Ed. 1976,
     14, 129.
20. Chatterjee, S.K.; Malhotra, A.; Pachauri, L.S. Angew.
     Makromol Chem., 1983, 116, 99.
21. Illiopoulos, I.; Audebert, R. Polym. Bull., 1985, 13, -
     171.
22. Papisov, I.M.; Branovskii, V. Yu; Kabanov, V.A. Vysokomol.
     soyed., 1975, A17, 2104. (Translated in Polymer Sci. USSR,
     1975, 17, 2428.
23. Iliopoulos, I.; Audebert, R. J. Polym. Sci. Polym. Phys.
     Ed. 1988, 26, 2093.
24. Anufrieva, Ye. V.; Gotlib, Yu. Ya.; Krakovyak, M.G.;
     Skorokhodov, S.S. Vysokomol. soyed., 1972, A14, 1430
     (Translated in Polymer Sci. USSR, 1972, 14, 1604.
25. Anufrieva, E.V.; Pautov, V.D.; Geller, N.M.; Krakoviak,
     M.G.; Papisov, I.M. Dokl. Akad. Nauk USSR, 1975, 220, 353.
26. Anufrieva, E.V.; Pautov, V.D.; Papisov, N.M.; Kabanov, V.A.
     Dokl. Akad. Nauk. USSR 1977, 232, 1096.
27. Chen, H.L.; Morawetz, H. Eur. Polym. J. 1983, 19, 923.
28. Bednar, B.; Morawetz, H.; Shafer, J.A. Macromolecules 1984,
     17, 1634.
29. Bednar, B.; Li, Z.; Huang, Y.; Chang, L.C.P.; Morawetz, H.
     Macromolecules, 1985, 18, 1829.
30. Ohno, H.; Tsuchida, E.; Makromol. Chem. Rapid Commun. 1980,
     1, 591.
31. Molyneux P.; "Water-Soluble Synthetic Polymers: Properties
     and Behavior" 1984, 2, 172.
32. Ikawa, T.; Abe, K.; Honda, K.; Tsuchida, E. J. Polym. Sci.,
     Polym. Chem. Ed., 1975, 13, 1505.
33. Berne, B.J.; Pecora, R. "Dynamic Light Scattering", Wiley-
     Interscience, New, York, 1976.
34. Pecora, R., ed. "Dynamic Light Scattering: Applications of
     Photon Correlation Spectroscopy", Plenum Press, New York,
     1985.

35.   Flamberg, A.; Pecora, R. J. Phys. Chem. 1984, 88, 3026.
36.   Provencher, S.W. Comp. Phys. Comm. 1982, 27, 229.
37.   Provencher, S.W. Comp. Phys. Comm. 1982, 27, 213.
38.   Provencher, S.W. "CONTIN" Users's Manual, European
      Molecular Biology Laboratory Technical Report #MBL-DA02,
      eidelberg, 1980.
39.   Provencher, S.W.; Hendrix, J.; De Maeyer, L.; Paulussen, N.
      J. Chem. Phys. 1978, 69, 4273.
40.   Hemker, D.J; Garza, V; Frank, C.W. Macromolecules, in
      press.
41.   Hemker, D.J.; Frank, C.W. Macromolecules, in press.
42.   Chu, D.Y.; Thomas, J.K. Macromolecules 1984, 17, 2142.
43.   Char, K.; Frank, C.W.; Gast, A.P.; Tang, W.T.
      Macromolecules 1987, 20, 1833.
44.   Israelachvili, J.; Pashley, R. Nature 1982, 300, 341.
45.   Dashevsky, V.G.; Sarkisov, G.N. Molec. Phys. 1974, 27,
      1271.
46.   Marcelja, S.; Mitchell, D.J.; Ninham, B. W.; Sculley, M. J.
      J. Chem. Soc. Faraday Trans 1977, 73, 630.
47.   Char, K.; Frank, C.W.; Gast, A.P. Macromolecules 1989, 22,
      3177.
48.   Oyama, H.T.; Hemker, D.J.; Frank, C.W. Macromolecules,
      1989, 22, 1255.
49.   Flamberg, A.; Pecora, R. J. Phys. Chem. 1984, 88, 3026.
50.   Witten Jr., T.A.; Sanders, L.M. Phys. Rev. Lett. 1981,
      47, 1400.
51.   Hurd, A.J.; Schaefer, D.W. Phys. Rev. Lett. 1985, 54, 1043.
52.   Von Smoluchowski, M. Phys. Z. 1916, 17, 593.
53.   Sonntag, H.; Strenge, K."Coagulation Kinetics and Structure
      Formation,"Plenum Press, New York, 1987, p 58.
54.   Weitz, D.A.; Lin, M.Y.; Huang, J.S. "Fractals and Scaling
      in Kinetic Colloid Aggregation", Exxon Monograph, 1987.
55.   Mandelbrot, B.B."The Fractal Geometry of Nature", Freeman,
      San Francisco, 1982.
56.   Meakin, P. Phys. Rev. Lett. 1983 51, 1119.
57.   Kolb, H.; Botet, R; Jullien, R. Phys. Rev. Lett. 1983, 51,
      1123.
58.   Botet, R.; Jullien, R.; Kolb, M.J. Phys. A: Math. Gen.
      1984, 17, 175.
59.   Weitz, D.A.; Huang, J.S.; Lin, M.Y.; Sung, J. Phys. Rev.
      Lett. 1985, 55, 1657.

RECEIVED July 2, 1990

Chapter 21

Roles of Molecular Structure and Solvation on Drag Reduction in Aqueous Solutions

Charles L. McCormick, Sarah E. Morgan, and Roger D. Hester

Department of Polymer Science, The University of Southern Mississippi, Hattiesburg, MS 39406-0076

Drag reduction performance of water-soluble copolymers tailored with specific structural features has been examined. These copolymers including polyelectrolytes, polyampholytes, and hydrophobically-modified polymers respond to changes in ionic strength as indicated by changes in hydrodynamic volume in aqueous solution. Drag reduction performance is greatly affected by polymer microstructure and by solvation. A new method of data representation in which drag reduction efficiency is shown as a function of volume fraction allows comparison of a large number of polymer types. Further normalization utilizing an empirical shift factor allows all data to fall on a single efficiency curve. Results of this study suggest that predictive dynamic extensional models might be improved by inclusion of parameters reflective of solvent and associative interactions as well as hydrodynamic volume.

The phenomenon of drag reduction (DR) was first reported by Toms four decades ago (1). Frictional resistance in turbulent flow can be reduced to as little as one-quarter of that of pure solvent by the addition of certain flow modifiers. Numerous studies of various polymers in both aqueous and organic solvents have shown that DR is affected by molecular weight, concentration, chain flexibility, and a number of other parameters. However, a quantitative understanding of the phenomenon has still eluded investigators; a number of conflicting theories and experimental results exist throughout the extensive literature on this subject.

Most theories suggest that polymer molecules interfere with production, growth, and transport of turbulent disturbances. Recent evidence points to the importance of molecular extension in energy dissipation. DR models may be classified somewhat arbitrarily into length scale, time scale and energy theories. Length scale models such as those of Virk (2,3) and Hlavacek (4,5) correlate with polymer chain length or radius of gyration. Time scale models of Lumley (6-8) and Ryskin (9,10) can be related to polymer relaxation time. Energy models generally deal with the ability of polymers to alter the energy balance in turbulent flow with major contributions from Virk (3), Walsh (11), Kohn (12),

0097–6156/91/0467–0320$06.00/0

and de Gennes (13). Berman (14) has suggested the operation of multiple mechanisms.

Despite conflicting explanations of DR, the direct correlation of coil size (or changes therein) to DR efficiency has been repeatedly demonstrated. However, it is not clear whether the parameter providing the best correlation is molecular weight, degree of polymerization, radius of gyration, or hydrodynamic volume. Correlations are complicated by solvent interactions and draining characteristics, molecular weight distribution, and substitutional patterns governing effective segment lengths. Recent reviews of molecular parameters in DR are found in references 15 and 16.

The purpose of our continuing research (16-19; McCormick, C. L. et al., Macromolecules, two articles in press; Safieddine, A. M., Ph.D. Thesis, University of Southern Mississippi, in press) is to systematically examine the interrelationships between structure and drag reduction performance of structurally tailored water-soluble copolymers. The effects of molecular parameters are examined in this work utilizing copolymer models whose dimensions rely on specific polymer/polymer or polymer/solvent interactions under given conditions of ionic strength. Copolymers include uncharged, hydrophilic macromolecules, polyampholytes, polyelectrolytes, and hydrophobically associating systems. Extensive studies of dilute solution properties including viscosity, hydrodynamic volume, rheology, and phase behavior have accompanied DR measurements.

Experimental

Materials.   Three molecular weight grades of poly(ethylene oxide) (PEO) were purchased from Union Carbide Corporation. Acrylamide (AM) and acrylic acid (AA) from Aldrich, diacetone acrylamide (DAAM) and 2-acrylamido-2-methylpropane sulfonic acid (AMPS) from Polysciences were purified by three recrystallizations from acetone or methanol. The monomer 3-acrylamido-3-methyl butanoic acid (AMBA) was synthesized via a Ritter reaction using a previously published procedure (20). 2-Acrylamido-2-methylpropanedimethyl-ammonium chloride (AMPDAC) synthesis was also reported previously (21).

Model copolymers and terpolymers utilized in drag reduction studies were synthesized and thoroughly characterized previously in our research group. These include: uncharged homopolyacrylamide and DAAM copolymers with AM (22); polyelectrolytes of AM with AMPS or AMBA (23); polyampholytes containing AMPS and AMPDAC (24,25).

Polymer Characterization.   Polymer compositions were determined from elemental analyses (M-H-W Laboratories, Phoenix) and ^{13}C-NMR. Intrinsic viscosities were determined on a Contraves Low Shear 30 Rheometer. Classical light scattering studies were performed using a Chromatix KMX-6 low angle laser light scattering spectrophotometer utilizing a 2mW He-Ne laser operating at 633 nm to obtain weight average molecular weights. Specific refractive index increment was determined with a Chromatix KMX-16 laser differential refractometer.

Quasielastic light scattering studies, yielding the translational diffusion coefficient were performed with the KMX-6 in conjunction with a Langley-Ford Model LFI-64 channel digital correlator. Hydrodynamic diameter was calculated from the diffusional coefficient using the Stokes-Einstein relationship.

Drag Reduction Measurements.   Polymer solutions were prepared by dissolving the required mass of polymer with 0.01% NaN_3 biocide in solvent in one liter flasks by gentle room temperature stirring for 48 hours. Solutions were then transferred to polypropylene tanks and diluted with solvent (deionized water or 0.514 M aqueous NaCl) to twenty liters, and gently stirred for an additional 24 hours before drag reduction measurements were made at 25°C.

Polymeric solutions were tested for drag reduction performance in both a rotating disk and a tube flow apparatus. The first system consisted of a modified Haake, Model RV3, rheometer equipped with a rotating disk. The stainless steel disk was 9 cm in radius, 2mm in thickness, and was machined to insure flatness and smoothness. The disk was centered in a chamber with the depth in the fluid being adjustable by changing the length of a stainless steel shaft to which the disk was attached. The chamber was a Pyrex jar 305 mm in diameter by 457 mm in height with a capacity of 33.4 liters.

The disk was driven by a variable speed motor. The motor drive and torque sensing unit are components from a Haake RV3 rotoviscometer. The torque applied to the rotating disk was determined using the stress measuring head which was calibrated using the method described by the manufacturer (26). The data acquisition system for recording the torque consisted of a Hewlett-Packard 41C calculator connected to an ADC 41 (Interface Instruments, Corvallis, OR) analog to digital interface. The experimental data, torque (τ_q) and disk angular velocity (ω), were converted to Reynolds number (Re) and friction factor (f) using Equations 1 and 2 which were developed for a disk of radius (R) rotating in an unbounded fluid of viscosity μ and density ρ (27,28).

$$f = \tau_q/(\pi\rho\omega^2R^5) \tag{1}$$

$$Re = \rho R^2\omega/\mu \tag{2}$$

The tube flow apparatus consisted of a smooth stainless steel tube 102 cm in length, L, with a diameter, D, of 0.210 cm. This single pass testing system was driven by a high pressure nitrogen gas source. Pressure taps were placed at 150 L/D and 350 L/D downstream from the tube entrance to determine pressure drop. Pressure was measured by a Validyne DP 15 differential pressure transducer. Flow rate was monitored by a load cell with strain gauges connected to a Hewlett-Packard plotter where weight was plotted as a function of time. From measured pressure drop and flow rate, friction factor and Reynolds number were calculated (27).

Results and Discussion

Tailored Copolymers. Synthetic copolymers (Table I) were prepared with structural features of particular interest when examining the drag reduction phenomenon. Acrylamide (AM) is a neutral, hydrophilic monomer with a high rate of propagation in aqueous solution. Copolymers of AM with sodium acrylate (NaA), sodium-2-acrylamido-2-methylpropane sulfonate (NaAMPS) and sodium-3-acrylamido-3-methylbutanoate (NaAMB) are anionic polyelectrolytes. The latter polymers are more electrolyte tolerant than the NaA copolymers due to the somewhat more hydrophobic character introduced by the geminal dimethyl groups and apparent intramolecular chain stiffening (29,30).

The uncharged diacetone acrylamide monomer (DAAM) is more hydrophobic than acrylamide. AM/DAAM copolymers have unusual intermolecular associations which increase at certain copolymer compositions with addition of electrolytes (22). High concentrations of DAAM result in intramolecular associations with micelle-like structures.

The polyampholytes of this study, the ADAS (24) and ADASAM (25), have unusual properties in that viscosity increases upon addition of simple electrolytes to the respective aqueous solutions. This "antipolyelectrolyte" behavior is due to disruption of strong ionic interactions between adjacent or closely spaced mers. It should be noted that unbalanced compositions of cationic and anionic units leads to traditional polyelectrolyte behavior. Additionally, the ADASAM terpolymers or low charge density ampholytes are much larger than the high charge density ADAS copolymers.

A number of solution parameters of the model copolymers of this study are listed in Table I along with structural data. Zero shear intrinsic viscosity $[\eta]$ from low shear rheometry, weight average molecular weights M_w and second virial coefficients A_2 from low angle laser light scattering, and translational diffusion coefficients D_o from quasielastic light scattering are given for each sample. Hydrodynamic diameter d_o was calculated from D_o using the Stokes-Einstein relationship.

Drag Reduction Studies. We continue to use a method of data reduction previously reported by our group (McCormick, C. L. et al., *Macromolecules*, two articles in press) which is quite instructive when examining the role of molecular parameters and solvation. Percent drag reduction is defined by Equation (1) in which f_s and f_p represent the friction factors of solvent and polymer solution, respectively.

$$\% \text{ DR} = [(f_s - f_p)/f_s] \times 100 \tag{3}$$

Values of friction factor can be obtained from plots of the type shown in Figure 1 at a given Reynolds number above the laminar/turbulent transition. Normalized DR is then plotted vs $[\eta]C$, a dimensionless parameter related to polymer volume fraction. Figure 2 is representative of this relationship, which we call an efficiency plot, for selected polymer models (Table I) tested in either tube flow or rotating disk geometries. Normalization procedures allow direct comparison of diverse polymer types with different degrees of polymerization and solvation.

Each branch of the family of curves represents a particular polymer type (symbols for legends are found in Table II). The most efficient have high values of $\% \text{ DR}/[\eta]C$ at low volume fractions. In each case a maximum in DR efficiency is obtained beyond which increasing values of $[\eta]C$ only serve to reduce efficiency. At particular abscissa values for each polymer type, a common slope is reached; eventually at high values of $[\eta]C$, the curves merge into a single line.

Effects of Composition. Close examination of Figure 2 reveals that the polymers of our study with the greatest drag reduction efficiency ($\%DR/[\eta]C$) at the lowest volume fraction are the diacetone acrylamide copolymers (DAAM). The uncharged homopolyacrylamide (PAM) and polyethylene oxide (PEO) yield moderate values; the homopolyelectrolytes are the least efficient.

Figure 3 is a plot of DR efficiency curves for the NaAMB copolyelectrolytes in 0.514 N NaCl. Those copolymers with the lowest mole percentages of the charged NaAMB comonomer exhibit the highest efficiencies. The NaAMB homopolymer is the least efficient. These trends parallel the hydrodynamic volume as measured by intrinsic viscosity and light scattering (Table I). The DAAM copolymers, on the other hand, show increasing DR efficiency with increasing incorporation of the hydrophobic comonomer (Figure 4); this trend does not parallel hydrodynamic volume or molecular weight. It

Table I
Structure and Solution Properties of Copolymer Models

Sample	Repeating Units [mole %]		Intrinsic Viscosity [η] (dl/g)			Light Scattering			
			H_2O	0.514 M NaCl	1 M Urea	$M_w \times 10^4$ (g/mol)[b]	$A_2 \times 10^4$ (mol cm³/g²)[c]	$D_0 \times 10^8$ (cm²/sec)[d]	d_0 Å[e]
WSR-301	+O—CH₂—CH₂+	[100]	—	16.0	—	5.3*	—	—	—
WSR-N-60K	"	[100]	—	13.0	—	4.0*	—	—	—
WSR-N-12K	"	[100]	—	6.4	—	1.7*	—	—	—
PAM-4	+CH₂—CH+ / C=O / NH₂	[100]	35.0	34.0	—	24.0	3.0	1.3	3300
PAM-MC	"	[100]	9.6	9.3	—	6.0	3.0	4.2	1600
	+CH₂—CH+ / C=O / NH₂	+CH₂—CH+ / C=O / NH / CH₃—C—CH₃ / CH₂ / COO⁻Na⁺							
NaAMB 5	[95]	[5]	—	22.0	—	24.0	2.4	1.9	2900
NaAMB 10	[90]	[10]	—	47.0	—	28.0	3.2	1.4	3300
NaAMB 10*	[90]	[10]	—	1.8	—	—	—	—	—
NaAMB 25	[78]	[22]	—	52.0	—	25.0	3.9	2.5	2200
NaAMB 25*	[76]	[24]	—	6.3	—	—	—	—	—
NaAMB 40	[66]	[34]	—	40.0	—	22.0	3.8	2.7	2000
NaAMB 40*	[64]	[36]	—	9.0	—	—	—	—	—
NaAMB 100	[0]	[100]	—	8.4	—	3.6	4.5	2.9	1600

Table I (continued)

	+CH₂—CH+ / C=O / NH₂	+CH₂—CH+ / C=O / NH / CH₂—C—CH₃ / CH₂ / C=O / CH₃						
DAAM 15	[83]	[17]	15.0	18.0	13.0	7.5(3.5)	1.3(3.0)	1.2(2.9)	3800(1700)
DAAM 20	[80]	[20]	11.0	16.0	14.0	6.9(4.3)	1.7(2.6)	1.8(3.1)	2600(1600)
DAAM 25	[74]	[26]	7.0	8.4	8.5	3.9(2.6)	2.1(2.1)	2.9(4.4)	1700(1200)
DAAM 30	[71]	[29]	6.2	8.0	—	6.1(5.9)	1.7(3.0)	2.8(3.3)	1700(1500)
DAAM 35	[65]	[35]	3.1	3.4	15.0	3.6(3.4)	4.0(2.7)	3.3(4.1)	1400(1200)

	+CH₂—CH+ / C=O / NH₂	+CH₂—CH+ / C=O / NH / CH₂—C—CH₃ / CH₃ / H₃C—N—CH₃ / H	+CH₂—CH+ / C=O / NH / CH₂—C—CH₃ / CH₃ / SO₃⁻						
ADASAM 2.5-2.5	[91.2]	[5.4]	[3.4]	4.5	14.0	—	7.4	2.9	1.7	2800
ADASAM 5-5	[85.6]	[8.2]	[6.2]	4.0	17.0	—	8.8	9.3	1.7	2700
ADASAM 5-10	[81.7]	[8.3]	[10.0]	24.0**	20.0	—	10.0	2.4	1.1	4200
ADAS 10	[0]	[27.0]	[73.0]	20.0**	8.0	—	9.5	1.6	2.5	2000
ADAS 50	[0]	[47.0]	[53.0]	4.0	9.0	—	2.7	4.5	4.0	1000

*Mole % in copolymer.
bWeight average molecular weight.
cSecond virial coefficient.
dTranslational diffusion coefficient.
eHydrodynamic diameter calculated from diffusional coefficient.

Data for b, c, d, and e were determined in 0.514 M NaCl.
Values in parentheses are for deionized water.

*Viscosity average molecular weight.
**Approximate values (polyelectrolyte effect).

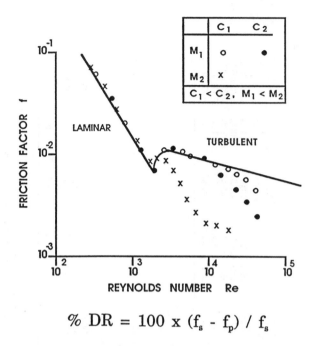

$$\% \; DR = 100 \; x \; (f_s - f_p) \; / \; f_s$$

Figure 1. Friction factor vs Reynolds number plot for three polymers at specific concentrations and molecular weights. The solid line represents the solvent response.

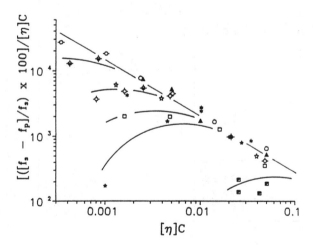

Figure 2. Copolymer solutions in 0.514 M NaCl solvent tested in disk flow at Re = 520,000. See Table II for symbol descriptions.

Table II
Symbol Table and Polymer Shift Factors

Polymer/Solvent	Symbol	Shift Factors in Disk Flow
Solvents:		
DI water	✳	—
0.514 M NaCl	✹	—
n-hexane	⊕	—
Homopolymers:		
PEO WSR-301	□	1.0
PAM-4	✱	2.0
Copolymers:		
DAAM 15	×	1.3
DAAM 20	⟡	2.2
DAAM 25	⬦	6.5
DAAM 30	◐	8.0
DAAM 35	⊸	10.0
NaA 5	☆	3.0
NaA 10	★	3.0
NaA 20	✦	8.0
NaA 35	✿	0.35
NaAMB 5	+	3.5
NaAMB 10	○	3.5
NaAMB 25	△	3.5
NaAMB 40	▲	3.5
NaAMB 100	◪	0.075
NaAMPS 5	■	2.1
NaAMPS 10	◊	2.1
NaAMPS 15	✛	2.1
NaAMPS 20	⌀	2.1
NaAMPS 35	●	2.1
NaAMPS 100	◢	0.09

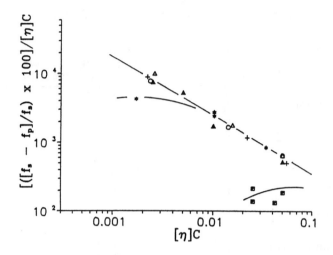

Figure 3. NaAMB copolymer solutions in 0.514 M NaCl solvent tested in disk flow at Re = 520,000. See Table II for symbol descriptions.

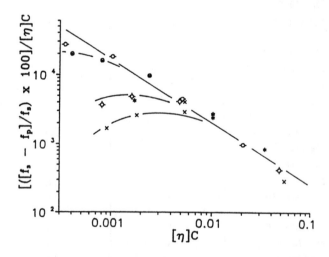

Figure 4. DAAM copolymer solutions in 0.514 M NaCl solvent tested in disk flow at Re = 520,000. See Table II for symbol descriptions.

is interesting to note that DAAM copolymers except DAAM 15 are more efficient than PAM-4 despite having much lower molecular weights. ADASAM 5-5 with d_o of 2700 Å and NaAMB 10 with d_o of 3300 Å in 0.514 NaCl exhibit higher DR efficiencies than the higher charge-density ADAS 50 (1000 Å) and NaAMB 40 (2000 Å). Solvent ordering in the vicinity of the relatively hydrophobic moieties, disordering by charged mers, and associations are likely responsible for observed behavior and will be discussed further in this chapter.

Effect of Molecular Weight. The effect of changes in molecular weight on DR behavior was studied using polymer samples with identical chemical structure but different molecular weights. Figure 5a is a plot of friction factor vs Reynolds number for 3 ppm solutions of the three PEO samples (Table I) measured in the rotating disk apparatus. As might be expected, when polymer concentration is held constant, friction reduction increases with molecular weight. If polymer concentration is adjusted to yield constant volume fraction (Figure 5b), the three samples show similar curves. Likewise (Figure 5c) plots of DR efficiency yield nearly superimposable curves for the three samples of differing molecular weight. The lowest molecular weight sample (WSR-N-12K) is slightly less effective than the other two, perhaps indicating that there may be a lower limit for significant drag reduction.

The NaAMB copolymers with nearly identical compositions (Table I) but different molecular weights exhibit similar DR trends to those above. In Figures 6a and 6b, friction factor vs Re plots of low and high molecular weight NaAMB samples are displayed at 3 ppm and at constant volume fraction, respectively. Volume fraction normalization in Figures 6b and in the DR efficiency curve 6c result in single curves for both samples.

The results of this portion of our study indicate that DR performance correlates well with volume fraction for polymers of singular composition but different molecular weight. The fact that a single efficiency curve results indicates that normalizing %DR for volume fraction also normalizes for molecular weight effects. This is extremely important for facile comparison of synthetic copolymers. It should be pointed out that no attempts were made to account for molecular weight distributions in this study.

Effect of Changes in Hydrodynamic Volume. The relationship between DR and hydrodynamic volume was further investigated using copolymer models of fixed composition and molecular weight. In these studies hydrodynamic volume was varied by changing the ionic strength of the solvent. The polyampholyte models of high (ADAS 50) and low (ADASAM 2.5-2.5) charge density show over twofold increases in intrinsic viscosity as the solvent is changed from water to 0.514 M NaCl (Table I). Both also demonstrated lower friction factors in NaCl solutions as illustrated in f vs Reynolds number plots of 5 ppm solutions in tube flow (Figure 7).

At equal concentrations, low charge density polymers are more effective than those of high charge density. This is best shown in DR efficiency plots in which comparisons are made based on equal volume fractions (Figure 8a). This trend is consistent with that predicted by light scattering studies (Table I). For polyampholytes of a particular structure, DR behavior in different solvents can be normalized by adjusting polymer volume fraction. Coincident curves of f vs Re are obtained via rotating disk for ADASAM 5-5 in deionized water and in 0.514 NaCl when concentration is adjusted to give constant volume fractions (Figure 8b). A single curve is likewise obtained for ADAS 50 in both solvents.

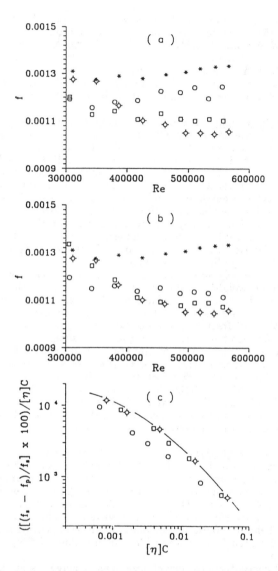

Figure 5. Drag reduction behavior of PEO polymers in 0.514 M NaCl tested by rotating disk. Solvent (*); WSR-N-12K, MW = 1.7 x 10^6 (O); WSR-N-60K, MW = 4.0 x 10^6 (□); WSR-301, MW = 5.3 x 10^6 (-O-).
  (a) Friction factor vs Reynolds number for 3 ppm PEO solutions.
  (b) Friction factor vs Reynolds number at constant polymer volume fraction, [η]C = 0.007.
  (c) Drag reduction efficiency measured at Re = 520,000 vs PEO polymer volume fraction.

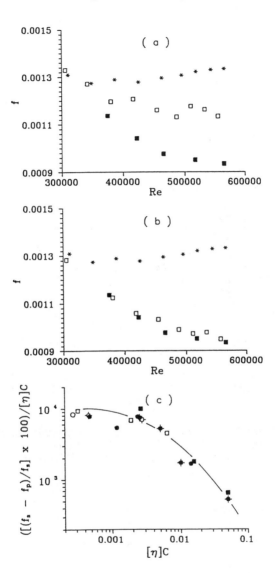

Figure 6. Drag reduction behavior of NaAMB copolymers in 0.514 M NaCl tested by rotating disk. Solvent (*); NaAMB 10, [η] = 1.8 dL/g (○); NaAMB 10, [η] = 47 dL/g (●); NaAMB 25, [η] = 6.3 dL/g (□); NaAMB 25, [η] = 52 dL/g (■); NaAMB 40, [η] = 9.0 dL/g (-◇-); NaAMB 40, [η] = 50 dL/g (-◆-).
(a) Friction factor vs Reynolds number for 3 ppm NaAMB 25 solutions.
(b) Friction factor vs Reynolds number for NaAMB 25 solutions at constant copolymer volume fraction, [η]C = 0.010.
(c) Drag reduction efficiency measured at Re = 520,000 vs NaAMB copolymer volume fraction.

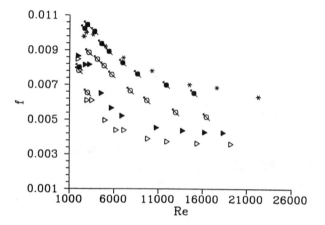

Figure 7. Friction factor vs Reynolds number for polymer solutions using deionized water (filled symbols) and 0.514 M NaCl (open symbols) solvents tested in tube flow. Deionized water (*), 5 ppm ADAS solutions (●, ◔), 5 ppm ADASAM 2.5-2.5 solutions (▶, ▷).

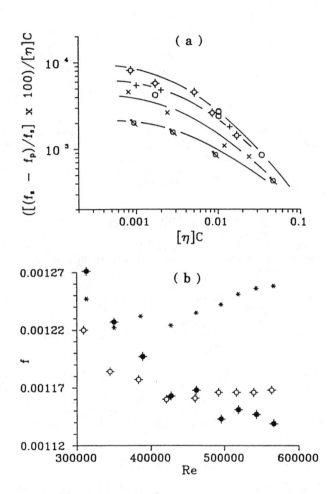

Figure 8. Drag reduction behavior of polyampholyte copolymer solutions using deionized water (filled symbols) and 0.514 M NaCl (open symbols) solvents tested by rotating disk.

(a) Drag reduction efficiency measured at Re = 520,000 vs polymer volume fraction for low and high charge density polyampholytes in 0.514 M NaCl. PAM-4 (O), ADAS 10 (×), ADAS 50 ($\widehat{Q}$), ADASAM 5-5 (-$\diamondsuit$-), ADASAM 5-10 (+).

(b) Friction factor versus Reynolds number for ADASAM 5-5 solutions (-$\diamondsuit$-, -$\diamondsuit$-) at constant copolymer volume fraction, $[\eta]C = 0.0015$.

Effect of Associations and Solvent Ordering. Associations of the relatively hydrophobic diacetone acrylamide units in the DAAM/AM copolymers appear to be responsible for the enhanced drag reduction efficiency observed for these systems (Figure 1). To further study such effects, viscosity, light scattering, and drag reduction experiments were conducted on each copolymer in the DAAM series utilizing three solvents which strongly affect the nature of inter- and intramolecular hydrophobic association. A comparison of the hydrodynamic volume data in Table I with the drag reduction studies in our rotating disk apparatus is quite revealing. Figure 9a is a plot of friction factor vs Reynolds number for a 15 ppm solution of DAAM 35 in deionized water, 1 M urea, and 0.514 M NaCl, respectively. Friction reduction is greatest in the saline solution in which intramolecular associations (and any existing intermolecular associations) would be enhanced and lowest in urea where hydrophobic associations are virtually eliminated. This behavior is not predicted by most drag reduction theoretical models which normally suggest a parallel between DR behavior and hydrodynamic volume. In this case DR is poorest in urea despite a 3-fold increase in intrinsic viscosity over that in saline solutions (Table I).

Universal DR Calibration for Diverse Polymer Types. Our method of plotting drag reduction efficiency clearly indicates that copolymer composition is an important consideration in designing optimal DR fluids. Significantly, a family of curves is generated for specified conditions with a distinct curve for each copolymer type as shown in Figure 2. Additional normalization can be accomplished by introduction of a shift factor Δ which allows all data to fall on a single efficiency curve (Figure 10). The polyethylene oxide curve for WSR-301 was chosen as a standard ($\Delta = 1$) to which all other curves were adjusted. Table II lists Δ values for all polymers tested in disk flow. Interestingly, shift factors range over two orders of magnitude with DAAM 35 giving the largest value of Δ and homopolyelectrolytes the lowest.

Agreement with Theoretical Models. Recently we applied our experimental results to a number of theoretical models found in the literature (McCormick, C. L. et al., Macromolecules, in press). One model showing particular promise is that of Ryskin (9,10) which involves polymer extension dynamics in what is referred to as a "yo-yo" model. Equation 2 yields the polymer effect on viscosity enhancement ζ_{turb}:

$$\zeta_{turb} \simeq 0.05\alpha^3 N_A a^3 N^2 C/M_a \qquad (4)$$

N_A = Avogadro's number M_a = mol. weight repeat unit
a   = length of a repeat unit α = ratio of chain length to that
C = polymer concentration   of a fully extended chain
N = degree of polymerization

For each copolymer model, α may be adjusted to yield acceptable correlation with experimental data. We are currently assessing experimental approaches to determine α. We believe that α may be related to our empirical shift factor Δ via the extensibility of an effective segment length or Kuhn segment. We also feel that changes in coil draining during extension and the related reordering of solvent are important contributions to the drag reduction phenomenon.

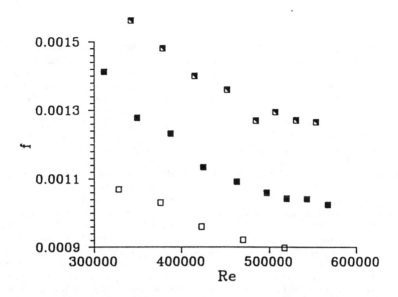

Figure 9. Friction factor vs Reynolds number for DAAM copolymers using various solvents tested by rotating disk.   15 ppm DAAM 35 in 1 M urea (▨), in deionized water (■), in 0.514 M NaCl (□).

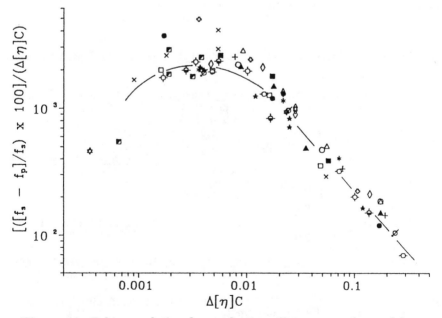

Figure 10. Polymer solution drag reduction efficiency vs polymer volume fraction employing shift factors.   See Table II for symbol descriptions and shift factor information.   Disk flow measurements were made at Re = 520,000.

Conclusions

The relationship between drag reduction behavior and chemical structure has been investigated for tailored water-soluble copolymers with measured values of intrinsic viscosity, molecular weight, and hydrodynamic volume. Studied were the effects of: structural changes at constant solvent and flow conditions, changes in molecular weight for constant structure, changes in hydrodynamic volume at constant molecular weight and structure, and changes in solvent nature and association for constant structure. Drag reduction studies were conducted both in a rotating disk and in a tube flow apparatus. Experimental data were presented by traditional f vs Reynolds number plots and in normalized %DR vs volume fraction plots at a given Reynolds number under specified conditions of solvent and concentration. Results clearly show the influence of polymer molecular structure on drag reduction efficiency; those polymers with the best efficiencies are polymers with the greatest potential for extension in turbulent flow. Polyelectrolytes which have often been reported to have the best absolute DR behavior are actually much less efficient than others in this study. An empirical shift factor may be introduced to collapse efficiency curves for families of copolymers onto a single universal curve. Initial correlations of experiment and theory utilizing extensional dynamic models such as that proposed by Ryskin have been promising. Of special interest is the possibility of correlating an extensibility factor with our empirical shift factor utilizing an effective segment length concept.

Acknowledgments

Research support from the Office of Naval Research and the Defense Advanced Research Projects Agency is gratefully acknowledged.

References Cited

1. Toms, B. A. Proceedings of International Congress on Rheology, 1949, Vol. 2, p 135.
2. Virk, P. S. A.I.Ch.E. J. 1975, 21, 625.
3. Virk, P. S. Biotechnology of Marine Polysaccharides; Colwell, R.; Pariser, E. R.; Sinskey, A J., Eds.; Hemisphere: Washington, 1985; p 149.
4. Hlavacek, B.; Rollin, L. A.; Schreiber, H. P. Polymer 1976, 17, 81.
5. Hlavacek, B.; Sangster, J. Can. J. Chem. Engng. 1976, 54, 115.
6. Lumley, J. L. J. Polym. Sci. Macromolec. Rev. 1973, 7, 263.
7. Fabula, A. G.; Lumley, J. L.; Taylor, W. D. Modern Developments in the Mechanics of Continua; Academic Press: New York, 1966.
8. Lumley, J. L. Phys. Fluids 1977, 20, Part II.
9. Ryskin, G. J. Fluid Mech. 1987, 178, 423.
10. Ryskin, G. Phys. Rev. Lett. 1987, 59, 2059.
11. Walsh, M. Ph.D. Thesis, Cal. Inst. of Technology, Pasadena, CA, 1967.
12. Kohn, M. C. J. Polym. Sci., Polym. Phys. Edn. 1973, 11, 2339.
13. de Gennes, P. G. Physica 1986, 140A, 9.
14. Berman, N. S. The Influence of Polymer Additives on Velocity and Temperature Fields; Gampert, B., Ed.; Springer-Verlag: Berlin, 1985; p 293.
15. Kulicke, W. M.; Kotter, M.; Grager, H. In Advances in Polymer Science; Springer-Verlag: Berlin, 1989; p 1.
16. McCormick, C. L.; Morgan, S. E. Prog. Polym. Sci. 1990, 15(3).

17. Morgan, S. E.   Ph.D. Thesis, The University of Southern Mississippi, Hattiesburg, MS, 1988.
18. McCormick, C. L.; Mumick, P. S.; Morgan, S. E.   Polymer Preprints 1989, 30(2), 256.
19. McCormick, C. L.; Hester, R. D.; Morgan, S. E.; Mumick, P. S.   Pacific Polymer Preprints 1989, 1, 147.
20. McCormick, C. L.; Blackmon, K. P.   J. Polym. Sci., Polym. Chem. 1986, 24, 2635.
21. Blackmon, K. P.   Ph.D. Thesis, The University of Southern Mississippi, Hattiesburg, MS, 1986.
22. McCormick, C. L.; Hutchinson, B. H.; Morgan, S. E.   Makromol. Chem. 1987, 188, 357.
23. McCormick, C. L.; Blackmon, K. P.; Elliott, D. L.   J. Polym. Sci., Polym. Chem. 1986, 24, 2619.
24. McCormick, C. L.; Johnson, C. B.   Macromolecules 1988, 21, 686.
25. McCormick, C. L.; Johnson, C. B.   Macromolecules 1988, 21, 694.
26. Instruction Manual, Model Rotovisco RV3, Haake Instruments: Saddle Brook, NJ.
27. Schlichting, H.   Boundary Layer Theory; McGraw-Hill: New York, 1979; p 647.
28. Bird, R. B.; Stewart, W. E.; Lightfoot, E. N.   Transport Phenomena; John Wiley & Sons: New York, 1980; p 181.
29. McCormick, C. L.; Blackmon, K. P.   Macromolecules 1986, 19, 1512.
30. McCormick, C. L.; Blackmon, K. P.; Elliott, D. L.   Macromolecules 1986, 19, 1516.

RECEIVED June 4, 1990

Chapter 22

Rheological Properties of Hydrophobically Modified Acrylamide-Based Polyelectrolytes

John C. Middleton, Dosha F. Cummins, and Charles L. McCormick

Department of Polymer Science, The University of Southern Mississippi,
Hattiesburg, MS 39406–0076

Associative polymers of acrylamide and n-decylacrylamide
with sodium-3-acrylamido-3-methylbutanoate, sodium acrylate, or
sodium-2-acrylamido-2-methylpropanesulfonate have been prepared
by a micellar technique. Low shear rheometry was used to
obtain plots of apparent viscosity as functions of ionic group type
and mole percent incorporation into the backbone, polymer
concentration and solution ionic strength. Results indicate that
these polymers maintain high viscosity in NaCl concentrations of
up to 0.514 M by intermolecular association of the hydrophobic
groups. The amount of aggregation is dependent on the type of
ionic group incorporated as well as the distance of the charged
group from the backbone. Maximum increases in apparent
viscosity are observed for the terpolymers containing carboxylate
groups close to the polymer backbone. A conceptual model based
on placement of both charged and hydrophobic groups along the
macromolecular backbone is proposed consistent with rheological
behavior.

Water-soluble polymers containing non-polar groups which aggregate through
hydrophobic interactions in a polar medium were first discovered while studying
the conformations of proteins (1). Such polymers contain both ionic or nonionic water-soluble groups and hydrophobic or amphiphilic groups. The unique
solution behavior of these hydrophobically modified, water-soluble
macromolecules has led to the synthesis and characterization of several
synthetic analogs. Novel rheological properties have made them attractive for
a variety of applications such as aqueous thickeners in latex coatings (2,3),
and mobility control agents in enhanced oil recovery (4,5). Synthetic,
hydrophobically modified polyelectrolytes may also serve as models of
biopolymers for studying how structure and activity are related in proteins and
biomembranes (6).
 The term "hydrophobic bond" was originally developed to describe the
grouping of non-polar side chains in proteins and has been found to be an
important factor in stabilizing folded conformations (1). Although no bond
actually exists the term "hydrophobic bond" or "hydrophobic interaction" has
received wide acceptance in the literature and several articles and books have

0097–6156/91/0467–0338$06.00/0

been devoted to the subject (7-9). When non-polar groups are introduced into a polar medium such as water, the water molecules hydrogen bond around the hydrophobic group creating what have been called "icebergs" (9). These icebergs are quasi-crystalline structures in which there is less randomness and slightly better hydrogen bonding than in ordinary liquid water at the same temperature. The ordering of water molecules around a hydrophobic group results in an energetically unfavorable decrease in entropy. If a sufficient number of non-polar groups are present, micellar structures will spontaneously form as the hydrophobic molecules are expelled from the solvent creating a more favorable entropic environment and an increase in free energy. The addition of external electrolytes such as NaCl promotes hydrophobic associations by increasing the polarity of the medium.

Studies in our laboratories have focused on developing macromolecules that can maintain or increase the viscosity of aqueous systems in the presence of mono- or multivalent electrolytes (10-19). Recent work concentrated on the synthesis and characterization of copolymers of acrylamide with n-alkylacrylamides with alkyl lengths of 8, 10, and 12 carbons. These polymers show unique solution behavior with the incorporation of less than 1 mol % of the n-alkylacrylamide group (20,21). Polymer association occurs intermolecularly through the pendent hydrophobic groups above a critical polymer concentration (C*). Above C* a rapid increase in apparent viscosity is observed as the polymers form networks with the hydrophobic groups acting as transient crosslinks (Figures 1 and 2).

While these macromolecules show increased viscosity in the presence of small molecule electrolytes above a critical concentration, they are slow to dissolve from the dry state (22). In order to enhance dissolution and provide electrolyte character, a series of terpolymers containing acrylamide (AM), 0.5 mole % of N-n-decylacrylamide (C-10 AM) as the hydrophobic group, and sodium-3-acrylamido-3-methylbutanoate (NaAMB), sodium acrylate (NaA), or sodium-2-acrylamido-2-methylpropanesulfonate (NaAMPS) have been synthesized (23,24). Rheological properties were determined by low shear viscometry in deionized water and sodium chloride solutions.

Experimental

Monomer Synthesis. Acrylamide and AMPS were obtained commercially from Aldrich Chemical Co. and purified by recrystallization from acetone. Acrylic acid was also obtained commercially from Aldrich and purified by vacuum distillation to remove inhibitor before use. N-n-decylacrylamide (20) and NaAMB (25) were synthesized by previously reported methods. The structures of all the monomers appear in Figure 3.

Polymer Synthesis. A series of terpolymers was prepared with monomer feeds of 0.5 mole percent of the N-n-decylacrylamide and 5, 10, 25, and 40 mole percent of each of the ionizable groups. The remaining polymer backbone was composed of acrylamide.

The incorporation of water-soluble and water-insoluble monomers into a polymer backbone was accomplished using a micellar polymerization method (4). This technique utilizes a surfactant to solubilize the hydrophobic monomer. Sodium dodecyl sulfate was the surfactant in this instance. A water-soluble initiator, potassium persulfate, was used to induce free-radical polymerization.

Viscometry. The appropriate amount of dried polymer was weighed into a glass container and solvent added. The polymers were dissolved by gentle shaking

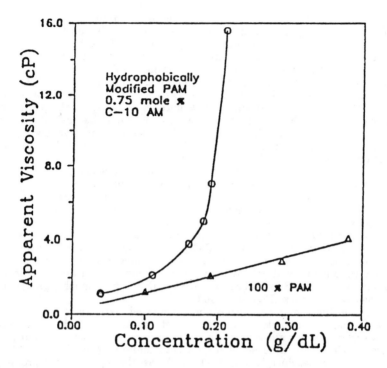

Figure 1. Apparent viscosity vs concentration for a polyacrylamide (PAM) polymer and PAM hydrophobically modified with 0.75 mol % n-decylacrylamide (C-10 AM).

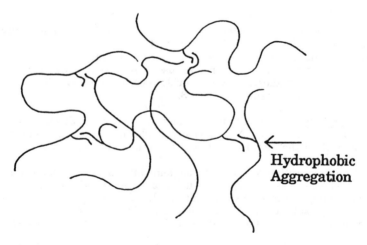

Figure 2. Proposed model for hydrophobically associating polymers.

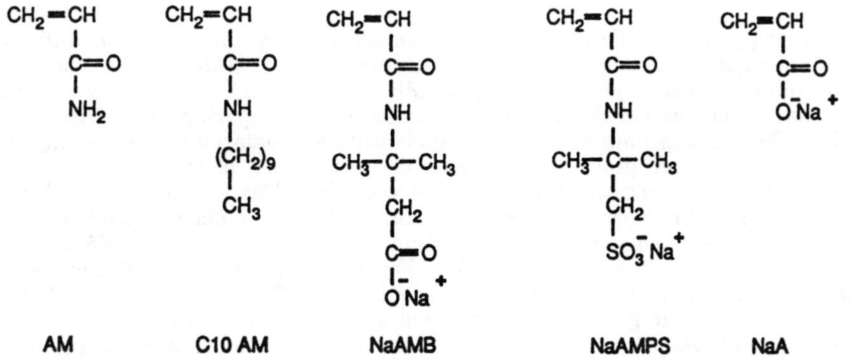

Figure 3. Monomer structures used to prepare terpolymers.

on an orbital shaker for 14 days to allow complete hydration before further dilutions of these stock solutions were made.   Viscosity experiments were conducted on the Contraves LS 30 low shear rheometer at a shear rate of 6 reciprocal seconds at 30°C.

Results and Discussion

Conceptual Model. Two opposing forces determine the solution behavior of hydrophobically modified polyelectrolytes in aqueous solution- electrostatic repulsions and hydrophobic associations. The electrostatic interactions of the anionic groups along the backbone tend to increase the hydrodynamic volume and repel polymer segments from another. Hydrophobic moieties aggregate in aqueous solution and may associate either intramolecularly in dilute solution or intermolecularly at higher concentration (above C*). The combination of these factors along with other molecular parameters such as molecular weight, polymer microstructure, pH and solvent ionic strength result in a complex, but technologically important system.

Solutions Studies. Three series of polymers were synthesized with differing ionic groups. One series contained sulfonate groups (NaAMPS) and the other two contained carboxylate groups (NaAMB and NaA). The effect of ionic group distance from the backbone was also evaluated by comparing the NaAMB and NaA. Ionic group content influence on viscosity was determined by varying the amount of charged group incorporation between 5 and 40 mol percent.

     Apparent viscosity (in centipoise) was plotted as a function of polymer concentration for each polymer in six different solvent ionic strengths. The solvents were deionized water and sodium chloride solutions of 0.085 M, 0.17 M, 0.259 M, 0.342 M and 0.514 M. For these studies, the polymers were grouped according to polymer ionic content. The effects of the three different types of electrolyte groups were compared as functions of solvent ionic strength vs apparent viscosity. The polymers containing 10 and 25 mole per cent electrolyte are shown in figures 4-9 as representative cases.

     In deionized water all of the polymers have high viscosities typical of polyelectrolytes in aqueous media (Figures 4 and 5). However, when small molecule electrolyte is added (Figures 6-9) the solution apparent viscosity is greatly reduced below the overlap concentration as the intramolecular ionic repulsions are shielded reducing the hydrodynamic volume of the polymer coils. Above the overlap concentration, however, which varies depending on the polymer composition and solution ionic strength, significant associative behavior is observed only for the NaA and NaAMB polymers.

Comparison of the NaAMPS, NaAMB and NaA Terpolymers. At low ionic strength and concentration, the electrostatic repulsive forces dominate the polymer solution behavior and all the polymers act as polyelectrolytes with similar viscosities. Ionic shielding by externally added NaCl both reduces the repulsive forces and increases the polarity of the medium.   Therefore hydrophobic association is favored as the non-polar n-decyl groups are excluded from the polar environment resulting in network formation. The NaAMPS polymers show an almost linear increase in viscosity with polymer concentration independent of ionic strength. However, for the NaAMB and NaA terpolymers, apparent viscosity increases linearly with sample concentration only in deionized water (Figures 4 and 5). In brine solutions, the viscosity increases exponentially above C*, indicative of intermolecular hydrophobic association

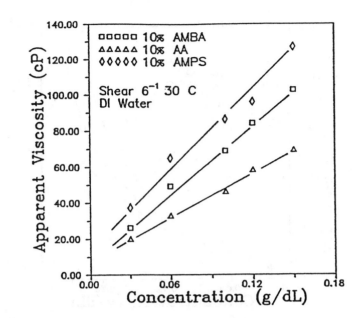

Figure 4. Apparent viscosity vs polymer concentration for the 10 mol % polymers in deionized water.

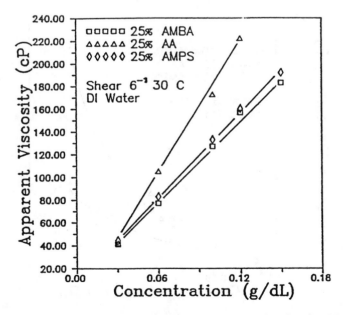

Figure 5. Apparent viscosity vs polymer concentration for the 25 mol % polymers in deionized water.

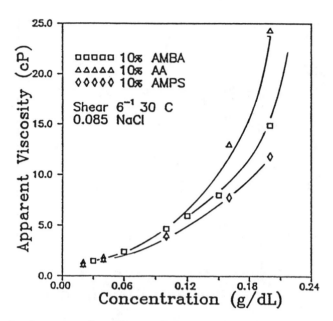

Figure 6. Apparent viscosity vs polymer concentration for the 10 mol % polymers in 0.085 M NaCl.

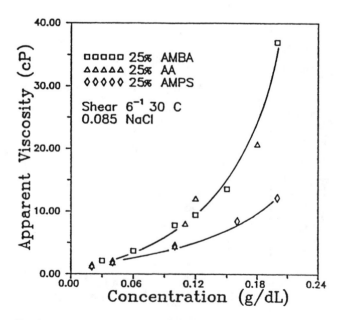

Figure 7. Apparent viscosity vs polymer concentration for the 25 mol % polymers in 0.085 M NaCl.

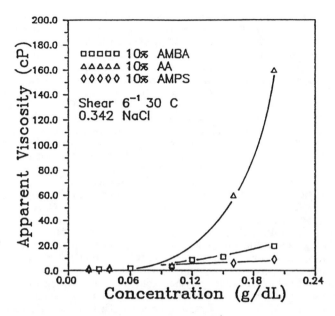

Figure 8. Apparent viscosity vs polymer concentration for the 10 mol % polymers in 0.342 M NaCl.

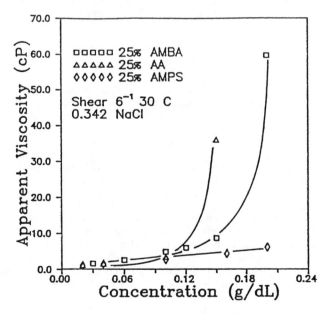

Figure 9. Apparent viscosity vs polymer concentration for the 25 mol % polymers in 0.342 M NaCl.

(Figures 6-9). The terpolymers containing NaA have lower overlap concentrations and higher viscosities than the other systems investigated. The terpolymers containing NaAMPS are the least affected by changing ionic strength because the sulfonate anion binds cations weakly and is not affected strongly by the added electrolyte. Polymer viscosity is actually lowered with increasing NaAMPS concentration in solvents of high ionic strength. As sulfonate group incorporation is increased from 10 to 25 mol %, all aggregation is prevented. The weaker carboxylate acid analogs, NaAMB and NaA, do show aggregation in brine solutions as charge is more effectively shielded through stronger site binding. Increasing the ionic group content from 10 to 25 mol % for the carboxylate polymers increases apparent viscosity. The presence of the carboxylyate groups and their counterions give the polymers added hydrodynamic volume through nearest neighbor steric and short-range electrostatic interactions without disrupting hydrophobic interactions.

The NaAMB and NaAMPS mers are farther from the polymer backbone and may interfere with hydrophobic association more than NaA mers which do not protrude as far, accounting for the much higher viscosities of the NaA terpolymers. Additionally, the *gem*-dimethyl groups of the NaAMB and NaAMPS may have sufficient hydrophobic character to intramolecularly stabilize the *n*-decyl groups through nearest neighbor interactions making them less thermodynamically driven to associate with other decyl groups (Figure 10).

The individual, flexible polymer coils may be represented as spheres containing charged groups in the interior as well as externally covering the surface (Figure 11). At low solution ionic strength the coils are expanded due to intramolecular electrostatic repulsions, but hydrophobic interactions are prevented by intermolecular repulsions. Introduction of NaCl ionic shielding lessens both intra- and intermolecular repulsions reducing the hydrodynamic volume of the individual coil. However, apparent viscosity increases due to hydrophobic aggregation in the absence of long range intermolecular repulsions.

Conclusions

Series of associative terpolymers with ionizable carboxyl and sulfonyl sites have been prepared and their solution properties evaluated. The NaAMPS terpolymers display typical polyelectrolyte behavior. The sulfonate groups are not well shielded by the Na counterions; the resulting ionic repulsions prevent hydrophobic aggregation. The NaAMB and NaA terpolymers, on the other hand, exhibit strong associative properties due to effective hydrophobic associations among the decyl groups. These polymers show increases in viscosity in brine solutions, apparently due to increased hydrophobic association and reduced intermolecular electrostatic repulsions. Distance of the ionic group from the backbone influences hydrophobic association. The NaAMB and NaAMPS monomers extend charged groups farther from the polymer backbone preventing associations among the hydrophobic decyl groups. This is not observed for the NaA monomer where the charge is much closer to the backbone. A hydrophobically modified, charged-sphere model is used to illustrate how the rheological behavior may be rationalized by the balancing the electrostatic repulsions and hydrophobic attractions in solutions of differing ionic strength.

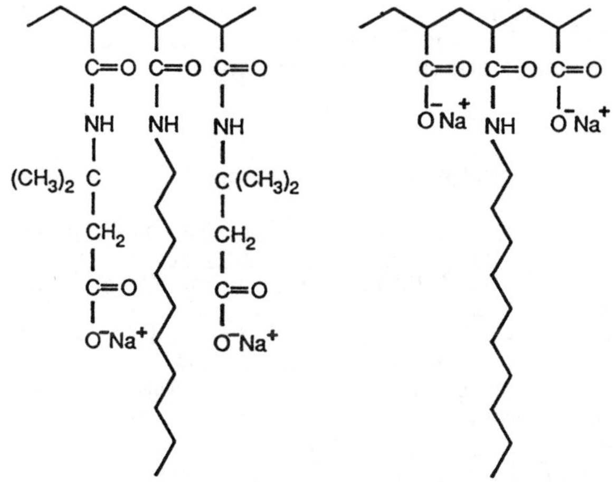

Figure 10. A schematic comparision of NaAMB and NaA relative to the C-10 group.

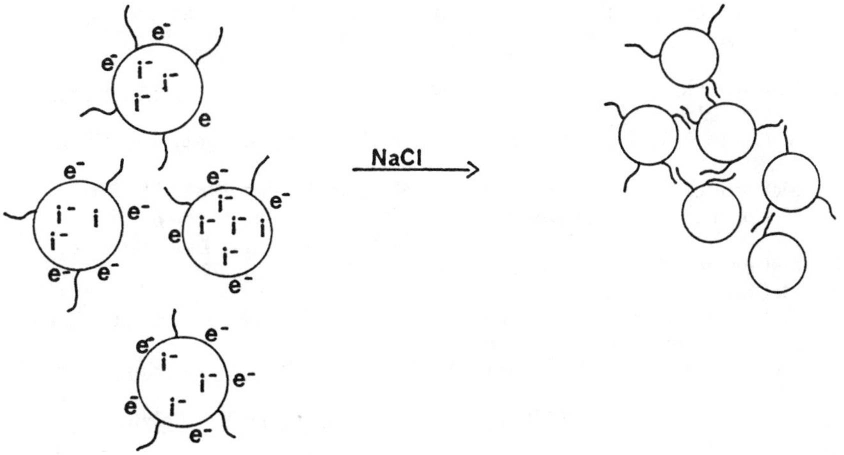

Figure 11. The shielding of electrostatic repulsions by added electrolyte results in polymer coil contraction due to reduced intramolecular repulsions (i) and polymer aggregation due to reduced intermolecular repulsions (e).

Acknowledgments

Financial support for this work was provided by the Department of Energy, the Office of Naval Research and the Defense Advanced Research Projects Agency and is gratefully acknowledged.

Literature Cited

1. Advances in Protein Chemistry; Kauzmann, W., Ed.; Academic Press: New York, 1959.
2. Landoll, L. M. J. Poly. Sci., Poly. Chem. Ed. 1982, 20, 443.
3. Glass, J. E. Advances in Chemistry Series No. 213; J. E. Glass.
4. Turner, S. R.; Siano, D. B.; Bock, J. (to Exxon Research and Engineering) U.S. Patent 4 520 182, 1985.
5. Evani, S. (to Dow Chemical Co.) U.S. Patent 4 432 881, 1984.
6. Miyamoto, S. Macromolecules 1984, 14, 1054.
7. Tanford, C. A. The Hydrophobic Effect: Formation of Micelles and Biological Membranes; Wiley-Interscience: New York, 1973.
8. Ben-Naim, A. Hydrophobic Interactions; Plenum Press: New York, 1980.
9. Shinoda, K. J. Phys. Chem. 1977, 81(13), 1300.
10. McCormick, C. L.; Elliott, D. L.; Blackmon, K. P. Macromolecules 1986, 19, 1516.
11. McCormick, C. L.; Blackmon, K. P. J. Macromol. Sci. Chem. 1986, A25, 1451.
12. McCormick, C. L.; Elliott, D. L. J. Macromol. Sci. Chem. 1986, A23, 1469.
13. McCormick, C. L.; Blackmon, K. P. Polymer 1986, 27, 1971.
14. McCormick, C. L.; Elliott, D. L.; Blackmon, K. P. Polymer 1986, 27, 1976.
15. McCormick, C. L.; Blackmon, K. P. Angew. Makromol. Chem. 1986, 144, 73.
16. McCormick, C. L.; Elliott, D. L.; Blackmon, K. P. Angew. Makromol. Chem. 1986, 144, 87.
17. McCormick, C. L.; Elliott, D. L. Polym. Sci., Polym. Chem. Ed. 1986, A25, 1329.
18. McCormick, C. L.; Johnson, C. B. Macromolecules 1988, 21, 686.
19. McCormick, C. L.; Johnson, C. B. Macromolecules 1988, 21, 694.
20. McCormick, C. L.; Johnson, C. B. Polym. Mater. Sci. Eng. 1986, 55, 366.
21. McCormick, C. L.; Johnson, C. B.; Tanaka, T. Polymer 1988, 29, 731.
22. McCormick, C. L.; Johnson, C. B. In Polymers in Aqueous Media; Glass, J. E., Ed.; Advances in Chemistry Series No. 223; American Chemical Society: Washington, DC, 1989.
23. McCormick, C. L.; Middleton, J. C. Polym. Mater. Sci. Eng. 1986, 55, 700.
24. McCormick, C. L.; Middleton, J. C.; Cummins, D. F. Polym. Preprints 1989, 30(2), 348.
25. McCormick, C. L.; Blackmon, K. P. J. Polym. Sci., Polym. Chem. 1986, A24, 2635.

RECEIVED June 4, 1990

BIOMEDICAL AND INDUSTRIAL APPLICATIONS

Chapter 23

Bioadhesive Drug Delivery

Sau-Hung S. Leung[1] and Joseph R. Robinson[2]

[1]Columbia Research Laboratories, 1202 Ann Street, Madison, WI 53713
[2]School of Pharmacy, University of Wisconsin, 425 North Charter Street, Madison, WI 53706

Bioadhesive drug delivery systems adhere to the mucin-network and/or underlying epithelial layer of mucosal surfaces. Mucus is a continuous network of cross-linked glycoproteins which, at physiological pH, carries a substantial negative charge due to the presence of sialic acid and sulfonic acid residues. The glycocalyx (polysaccharide-containing structures) on the external surface of cells is partly responsible for the adhesive properties of the cell and carries a net negative charge as well.
     Bioadhesion begins with establishment of intimate contact between the polymer and substrate. This is followed by bond formation and interpenetration/interdiffusion of polymer and substrate. Any increase in contact area and establishment of physical entanglement at the interface strengthens the force of bioadhesion. Thus, factors that favor intimacy of contact, increased bond formation and enhanced physical entanglement will potentially increase bioadhesive strength.
     Bioadhesive polymers can be used in drug delivery to localize delivery systems at specific sites for either local treatment or for prolonged delivery of drug in a route of administration that would otherwise cause rapid removal of the system.

Bioadhesive drug delivery is an important means of drug delivery which has recently received considerable attention by pharmaceutical scientists.(1-11) The reasons for this attention include the fact that bioadhesives can be used to improve intimacy of contact between the drug

0097–6156/91/0467–0350$06.00/0
© 1991 American Chemical Society

delivery system and the absorptive surface, as well as to increase dosage form residence time. Localization of the dosage form can improve local therapy and can allow local modification of tissue permeability to improve bioavailabliity of the drug. The focus of this review is on structure-adhesion relationship and strategies for bioadhesive drug delivery.

In bioadhesive drug delivery, two different components are involved (Figure 1): 1) the bioadhesive drug delivery system, which acts as a platform for delivery of drug as well as containing a bioadhesive polymer on its surface; and 2) the underlying substrate. The substrate may be the mucin layer or the epithelium layer, or perhaps some of each.

Over the past decade there has been considerable effort devoted to establishing suitable screening techniques to identify bioadhesive polymers and to explore the mechanism(s) of bioadhesion. Most of the exploratory work, aimed at understanding the mechanisms of bioadhesion, utilize the various screening techniques that have been reported in the literature.

Typically, potential bioadhesive screening techniques can be divided into those involving tensile or shear stress. A tensile strength approach, between mucus and water-soluble polymers, was reported using a Wilhelmy plate method,(3) as well as a modified tensiometer method.(11,12) Shear stress was measured using a dual tensiometer method.(11) The static and dynamic bioadhesive properties of polymer particles was also studied using a thin channel, filled with artificial mucus gel or natural mucus.(10,14) Several aspects of the mechanism of bioadhesion have been reported,(10) and the bioadhesive potential and structural requirements of natural and synthetic polymers for bioadhesion were studied.(3,6,10,13) The results of these studies showed that highly charged carboxyl polyanions are good bioadhesives. A number of bioadhesive drug delivery systems, for different routes of administration, have been developed.(1,4,5,7,15-20)

In order to design a specific dosage form for drug delivery across the mucin/epithelial layer, a better understanding of the physico-chemical characteristics of the mucus and surface epithelial layer is required. An improved light microscopy method namely, laser confocal fluorescence microscopy, may allow biological structures to be visualized with a minimal amount of disturbance.(29) Indeed, the confocal imaging technique has great potential in studying the interactions of bioadhesives with biological substrate surfaces.

BIOADHESIVE PARAMETERS

<u>Charges</u> The presence of charge groups or at least those

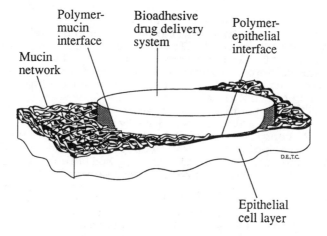

Figure 1: Schematic Diagram at the site of Bioadhesion

functional groups capable of strong hydrogen bonding favors bioadhesion. In studying the bioadhesive strength of cross-linked polyacrylic acid, (21) with an apparent pka of approximately 4.75, the adhesive force drops sharply at around pH4, which suggests that bioadhesion is favored when carboxylate groups are in an undissociated form. Thus, hydrogen-bonding through polar groups appears to be involved in bioadhesion.

<u>Hydration and Expanded Nature of the Polymer</u>   When bioadhesives hydrate in an aqueous media, they swell and form a gel. Besides pH and ionic strength of the aqueous medium, the type and number of hydrophilic functional groups in the polymer structure affect the rate and extent of hydration. As the percent of charged groups decrease, the degree of hydration and hence expanded nature of the polymer decreases in a similar fashion. (11) The effect of the expanded nature of the mucin network and adhesive network have been studied. (11) Results showed that bioadhesive strength increased with increased openness of the mucin and polymer networks.

<u>Chain Segment Mobility</u> The ability of the polymer chains to interpenetrate and thus increase the area for interaction can be approximated by their ability to diffuse. Over a sufficiently restricted temperature range, the experimental diffusion coefficient, D, shows an exponential temperature dependence of the Arrhenius type (22)

$$D=D_o(-E/RT) \tag{1}$$

where the pre-exponential factor D_o is a constant and is independent of temperature over a given temperature range, and E is the experimental activation energy for diffusion or for mobility of the segment chains.

Bioadhesion of a hydrating polymer to a mucin network or underlying epithelia was found to be a time dependent process. (23) As the contact time increases, depth of interpenetration increases and results in an increase in strength of adhesion. Furthermore, the bioadhesive process was found to be temperature dependent. (23) In other word, interdiffusion/interpenetration increases with temperature, which results in an increase of contact area and adhesive strength. Thus, an increase in the depth of interpenetration increases strength of bioadhesion. (23)

MUCUS LAYER

Mucus covers all the orifices of the body with the exception of the ears, presenting a continuous, unstirred gel layer over the mucosa. Thus, the mucus layer is

normally the first layer that the bioadhesive dosage form
will encounter. Mucins are synthesized either by goblet
cells lining the mucus epithelium or by special exocrine
glands with mucus cell acini.(49) The basic component of
all mucus is the mucin glycoprotein with oligosaccharide
side-chains,(24,25) and terminal sialic acids(26,27) with
pKa of 2.6.(28)   Thus, the mucin network carries a
substantial negative charge at a physiological pH of 7.4,
which affects bioadhesion significantly.
    Besides   the charge of the mucus layer,the thickness
may play a role in bioadhesion. The thickness of the
mucus layer is the distance between the solution interface
and the mucus-mucosa interface. The mean mucus thickness
is 192um in the human stomach, 77um in the rat stomach,
81um in the rat duodenum,(30) and 1.4 um in the human
conjunctiva.(31)
    The continuity and thickness of the mucus layer is
important for its protective function against physical and
chemical insults. In the gastrointestinal tract, gastric
mucin, together with bicarbonate secretion, provides a
diffusion barrier for hydrogen ion and pepsin, and
maintains a near-neutral pH at the mucosal surface. Once
the continuity of the mucus layer is interrupted or its
thickness is reduced, bioadhesion at the epithelial
surface may occur.

EPITHELIUM

Reduction of the mucus thickness and interruption of the
continuity   of   mucin   coverage   can   be   accomplished
physically or chemically.   It is possible that a good
bioadhesive can adhere to the epithelium directly or
penetrate across the reduced mucus network and attach to
the epithelial surface. Attachment of adhesives to the
exposed epithelial surface can serve three different
functions: form a continuous layer with the mucus network
and thus reduce exposed cell surface area; act as a
protective covering for the underlying epithelium against
physical and chemical insult; and act as a drug reservoir
for sustained drug delivery and allow recovery of the
damaged or diseased cell layers.
    It is thought that adhesion of bioadhesives onto the
epithelial cell layer depends significantly on the
interactions between the bioadhesives and specialized
structures and/or macromolecules. One of the specialized
structures is the glycocalyx. The term glycocalyx
includes all polysaccharide-containing structures on the
external surface of cells, and is maintained and
synthesized continuously by the underlying cell.(32) Some
examples of specialized macromolecules synthesized by
cells are fibronectins,(33) lectins,(34) and cell adhesion
molecules (CAMs).(35)   Fibronectin is a component of the
extracellular matrix, which binds certain forms of

collagen and glycosaminoglycans and mediates the adhesion and spreading of cells in culture.(36)   Lectins are carbohydrate-binding proteins and glycoproteins derived from both plants and animals.(37)   Since the surface of virtually all cells have a number of glycosylated moieties, the effect of these specialized structures on bioadhesion should not be ignored.

STRATEGIES FOR BIOADHESIVE DRUG DELIVERY

Ocular Drug delivery   In general, mucus forms the inner layer of the tear film, which is in contact with the microvilli of the corneal and conjunctival epithelium. Blinking spreads the glycoprotein along the corneal surface and permits continuous replenishment of the mucin layer.(38)   Normal tear flow would facilitate drug release from a precorneal dosage form by providing a constant flow of bathing fluid. However, the major reason for failure of conventional ocular drug delivery systems is excessive drainage of the drug before adequate corneal absorption can occur. Thus, a good ocular bioadhesive dosage form needs to reduce drug loss via the ocular drainage system. One way to achieve this goal is by reducing the drug dissolution rate. A film of polyvinyl alcohol was used to deliver pilocarpine ocularly.(65,66)   The release of pilocarpine was controlled by slow dissolution of the insert.   The ocular bioavailability of pilocarpine with this system was found to increase two-fold over an aqueous formulation.

Besides excessive ocular drainage, another problem for ocular drug delivery system is the rapid response of the eye to external stimuli.   The eye is an extremely sensitive organ, which responds quickly to external insults. The first encountered problem with instillation of eye drops or placement of ocular inserts is an increased tear secretion stimulated by the drug, dosage form, excipients, tonicity, pH or physical irritation. The increase in tear secretion affects liquid ocular drug delivery systems more by diluting the drug concentration and reducing drug availability.

The pH in the precorneal area is 7.3 to 7.7. At this neutral pH, protonation of the carboxylic groups is minimum and the hydration volume of an anionic bioadhesive, cross-linked polyacrylic acid, is maximum. This increase in hydration results in an increase in the expanded nature of the polymer network. Which favors the interpenetration/interdiffusion process and increases physical entanglement between the polymer and mucin networks with a subsequent increase in bioadhesive strength.

Lightly cross-linked (0.3%) polyacrylic acid was used in ocular delivery of progesterone in rabbits and was found to increase the area under the curve 4.2 times greater than a conventional ocular suspension. (64) This increase in bioavailability of progesterone is probably due to an increase in retention of the dosage form in the precorneal area via bioadhesion. Thus, prolonged residence time of the ocular formulation in the preocular area increased ocular absorption.

Another ocular product consisting of partially esterified acrylic acid, Piloplex, was designed to prolong the therapeutic effect of pilocarpine. (67,68) It is an aqueous emulsion (dispersion) with limited water solubility. Upon instillation of the emulsion into the eye, the apparent opaque mass adhered to the preocular tissue and stayed in the lower fornix for extended periods of time, releasing drug. The release of pilocarpine is probably controlled by slow dissolution of the polymer and diffusion of pilocarpine from the polymer.

Nasal Drug Delivery The nasal route is a potentially good route for systemic drug delivery, especially for drugs that have extensive first-pass metabolism or gastrointestinal degradation. In terms of nasal drug absorption, hydrophobic drugs seem better absorbed than hydrophilic drugs. (44) Absorption of insulin by the nasal mucosa increased substantially by addition of hydrophobic salts, (40-42) and nonionic surfactants. (43)

In a normal nose, the nasal mucosa is covered by a thin layer of mucus, which contains 90 to 95% water, 1 to 2% salt and 2 to 3% mucin. The human nasal mucosa is about 150 cm^2 in area,[5] which provides an excellent adsorptive surface. Ciliated cells are present on the surface of the mucosa, which moves the mucous fluid at a rate of 5mm/min toward the throat. (69) Thus, a good nasal bioadhesive has to hydrate rapidly in the nasal mucus layer in the presence of a limited supply of water. The adhesive must adhere reasonably well to the mucus layer or the underlying epithelial surface after hydration, and resist nasal clearance by the cilia.

Carbopol 934 was used in development of a powder nasal drug delivery system of insulin. (39) It was found that incorporation of sodium carbopol into the nasal formulation results in a sustained release of insulin, which increases with carbopol concentration.

An adhesive powder spray for nasal allergy was also developed to deliver beclomethasone dipropionate. (71) Hydroxypropyl cellulose was used as the bioadhesive base. Once administered, the powder swelled and adhered to the mucosal membrane, up to 6 hours after application. (72)

Furthermore, a bioadhesive microsphere was designed for nasal delivery of sodium cromoglycate. (70) The bioadhesive microsphere consisted of albumin, starch, and DEAE-dextran. The slow clearance of the microspheres from the nasal cavity should prolong contact between the delivery system and the mucin/epithelial layer and increase availability of the drug.

<u>Buccal Drug Delivery</u> The buccal cavity is a convenient and easily accessible site for drug delivery, and was found to have similar permeability characteristics to the dermis in skin.<u>(45)</u> The buccal cavity has sufficient water secretion from the salivary gland (1 to 2 liters per day) to dissolve active ingredients and properly hydrate any polymer that functions as a bioadhesive. Futhermore, the buccal route offers the advantages of a lower enzymatic barrier, avoiding first-pass metabolism and gastric enzymatic degradation. However, some metabolic barrier still exists,<u>(47)</u> especially in certain disease states, e.g., psoriasis.<u>(48)</u>

The oral cavity permits precise localization of a bioadhesive drug delivery system and allows removal of the dosage form when irritation to the mucosal area occurs. Localization of the dosage unit allows modification of the underlying absorbing tissue to enhance drug availability. For example, addition of sodium glycocholate, an absorption enhancer, increases the effect of insulin by 40%.<u>(56)</u>

There are two kinds of mucosal tissues, keratinized and non-keratinized, in the oral cavity. The flexible mucosal tissues in the oral cavity, e.g., the soft palate, ventral surface of the tongue, floor of mouth, alveolar mucosa, vestibular lips and cheek, are characterized by a non-keratinized epithelium. Whereas, the hard palate, the dorsum of the tongue and gingiva, are characterized by a keratinized epithelium.<u>(46)</u> The dosage unit should be placed on the non-keratinized mucosal tissue for better systemic absorption. Thus, an understanding of the distribution of non-keratinized tissue in the buccal cavity is essential.

There is one additional distinct feature of the buccal mucosa. The mucus of the buccal mucosa is derived primarily from salivary glands. Thus, there is no build-up of the mucus underneath any bioadhesive drug delivery system, to interface with interactions between the adhesive and the underlying cell layer.

An example of adhesive tablet for buccal or sublingual administration was developed using hydroxypropyl cellulose.<u>(73,74)</u> Once administered the surface of the tablet swelled, and the tablet slowly eroded and released entrapped medicament. The adhesive buccal nitroglycerin tablets were found to have pharmacodynamic effects for up to 5 hours.

Another buccal adhesive dosage form is adhesive gels,<u>(16,75)</u> using hydroxylpropyl cellulose, polyacrylic acid and polymethylmethacrylate as gel-forming polymers. As compared to solution formulations, adhesive gels significantly prolong residence time of the formulation to the oral mucosa.

Buccal tablets and gels increase contact time between
the formulation and buccal mucosa and increase
availability of the drug. However, the direction of drug
release is not controlled in these systems and large
amounts of drug will be released and swallowed. Thus,
better control of drug release is desirable.

Bioadhesive buccal patches can be designed to release
drug unidirectionally to the mucosa for systemic
absorption or unidirectionally to the buccal cavity for
local effect, or bi-directionally to both mucosa and
buccal cavity.(76,77) The direction of drug release can
be controlled by incorporation of a nonpermeable membrane
to act as a diffusion barrier. Thus, bioadhesive buccal
patches can be mono or multilaminated in design.

Stomach and Intestinal Drug Delivery In different regions
of the stomach, there are different types of mucosa.(50)
The cardiac glands, located at the cardia region are
coiled and lined by mucus-producing epithelium. The
gastric or fundic glands, located in the fundus, secrete
hydrochloric acid and pepsinogen. The pyloric glands,
located in the pyloric region, synthesize and secrete
gastrin.

The small intestine in adults is approximately 12 feet
in length and is composed of the duodenum, jejunum and
ileum.(50) The villi and microvilli in the small
intestine provides a tremendous surface area for
absorption. It is noted that important nutrients have
specific sites of absorption. Iron and calcium are
absorbed primarily in the duodenum. Fat and sugar are
absorbed primarily in the proximal region of the small
bowel. For a diseased intestine, absorption is sometimes
impaired. Thus, identification of a absorption window, if
any, of a particular drug is important, especially for
drugs with a narrow absorption window. The use of a
specially designed bioadhesive may localize the dosage
unit at or around the absorption window.

Another important factor in oral drug delivery systems
is the motility pattern of the gastrointestinal (GI)
tract. There are two modes of GI motility patterns, the
digestive (fed) mode and interdigestive (fasted) mode. In
the fed state, the stomach is in a state of continuous
motility with substantial retropulsive forces. There are
four distinct phases in the fasted mode.(53) Phase I is
a quiescent period; phase II has random spikes of
electrical activity; phase III (housekeeper wave) is a
period of regular spike bursts of regular contractions at
maximal frequency that migrate distally, and phase IV is
the transition period between phase I and phase III. The
above cycle in the fasted mode is commonly known as the
interdigestive migrating motor complex (IMMC) and the
average duration range is 90 to 120 minutes.

When there is continuous contraction, especially during

the housekeeper wave, in the stomach and small intestine, the bioadhesive drug delivery system may be dislodged or loosened by the distally migrating contractile forces. Thus, a good bioadhesive drug delivery system needs to overcome the cleaning effect of the housekeeper wave. Furthermore, there is substantial mucin turnover in the stomach. Once a bioadhesive dosage form is swallowed, the external surface of the dosage form will be covered with "soluble" mucin. Thus, all the bioadhesive sites on the surface of the dosage form will be covered, and the apparent bioadhesive strength will be reduced to that of mucin-mucin interaction, and swept down the GI tract during the phase III housekeeper wave.

Colon and Rectal Drug Delivery The colon is divided into 4 parts; the ascending, transverse, descending and sigmoid sections. The rectum is continuous with the sigmoid colon, and is about 12 cm in length. (51) The mucous coat of the large intestine is smooth and devoid of villi. (51) The surface is covered with a large number of goblet cells.

Bioadhesive drug delivery systems designed to target the colon area encounter similar problems as gastric and intestinal bioadhesives. The two important factors controlling residence time are GI motility and mucin turnover. Furthermore, maintenance of the bioadhesive surface of the dosage form after it has passed through the stomach and small intestine is another major concern. One possible approach is to coat the bioadhesive particles with hydrophobic but erodible polymer. The rate of erosion of the protective coat is controlled so that the bioadhesives will not be hydrated until the dosage form reaches the colon. Once it reaches the colon, bioavailability of the drug can be improved by using enhancers, (58) or dextran ester prodrugs. (59,60)

The mucous membrane in the rectum is thicker and more vascular than the colon. There is one distinctive feature of the rectal blood supply, (50) namely, the proximal one-third of the rectum is drained by the superior hemorrhoidal vein, which flows into the inferior mesenteric vein to the hepatic portal system. Whereas, the middle and distal one-third of the rectum are drained by the middle and inferior hemorrhoidal veins respectively, and enter the systemic circulation via the iliac veins. Thus, in order to by-pass hepatic first-pass metabolism, the rectal dosage form has to be localized in the lower two-thirds of the rectum.

Normally, after insertion, a rectal suppository has a tendency to migrate upward toward the upper two-thirds of the rectum. The use of bioadhesives that attach to the mucin and/or epithelial layer can localize the dosage form in the lower two-thirds of the rectum and delay its upward migration. However, defecation removes the dosage form irrespective of bioadhesion.

Cervical and Vaginal Drug Delivery   The cervix is the
distal portion of the uterus. (51)   The vaginal portion of
the cervix (exocervix) projects into the vaginal cavity,
and many cervical drug delivery systems are applied to the
vaginal portion of the cervix. The cervical canal is
covered by a columnar epithelium which is responsible for
mucus secretion. (52)
     During a woman's lifetime, there are marked changes and
variations in endogenous estrogen level, which affects the
environment of the cervix and vagina. During reproductive
years, the vaginal squamous epithelium is thick and
stratified, and produces an acidic environment via
metabolism of glycogen. (61) When estrogen dominates, the
cervical mucus appears clear and with low viscoelasticity,
whereas when progesterone dominates, the cervical mucus is
thick and has high viscoelasticity.
     For post-menopausal women, the endogenous estrogen
level decreases with resulting shrinkage of the cervix.
There is a decline in production of cervical mucus, and
the vaginal mucosa may be only 3 to 4 cells thick with a
decrease in vascularity. As the vaginal squamous
epithelium atrophies, the glycogen content decreases,and
the vaginal pH will become alkaline to slightly acidic
with resulting bacterial growth and infection. Post-
menopausal vaginal dryness was successfully treated by
using a bioadhesive moisturizer. (62) The moisturizer was
applied as a cream and the hydrated polycarbophil
bioadhesive attaches to the vaginal mucosa for about 48
hours and maintains the dessicated vaginal tissue at an
acidic pH, in addition to moisturization.
     A soluble hydroxypropyl cellulose cartridge impregnated
with drug was used for vaginal drug delivery. (63) The
cartridge hydrates to a high viscosity gel and releases
drug over an extended period of time. Bioadhesive tablets
consisting of a combination of hydroxpropyl cellulose and
Carbopol   934,   were   used   to   deliver   bleomycin
vaginally. (54)   There   was   a   high percentage   of
disappearance of cancerous focus in the earliest stage of
uterine cervix cancer without extirpation of the uterus,
when this dosage form was used.
     A bioadhesive stick was used to deliver drug
cervically. (64)   Drugs entrapped in a mixture of
hydroxypropyl cellulose and carbopol were released slowly.
A double layered stick-type bioadhesive formulation was
also used to deliver bleomycin cervically. (55) The double
layered stick consists of an outer layer and a center core
containing 20mg and 30mg of bleomycin respectively. The
double layered stick was found to be able to delivery
bleomycin continuously for 1 week.

SUMMARY

The bioadhesive process is time and temperature dependent. It begins with the establishment of intimate contact between the bioadhesive and substrate, followed by interpenetration and bond formation. After selection of a good bioadhesive for a specific administrative route, the factors that affect bioadhesion need to be identified. Proper modification of the dosage form is required to satisfy particular need and conditions of the chosen route. Bioadhesive drug delivery has great potential in localizing the dosage form in specific absorptive surfaces and in improving the bioavailability of a chosen drug.

LITERATURE CITED
1) Ishida, M., Machida, Y., Nambu, N., and Nagai, T., New mucosal dosage form of insulin, <u>Chem. Pharm. Bull., 29,</u> 810, 1981.
2) Gurny, R., Meyer, J.M., and Peppas, N.A., Bioadhesive intraoral release systems: design, testing, and analysis, <u>Biomaterial, 5,</u> 336, 1984.
3) Smart, J.D., Kellaway, I.W., and Worthington, H.E.C., An in-vitro investigation of mucosa-adhesive materials for use in controlled drug delivery, <u>J. Pharm. Pharmacal., 36,</u> 295, 1984.
4) Nagai, T., and Machida, Y., Advances in drug delivery. Mucosal adhesive dosage forms, <u>Pharm. Internal., Aug.,</u> 196, 1985.
5) Nagai, T., Topical mucosal adhesive dosage forms, <u>Medicinal Res. Revs., 6,</u> 227, 1986.
6) Park, K., and Robinson, J.R., Bioadhesive polymers as platforms for oral-controlled drug delivery; method to study bioadhesion, <u>Int. J. Pharm., 19,</u> 107, 1984.
7) Ishida, M., Nambu, N., and Nagai, T., Highly viscous gel ointment containing carbopol for application to the oral mucosa, <u>Chem. Pharm. Bull., 31,</u> 4561, 1983.
8) Park, K., Ch'ng, H.S., and Robinson, J.R., Alternative approaches to oral controlled drug delivery: Bioadhesives and in-situ systems, in <u>Recent Advances in Drug Delivery Systems,</u> Anderson, J.M., and Kim, S.W., eds., Plenum Press, 1984, p. 163.
9) Longer, M.A., Ch'ng, H.S., and Robinson, J.R., Bioadhesive polymers as platforms for oral controlled drug delivery III: Oral delivery of chlorothiazide using a bioadhesive polymer, <u>J. Pharm. Sci., 74,</u> 406, 1985.
10) Peppas, N.A., and Buri, P.A., Surface, interfacial and molecular aspects of polymer bioadhesion on soft tissues, <u>J. Controlled Release, 2,</u> 257, 1985.

11) Leung, S.H.S., and Robinson, J.R., The contribution of anionic polymer structural features to mucoadhesion, J. Controlled Release, 5, 223, 1988.

12) Ch'ng, H.S., Park, H., Kelly, P., and Robinson, J.R., Bioadhesive polymers as platforms for oral controlled drug delivery II: Synthesis and evaluation of some swelling, water-insoluble bioadhesive polymers, J. Pharm. Sci., 74, 399, 1985.

13) Chen, J.L., and Cyr, G.N., Compositions producing adhesion through hydration, in Adhesive Biological System, Manly, R.S., ed., Academic Press, New York and London, 1970, chap. 10.

14) Mikos,A.G., and Peppas,N.A. Comparison of experimental techniques for the measurement of the bioadhesive forces of polymeric materials with soft tissues, in Proc. 13th Int. Symp. Controlled Release of Bioactive Materials, Chaudra,I.A. and Thies,C.,Eds.,Controlled Release Society, Inc. Lincolnshire,III., 1986, 97

15) Ishida, M., Nambu, N., and Nagai, T., Mucosal dosage form of lidocaine for toothache using hydroxypropyl cellulose and carbopol, Chem. Pharm. Bull., 30, 980, 1982.

16) Bremecker,K.D.,Strempel,H.,and Klein,G.,Novel concept for a mucosal adhesive ointment, J.Pharm. Sci., 73, 548, 1984.

17) Yotsuyanagi, T., Yamamura, K., and Akao, Y., Mucosa-adhesive film containing local analgesic, The Lancet, Sept. 14, 613, 1985.

18) Nagai,T.,Nishimoto,Y.,Namby,N.,Suzuki.,and Sekine,K., Powder dosage form of insulin for nasal administration, J. Controlled Release, 1, 15, 1984

19) de Leede, L.G.J., de Boer, A.G., Portzger, E., Feijen, J., and Breimer, D.D., Rate-controlled rectal drug delivery in man with a hydrogel preparation, J. Controlled Release., 4, 17, 1986.

20) Harris,A.S.,and Stenberg,P.,Pharmaceutical aspects of prostaglandin formulations for local administration of obstetrics, Pharms.Int. Engl. Ed.,May, 133, 1981.

21) Park, H., On the mechanism of bioadhesion, Ph.D. Thesis, University of Wisconsin-Madison, 1986.

22) Peppas, N.A., and Reinhart, C.T., Solute diffusion in swollen membranes. Part I. A new theory, J. Membrane Sci., 15, 275, 1983.

23) Leung,S.H.S., and Robinson,J.R., The contribution of anionic polymer structural features to mucoadhesion II., J. Controlled Release,in press

24) Allen, A., and Garner, A., Progress report: Mucus and bicarbonate secretion in the stomach and their possible role in mucosal protection, Gut, 21, 249, 1980.

25) Silberberg, A., and Meyer, F.A., Structure and function of mucus, in Mucus in health and disease-II, Advances in Experimental Medicine and Biology, 144, 53, 1982.

26) Gottschalk, A., in The Chemistry and Biology of Sialic Acid and Related substances, Cambridge University Press, London, 1960.

27) Jeanloz, R.W., in Glycoprotein, Their Composition, Structure and Function, Gottschalk, A., ed., Elsevier, Amsterdam, 1972.

28) Johnson, P.M., and Rainsford, K.D., The physical properties of mucins, preliminary observations on the sedimentation behavior of porcine gastric mucin, Biochim. Biophys. Acta, 286, 72, 1972.

29) Yongyut Rojanasakul,Ph.D. Thesis, University of Wisconsin-Madison, WI, 1989.

30) Allen, A., Hutton, D.A., Pearson, J.P., and Sellers, L.A., Mucus glycoprotein structure, gel formation and gastrointestinal mucus function, in Mucus and Mucosa, Ciba Foundation Symposium 109, Pitman, London, 1984, p. 137.

31) Nichols,B.A., Chiappino,M.L., and Dawson,C.R., Demonstration of the mucus layer of the rear film by electron microscopy, Invest. Ophthalmol. Vis. Sci., 26, 464, 1985

32) Ito,I., Structure and function of the glycocalyx, Fed. Proc. Fed. Am. Soc. Exp. Biol.,28, 12, 1969.

33) Yamada,M., and Okigaki,T., Promotion of epithelial cell adhesion on collagen by proteins from rat embryo fibroblasts,Cell Biol. Int. Rep., 7, 1115, 1983

34) Massa,S., and Bosmann,H.B., Cellular adhesion; description, methodology and drug perturbation, Pharm. Ther., 21, 101, 1983.

35) Edelman,G.M., Cell adhesion molecules, Science, 219, 450, 1983.

36) Dessau,W.,Jilek,F.,Adelman,B.C., and Hormann,H., Similarity of antigelatin factor and cold insoluble globulin, Biochim. Biophys. Acts, 533, 227, 1978

37) McCoy,J.P.,Jr., Contemporary laboratory applications of lectins, Biotechniques, 4, 252, 1986.

38) Adams,A.D., The morphology of human conjunctival mucus, Amb Ophthalmal 497, 930, 1979.

39) Nagai, T., Nishimoto, Y., Nambu, N., Suzuki, Y., and Sekine, K., Powder dosage form of insulin for nasal administration, J. Controlled Rel., 1, 15, 1984.

40) Gordon, G.S., Moses, A.C., Silver, R.D., Flier, J.S., and Carey, M.C., Nasal absorption of

insulin: Enhancement by hydrophobic bile salts,
Proc. Natl. Acad. Sci. USA, 82, 7419, 1985.

41) Moses,A.C.,Gordon,G.S.,Carey,M.C.,andFlier,J.S.,
Insulin administered intranasally as an insulin-
bile salt aerosol, Diabetes, 32, 1040, 1983.

42) Hirai,S.,Yashiki,T.,   and Mima,H.,   Effect   of
surfactants on the nasal absorption of insulin in
rats, J. Pharm. Sci. 9, 165, 1981.

43) Salzman,R.,Manson,J.E.,George,G.T.,Griffing,T.,
Kimmerle,R., Ruderman,N., McCall,A., Staltz,E.I.,
Mullin,C., Small,D., Armstrong,J., and Melby,J.,
Intranasal   aerosolized   insulin.   Mixed-meal
studies and long-term use in type I diabetes. New
Engl. J. Med., 312, 1078, 1985.

44) Duchateau, G.S.M.J.E., Zuidema, J., Albers, W.M.,
and   Merkus,   F.W.H.M.,   Nasal   absorption   of
alprenolol and metaprolol, Intern. J. Pharm., 34,
131, 1986.

45) Galey,W.R.,Lonsdale,H.K., and Nacht,S., The in
vitro permeability of skin and buccal mucosa to
selected drugs and tritiated water, J. Invest.
Dermatology, 67, 713, 1976.

46) Chen,S.Y., and Squier,C.A., The ultrastructure of
the oral epithelium.In the structure and function
of   oral   mucosa   (Meyer,J.,   Squier,C.A.,
Garren,K.W.,eds.) Pergamon, New York, 1984, pp7-30

47) Garren,K.W., Topp,E.M., and Repta,A.J., Buccal
absorption   III.   Simultaneous   diffusion   and
Metabolism of an aminopeptidase substrate in the
hamster cheek pouch, Pharm. Res., 6, 966, 1989.

48) Hammar,H.,Enzymes   of   buccal   epithelium   in
psoriasis, Br. J. Dermatology, 89, 619, 1973.

49) Schachter, H., and Williams, D., Biosynthesis of
mucus glycoproteins, AEMB, 144, 3, 1982.

50) Greenberger,N.J., Gastrointestinal Disorders. Year
Book Medical Publishers, Inc., 4th ed. 1986.

51) Gray,H.,   Anatomy   of   the   Human   Body,
Clemente,C.D.,ed., Lea and Febiger, Philadephia,
30th ed.,1985.

52) Jones,H.W.,III.,Wentz,A.C.,and   Burnett,L.S.,
Novak's   Textbook   of   Gynecology,
Brown,C.L.,ed.,Williams and Wilkins, Baltimore,
11th ed.,1988.

53) Szurszewski, J.H., A migrating electric complex of
the canine small intestine, Am. J. Physiol., 217,
1757, 1969.

54) Machida, Y., Masuda, H., Fujiyama, N., Ito, S.,
Iwater, M., and Nagai, T., Preparation and phase
II clinical examination of a topical dosage form
for treatment of carcinoma colli containing
bleomycin with hydroxypropyl cellulose, Chem.
Pharm. Bull., 27, 93, 1979.

55) Iwate,M.,Machida,Y., and Nagai,T., Double-layered
Stick-type formulation of bleomycin for treatment

of uterine cervical cancer, <u>Drug Design Delivery,</u> <u>1,</u> 253, 1987.

56) Matthews, D.M., and Adibi, S.A., Peptide absorption, <u>Gastroenterology, 71,</u> 151, 1976.

57) Machida, Y., Masuda, H., Fujiyama, N., Iwater, M., and Nagai, T., Preparation and phase II clinical examination of topical dosage forms for the treatment of carcinoma colli containing bleomycin, carboquone, or 5-fluorouracil with hydroxypropyl cellulose, <u>Chem. Pharm. Bull., 28,</u> 1125, 1980.

58) Tomita,K.,Shiga,M.,Hayashi,M., and Awazu,S., Enhancement of colonic drug absorption by the paracellular permeation route, <u>Pharm. Res., 5,</u> 341, 1988.

59) Harboe,E.,Larsen,C.,Johansen,M., and Olesen,H.P., Macromolecular prodrugs, XV. Colon-targeted delivery-bioavailability of naproxen form orally administered dextran-naproxen ester prodrugs varying in molecular size in the pig, <u>Pharm. Res., 6,</u> 919, 1989.

60) Larsen,C.,Harboe,E.,Johansen,M., and Olesen,H.P., Macromolecular prodrugs, XVI. Colon-targeted delivery-Comparison of the rate of release of naproxen from dextran ester prodrugs in homogenates of various segments of the pig gastrointestinal (GI) tract, <u>Pharm. Res., 6,</u> 995, 1989.

61) Bergman, A., and Brenner, P.F., Alterations in the urogenital systems, in <u>Menopause-Physiology and Pharmacology,</u> Mishell, D.R. Jr., ed., Year Book Medical Publishers, Inc., Chicago, London, 1987, chap. 5.

62) Replens by Columbia Laboratories, Miami, Florida.

63) Williams, B.L., Soluble medicated hydroxypropyl cellulose cartridge, U.S. Patent 4, 317, 447, March 2, 1982.

64) Hui, H.W., and Robinson, J.R., Ocular delivery of progesterone using a bioadhesive polymer, <u>Int. J. Pharm., 26,</u> 203, 1985.

65) Saettone,M.F., Giannaccini,B.,Chetoni,P., Vehicle effects in ophthalmic bioavailability: an evaluation of polymeric inserts containing pilocarpine, <u>J. Pharm. Pharmacol., 36,</u> 229, 1984.

66) Grass,G.M.,Cobby,J.,Makoid,M.C., Ocular delivery of pilocarpine from erodible matrices, <u>J. Pharm. Sci., 73,</u> 618, 1984.

67) Ticho,U.,Blumenthal,M.,Zones,S.,Gal,A.,Blank,I., and Mazor,Z.W., A clinical trial with piloplex- A new long-acting pilocarpine compound: preliminary report, <u>Ann. Ophthal., April,</u> 555, 1979.

68) Robinson,J.R., and Li,V.H.K., Ocular disposition and bioavailability of pilocarpine from piloplex and other sustained release drug delivery system,

in,   Recent   Advances   in   Glaucoma,
Ticho,U.,David,R.,eds. Excerpta Medica, Amsterdam,
1984, 231.
69) Mygind,N., Nasary Allergy, Blackwell Scientific,
Oxford, 1978.
70) Illum,   L.,   Jorgensen,   H.,   Bisgaard,   H.,
Krogsgaard, O., and Rossing, N., Bioadhesive
microspheres as a potential nasal drug delivery
system, Intern. J. Pharm., 39, 189, 1987.
71) Suzuki,Y.,Ikura,H.,Yamashita,G., and Nagai,T.,
Powdery pharmaceutical preparation and powdery
preparation to the nasal mucosa and method for
administration there of, Japanese Patent 1, 286,
881, October 31,1985; U.S. Patent 4, 294, 829,
October 13, 1981; E.P.C. Patent 23, 359, July 29,
1980.
72) Kuroishi,T.,Aska,H., and Okamoto,M., Phase I study
of beclomethasone dipropionate powder preparation
(TL-102)-single and repeat administration, Jpn.
Pharmacol Ther., 27, 4055, 1984.
73) Schor,J.M.,Davis,S.S.,Nigaloyem,A., and Bolton,S.
Sustain transmucosal tablets, Drug Dev. Ind.
Pharm., 9, 1359, 1983.
74) Erb,R.J., Bioavailability of controlled release
buccal   and   oral   nitroglycerin   by   digital
plethysmography,   in   Controlled   Release
Nitroglycerine in Buccal and Oral form, Advances
in   Pharmacotherapy,   Vol.I,   Bussmann,W.D.,
Drics,R.R., and Wagner,W., eds.,S.Karger,Basel,
1982, 35.
75) Ishida, M., Nambu, N., and Nagai, T., Mucosal
dosage form of lidocaine for toothache using
hydroxypropyl cellulose and carbopol, Chem. Pharm.
Bull., 31, 980, 1982.
76) Gupta,P.K.,Leung,S.H.S.,   and   Robinson,J.R.,
Bioadhesive/Mucoadhesives in drug delivery to the
gastroinestinal   tract,   in   Bioadhesive   Drug
Delivery Systems, (V. Lenaerts and R. Gurny, eds.)
CRC Press Inc., Baca Roton, Florida, 1990, ch.4.
77) Merkle,H.P.,Anders,R.,   and   Wermerskirchen,A.,
Mucoadhesive buccal patches for pepticle delivery,
in   Bioadhesive   Drug   Delivery   Systems,
(Lenaerts,V., and Gurny,R., eds.) CRC Press Inc.,
Baca Roton, Florida, 1990, ch.6.

RECEIVED June 4, 1990

Chapter 24

Heparin, Heparinoids, Synthetic Polyanions, and Anionic Dyes

Opportunities and New Developments

William Regelson

Medical College of Virginia, P.O. Box 273, Richmond, VA 23298

Review of native heparin action provides new applications for synthetic polyanions. Furor over dextran sulfate or suramin, anionic dyes as anti-AIDS or anti-tumor drugs forgets the past history of polyanions: As physicochemical combinants, reverse transcriptase inhibitors and immunoadjuvants. This goes back to Ehrlich and anionic dyes, S. S. Cohen and heparin, ethylene maleic anhydride copolymers, and the macroanionic tungstates and heteropolymolybdates. Unfortunately, in this AIDS era, vaccine enhancing action of the polycarboxylate polymers Diveema (MVE, pyran copolymer) is still ignored. Their clinically effective anti-tumor potential requires intraperitoneal not intravenous use. Preclinical experience with newer sulfated native glycosaminoglycan heparin fractions should re-evaluate the place of synthetic polyanions for anti-coagulant, lipolytic (anti-atherosclerosis action), anti-inflammatory effects and DNA modulation. Inhibition of angiogenesis and smooth muscle proliferation via modulation of cytokines as well as oral absorption and endothelial localization provides new roles for polyanions of clinical value.

In recent years, there has been a veritable explosion of interest in both native heparin or heparin related, glycosaminoglycans; synthetic polyanions, or anionic dyes. It is the purpose of this review to discuss pertinent observations regarding the structural and physiologic actions of these negatively charged polymers because of new interest in their clinical application, independent of uses involving surface modification of implantable polymers.

Since Jaques' (1,2) and Engelberg's (3,4) last reviews, Lane & Lindhal (5) have provided a new text that focuses on heparin related proteoglycans where there are excellent sections discussing structure and function (6), but the text is focussed on the naturally derived compounds and completely ignores the water soluble synthetic polyanions whose development stemmed from these native compounds.

0097–6156/91/0467–0367$07.75/0
© 1991 American Chemical Society

Since our last reviews (7-11) dealing with the systemic action of pyran copolymer (Diveema, vinyl ether maleic anhydride, MVE2), and the biologic action of polyanions (9-11), the clinical value of these polymers is no longer clinically focussed on their anti-coagulant, immunostimulatory action or coulombic electrostatic interactions (6-8), but has extended to their role as carriers of chemotherapy, and most important, as enzyme carriers to targeted sites of pathology (12-14). The biologic use of enzymes linked to polymers as "solid enzymes," now transcends extra-corporeal action or commercial detergent protease delipidation but extends to systemic use in the treatment of tumors and infection.

DRUG AND ENZYME CONJUGATES

The application of polyanionic polymers as drug carriers and enhancers of pharmacokinetics or drug localizing activity has been discussed by Maeda et al. (12,13). The most recent successful example of the above is the application of pyran copolymer as a carrier of superoxide dismutase (SOD) to areas of inflammation induced by the influenza virus (14,15). The anti-inflammatory enzyme SOD conjugated to pyran (mw 5600) by carbamide linkages protected mice from lethal influenza infection from 5 to 8 days after viral inoculation.

Prior to the use of SOD coupled to pyran, polyethylene glycols (PEG) have been used as carriers of SOD, to prolong half-life and decrease immunogenicity of the enzyme. However, the clinical application of SOD alone, or with PEG as the polymeric carrier, has previously been focussed on the treatment of rheumatoid disease, ischemia or trauma. Pyran-SOD should be useful in preventing injury mediated by any macrophage or leukocyte induced host reaction as SOD tissue protection works by minimizing free radical damage induced by white blood cell enzymes such as xanthine oxidase (15). The enzyme SOD, by itself, has a very short in-vivo half life, which limits therapeutic value although it is available in veterinary medicine (orgatein) as an anti-inflammatory agent. If immunogenicity will not prove to be a problem, the success with pyran-SOD in the Japanese influenza model suggests that it may be applicable to any acute and latent forms of tissue injury where free radicals are the initiating pathologic factor.

The route of administration of pyran SOD should be critical in that, as in the case of influenza where the primary pathology is in the lungs, giving pyran intravenously optimizes concentration in the area of primary pathology. Intravenous pyran also localizes in liver, spleen, and kidney (16). This also relates to molecular weight, and one would assume that pyran-SOD could have similar organ focussed concentration enabling the therapeutic value for pyran-SOD to be useful in liver injury for treatment of inflammation from viral hepatitis or toxic agents.

Of interest to the above, Karlsson and Marklund (17) have shown that heparin induces the release of extra-cellular SOD, and heparin has an affinity for SOD. In other areas of medicinal chemistry, the development of polyanionic polymers for the focussed transport of proteins or peptides has been successfully developed by Maeda et al. (12,13,18,19). They linked polystyrene-co-maleic acid to the anti-tumor antibiotic, neocarzinostatin (SMANCS) for the delivery of cancer chemotherapy.

Neocarzinostatin is an antibiotic with nuclear translational inhibiting activity blocking tumor growth. The SMANCS conjugation produces a polymer of 16,000 mw with a ten fold increase in the biologic half life of the polymer complex as compared to neocarzinostatin alone. What is most important, is that the physical characteristics of the polymer coupled neocarzinostatin is changed: there is marked decrease in its cytotoxicity and it shows enhanced solubility in lipid emulsions so that it can be administered in such a form (e.g. Lipiodol, an iodinated lipid emulsion). This serves to present SMANCS in lipid droplets that leak out of damaged blood vessels for prolonged retention, up to weeks, within the tumor bed. When SMANCS is given intra-arterially to the area of regional tumor involvement (18-22) local intra-tumoral concentration permits prolonged regional anti-tumor action, with minimal toxicity to bone marrow and other sites. Using this technique, giving SMANCS/Lipiodol through the hepatic artery, primary inoperable cancer of the liver (hepatoma) may be cured. This is because primary hepatoma normally has a sinusoidal circulation that permits macromolecules to enter the tumor bed through blood vessel walls. With intra-arterial SMANCS, success has also been seen in treating selected adenocarcinoma of the lung, via the bronchial artery. The Kumamoto group are now developing methods to enhance local intra tumoral SMANCS concentration by radiating regional tumor sites with doses of radiation that cause local vascular damage, leading to the vascular penetration of SMANCS into the tumor utilizing systemic intravenous administration. Another advantage to the clinical use of these polyanionic conjugates is that they are capable of inducing interferon (21) for host mediated enhanced anti-viral or anti-tumor action.

The use of anionic polymers coupled to enzymes or chemotherapeutic agents provides a complex carrier system that behaves in similar fashion to blood proteins. What makes the use of these agents feasible is that regional tissue entrapment of these polymers is now possible in tumors via intra-arterial administration or systematically following endothelial injury (12,13,18,19). The pharmacologic use of macromolecule carriers with particular reference to tumor inhibition has recently been reviewed by Maeda (12,13). The recent success of Maeda's group (18-22) follows a long history of failure by Ringsdorf's group (23) and others (7) because of our lack of understanding, at the time, of endothelial vascular barriers that prevented the entry of polymeric complexes into the tumor bed. This same problem exists today for liposomal drug delivery (16) which could become more effective with selective vascular injury to produce protein leakage.

Dumitriu et al. (24) have reviewed the literature regarding enzyme and drug carriers linked to polymers which have growing validity when applied to direct regional intratumoral injection for local control (12,13,18-22,25).

DETOXIFICATION

Relative to the use of polyanionic conjugates, one observation which has never been adequately exploited is that of Higginbotham in 1960 (27). He found that heparinization of mice could protect from toxicity due to cationic antibiotics: polymyxin B, neomycin, viomycin, and dihydrostreptomycin. In the case of neomycin-heparin, antibiotic potency was maintained despite its administration as a complex. Unfortunately, there is no data regarding chronic toxicity, and this approach was never developed clinically. Instead, less toxic antibiotics were developed by selection or modification of endogenous structure. However, this approach has relevance to the concept that the polyanionic nature of stromal glycosaminoglycans serves as broad non-specific detoxifying system to cationic products such as polyamines (28). In another area, as discussed by us previously (7), and in a recent review by Levy (29) the complexing of cationic DEAE dextran to polyanionic polynucleotides to enhance interferon induction and improve therapeutic index was a promising development that never achieved clinical enthusiasm.

Coupling or charge complexing of antibiotics to copolymers is worthy of clinical interest for both prolonged action and regionalization, and there is a new need to rethink the practical applications inherent in anionic macro-molecules. Historically, this goes back to the period of Paul Ehrlich and anionic dyes which has not advanced very far since it was reviewed by Jaques (1,2) and Regelson (9,10). However, the trypanocide suramin is having a clinical rebirth, as discussed below.

ANIONIC DYES

The anionic dye suramin, has been shown to inhibit the HIV virus and other viruses in in-vitro models (30), but the clinical results have been disappointing because of problems of hemorrhage and adrenal insufficiency complicating clinical trial. More recently, suramin has been revived for clinical testing as an anti-tumor drug, where it may have clinical activity against a wide range of tumors including carcinoma of the prostate and adrenals (31). Suramin has also been shown to modulate NK and monocytic cell mediated toxicity (32) with increased cytotoxicity engendered in human monocytes and mouse peritoneal macrophages upon exposure to the anionic dye. Similar effects have been seen for synthetic polyanions such as pyran copolymer (8) which led to clinical anti-tumor trial of MVE2, the 15,000 mw fraction of this divinyl-ether maleic anhy-dride copolymer, as a bioresponse modifier. These results are variable depending on preincubation, source of effector cells, and route of administration (16).

Suramin utilization represents a revival of the clinical application of polyanions, and there is intense activity evaluating anionic dyes and polyanions for their potential usefulness in the treatment of AIDS. Unfortunately, the work of Ray & Agrawal (33), who showed that aminofluorene sulfonic acids had high selective concentration in mouse tumors, has been lost, but should be revisited.

ANTI-TUMOR

While the use of anionic dyes is in clinical ascendancy, the clinical anti-tumor use of pyran copolymer (MVE2) (8) and related polyanions like polyethylene sulfonate (PES) (34) has halted prematurely because of fear of toxicity. The exception is carbetimer, a low molecular weight ethylene maleic anhydride copolymer (35). The same has been true of the synthetic interferon inducing polynucleotides, although ampligen, an asymmetric polynucleotide interferon inducer, is reported to have clinical value (36-38). Levy (29) has reviewed the history of the polynucleotides as interferon inducing immunomodulators for their anti-tumor and antiviral applications. He suggests that clinical dosage selection was too high for optimal clinical results. It is this author's opinion that the route of administration is the critical feature of disappointing clinical results which indicate that the intraperitoneal route must be developed (16).

It is important to stress that there has been lack of critical clinical understanding of the use of polyanions as immunostimulators or bioresponse modulators. As an example, for immunostimulation, the best clinical route is intraperitoneal (IP) (16,39) where peritoneal exudate cells can play a therapeutic role. With subcutaneous ports and indwelling peritoneal catheters, the IP route is clinically feasible.

Polyanions are potent immunoadjuvants but their clinical study was not based on this model. MVE2, when given after tumor burden reduction by chemotherapy was curative in mouse models, but this was never developed clinically despite sponsorship of MVE2 studies by the NCI Bioresponse Modifier Program (40).

As mentioned previously, carbetimer, derived from our Monsanto collaborative tumor screen of the 60's (10), is now in phase II/III clinical study where it is apparently showing clinical anti-tumor effects (35). Of interest, it is not an immunostimulator or anti-coagulant. It is thought to be a pyrimidine antagonist (41), but one must determine if its mechanism is based on direct or indirect competitive effects on cytokines or phosphokinases such as have been shown to be influenced by native heparin or heparitan sulfate which is discussed later on in this paper. Of clinical interest, carbetimer produces hypercalcemia and CNS effects that are similar to those seen for pyran copolymer and suramin.

HEPARANASE INHIBITION: ANTI-METASTATIC EFFECTS

Heparan sulfate, which is a component of the vascular wall is cleaved by metastatic murine B16 melanoma cells (42), while heparin, but not other glycosaminoglycans inhibited heparan degradation. Similar effects were reported by Bar-Ner et al. (43) wherein a methylcholanthrene induced T-lymphoma produced an endoglycosidase (Heparinase) which was also inhibited by heparin. Based on these observations, Irimura et al. (44) explored the structure activity relationship of heparin's inhibition of endo-β-glucuronidase: N-sulfate,O-sulfate and glucosamine residues and carboxy groups on uronic residues were responsible for the inhibition of heparanase.

Heparin fragments containing these residues reduced B16 intravenous tumor lung metastatic capacity on in-vitro incubation. The additional significance of heparanase will be discussed in the atherosclerotic section as it has bearing on atherosclerosis wherein heparan/ heparin content of blood vessels is an age related atherosclerotic event.

IMMUNOADJUVANCY AND ANTI-VIRAL ACTION

Pertinent to our search for HIV vaccines is the observation by Campbell & Richmond (45) that pyran has vaccine adjuvancy. When an ineffective hoof and mouth vaccine was given intraperitoneally simultaneously with pyran it resulted in 100% protection of mice; other work has shown anti-viral action even without vaccine adjuvancy: Morahan et al. (46) demonstrated that pyran could protect mice in-vivo from herpes type 1 and type 2 (HSV) infection with reduction of both mortality and the severity of lesions. This use of IV pyran post herpes infection reduced virus titers and resultant pathogenesis suggesting that it may have value in early or latent herpes infection. The use of pyran or related polyanions for vaccine adjuvancy, should be renewed with particular reference to retrovirus (HIV, AIDS) immunization or therapy particularly in view of the associated finding of a herpes related virus infection in patients with HIV.

This clinical potential is supported by mouse data showing that pyran protects against the expression of Friend leukemia up to 3 months after a single injection (47). Friend leukemia is an RNA retro-virus with immunodepressing properties similar to that seen for HIV.

In further support of the above, the clinical topical application of heparin related polyanions has been suggested long ago by Nahamias et al. (48) who showed that in-vitro herpes simplex infectivity could be reduced by a wide range of polyanions. This action was dependent on branching of the molecule, the type of glycosidic bond, the degree of sulfation, and the size of the molecule. In their study, heparin was the only effective native polyanion. This work relates to the early observations of Duran-Reynals (49) that vaccinia infection can be held in place by the binding action of tissue hyaluronate (ground substance).

The anti-viral action of polyanions can express itself through four independent modes of selective therapeutic value:

1) As a direct charge mediated interaction which has been described for both polyanion and anionic dye neutralization of viruses (1,2,9-11). This is an old story going back to S.S. Cohen with tobacco mosaic virus (TMV) in 1942 (50) and with Vaheri and others with herpes and other viruses (51). This was reviewed by Jaques (1) and myself (9,10) and interest in this is now being revived with the observation that suramin, fuchsin, aurintricarboxylic acid, heparin, and dextran sulfate inhibit HIV infected cells (52,53).

In a good review of current and past data, the anionic dye, suramin, has been shown to inhibit hepatitis B, rous sarcoma, and hepatitis delta virus by the above mechanism (30). The authors speak of patterns of enzyme inhibition induced by suramin, but one would guess, as has been shown much earlier by Vaheri et al. (51) and in our institution (9,10,54), that effects are electrostatic and probably not related to specific enzyme blockade.

2) The other anti-viral mechanism provided by a polyanions places the emphasis on current attempts to control HIV-1 infection via inhibition of reverse transcriptase, the key enzyme involving HIV viral proliferation (52,53,55). Reverse transcriptase inhibition for pyran copolymer was shown by Papas et al. in 1974 (56). Similar actions are seen for suramin, but not necessarily related to antiviral effects (30). Alternatively, these anionic compounds can also inhibit DNA polymerases (30,53) necessary for viral production.

3) A third potential anti-viral action relates to the fact that polyanions inhibit aspartic related peptic proteases that are a vital part of the HIV infectious viral sequence (57).

4)   Finally, polyanions induce interferon, and stimulate macrophage function involved in host mediated anti-viral action, as well as antigenic processing (8,29). Interferon inhibits viral proliferation, including HIV, and polyanions like pyran copolymer and the polynucleotides (8,29,36-38), enhance immune responsiveness leading to the development of effective vaccines (8,45).

5)   In the synthesis of antivirals, it should be remembered that olefin-maleic anhydride copolymers were used as oral antidiarrheal agents (58). When cross linked with vinyl crotonate they were found to remove intestinal water and adsorb bacterial endo and exo enterotoxins from the gastrointestinal tract (59). However, the action of these polymers for the prevention or attenuation of intestinal bacterial and viral infection has been neglected in recent years, although there is a real place for their application as adsorbents in tampons for prevention of toxic shock syndrome during menstruation.

The charge related or adsorption action of polyanionic polymers has been well reviewed (1,2,9,10,51). No clinical or commercial exploitation developed from this work although Melnick's group working with Monsanto (60,61) gave serious consideration to the use of ethylene maleic anhydride copolymers in water purification.

Polyanions can also play a role in diagnostic pathology techniques as in-vitro, they alter the cytopathogenicity of encephalomyocarditis viruses (EMC) and plaquing is modulated in the presence of negatively charged agar overlays which could be reversed by polycations (62). As discussed by Duran-Reynals (49), these in vitro observations should be pertinent to the clinical pathology of viral or bacterial infection.

There should be renewed interest in Duran-Reynals finding (49) that local hyaluronidase action enhanced cutaneous vaccinia pathology beyond local cutaneous binding sites. These old observations provide us with the opportunity to develop polyanions, or glycosaminoglycan charged heparinoids that could neutralize the presence of cutaneous released viruses. Of recent interest to this, interleukin I (IL I) has been shown to generate hyaluronate in dermocytes in tissue culture (63) suggesting a possible cutaneous place and a mechanism for IL I application in control of cutaneous HSV manifestation.

LYMPHOCYTE MOBILIZATION

In relation to immune responsiveness, lymphocytes are dramatically mobilized from spleen and lymph nodes by polymethacrylic acid (64). We have seen similar phenomena occurring in dogs related to bone marrow mobilization on chronic intraperitoneal administration of ethylene maleic anhydride copolymers (65).

In view of the interaction of heparinoids with lymphokines or leukokines, this represents a possible explanation of the above phenomenon (66). Alternatively, one should investigate this in relation to the function of thyroid hormones which in hyperthyroidism induces a similar lymphocytosis. Increased thyroid function correlates with resistance to tuberculosis (67) and Duran-Reynals (49) has shown that resistance to TB in animal models relates to ground substance permeability, an observation we have forgotten in regard to cutaneous pathology.

IMMUNITY

We have thoroughly reviewed the immunostimulating action of pyran copolymer in 1985 (8) which is a model for the immunostimulating action of other polyanions. Their anti-tumor activity involves activation of macrophages as effector cells which can distinguish the tumor cell's surface from that of normal cells. This results in selective tumor cell killing, sparing normal cells, when these polyanionic activated macrophages are in contact with the tumor cell surface (8,68-70). Still unexplained, Kaplan's laboratory found that this could be mediated by a reduction division decreasing cellular DNA (71,72).

Of major clinical interest, pyran can be given as a curative immunoadjuvant in conjunction with chemotherapy that reduces the tumor burden (56) provided the chemotherapy does not reduce host response (7,8). As in the case of vaccine adjuvancy, this use of pyran makes it also potentially curative in post-operative situations. Unfortunately, none of the clinical studies that were prompted by observations of the Chirigos' group (56) were applied in this manner. Studies with the MVE2 fraction of pyran were conducted in advanced inoperable patients where the drug was given intravenously instead of intraperitoneally where best immunoadjuvant effects are seen (8,16). Perhaps the renewed interest in the imidazole, levamisole-5 fluorouracil adjuvant combination enhancing survival in Dukes C bowel cancer will give new impetus to the appropriate clinical use of pyran copolymer (MVE2) as an immunoadjuvant.

There is still activity involving the clinical application of polynucleotides (29,36-38), and Akashi et al. (73) have combined polyadenine with a maleic anhydride copolymer which is under development as an immunoadjuvant. Alternatively, Ottenbrite et al. (74) are encapsulating their immunostimulating polyanion in polysaccharide encapsulated liposomes for macrophage activation.

In another area dealing with immunity, native heparin, inhibits formation of the classical C convertase complement pathway. This has been reviewed by Lane (75) in relation to protein binding. The physical chemical nature of the inhibitory heparin fractions relate to enhanced activity with increasing molecular weight and the degree of sulfation (76). This work has defined the structural relationships between anti-coagulant activity and complement binding capacity. Similar studies have been conducted defining the anti-complement activity of dextran derivatives with varied carboxylic and benzylamide sulfonate groups (7,9,10,77).

Heparin inhibits sialyltransferase activity and these patterns of inhibition block lymphokine IL2 expression interfering with active killer cell proliferation by 50%. (78). In contrast, Dziarski (79), has shown that heparin enhances the mixed lymphocyte reaction and cytotoxic responses against H_2-histocompatible tumors.

Heparin and related compounds like dextran sulfate, or those dextran derivatives that have heparin-like action as discussed earlier, inhibit serum complement (77). This action of heparinoids has also been discussed in an early paper dealing with the action of heparin as an agent producing thrombocytopenia and anaphylaxis (80). Since then, heparin induced thrombocytopenias as reviewed by Godal (81), has been discussed in almost 100 papers. Thrombocytopenia is also a toxic effect of synthetic polyanions, and although many papers discuss thrombocytopenia, little has been concluded with the exception of Horne (82) who discusses heparin binding to normal and abnormal platelets and concludes that polymer size and charge density are the determinants of platelet heparin binding. It should be remembered that in that situation heparin action can both prevent and/or activate fibrinolysin (83).

ANTIGEN OR ANTIBODY LOCALIZATION

The modulation of ground substance permeability by polyanions was seen many years ago in the ability of sulfated polysaccharides to block the penetration of mucin by pepsin (84). This is relevant to past observations regarding the electrostatic binding of viruses to polyanions and is now relevant to the clinical expression of auto-immune syndromes. As an example, monoclonal anti-DNA antibodies have been shown to bind to cardiolipin and other negatively charged phospholipids as well as hyaluronic acid and chondroitin sulfate (85,86). Most recently, Aotsuka et al. (87) have shown that anti-DNA or heparin sulfate antibodies found in systemic lupus (SLE) patients bind to the anionic dye cibacron blue when it is cross-linked to agarose. The levels of antibody activity correlate with the intensity of clinical disease.

Based on the above, we now have the clinical possibility that antibody localization or its expression may relate to the presence of heparin or other glycosaminoglycans acting locally or systemically. Apart from the significance of this in clinical pathology, these observations provide a technically feasible method for binding or eluting antibodies through extracorporeal exposure of plasma from patients with SLE or related disease. Alternatively, it provides an approach to removing blocking antibody which may be pertinent to endogenous tumor growth regulation.

PROTEINASE INHIBITION

Macroanionic and macrocationic inhibition of proteinases has been described in 1954 (88). Delicort and Stahman (89) showed that polylysine and polyglutamic acid produce activation or inhibition of pepsin, depending on concentration. This may be pertinent to the fundamental mechanisms of blood clotting which involve proteinase interactions which are governed by sequential activation of serine proteinases (90).

A new super family of plasma serine inhibitors called serpins have been described. Serpins are small glycoproteins important to the inhibition of elastase, thrombin, fibrinolysin, trypsin, chymotrypsin, and C_1-esterase where protein substrates are more readily digested following oxidation resulting in propagation of tissue injury from coagulant or elastase action, or the release of tissue phospholipases. These serpins possess structural peptide homology that shows their significant place in evolutionary history (91). An example of their clinical importance is seen in parallel structural affinities between alpha 1-antitrypsin and antithrombin which is modulated by heparin through cofactor II (92).

Serpins control proteolytic activity governing coagulation and inflammation (93). Serpin relationships to fibrinolysins are pertinent to an observation we made in 1977 (94) that splenic serine protease activity correlated with viral susceptibility in specific mouse strains. Those mouse strains resistant to N-tropic Friend leukemia had high levels of fibrinolytic protease action (TAME and BAME esterase), while sensitive strains had low activity. The IV administration of pyran copolymer enhanced susceptibility to the virus and lowered proteolytic activity. In contrast, IP pyran injection had the opposite effect and enhanced viral resistance. There was a similar correlate between BAME and TAME esterase activity and susceptibility to herpes (HSV-1) infection in mouse spleens of different strains.

The above may be pertinent to serpin action as a protease inhibitor, as BAME, TAME esterase activity was present in splenic homogenates which was affected by IP pyran injection. These observations now take on new meaning because of the importance of proteases in virus infection that is pertinent to resistance to HIV infection and other infections where retro-viruses are involved. As an example, Blundell & Pearl (57) have reviewed data showing the importance of aspartic proteinases which are related to retroviral proteinases, a key factor in HIV viral infection. The aspartic proteinase is also a structural feature of gastric proteases and lysosomal cathepsin D (95).

Pyran's modulation of Friend leukemia resistance is clinically pertinent (47) as the Friend virus is a retrovirus that produces immune paralysis in mice similar to that seen in HIV infection. Based on this, we should screen the inhibiting action of polyanions on proteases as there may be more than one possible therapeutic application of synthetic or native polyanions in the control of viral pathology.

In another area related to proteolysis, heparin fragments also show significant action as elastase inhibitors and long term effects on chronic administration suggest that heparin fragments might, as discussed earlier, have potential value in the treatment of emphysema (96) where elastase is a key etiologic factor.

CATHEPSIN AND VASCULAR EFFECTS

Neutrophils contain cationic proteins of proteolytic capacity such as cathepsin G. (97). This enzyme released by neutrophil adhesion to the endothelial vascular lining enhances vascular permeability to proteins resulting in leak outside the vessel. This is associated with endothelial retraction and intercellular gaps permitting serum albumin to leak across blood vessels. The polyanion heparin or dextran sulfate protects the vascular endothelium from this effect. Heparinoids and heparin bind to vascular endothelial intracellular cement (98) and play a role in atherosclerosis or myocardial infarction (3,4,99-101).

VASCULAR AND GLOMERULAR EFFECTS

Heparin has been found to inhibit renal mesangial growth independent of its anticoagulant effects (102,103) and has had a long history as a possible inhibitor of the clinical development of glomerularnephritis or other forms of renal injury

(104,105). The focus of this action in the past was placed on heparin's anticoagulant and anti-platelet or anti-complement activity. However, glomerularnephritis is a disease of proliferated epithelial cells (mesangial cells) which greatly diminishes kidney function. Of interest, interleukin 1 (IL-1), platelet derived growth factor (PDGF) are macrophage secreted factors which can stimulate mesangial cells to grow. In addition, glomerular vascular epithelial cells respond to epidermal growth factor (EGF), fibronectin, and glycoproteins, and can in autocrine fashion, be stimulated to produce both IL-1 and PDGF (66).

An additional place for heparin is that it also blocks nephrosclerosis in in-vivo model systems. This has been well reviewed by Adler (66) who has studied the direct inhibition by heparin of renal vascular cells in-vitro. There was no correlate with antithrombin effect, but with O-sulfation and n-substitution requirements Adler concluded that heparin's action may be via its bonding to cell growth factors. He sites heparin's affinity to histidine-rich glycoprotein, platelet factor 4, fibronectin, thrombospondin, antithrombin, heparin cofactor II, PDGF, fibroblast, endothelial, and other growth factors. In a sense this is the rediscovery of Albert Fischers work in the 1930's wherein heparin inhibited tissue culture growth by interfering with the isoelectric point of proteins (9,10). This was studied by Gorter and Nanninga in 1953 (106) and similar action has been described for the anionic dye, suramin in a recent review by Stein et al. (31).

Adler (66) gives us a new view of pathophysiology: He suggests that the action of the matrix, surface related glycosaminoglycan, heparan sulfate plays the same role of heparin. However, in contrast, while heparin interacts with basic endothelial growth factor (ECGF) and fibroblast growth factor (bFGF) to inhibit mitosis, heparan and related compounds can also selectively potentiate mitogenic activity as seen for lung diploid fibroblasts, an action which is associated with plasminogen activation (107) resulting in proliferative events. The complexities of this are reported below:

Castellot et al. (108,109) describe the inhibition of vascular smooth muscle cells by heparin via a block to G_o-S in the mitotic cycle. This group has examined protein induction effects of heparin along with the pattern of growth inhibition (108). All of this is pertinent to the action of heparin fractions on blood vessel walls where it blocks angiogenesis in the presence of unique steroids (110,111). The blocking action of heparin on angiogenesis may reflect on its affinity for copper as well as fibroblastic growth factor.

GROWTH

The literature in regard to heparin effects on proliferation is still growing since our in depth reviews in 1968 (9,10). Bikfalvi et al. (112), updated this in their recent paper showing the binding, internalization, and growth inhibition of human omental endothelium by heparin. This is of interest because of the angiogenic action of certain glycolipids found in omental fat (113).

In summary, heparin inhibits migration and proliferation of smooth muscle cells and stimulates endothelial migration in the presence of endothelial cell growth factor (ECGF). In contrast, heparin potentiates the mitogenic activity of growth factors that bind heparin i.e., ECGF and acidic fibroblast growth factor (aFGF), where it stabilizes tertiary structures that effect cell surface binding. However, basic fibroblast growth factor (bFGF) is inhibited by heparin.

Heparin forms complexes with both basic and acidic fibroblast growth factors (FGF) which have angiogenic action. Heparin activates bFGF cell surface binding as can the anionic dye, suramin, while protamine, a cationic heparin antagonist, blocks FGF receptor binding (114). Suramin blocks EGF binding to renal carcinoma cells (115).

In addition, bovine brain endothelial cell growth factor (ECGF) stimulates the synthesis of plasminogen activators. As discussed previously, this action is enhanced by heparin on lung diploid human fibroblasts (107). Heparin also potentiates the mitotic stimulatory action of ECGF on endothelial cells (116).

In contrast, inhibition of arterial smooth muscle intimal hyperplasia after arterial injury has been reported for heparin and heparinoids (117). Heparin stimulates thrombospondin, a cell attachment protein and arrests smooth muscle at the G_0/G_1 phase of the cell cycle.

In regard to the above mechanisms: Myoinositol-PO_4 governs the endocytosis or synthesis and/or movement of heparin SO_4 into the nucleus. Heparan SO_4 concentration is increased in tissue culture with contact inhibition and can modulate the rate of cell division. Cell proliferation decreases with a decline in nuclear heparan SO_4. The cell matrix proteins that bind heparan SO_4 to the plasma membrane also bind alkaline phosphatase, acetylcholine esterase, myelin basic protein, thy-1 glycoprotein and the variable surface glycoprotein of trypanosomes. Lithium ions along with myoinositol phosphate or inositol deficiency lowers nuclear heparin SO_4 whose surface binding can be lost following insulin activated phospholipase C (118).

TUMOR INHIBITION

The anti-mitotic effect of polyanions and their application to the clinical control of cancer has been the major focus of our research and that of a wide range of clinical investigators (7-10). As mentioned previously, the recent defined clinical action of the ethylene maleic anhydride copolymer, carbetimer has been an outgrowth of this. Unfortunately, polyethylene sulfonate, and pyran copolymer (MVE2) have failed clinically because of toxicity despite observations suggestive of clinical anti-tumor activity (8,9,10). As discussed previously, one reason for failure, as seen in the experience with MVE2 has been the clinical focus on intravenous MVE2 administration to advanced cancer patients while intraperitoneal administration for this polyanion offers better therapeutic alternatives now that IP drug administration is relatively safe (16). Attempts at the development of synthetic polyanions, with the best therapeutic index have been made by us and others (9,119,120).

Relative to native polyanionic growth control, Iozzo's (121) review discusses the role of cell surface heparin proteoglycan on the expression of the neoplastic phenotype where glycosaminoglycan compounds differentially effect cell recognition adhesion, migration, and growth. Most recently Robertson et al. (122) have shown that heparin and heparin sulfate can promote tumor cell invasion of experimental collagen matrices. The effects of glycosaminoglycans are contradictory depending on the system studied: As an example, heparin can enhance capillary endothelial motility, while inhibiting endothelial proliferation and lung metastases (123). The results can be contradictory depending on the model and the character of the heparin studied (8,16).

TUMOR NECROSIS FACTOR

The action of pyran copolymer in enhancing endotoxin lethality (124) must be seen in the light of the action of heparinoids as factors in the isolation of tumor necrosis factor (TNF). In this regard, we have observed that pyran copolymer can reveal the clinical presence of endotoxin based on the in-vitro presence of serum precipitation bands. This is not surprising in view of Thomas et al. (125) observation that synthetic polyanions can induce the generalized Schwartzman reaction (8,9,80). The latter is characterized by two appropriately spaced intravenous injection of gram negative endotoxin to rabbits. This results in fibrinoid deposition within capillaries that can also be precipitated by heparinoids (80). Paradoxically, this phenomenon can be prevented by pretreatment with heparin at the time of the second precipitating injection. The focus on this has related to precipitation of in situ fibrinogen complexes, but there is also a contribution from neutrophils that may play a role in the Schwartzman pathology as leukopenia is protective (125).

CONNECTIVE TISSUE

Proteoglycans related to heparin are produced by mast cells which show heterogeneity related to the sulfated polysaccharides that are produced (126). Inhibition of extracellular matrix degradation by tumor protease extracts can be inhibited by heparin and related glycosaminoglycans as well as protease inhibitors (127). The major degradation of glycosaminoglycans can be lysosomal or surface related and displacement can occur by electrostatic interaction as is seen in displacement by heparin of surface related sulfated molecules in granulosa cells in tissue culture. There are both internal and external degradative pathways that differ depending on the cell line investigated (127,128). Gallagher & Lyon (129) discuss in detail the heparan related surface binding affinities which include neural affinities governing acetyl choline interaction.

Type IV collagen is a basement membrane protein that interacts with both heparin and heparan (130) which as discussed previously is important to tumor metastasis, the adhesion of cells and filtration governing capillary blood supply and interstitial cell maintenance. In cultured fibroblasts, heparin sulfate binds fibronectin and can modulate actin-containing stress fibers which govern cell morphology (131). These observations are important to diabetes where there are abnormalities in the biosynthesis of collagen, proteoglycans, and structural glycoproteins with the loss of anionic sites in blood vessels (132).

Treatment of diabetic KK mice with low molecular weight heparin fragments (CY222) decreases the synthesis of type III collagen and fibronectin which is abnormally produced in the diabetic mice (133). In streptozotocin induced diabetes there is a decrease in glomerular heparan sulfate binding which may result in changes in anionic charge and protein renal loss (134). The changes in intrinsic synthesis of these anionic biopolymers may explain the defects in kidney function and blood vessels seen in diabetes.

BRAIN: AMYLOID

As reviewed by Gallagher & Lyon (129), heparan sulfate proteoglycans are a component of cholinergic synaptic vesicles. In addition, the presence of heparan sulfate and heparin in large concentrations in the brain as a structural component indicate that changes in their character or content could alter brain function. Pertinent to the place of these proteoglycans in CNS pathology, Magnus et al. (135) have found that proteinase treated amyloid fibrils reveal the presence of glycosaminoglycans which include chondroitin sulfate, dermatan and heparan/ heparin sulfate. These workers suggest that the accumulation of glycosamino-glycans can precede the deposition of amyloid fibrils. This is not surprising in view of the modulation of collagen by glycosaminoglycans (136).

Heparan sulfate is an important constituent of the brain. It is found complexed with chondroitin sulfate with a 62/21% ratio of the former to heparan, with the latter having structural affinities to heparin (137). The role of these glycosaminoglycans in brain structural metabolism remains to be elucidated but retinal neuronal cells as well as glial cells have the capacity (138) to make heparitan.

OSTEOPOROSIS: Ca⁺⁺FLUX

The action of heparin and polyanions on bone matrix formation has been reviewed in 1968 (9,10) and has direct relevance to their potential role in calcium metabolism. The demonstrations of this were made by Adams et al. (139) with heparin, dextran and laminarin sulfate induction of osteoporosis. Polyethylene sulfonate, which has similar properties, increases alkaline phosphatase levels (34). In this regard as mentioned previously, it is of interest that the ethylene maleic anhydride copolymer carbetimer, now in clinical anti-tumor study, produces hypercalcemia (35). However, there are no reports of the effects of this polycarboxylate polyanion on bone matrix.

These effects on calcium go back to Victor Heilbrunn (140) in the late 1940's who in his Dynamics of Living Protoplasm discussed the role of heparin and related glycosaminoglycans as growth controlling substances through alteration of ionic calcium balance. In that period, Chaet (141) showed that heparin could produce diastolic arrest in isolated frog hearts, a Ca⁺⁺ mediated event, and this was used by Upjohn in 1956 as a model for heparinoid selection in the search for anti-mitotic heparin related compounds. Independently, there is evidence that polyanions alter osteoblast sensitivity to vitamin C (142) and parahormone, and we have the clinical evidence of heparin or heparinoid induction of osteoporosis and interruption of bone and cartilage growth (7,9,10,139).

More recently, Mellon (143) has shown that anionic dyes, and to a lesser extent heparin, compete for 1,25-dihydroxyvitamin D_3 polynucleotide receptor binding suggesting that polyanions can alter sterol metabolism to produce a rachitic bone picture. This effect of polyanions on vitamin D_3 binding should not be surprising in that these substances can compete for nucleotide binding of proteins such as RNA and DNA polymerases and for the estrogen receptor binding site. The latter may be important to the syndrome of menopausal osteoporosis, and although polyanions probably produce bone alterations through diverse mechanisms, i.e. enhancement of ascorbic acid deficiency and enhancement of parahormone and vitamin A and D effects. These polyanionic actions have never been defined or translated into therapeutic approaches that would have an impact on clinical disease. Pertinent to this, most recently, heparitan sulfate has been shown to precipitate type IV collagen (130), and the role of heparitan sulfate polyanions in endothelial basement membrane formation is becoming relevant. Alternatively, heparin may activate collagenases, and as in the case of the heteropolymolybdates, which are macroanionic, we should look at sulfide oxidase inhibition which is a key factor in heteropolymolybdate skeletal osteoporotic toxicity (144).

CELL MEMBRANE, TRANSDUCTION
(INOSITOL 1,4,5 TRIPHOSPHATE [IP$_3$])

Heparin blocks calcium flux mediated by inositol-1,4,5 triphosphate (IP$_3$). Thus it affects membrane related message transduction utilizing Ca^{2+}. This is seen for a variety of tissue systems including the brain, rat liver microsomes, and the adrenal cortex. Heparin is a potent non-competitive inhibitor of IP$_3$-kinase, IP$_3$-binding, and IP$_3$ induced Ca^{2+} release (145) governing signal transduction.

CHELATION

The binding of calcium to heparin and other cations is reviewed by Niedoszynski (146) but he makes no reference to copper affinity as discussed by Folkman & Ingber (111). Polyanions have chelating action in that they have shown systemic plutonium decorporation (147). In addition, Phillip & Bock (148) showed that sulfated dextran binds iron, and heparin may remove zinc from sperm (149,150). The binding of polyanions to small chelating molecules could extend circulation, half-life, or control localization of chelators in tumors or abscess sites as has been done by Maeda's group for SOD and neocarzinostatin (12-14,18-20). A viable place for polyanions clinically would be to attach them to chelators with binding capacity for iron which would allow them to compete with bacterial growth factor siderophores or to inhibit urease splitting bacteria which could control ammonia formation, the latter would be of value in hepatic insufficiency, renal infection, stone formation and/or enhancement of carcass weight which may depend on inhibition of bacterial urease splitters.

NUCLEUS

Currently, there is once more repeated emphasis on morphologic nuclear changes induced by heparin (151,152) which relate to euchromatization and extrusion of nuclear contents.

Acid mucopolysaccharides have been invoked as playing a role in the transition from heterochromatic to the euchromatic state of the nucleus. Polyanions increase the template availability of nuclear DNA (153,154).

Dependent on molecular weight, the polyanion polyethylene sulfonate inhibits deoxyribonucleotidyl transferase, DNAase and RNAase (155). This area of polyanionic action regarding DNAse inhibition has been reviewed by us in our previous papers, and this activity was not found to correlate with anti-tumor action (9,10).

Heparin has a decondensing action on sperm (152) or isolated nuclei (156,157). This relates to heparin binding sites related to the degree of glycosaminoglycan sulfation (158). Human sperm preincubated with heparin show increased capacity for fertilization which may relate to zinc liberation where zinc is thought to be a nuclear chromatin stabilizer (149,150). Similar effects of cell nuclei are shown by synthetic polyanions as well as native heparin. (7,54,153,154,159).

INHIBITION OF TRANSCRIPTION BY HEPARIN

Wheat germ polymerase binds both to heparin and DNA in affinity columns where heparin competes for DNA-binding sites and blocks RNA transcription with inhibition of the polymerase (160).

There is a specificity for polyanions which activate the template properties of nuclei and alter chromatin configuration. This is a function of the molecular weight and charge distribution, and is seen for heparin, polyaspartic acid and dextran SO_4. Other polyanions such as chondroitin SO_4, casein, polygalacturonic acid and hyaluronic acid are without effect (157). The balanced release of template restriction/vs. DNA competition with polyamines may be the focus of this action. In this regard, we have shown polyanions to derepress heterochromatic condensed chromosomes (153,154).

Protein kinases (casein protein kinases) that govern the phosphorylation of proteins derived from the liver are distinct in their affinities for heparin. Heparin blocks nuclear protein phosphorylation in isolated liver tumor nuclei which is associated with a decline in GTP with a resultant block to casein kinase (161). These enzymes are key factors in cell division (162) and post translational events governing protein synthesis (163). In addition, heparin can inhibit specific initiation by RNA polymerases I and II in isolated nuclei (164).

The nucleotide analog dichlororibofuranosil benzimidazole blocks in-vitro transcription by RNA polymerase II (DRB) which inhibits protein kinase which is inhibited by heparin (165). In viral models, anionic polyelectrolytes like pyran copolymer inhibit deoxyribonucleic acid polymerases from human cells and from simian sarcoma virus. These effects depend on the cationic environment of the template primer and are thought to relate to the capacity of polyanions to complex bivalent cations (166).

Papas et al. (56) showed that pyran copolymer was a potent inhibitor of DNA polymerase that was isolated from the avian myeloblastosis virus. Inhibition by pyran was shown for all oncorna virus polymerases and for mammalian polymerases. A 92% inhibition was found for the RNA dependent DNA polymerase suggesting once more that pyran may have value in the treatment of HIV infection.

The action of polyanions on inhibiting DNAse I and II is reported in our earlier reviews (9,10).

LIPID

Heparin, in intestinal brush border epithelium, binds pancreatic cholesterol esterase, and triglyceride lipase and plays a role in the binding and uptake of neutral lipids (167). This observation may explain the role of membrane bound heparin in neutral fat absorption and may conceivably relate to intestinal pathology if heparin bound activated lipases produce free fatty acids in an environment that permits free radical formation via prostaglandin, leukotriene interaction.

The lipolytic action of heparin and its interaction with cofactors is reviewed in detail by Olivecrona & Bengtsson (168) but they do not discuss this in the light of past clinical experience of heparin use in atherosclerosis.

INFLAMMATION

A key enzyme involved in inflammation is phospholipase A_2 which is released by phagocytic cells or tissue injury. Certain pathogenic organisms such as the protozoan Naeglaria or staphylococci release this enzyme as part of their pathogenic action. Heparin effectively inhibits this enzyme derived from neutrophils depending on the pH of the environment, with heparin most effective at pH 5.5. This is thought to be a charge mediated effect, but heparin fractions may differ as to pH optima and phospholipase A_2 inhibiting activity (169). These observations are important to inflammation and vascular injury as they may explain the anti-inflammatory activity of heparin.

ATHEROSCLEROSIS

The pathophysiology of atherosclerosis and the protective effect of heparin has been reviewed extensively by Engelberg in several recent papers (3,4).

As vascular injury and platelet aggregation are key mediating factors in atherosclerosis, there is a renewed interest in heparin as a therapeutic factor. Heparin has been used with varied clinical enthusiasm as an agent to control atherosclerosis and coronary disease (99-102) going back to Engelberg (99) and confirmed enthusiastically in 1987 where its prophylactic subcutaneous use reduced reinfraction by 63% and general mortality by 48% in a selected group of post myocardial infarct patients, as compared to controls.

Its early place in this area was governed by its lipemic or lipolytic action. Given subcutaneously, as described above, it can significantly protect post myocardial infarct patients with coronary disease but this area of its chronic prophylactic application has lost adherents in recent years although it still has a primary place in the treatment of myocardial infarction (101).

In addition to initial injury with plaquing and thrombosis, arteriosclerosis is primarily based on smooth muscle proliferation where arterial smooth muscle cells secrete proteoglycans which can interact with lipoproteins (170-171). These proteoglycans contain chondroitin sulfate, dermatan sulfate, and heparin sulfate, all polyanionic glycosaminoglycans of negative charge. These compounds may create a vascular visceoelastic gel that regulate plasma protein constituent entry in the vascular wall (172,173) and heparan sulfate or heparin derived from

vascular smooth muscles can control arterial vessel smooth muscle hyperplasia (108,109,173).
The concentration of dermatin and chondroitin sulfate with cholesterol increases in arteriosclerotic plaques with age while heparin sulfate concentration decreases (173,174). This is pertinent to the work of Nakajima et al. (42), Barner et al. (43) and Irimura et al (44) who have shown heparin or its fragments to inhibit heparanase activity in tumors. They reported that tumor containing endo-β glucouronidases destroy blood vessel walls. Similar activity may occur with age as a function of arterial plaque formation. Heparin potentiates the production of endothelial cell growth factor and plasminogen activators responsible for clot lysis (107) which may be factors in vascular occlusion. These observations along with renewed interest in lipoprotein lipase will move heparin related compounds back into the main stream of therapy for atherosclerosis.

STEROID METABOLISM, ANGIOTENSIN

The antihormone action of heparin has been observed as far back as 1957, where it has been shown to block aldosterone production and possess natriuretic action (9,10) and can block the lymphopenic action of glucocorticoids. As mentioned earlier in conjunction with steroids, heparin blocks angiogenesis (111). However, the action of polyanions can be mediated through their inhibition of adrenal steroidogenic enzyme action. This is seen for suramin, heparin, and heparinoids. In the case of suramin fully, 25% of patients develop adrenal insufficiency related to significant inhibition of adrenocortical enzymes (175). These effects may be mediated by suramin's inhibition of P-450 hydrolytic enzymes, and probably explain why these compounds sensitize to bacterial endotoxin lethality (124).
Heparin treated rats show a decrease in angiotensin II receptors in adrenal glomerulosa cells with a resultant decrease in aldosterone production (176-178). Heparin treatment has shown a clinical atrophy in the adrenal glomerulosa (178,179) and can block ACTh adrenocortical hyperplasia. In addition, heparin has been shown to have a direct effect on steroid synthesis and can inhibit glycolipid synthesis by T lymphocytes. This includes asialo GM1, GM2, and a neutral lactosylceramide. In addition ganglioside synthesis can be inhibited.
In another area, Lerea et al. (180), have characterized a uterine cytosol inhibitor of estrogen receptor binding to nuclei as having properties similar to heparin. This is not surprising in view of estrogen receptor binding to a variety of polyanions (181). Similar binding of glucocorticoid receptors to polyanions have been reported (182,183) with isolation of estrogen, progesterone and androgen receptors by immobilized heparin as well as glucocorticoid receptors.

GASTRIC AND INTESTINAL MUCOSA

While heparinoids inhibit peptic activity (7,184). This can be concentration related to charge effects causing enhancement of protease effects (84). Heparin is responsible, regardless of its route of administration for the release of diamine oxidase from the tip of intestinal villi. Diamine oxidase serum levels reflect on the degree of intestinal mucosal injury, and the recovery of enzyme levels post heparin administration reflect on renewal of intestinal lining cells (185). This is of interest because of the diarrheal toxic action of heparinoids (34).

RESPIRATION

Since our reviews of 1968 (9,10) Konig et al. (186) have shown moderate polyanionic inhibition of mitochondrial respiration between cytochrome b and c, with strong inhibition of oxidative phospholyation and ATPase.

HEPARIN AND POLYANION STRUCTURE

The anticoagulant, anti-thrombotic action has been reviewed extensively, most recently by Lane & Lindahl's new text on heparin (5). Braud & Vert (187) discuss this in relation to sulfation and anticoagulant activity. New low molecular weight heparins are available which apparently demonstrate better systemic absorption on subcutaneous administration for anti-thrombotic action. These new heparins, like enoxaparine, have less in-vivo hemorrhagic action and apparently have minor effects on platelets and lipolysis (189). In this regard, there are distinct differences between the anticoagulant effects of heparin and synthetic polyanions like pyran copolymer. Pyran is superior to clinical heparin as an inhibitor of the thrombin-fibrinogen clotting reaction (189,190).

The anti-thrombotic and bleeding effects of glycosaminoglycans is governed by the degree of sulfation. Higher levels of sulfation enhance inhibition of platelet aggregation which increases blood loss (191). Petitou et al. (192) have shown structure activity correlates for anti-thrombin III action of heparin pentasaccharide sequences which was dependent on O-sulfate residues. These relationships have also been examined in structural detail by Oscarsson et al. (193), and Linhardt et al. (194).

Efforts are still underway to make better synthetic anti-tumor polyanions (195-197). The current and perhaps more productive application of these agents is to test them as reverse transcriptase inhibitors for anti-viral, anti-HIV (AIDS) application.

ORAL ABSORPTION AND ENDOTHELIAL ACTION

Most recently, Jaques group (198) has found that dextran sulfate and heparin are orally absorbed, as determined by plasma and endothelial localization. This is not surprising as A. H. Robins attempted to develop an oral dextran sulfate anti-ulcer peptic inhibitor which was never marketed because of the induction of mesenteric lymphadenopathy. What makes the Jaques observation of major clinical relevance is the endothelial localizing action of these orally absorbed polyanions suggesting they will have major clinical value in the treatment of atherosclerosis or vascular thrombosis (199). It also suggests that we reevaluate the oral use of dextran sulfate, or related polyanions as anti-virals in the treatment of AIDS.

SUMMARY

The reader is referred to other reviews of biologic action of synthetic polyanions (1-6) with major emphasis on anti-coagulant, anti-atherosclerosis and anti-viral action. Anti-viral, anti-tumor, and immunologic effects have been summarized by us in 1968, 1979 and 1985 (7-10). These also include metabolic polyanion effects and effects on cation and nuclear relationships.

Our review of pyran's (Diveema, MVE2) action (8) describes in detail the clinical and immunologic action of these polyanionic compounds which have not received a fair clinical anti-tumor trial as their best place is intraperitoneal when used as adjuvants when tumor burden is reduced (16). However, a new focus on their effects on cytokine action presents a new opportunity for their growth modulator action (66,107-111). The effects of polyanions on enzymes, nuclear function, and as antivirals and vaccine adjuvants is updated in this paper, where the most cogent place for their current clinical application is in the treatment of HIV infection (AIDS) where interferon induction and reverse transcriptase inhibition adds to validity. Their place as vaccine adjuvants deserves major consideration (8,45).

We also predict that heparinoids will have a rebirth both in their treatment of atherosclerosis where clinical data reconfirms earlier benefit (3,4). Heparinoids will also have renewed interest in acute and chronic inflammation (7,9,169). The inhibition of vascular smooth muscle proliferation, (66,102-106) the activation of lipoprotein lipases (167,168) and the inhibition of phospholipase A_2, (169) as well as anticoagulant action provide for a clinical return in these areas, particularly in view of their oral absorption and endothelial localization (198,199).

Finally, the exo or endotoxin gastrointestinal adsorbing action of polyanions as anti-diarrheals (58,59) should have application for adsorbent topical use such as a regional vaginal use in the prevention of the toxic shock syndrome.

We have touched on chelation and other actions, but most importantly, we feel that the greatest immediate clinical impact will be in the role of polyanions as drug or enzyme carriers which has opened up to us, thanks to the excellent work of Maeda's group (12-15,18-22). Whatever natural polyanions (heparin) can do, synthetics or natural derivatives (heparinoids) can accomplish. What is needed is an understanding of related physiologic and toxicologic effects for broad application of polyanionic heparinoids to bio-medicine.

Literature Cited

1. Jaques, L. B. Prog. Med. Chem. 1967, 5, 139-181.
2. Jaques, L. B. Pharmacol. Rev. 1980, 31, 97-166.
3. Engelberg, H. Pharm. Rev. 1984, 36, 91-110.
4. Engelberg, H. Sem. Thromb. Hemostasis. 1985, 11, 48-55.
5. Lane, D. A. In Heparin: Clinical and Biological Properties. Clinical Applications; Lindahl, L. I., Ed; CRC Press, Boca Raton, 1989.
6. Fransson, L-A. In Heparin; Lane, D. A.; Lindhal, U., Eds; CRC Press: Boca Raton, 1989. p 115.

7.   Regelson, W. In Microsymposium Volume of J. Polymer Sci. Symposium; Sedlacek, B., Ed; 1979, Vol. 66, p. 483.
8.   Regelson, W. In International Encyclopedia of Pharmacology and Therapeutics; Mitchell, M. Ed; Pergamon Press: New York, 1985, Vol. 115, p. 1-44.
9.   Regelson, W. Adv. In Cancer Res. Academic Press: New York 11, 1967; 223-303.
10. Regelson, W. In Chemotherapy; 1968, Vol.3, p 303-357.
11. Regelson, W. J. Med. 1974, 5, 50-68.
12. Maeda, H.; Oda, T.; Matsumura, Y.; Kimura, M. J. Bio. Comp. Polymers 1988, 3, 27-43.
13. Maeda, H.; Matsumura, Y. In CRC Reviews in Therapeutic Drug Carrier System; CRC Press: Boca Raton, In Press.
14. Oda, T.; Akaike, T.; Hamamoto, T.; Suzuki, F.; Hirano, T.; Maeda, H. Science 1989 244, 974-976.
15. Akaike, T.; Ando, M.; Oda, T.; et al. J.C.I., 1989 In press.
16. Regelson, W. J. Bioactive and Compatible Polymers 1986, Vol. 1, 84-107.
17. Karlsson, K.; Marklund, S. L. J. Clin. Invest. 1988, Sep; 82(3), 762-6.
18. Maeda, H.; Matsumoto, T.; Konno, T.; Iwai, K.; Lieda, M. J. Protein Chem. 1984, 3, 181-183.
19. Maeda, H.; Ueda, M.; Morinaga, T.; Matsumoto, T. J. Med. Chem. 1985, 28, 455-461.
20. Konno, T.; Maeda, H.; Iwai, K.; Tashiro, S.; et al. Eur. J. Cancer Res. 1983, 19: 1055-1065.
21. Suzuki, F.; Munakata, T.; Maeda, H. Anticancer Res. 1988, 8: 97-104.
22. Kobayashi, A.; Oda, T.; Maeda, H. J. Bio. Comp. Polymers 1988, 3, 319-322.
23. Przybylaski, M.; Zaharko, D. S.; Chirigos, M. et al. Cancer Treat. Rep. 1978, 62, 1837-1843.
24. Dumitriu, S.; Popa, M.; Dumitriu, M. J. Bio. Comp. Polymers 1988, 3, 243-312.
25. Kaetsu, I. J. Bio. Comp. Polymers 1988, 3, 164-183.
26. Higginbotham, R. D. Tex. Rpts. Biol. Med. 1960, 18, 408-417.
27. Higginbotham, R. D. Int. Arch. Allergy 1956, 15, 195-198.
28. Campbell, R. A. In Polyamines In Health and Disease; Hunt-Retzlaff, Z., Russi, J. B., Eds.; CRC Press (place?), 1989; Vol. 2, p 164-176.
29. Levy, H. B. J. Bio. Comp. Polymers 1986 348-385.
30. Petcu, D. J.; Aldrich, C. E.; Cortes, C. et al. Virol. 1988, 167, 385-392.
31. Stein, C. A.; LaRocca, R. U.; Thomas, R.; et al. J. Clin. Oncol. 1989, 7, 499-508.
32. Zhang, H-X; Sozzani, S.; D'Alessandao, I; et al. Int. J. Immunopharmac. 1988, 10, 695-707.
33. Ray. F. E.; Argrawal, K. C. Cancer Res. 1966, 26, PTI: 1740-1744.
34. Regelson, W.; Holland, J. R. Clin. Pharm. Therap. 1962, 3, 730-749.
35. Dodion, P.; DeValeriola, D.; Body, J.J.; et al. Eur. J. Cancer Clin.Oncol. 1989, 25, 279-286.
36. Brodsky, I.; Strayer, D. R.; Krueger, L. J.; Carter, W. A. J. Biol. Response Mod. 1985, 4(6), 669-75.

37. Pinto, A. J.; Morahan, P. S.; Brinton, M. A. Int. J. Immunopharmacol 1988, 10(3), 197-209.
38. Carter, W. A.; Strayer, D. R.; Brodsky, I; et al. Lancet 1987, 1(8545), 1286-92.
39. Carrano, R. A.; Iullucci, J. D.; Luce, J. K.; et al. Marcel Dekker: New York, 1984; pp. 243-260.
40. Dean, J. H.; Padarathsingh, B. L.; Keys, L. Cancer Treat Res. 1978, 62, 1807-1816.
41. Ardalan, B.; Paget, G. E. Cancer Res. 1986, 46, 5473-7463.
42. Nakajima, M.; Irimura, T.; DiFerrante, N.; et al. J. Biol. Chem. 1984, 259, 2283-2290.
43. Bar-Ner, M.; Kramer, M. D.; Schirrmacher, V.; et al. Int. J. Cancer 1985, 35, 483-91.
44. Irimura T.; Kakajima, M.; Nicolson, G. L. Biochem. 1986, 25, 5322-5328.
45. Campbell, C. H.; Richmond, J. W. Infection and immunity 1973, 7, 199-204.
46. Morahan, P. S.; Cline, P. F.; Breinig, M. C.; et al. Antimicrob. Agents and Chemoth. 1979, 15, 547-553.
47. Schuller, G. B.; Morhahan, P. S.; Snodgrass, M. Cancer Res. 1975, 35, 1915-1920.
48. Nahamias, A. J.; Kibrick, S.; Bernfeld, P. Proc. Soc. Exp. Biol. Med. 1964, 115, 993-996.
49. Duran-Reynals, F. Bact. Rev. 1942, 197-252.
50. Cohen, S. S. The isolation and caystallization of plant viruses and other protein macro molecules by means of hydrophilic colloids. 353-362, 1942.
51. Vaheri, A. Acta Pathol. Microbiol. Scand. Suppl. 1964, 32-51.
52. Baba, M.; Schols, D.; Pruwels, R.; Balzarini, J.; deClercq, E. Biochem. & Biophys. Res. Commun. 1988, 155, 1404-11.
53. Nakane, H.; Balzarini, J.; DeClercq, E.; Ono, K. Eur. J. Biochim. 1988, 177, 91-96.
54. Tunis, M.; Regelson, W. Arch. Bioche. Biophys. 1963, 101, 448-455.
55. Jeffries, D. J. J. Infect. 1989, 18: Suppl.I, 5-13.
56. Papas, S.; Pry, T.W.; Chirigos, M.A. Prot. Nat. Acad. Sci. USA. 1974, 71, 367-70.
57. Blundell, T.; Pearl, L. Nature 1989, 337, 596-7.
58. Nash, J. F.; Lin, T. M. U.S. Patent 3 224 941, 1965.
59. Tust, R. H.; Lin, T. M. Proc. Soc. Exp. Biol. & Med. 1970, 135, 72-6.
60. Grinstein, S. J.; Melnick, L.; Wallis, C. Bull. Who. 1970, 42, 291.
61. Wallis, C.; Homa, A.; Melnick, J. L. Appl. Microbiol. 1972, 23, 740-4.
62. Liebhaber, H.; Takemoto, K. K. Virology 1961, 14, 502-5.
63. Postlewaite, A. E.; Smith, G. N., Jr.; Lachman, L. B. et al. J. Clin. Invest. 1989, 83, 629-36.
64. Ormai, S.; Hagenbeek, A.; Palkovits, M.; Van Bekkum, D. W. Cell Tissue Kinet. 1973, 6, 407-23.
65. Mihich, E.; Simpson, C. L.; Regelson, W. et al. Fed. Proc. 1960, 19, 142.
66. Adler, S. A.M.J. Physiol. Renal. Fluid Electrolyte Physiol. 1988, 24, F781-F786.

67. Lurie, M. B. <u>Resistance to tuberculosis in experimental studies in active and acquired defensive mechanisms</u>; Harvard Univ. Press: Cambridge, MA, 1964.

68. Kaplan, A. M. In <u>Anionic Polymeric Drugs</u>. Donnaruma, L. G.; Ottenbrite, R.M.; Vogl., O., Eds.; John Wiley & Sons: New York, 1980, pp. 227-53.

69. Baird, L. G. Ph.D. Medical College of Virginia, Richmond, Virginia, 1976.

70. Kaplan, A. M.; Walker, P. L.; Morahan, P. S. <u>Fogarty Int. Center. Proc.</u> 28: 277-86. U.S. Gov't. Print Off, Wash., D.C., 1977b.

71. Kaplan, A. M.; Collins, J. M.; Morahan, P. S.; Snodgrass, M.J. <u>Cancer Treat. Rep.</u> 1978, <u>62</u>, 1823-9.

72. Kaplan, A. M.; Connoly, K.; Regelson, W. In <u>Proc. on the host invader interplay</u>. Van Den Bossche, H., Ed.; Elsevier: North Holland Netherlands, 1980, p. 429.

73. Akashi, M.; Iwasaki, H.; MiYauchi, N. <u>J. Bio. Comp. Polymers</u> 1989, <u>4</u>, 124-36.

74. Ottenbrite, R. M.; Sunamoto, J.; Sato, T. et al. <u>J. Biocomp. Polymers</u> 1988, <u>3</u>, 184-90.

75. Lane, D.A. In <u>Heparin</u>; Lane, D.A.; Lindahl, U., Eds.; CRC Press: Boca Raton, 1989, p. 363.

76. Maillet, F.; Petitou, M.; Choay, J.; Kazatchkine, M.D. <u>Molec. Immunol.</u> 1988, <u>25</u>, 917-23.

77. Mauzac, M.; Maillet, F.; Jozefonvicz, J.; Kazatchkine, M.D. <u>Biomaterials</u> 1985, <u>6</u>, 61-4.

78. Schwarting, G. A.; Gajewski, A. <u>Bioscience Reports</u> 1988, <u>8</u> 389-98.

79. Dziarski, R. <u>J. Immunol.</u> 1985, <u>143</u>, 356-365.

80. Regelson, W. <u>Hemat. Rev.</u> 1968, <u>1</u>, 193-227.

81. Godal, H.C. In <u>Heparin</u>; Lane, D.A.; Lindahl, U., Eds.; CRC Press: Boca Raton, 1989; p. 53.

82. Horne, M.K.,III <u>Thromb. Res.</u> 1988, <u>51</u>, 135-144.

83. Ungar, G.; Mist, S.H. <u>J. Exp. Med.</u> 1949, <u>90</u>, 39-51.

84. Anderson, N. <u>J. Pharm. Pharmacol.</u> 1961, <u>13</u>: 122T-123T.

85. Fraber, P.; Capel, P. J. A.; Rijke, G. P. M.; Viewinden, G. et al. <u>Clin. Exp. Immunol.</u> 1984, <u>55</u>, 502-07.

86. Fraber, J. H. M.; Rijke, T. P. M.; Van dePutte, L. B. A.; Capel, P. J. A.; Berden, J. H. M. <u>J. Clin. Invest.</u> 1986, <u>77</u>, 182.

87. Aotsuka, S.; Okawa-Takatsuji, M.; Kinoshita, M.; Yokohari, R. <u>Clin. Exp. Immunol.</u> 1988, <u>73</u>, 436-442.

88. Spensley, P. C.; Rogers, H. J. <u>Nature</u> 1954, <u>173</u>, 1190-2.

89. Delicort, E. E.; Stahmann, M. A. <u>Nature</u> 1955, <u>176</u>, 1028-9.

90. Bjork, I.; Olson, S. T.; Shore, J. D. In <u>Heparin</u>; Lane, D.A.; Lindahl, U., Eds.; CRC Press, Boca Raton, 1989; p.229.

91. Carrell, R. W., Boswell, D. R. In <u>Proteinase Inhibitors</u>; Barret; Salvesson, Eds.; Elselvica Sci. Pub, BV; 1986; p. 403.

92. Tollefson, D. M. In <u>Heparin</u>; Lane, D.A.; Lindahl, U., Eds.; CRC Press: Boca Raton, 198p; p.257.

93. Pizzo, D. V.; Mast, A. E.; Feldman, S. R.; Salvesen, G. <u>Biochim Biophis. Acta.</u> 1988, <u>967</u>, 158-62.

94. Roberts, P. S.; Regelson, W. In Control of Neoplasia by Modulation of the Immune Response; Chirigos, M.A., Ed.; Raven Press: New York, 1977; p. 549.
95. Tang, J.; Wong, N.S. J. Cell. Biochem. 1987,33, 53-63.
96. Redini, F.; Lafuma, C.; Hornbeck, W.; Chory, J.; Robert, L. Biochem. Pharm. 1988, 37, 4257-61.
97. Peterson, M.W. J. Lab. Clin. Med. 1989, 113, 297-308.
98. Ohta, G.; Sasaki, H.; Matsubara, F.; et al. Proc. Soc. Exp. Biol. Med. 1962, 109, 298-300.
99. Engelberg, H.; Kuhn, R.; Steinman, M. Circulation 1956, 13, 489-98.
100. Schmidt, A.; Bunte, A.; Buddecke, E. Biol. Chem. Hoppe-Selyer 1987, 368, 277-284.
101. Serneri, G. G.; Rovelli, F.; Gensini, G. F.; et al. Lancet 1987, I, 937-42.
102. Wright, T. C., Jr.; Castellot, J. J.; Diamond, J. R.; et al. In Heparin; Lane, D. A.; Lindahl, U., Eds.; CRC Press: Boca Raton, 1989; pp. 295-316.
103. Karnovsky, M. J.; Hoover, R. L.; Castellot, J. J. In Prespectives in Inflammation, Neoplasia and Vascular. Cell Biology. Edgington, T. S.; Ross, R.; Silverstein, S. C., Eds.; Alan R. Liss: New York, 1987; pp.175-194.
104. Olson, J. L. Kidney Int. 1984, 25, 376-382.
105. Diamond, J. R.; Karnovsky, M. J. Renal Physiol. Basel. 1986, 9, 366-374.
106. Gorter, F.; Nannignga, L. Discussions of the Faraday Soc. 1953, 13, 205-217.
107. Rappaport, R. S.; Ronchetti-Blume, M.; Vogel, R. L.; Hung, P. P. Thromb. Haemostasis 1988, 59, 514-522.
108. Castellot, J. J.; Frureau, L. V.; Karnovsky, M. J.; Rosenberg, R. J. Biol. Chem. 1982, 257, 11256-11260.
109. Castellot, J. J.; Cochran, D. L.; Karnovsky, M. J. J. Cell. Physiol. 1985, 124, 21-28.
110. Crum, R.; Szabo, S.; Folkman, J. Science 1965, 230, 1375-1377.
111. Folkman, J.; Ingber, D. E. In Heparin; Lane, D. A.; Lindahl, U., Eds.; CRC Press: Boca Raton, Florida, 1986, pp. 317-333.
112. Bikfalvi, A.; Dupuy, E.; Ruan, C.; et al. Cell Biol. Int. Rpts. 1988, 12, 931-942.
113. Goldsmith, H. S.; Griffith, A. L.; Kupferman, A.; et al. JAMA 1984, 252, 2034-36.
114. Neufeld, G.; Gospodarowicz, D. J. Cell Physiol. 1988, 136, 537-552.
115. Sovova, V.; Vydra, J.; Cerna, H.; et al. Folia Biol. (Papha). 1988, 34, 233-9.
116. Schreiber, A. B.; Kenney, J.; Kowalski, W. J.; Friesel, R.; Mehlman, T.; Macing, T. Proc. Natl. Acad. Sci. USA. 1985, 82, 6138-42.
117. Dryjski, M.; Mikat, E.; Bjornsson, T.D. J. Vasc. Surg. 1988, 8, 623-33.
118. Ishihara, M.; FeDarko, N.; Conrad, E. H. J. Biol. Chem. 1977, 262, 4708-16.
119. Hodnett, E. M.; Amirmozzami, J.; Tien Hai Tai, J. J. Med. Chem. 1978, 21, 652-7.
120. Ottenbrite, R. M.; Regelson, W.; Kaplan, A.; et al. In Polymer Drugs; Gonaruma, G., Ed.; John Wiley & Sons: New York, 1978, pp. 263-304.
121. Iozzo, R.V. J. Cell Biochem. 1988, 37, 61-78.

122. Robertson, N. P.; Starkey, J.R.; Hamner, S.; Meadows, G.C. Cancer Res. 1989, 49, 1816-23.
123. Coombe, D. R.; Parish, C. R.; Ramshaw, I. A.; Snowden, J. M. Int. J. Cancer 1987, 39, 82-8.
124. Munson, A. E.; Regelson, W. Proc. Soc. Exp. Biol. Med. 1971, 137, 553-7.
125. Thomas, L.; Brunson, J.; Smith, R. T. J. Exp. Med. 1955, 99, 249-261.
126. Forsberg, L. S.; Lazarus, S. C.; Seno, N.; et al. Biochim. Biophys. ACTA 1988, 967, 416-428.
127. Keren, Z.; Leland, F.; Nakajima, M.; LeGrue, S. J. Cancer Res. 1989, 49, 295-300.
128. Yanagishita, T. A.; Hascall, V. C. J. Biol. Chem. 1988, 259, 10270-83.
129. Gallagher, J. T.; Lyon, M. In Heparin; Lane, D. A.; Lindahl, U., Eds.; CRC Press: Boca Raton, Florida, 1989, pp. 135-158.
130. Koliakos, G. G.; Gouzk-Koliakos, K.; Burcht, L. T.; et al. J. Biol. Chem. 1989, 264, 2313-23.
131. Woods, A.; Hook, M.; Kjellen, L.; et al. J. Cell Biol. 1984, 99, 1743-53.
132. Rohrbach, R. Virchows Arch. 1986, 51, 127-135.
133. Asselot, C.; Labet-Robert, J.; Kern, P. Biochem. Pharm. 1989, 38, 895-9.
134. Klein, D. J.; Brown, D. M.; Oegema, T. R. Diabetes 1989, 35, 1130-42.
135. Magnus, J. H.; Husby, G.; Kolset, S. O. Ann. Rheum. Dis. 1989, 48, 215-9.
136. Vogel, K. G.; Trotter, J. A. Coll. Relat. Res. 1987, 7, 105-14.
137. Ripellino, J. A.; Margolis, R. U. J. Neurochem. 1989, 52, 807-12.
138. Threkeld, A.; Adler, R.; Hewitt, A. T. Dev. Biol. 1989, 132, 559-68.
139. Adams, S. S.; Thorpe, H. M.; Glynn, L. E. Lancet 1958, II, 618-20.
140. Heilbrunn, L. V. Dynamics of Living Protoplasm: Academic Press, 1956, New York.
141. Chaet, A. B. Physiol. Zool. 1955, 28, 315-21.
142. Ohlwiler, D. A.; Jurkiewicz, M. J.; Butcher, H. R.; et al. Surg. Forum 1959, 10, 301-3.
143. Mellon, W. S. Molec. Pharm. 1984, 25, 86-91.
144. Carruthers, C.; Regelson, W. Oncologia 1963, 16, 101-8.
145. Guillemette, G.; Lamontagre, S.; Boulay, G.; Mouillac, B. Mol. Pharm. 1989, 35, 349-344.
146. Niedoszynski, I. In Heparin; Lane, D. A.; Lindahl, U., Eds.; CRC Press: Boca Raton, Florida, 1989, pp. 51-63.
147. Lindenbaum, A.; Rosenthal, M. W.; Guilmette, R. A.: Proc. I.A.E.A. Syoposium on diagnosis and treatment of internally deposited radionucleotides: Vienna, Austria,IAEA SE 6/22, 1975.
148. Phillip, B.; Bock, W. J. Polym. Sci. Polym. Symp. 1979, 66, 83-94.
149. Reyes, R.; Magdaleno, M.; Hernandez, O.; et al. Arch. Androl. 1983, 10, 155-60.
150. Valencia, A.; Wens, M. A.; Merchant, H.; et al. Arch. Androl. 1984, 12, 109-113.
151. Zirne, R. A.; Erenpreiss, J. G. Neoplasma 1986, 33, 39-48.
152. Delgado, N. M.; Huacuja, L.; Merchant, H.; et al. Arch. Androl. 1980, 4, 305-13.

154. Miller, G.; Berlowitz, L.; Regelson, W. Chromasoma 1971, 32, 251-61.
155. Bach, M.K. Biochim/ Biophys ACTA 1964, 91, 619-26.
156. Bornens, M. Nature 1973, 244, 28-39.
157. Kraemer, G. J.; Coffey, D. Biochim. Biophys. Acta. 1970, 224, 553-67.
158. Delgado, N. M.; Reyes, R.; Huacuja, L.; et al. Arch. Androl. 1987, 8, 87-95.
159. Capatini, S.; Cocco, L.; Matteucci, A.; et al. Cell Biol. Int. Rpts. 1984, 8, 289-96.
160. Dynan, W. S.; Burgess, R. R. Biochemistry 1979, 18, 4581-88.
161. Friedman, D. L.; Kleiman, N. J.; Campbell, F. E.; Jr. Biochim. Biophys. Acta. 1985, 847, 165-76.
162. Baydoun, H.; Feth, F.; Hoppe, J.; et al. Arch. Biochem. Biophys. 1986, 245, 504-11.
163. Renart, M. F.; Sastre, L.; Sebastian, J. Eur. J. Biochem. 1984, 140, 47-54.
164. Zhang-Keck, Z-Y; Stallcup, M. R. J. Biol. Chem. 1988, 263, 3513-20.
165. Zandomeni, R.; Weinmann, R. J. Biol. Chem. 1984, 259, 14804-11.
166. Dicioccio, R.; Srivastaud, B. I. Biochem. J. 1978, 175, 519-24.
167. Bosner, M. S.; Gulick, T.; Riley, D. J. S.; Spilburg, C. A.; Lange III, L. G. Proc. Natl. Acad. Sci. 1988, 85, 7438-42.
168. Olivecrona, T.; Bengtsson-Olivecrona, G. In Heparin; Lane, D. A.; Lindahl, U., Eds.; CRC Press: Boca Raton, Florida, 1989, pp. 335-361.
169. Franson, R.; Patriarca, P.; Elsbach, P. J. Lipid Res. 1974, 15, 380-8.
170. Srinivasan, S. R.; Dolan, P.; Radhakrishnamurthy, B.; Berenson, G. S. Biochim. Biophys. Acta. 1975, 388, 58-70.
171. Srinivasan, S. R.; Radhakrishnamurthy, B.; Dalferes, E. R., Jr.; Berenson, G. S. Atherosclerosis 1979, 34, 105-18.
172. Salisbury, B. G. J.; Wagner, W.D. J. Biol. Chem. 1981, 256, 8050-7.
173. Wright, T. N. In Progress in Hemostasis and Thrombosis; Spaet, T. H., Ed.; Grune and Stration: New York, 1980; Vol. 5, p. 1-39.
174. Hollman, J.; Schmidt, A.; VonBassewitz, D. G.; Buddecke, E. Arteriosclerosis 1989, 9, 154-8.
175. Ashby, H.; DiMattina, M.; Linehan, W. M.; Robertson, C. N.; Queenan, J. T.; Albertson, B. D. J. Clin. Endocrinol. Metabl. 1989, 68, 505-8.
176. Azukizawa, S.; Iwasaki, I.; Kigoshi, T.; Uchida, K.; Morimoto, S. ACTA Endocrinol (Copenh.) 1988, 119, 367-72.
177. Levine, O. H.; Laidlaw, J. C.; Ruse, J. L. Can. J. Physiol. Pharmacol. 1972, 50, 270-5.
178. Abbott, E. C.; Gornall, A. G.; Sutherland, D. J. A.; et al. Can. Med. Assoc. J. 1966, 94, 1155-64.
179. Wilson, I. D.; Goetz, F. C. Am. J. Med. 1964, 36, 635-9.
180. Lerea, C. C.; Klinge, E. M.; Bambara, R. A.; et al. Endocrinol. 1987, 121, 1140-54.
181. Yamamoto, K. R.; Alberts, B. M. Annu. Rev. Biochem. 1976, 95, 751-5.
182. Dahmer, M. K.; Gienrungroj, W.; Pratt, W. B. J. Biol. Chem. 1985, 260, 7705-09.
183. Blanchardie, P.; Lustenberger, P.; Orsonneau, J. L.; Bernard, S. Biochemie. 1984, 66, 505-11.
184. Bianchi, R. G.; Cook, D. L. Gastroenterology 1964, 47, 409-12.

185. D'Agostino, L.; Daniele, B.; Pignata, S.; et al. Biochem. Pharm. 1989, 38, 47-9.
186. König, T.; Kocsis, B.; Meszaros, L.; et al. Biochim. Biophys. Acta 1977, 462, 380-9.
187. Braud, C.; Vert, M. J. Bioactive and Comp. Polymers 1989, 4, 269-80.
188. Frydman, A. M.; Bara, L.; LeRoux, Y.; Woler, M.; Chaliac, F.; Samama, M. M. J. Clin. Pharmacol. 1988, 28, 609-18.
189. Shamash, Y.; Alexander, B. Biochim. Biophis. Acta 1969, 194, 449-61.
190. Roberts, P. S.; Regelson, W.; Kingsbury, B. J. Lab Clin. Med. 1973, 2, 822-8.
191. Van Ryn-McKenna, J.; Ofosu, F. A.; Hirsh, J.; Buchanan, M. R. Br. J. Haematol. 1988, 71, 265-79.
192. Petitou M.; Lormeau, J. C.; Chory, J. Eur. J. Biochem. 1988, 176, 637-40.
193. Oscarsson, L-G.; Pejler, G.; Lindahl, U. J. Biol. Chem. 1989, 264, 296-304.
194. Linhardt, R. J.; Rice, K. G.; Kim, Y. S.; et al. Biochem. J. 1988, 254, 781-7.
195. Ottenbrite, R. M.; Takatsuka, R. J. Biocom. Polymers 1986, 1, 461-6.
196. Takatsuka R.; Kondo, S. J. Biocomp. Polymers 1987, 2, 31-48.
197. Butler, G. B.; Tollefson, N. M.; Gifford, G. E.; Flick, D. A. J. Bio. Comp. Polymers 1987, 2, 206-21.
198. Jaques, L. B.; Hiebert, L. M.; Wice, S. M. Sub. J. Lab. Clin. Med. 1990.
199. Herbert, L., Liu, J-M. In Press, 1990.

RECEIVED March 14, 1990

Chapter 25

Amine-Functionalized, Water-Soluble Polyamides as Drug Carriers

Eberhard W. Neuse and Axel G. Perlwitz

Department of Chemistry, University of the Witwatersrand, WITS 2050, Republic of South Africa

Solubility in aqueous media is a prerequisite for the efficacious drug carrier action of polymeric carrier molecules designed for the reversible binding of certain pharmacologically active agents requiring intravenous or intracavitary administration in clinical use. The macromolecular carriers discussed in this communication are aliphatic polyamides possessing intrachain–type or side chain–attached, primary or secondary amine functions capable of drug binding. The polymers are perfectly soluble in water, which permits a rough fractionation by dialysis. The products retained in membrane tubing with 12000 – 14000 molecular–mass cut–off have inherent viscosities of 5–20 mL g^{-1}. Several side–chain modification and model drug anchoring reactions are described, all leading to water–soluble product polymers. Notable among these are conjugates with organoiron (ferrocene) or platinum coordination complexes as examples of the pharmacologically important class of metal–containing polymeric drugs.

For the efficacious administration of polymer–bound drugs by intra–venous or intracavitary (e.g. intraperitoneal) methods, smooth solubility of the conjugate in water is a desirable and frequently crucial prerequisite. Such solubility behavior will enable the conjugate to be distributed rapidly in the body's central circulation system and thus will facilitate transport to the target tissue. This feature will be of particular benefit with conjugates comprising lipophilic drug molecules unable per se to dissolve efficaciously in aqueous media. Polyethyleneimines, polyacrylates and other vinyl–type polymers have been used in numerous investigations as water–soluble drug carriers. However, problems of nonbiodegradability, frequently coupled with toxicity, would tend to militate against the use of these polymer types as components of pharmaceutical preparations. Heterochain–type polymers comprising amide, ester, or carbohydrate ether links in the backbone may provide the required biodegradability and thus would appear to offer advantages in the biomedical field. Polysaccharides, polypeptides and several

0097–6156/91/0467–0394$06.00/0

types of protein have indeed been used for drug anchoring, and there is room for challenging future developments, notably those concerning drug conjugates with immunoglobulins and antibodies possessing targeting functions. Some exellent discussions of these timely topics are on record (1–5). Problems of a different kind, however, have been associated with proteins as drug carriers insofar as enzymatic backbone degradation may now turn out to be too rapid for the carrier chain to maintain its operational integrity. In addition, most proteinaceous polymers are strongly immunogenic, in part owing to the large number of different amino acid constituents in the main chain, and are swiftly scavenged by the body's reticulo–endothelial system. Lastly, the reactive side groups in a protein intended to be utilized as drug–binding sites tend to be poorly described in terms of location and accessibility; as a result, the drug–coupling chemistry is frequently ill–defined and speculative, and well characterized, water–soluble protein–drug conjugates have rarely been isolated in the solid state.

As part of a major program to develop more readily manageable, biologically functional polymer–drug conjugates distinguished by solubility in aqueous media, a large number of macromolecular carriers possessing amino, formyl, or carboxyl groups capable of reversible bond formation with biologically active agents have been synthesized in our laboratory. In the present article we describe three different synthetic approaches leading to linear polyamides with incorporated intra– or extrachain amino groups and discuss side–chain modification and model coupling reactions demonstrating the drug binding facilities of these target polymers.

Structural Prerequisites

For the structural design of macromolecular carriers intended for the synthesis of water–soluble polymer–drug conjugates a number of critical requirements must be considered, including the following:

1. The polymeric backbone should be highly flexible as this will lead to an increase in the (positive) entropy of solution and thus render the dissolution process thermodynamically more favorable. In addition, it should incorporate an adequate number of groups or segments susceptible to solvation in aqueous medium, thus providing solubility in water.
2. The carrier molecule must comprise reactive functional groups as suitable binding sites for drug attachment. The groups should be separated from the main chain by short (5–15 constituent–atoms) side chains or spacers to diminish the steric inaccessibility caused by the polymeric backbone and should be of such a nature as to permit biological (i.e. hydrolytic and enzymatic) cleavage of the bonds generated in the drug coupling reaction.
3. Additional biofissionable functions required to facilitate drug release in the biological environment should be inserted into the spacer segments. At least one of these should be sufficiently remote from the main chain to allow for approach by proteolytic enzymes.
4. The carrier backbone should be biodegradable to permit catabolic elimination in the 'spent' state. To this end, amide or ester links susceptible to slow hydrolytic or enzyme–mediated cleavage should be incorporated into the main chain.
5. The carrier polymer must be nontoxic and should exert minimal immunogenicity. Synthetic oligo– and polypeptides are generally superior to natural proteins in this respect, notably if the number of amino acid monomers utilized in the synthesis is kept to a minimum.

The structural prerequisites enumerated in the foregoing were taken into account in the choice of the polyamide structures **1 – 3** here presented. The amino groups incorporated into these carrier polymers as drug binding sites offer several distinct preparative and functional advantages. Specifically:

1. Amino groups lend themselves to coupling reactions with carboxyl groups under mild conditions. Such coupling can often be carried out in aqueous or aqueous–organic solvents with the aid of techniques developed in amino acid chemistry.
2. Being susceptible to protonation and hydrogen bonding, amino groups can function as hydrosolubilizing entities whenever present in excess over the number required for drug binding.
3. Amino groups will render the polymers cationic by virtue of their susceptibility to protonation. In the treatment of malignancies, this may be of potential therapeutic benefit, many cationic polymers being known to show antineoplastic activity (6). In addition, improved pharmacokinetics may result from the conjugate's facilitated pinocytotic cell entry (7) and its enhanced attraction to cancerous cell tissue, the membrane surfaces of which are oftentimes more negatively charged than those of resting cells (8,9).

Synthesis of Carrier Polymers

In the first synthetic approach, building up on the known (10) poly–D,L–succinimide synthesis by polycondensation of aspartic acid, copoly–merizations of D,L–aspartic acid with Nδ–phthaloylornithine in various monomer feed ratios m/n were conducted in orthophosphoric acid at 185°C. This gave copoly(imide–amides), which upon hydrazinolysis in DMF suffered N–deprotection concomitantly with imide ring opening (Scheme 1). The resulting water–soluble target polymers, of the poly(α,β–D,L–aspart–hydrazide–co–D,L–ornithine) type **1** (α form only shown here), possessed backbone compositions with the (randomly distributed) asparthydrazide and ornithine units present in x/y ratios higher by some 20–30% than the corresponding monomer feed ratios m/n (see Experimental Part). The powerful thermodynamic driving force for imide ring formation from the aspartic acid monomer can be expected to lead to preferential incorporation of succinimide units at the high polycondensation temperature of these experiments and thus may account for the observed differences in polymer composition and feed ratio.

In the second approach, aliphatic diamines of the type NH_2–CH(CH$_3$)–CH$_2$O(CH$_2$CH$_2$O)$_n$CH$_2$CH(CH$_3$)–NH$_2$ ($n \simeq 20$, 45) were copolymerized with iminodiacetic acid in polyphosphoric acid (PPA) at 140–170°C. This yielded polyiminodiacetamides of the general structure **2** (Scheme 2). The intrachain poly(ethyleneoxy) blocks in these carrier molecules act as the principal contributors to the compounds' excellent water solubility, and the secondary amino groups introduced as constituents of the iminodiacetic acid monomer provide the drug anchoring sites.

The susceptibility of poly–D,L–succinimide to nucleophilic imide ring opening mediated by amines, affording N–substituted α,β–D,L–aspartamide polymers, was utilized in the third synthetic approach. This ring opening reaction has been the subject of extended investigations (11–13), notably by Drobník and Saudek with coworkers, and the drug carrier potential of polyaspartamides has been proficiently explored (13–16). In our laboratory, the reaction was found to proceed not only with monofunctional, but also, under carefully maintained experimental conditions, with difunctional amines

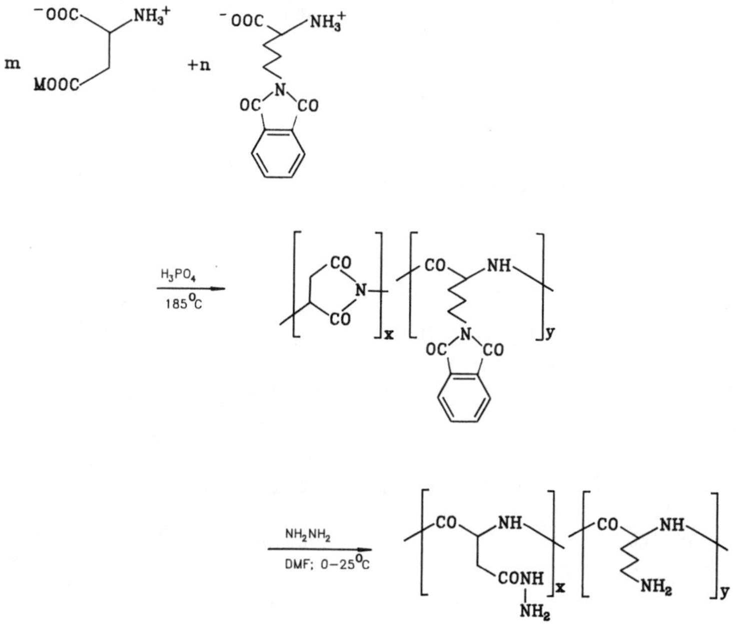

(1)

1

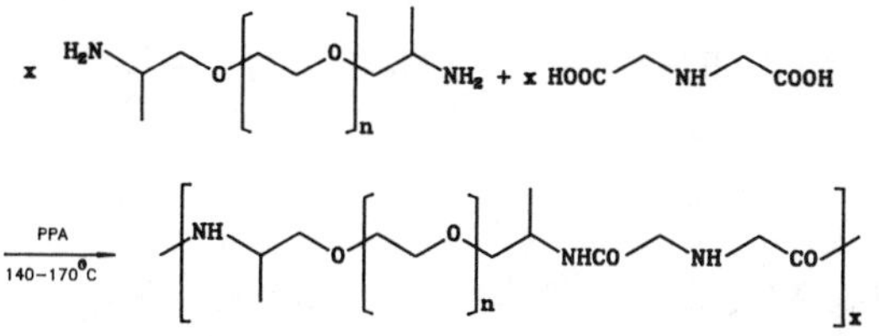

(2)

2

as nucleophiles. This permitted the direct synthesis of polymers bearing free amino side groups without involvement of cumbersome N–protection and deprotection techniques (Scheme 3; α forms shown only). Diamines used included propylenediamine, N–(2–hydroxyethyl)ethylenediamine, diethylenetriamine, and hydrazine. The reactions were performed by treating the polysuccinimide with an excess of diamine in N,N–dimethylformamide solution. In a similar fashion, sequential treatment of the polyimide with a monofunctional and a difunctional nucleophile in given proportions afforded copolyamides composed in random sequence of one type of repeat units possessing a drug–binding primary or secondary amino group and another type of units bearing tertiary amino groups capable of providing hydro–

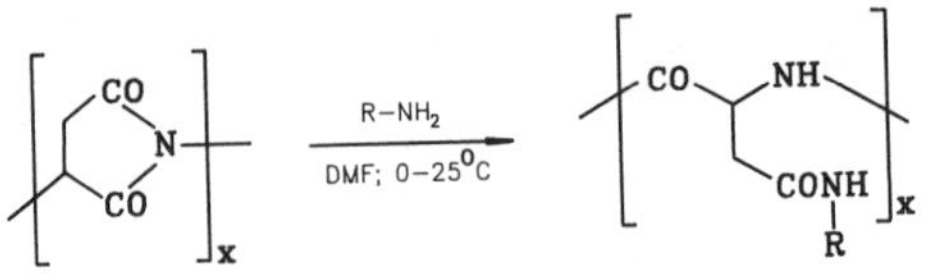

$$(3)$$

3

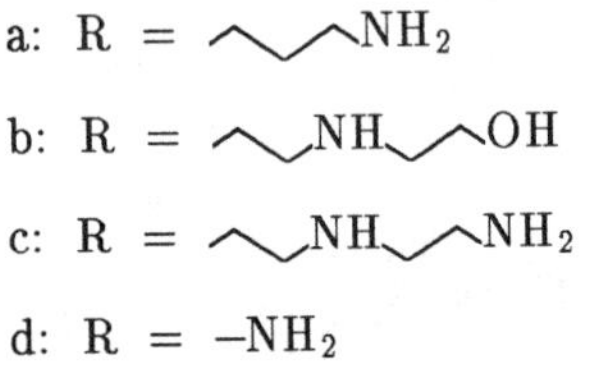

a: R = $\wedge\wedge$NH$_2$

b: R = $\wedge$NH$\wedge$OH

c: R = $\wedge$NH$\wedge$NH$_2$

d: R = $-$NH$_2$

solubility but unable to undergo simple drug coupling reactions (Scheme 4; α forms only).

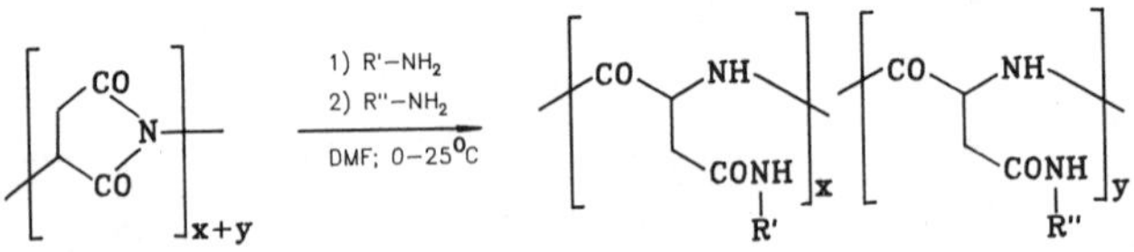

$$(4)$$

4

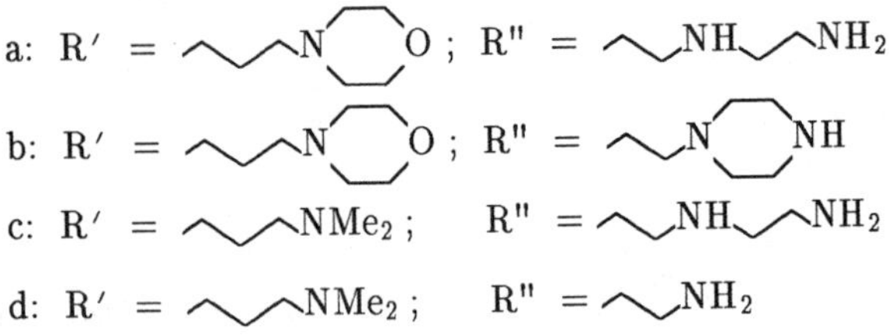

a: R$'$ = $\wedge\wedge$N$\bigcirc$O ; R$''$ = $\wedge$NH$\wedge$NH$_2$

b: R$'$ = $\wedge\wedge$N$\bigcirc$O ; R$''$ = $\wedge$N$\bigcirc$NH

c: R$'$ = $\wedge\wedge$NMe$_2$;   R$''$ = $\wedge$NH$\wedge$NH$_2$

d: R$'$ = $\wedge\wedge$NMe$_2$;   R$''$ = $\wedge$NH$_2$

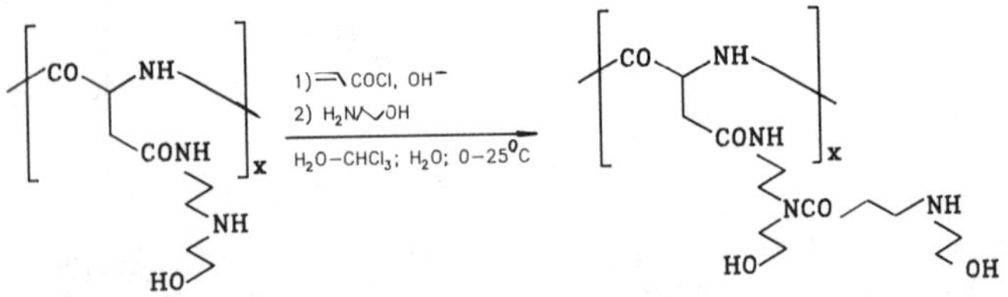

$$(5)$$

3b   **5**

Polyamides **2**, **3a**, **3b**, and **3d** were further modified by side–chain N–acryloylation, and the acryloylated polymers, on treatment with ethanolamine, ethylenediamine, or propylenediamine in water, converted to side chain–extended, hydroxyl– or amine–functionalized derivatives possessing the same solubility properties as their polymeric precursors. Polymer **5** derived from **3b** exemplifies the products so obtained (Scheme 5; α form only). The polyamides **1–5** were purified by dialysis in aqueous solution. Freeze–drying provided the polymers in 25–70% yield as water–soluble solids.

Anchoring Reactions

The drug–binding (anchoring) capabilities of the carrier polymers **1–4** were tested in a large variety of coupling reactions with suitably functionalized model compounds. In exemplifying experiments, copolymer **1** $(x/y = 3.7)$ was treated with an excess of 5–(acryloylamino) salicylic acid in aqueous solution; this gave the water–soluble, polymeric salicylic acid derivative **6** $(x/y = 3.7)$ *via* Michael addition of primary amino groups in **1** across the activated double bond in the acryloyl compound. The same educt polymer **1** $(x/y = 3.7)$, upon coupling with phenylacetic acid *via* the active N–hydroxysuccinimide ester of this acid in acetonitrile–water, converted to the conjugate **7** $(x/y = 3.7)$. In an analogous reaction, **1** $(x/y = 6.2)$ gave the conjugate **7** $(x/y = 6.2)$. Treatment of copolyamide **4a** $(x/y = 2.34)$ with potassium tetrachloro–

6   **7**

platinate in aqueous solution led to Pt complexation with the ethylenediamine ligand in the carrier polymer, affording the polymeric platinum complex **8** $(x/y = 2.34)$. In another exemplifying experiment conducted analogously, **4a** $(x/y = 3.0)$ gave the polymer complex **9** $(x/y = 3.0)$. Reaction of 2–ferrocenylpropanoic acid, *via* its active hydroxy–

8   **9**

succinimide ester, with copolyamide **4d** $(x/y = 1.5)$ in aqueous tetrahydrofuran gave the ferrocene–containing conjugate **10** $(x/y = 1.5)$, in which the proportion of z varied according to the employed reactant ratio. All conjugates were purified by dialysis and isolated by freeze–drying in yields of 35–70%. Conjugates with y (z in **10**) <0.4x dissolved completely in water. However, solid–state interactions upon extended storage at ambient temperature tended to cause crosslinking with concomitant loss of solubility, and the conjugates were therefore routinely stored at −25°C. The compositions of polymers 1–10 were determined microanalytically and by ^{1}H NMR spectroscopy. Inherent viscosities (in H_2O) were typically in the range of 5–20 mL g^{-1}. In light of the antineoplastic effects shown by numerous compounds of the iron (17–20) and platinum (21) group elements and certain other metals (22–24), the three conjugates 8–10 are of interest as prototype metal–containing, water–soluble polymers in biomedical applications.

10

In summary, this investigation has demonstrated the practicability of synthesizing linear, water–soluble polyamides fitted with amine–functionalized side chains and their conjugation products with drug model compounds. The results invite extended work in this field with emphasis on the incorporation of inherently hydrophobic, organometallic or inorganic coordination compounds showing biological activity into biocompatible, water–soluble conjugates.

Experimental Part

Solid–state IR spectra were taken on KBr pellets. ^{1}H NMR spectra (60 MHz) were recorded on D_2O solutions; chemical shifts δ, in ppm, are given relative to sodium 3–(trimethylsilyl)–2,2,3,3–d$_4$–propionate as internal standard. Inherent viscosities, η_{inh}, were determined on aqueous solutions (c = 0.2% w/v) in Cannon–Fenske tubes at 30.00 ± 0.05°C. Dialysis was performed in cellulose tubing (Spectrum Industries, Inc.) with 3500–4000 or 12000–14000 molecular–mass cut–off. Unless stated otherwise, the operation was conducted against running water, and the dialysis period was 24 h. N,N–Dimethylformamide (DMF) was dried over Molecular Sieves 4A and distilled under reduced pressure in a stream of N_2. All manipulations inolving this solvent were performed under anhydrous conditions.

Poly(α,β–D,L–asparthydrazide–co–D,L–ornithine) 1. The mixture of D,L–aspartic acid (150 mmol), Nδ–phthaloylornithine (50 mmol) (25), and H_3PO_4 (85%, 10 g), homogenized and saturated with N_2 at 60°C, was heated for 3h at 185°C (200°C bath temperature) under reduced pressure in a rotating evaporator, and the crude product was worked up as described (10) for aspartic acid homopolymerization. Following chain extension in DMF solution mediated by N,N'–dicyclohexylcarbodiimide (DCC) as in the homopolymer synthesis (10), the polymeric product was precipitated with H_2O (100 mL). The washed and dried (2d at 85°C, 0.5 torr) polymer was collected in 70% yield. Anal. found: C/N, 4.8; calcd. for poly[D,L–succinimide(3.7)–co–Nδ–phthaloylornithine(1.0)]: C/N, 4.9. In analogously conducted runs, monomer feed ratios (Asp/PHT–Orn) were 1, 2, and 5, and the resulting polymers possessed compositions corresponding to ratios of 1.3, 2.5, and 6.2 of succinimide to phthaloylornithine units in the main chain.

Hydrazine hydrate (98%, 11.75 mmol) was added dropwise to the stirred and cooled (0°C) solution of poly[D,L–succinimide(3.7)–*co*–phthaloylornithine(1.0)] (1 mmol, base unit) in DMF (8 mL). Stirring of the resulting suspension was continued for 12 h at room temperature, pH 8. After the addition of EtOH–BuOH (1:1; 20 mL), the mixture was centrifuged, and the collected solid phase was dissolved in H_2O. The pH was adjusted to 5, and the filtered polymer solution was dialyzed (3500 − 4000 molecular–mass cut–off) and freeze–dried, to give water–soluble poly(asparthydrazide–*co*–ornithine) in 48% yield; η_{inh}, 15 mL g^{-1}. Anal found: C/N, 1.6. Calcd. for 1 (x/y = 3.7): C/N, 1.5. Copolymers 1 (x/y = 1.3, 2.5, and 6.2), analogously obtained by hydrazinolysis of the respective poly(succinimide–*co*–phthaloylornithine) precursors, were equally soluble in H_2O.

Poly[imino–2,1–propyleneoxy–poly(ethyleneoxy)–1,2–propyleneimino–carbonylmethyleneiminomethylenecarbonyl] 2. The well ground mixture of bis(2–aminopropyl)poly(ethyleneglycol) 800 (Fluka AG; molecular mass 900) (10 mmol) and iminodiacetic acid (10 mmol) was briefly homogenized with warm (60 C) polyphosphoric acid (20 g) and was heated for 16 h at 165°C (180°C bath temp.) under reduced pressure on a rotating evaporator. The cooled melt was dissoled in H_2O (200 mL), and the solution, after pH adjustment to 5 − 6 and filtration, was dialyzed and freeze–dried as in the preceding experiment. The crude solid, collected in 80% yield, was redialyzed in aqueous solution (12000 − 14000 molecular–mass cut–off). The poly[imino–2,1–propyleneoxy–icosa(ethyleneoxy)–1,2–propyleneimino–carbonylmethyleneiminomethylenecarbonyl] so purified, and isolated by freeze–drying in 42% yield, was completely soluble in H_2O; η_{inh}, 6 mL g^{-1}. Anal. found: C/N, 18. Calcd. for 2 (x = 20): C/N, 17. In a similar fashion, use of bis(2–aminopropyl)poly(ethylene glycol) 1900 (molecular mass 2000), with heating performed for 8h at 150°C (160 C°bath temp.), gave crude 2 (x = 45), which, upon redialysis as above, was obtained in 28% yield as a hygroscopic solid completely soluble in H_2O; η_{inh}, 8 mL g^{-1}. Anal. found: C/N, 36. Calcd. for 2 (x = 45): C/N, 33.

Poly(α,β–D,L–aspartamides) 3. The solution of poly(D,L–succinimide) (10) (50 mmol) in DMF (60 mL) was added over a 45–min period to the stirred solution of propylenediamine (500 mmol) in DMF (60 mL) at 0°C. The resulting clear solution was stirred briefly (15 min) at 0°C and for 5 h at ambient temperature. The polymeric product was precipitated by EtOH–Et$_2$O (1:2; 250 mL), dialyzed in aqueous solution (48 h; 12000 − 14000 molecular–mass cut–off) and freeze–dried. The solid thus obtained in 62% yield dissolved completely in H_2O; η_{inh}, 15 mL g^{-1}. Anal. found: C/N, 2.4. Calcd. for 3a: C/N, 2.3. In a similar fashion, reactant ratios (amine/poly–succinimide base unit) being 8, 8, and 6, the aminolysis of polysuccinimide with N–(2–hydroxyethyl)ethylenediamine (5 h), diethylenetriamine (5 h), and hydrazine (6 h) at ambient temperature gave the water–soluble poly–aspartamides 3b, c and the polyhydrazide 3d, respectively, in yields of 40 − 70%, η_{inh}, 5 − 20 mL g^{-1}. ^{1}H NMR data were in agreement with the proposed structures 3.

Copoly(α,β–D,L–aspartamides) 4. The following experiment, affording the ethylenediamine–functionalized 4d, exemplifies the two–step preparation of copolyamides 4. The solution of poly(D,L–succinimide) (25 mmol) in DMF (20 mL) was added dropwise, with stirring, to 4–(3–aminopropyl)morpholine

(17.5 mmol) dissolved in the same medium (10 mL) and precooled to 0°C. The resulting solution was stirred for 6 h at room temperature and was then added dropwise to diethylenetriamine (15 mmol) in DMF (20 mL). Stirring was continued for another 6 h at ambient temperature. The polymeric product was precipitated, redissolved in H_2O, dialyzed, and freeze–dried as in the preceding experiment, giving water–soluble, solid **4d** (x/y = 2.34) in 53% yield; η_{inh}, 9 mL g^{-1}. Anal. found: C/N, 3.2. Calcd. for **4d** (x/y = 2.34): C/N, 3.1. ^{1}H NMR/δ: 4.0 − 3.7 (9H, $-CH_2OCH_2-$), 3.5 − 1.5 (38H, remaining CH_2).

Side Chain Modification of 3. The derivatization of polyaspartamides **3** at the side chains is exemplified by the preparation of **5** from **3b**. Polymer **3b** (10 mmol, base unit) was acryloylated with acryloyl chloride (30 mmol) in $CHCl_3-H_2O$ at 0°C, pH 6 − 7 (adjusted with NaOH), and the N–acylated derivative was collected from the dialyzed (12000 − 14000 molecular–mass cut–off) aqueous phase by freeze–drying in 38% yield as a water–soluble solid; η_{inh}, 17 mL g^{-1}. ^{1}H NMR/δ: 6.9 − 5.8 (3H, vinylic), 3.8 − 2.9 (9H, CH_2).

The polymer (1.0 mmol), dissolved in H_2O (5 mL), was treated with ethanolamine (3.0 mmol) in H_2O (3 mL) for 2 h at 0°C and for another 20 h at 20 − 25°C. Following pH adjustment to 8 with aqueous 1M HCl, the filtered solution was dialyzed and freeze–dried. The hygroscopic solid thus obtained in 45% yield was completely soluble in H_2O; η_{inh}, 12 mL g^{-1}. Anal. found: C/N, 3.3. Calcd. for **5**: C/N, 3.2$_5$.

Polymer–Bound Aminosalicylic Acid 6. The solution of copolyamide **1** (x/y = 3.7) (1.0 mmol, base unit) and 5–acryloylaminosalicylic acid (5.0 mmol; by acryloylation of 5–aminosalicylic acid with acryloyl chloride in $CHCl_3-H_2O$ at pH 7 − 8) in H_2O (8 mL) was stirred for 20 h at room temperature and for another 6 h at 80°C. Throughout this period, the pH of the solution was maintained at 7.5 − 8.0 ($NaHCO_3$). After pH adjustment to 2, the product polymer was precipitated with EtOH (20 mL) and redissoled in H_2O (20 mL). The solution was dialyzed (3500 − 4000 molecular–mass cut–off) and freeze–dried, giving **6** (x/y = 3.7) in 32% yield as a water–soluble solid; η_{inh}, 7 mL g^{-1}. Anal. found: C/N, 2.1. Calcd. for **6** (x/y = 3.7): C/N, 2.0. ^{1}H NMR/δ: 7.7 − 6.8 (3H, arom.), 3.6 − 1.6 (18H, CH_2).

Polymer–Bound Phenylacetic Acid 7. A solution of N–(phenylacetoxy)–succinimide was prepared by treatment of phenylacetic acid with DCC and N–hydroxysuccinimide (3.0 mmol each) in MeCN at 0°C (2 h), then ambient temperature (6 h), followed by filtration from the precipitated urea. The solution, after volume reduction to ca. 5 mL, was added to the stirred solution of **1** (x/y = 3.7) (1.0 mmol, base unit) and $NaHCO_3$ (1 mmol) in H_2O (3 mL). The suspension was stirred for 15 h and, after the addition of more $NaHCO_3$ (2 mmol) and dilution with H_2O and MeCN (2 mL each), for another 6 h at ambient temperature. After removal of the organic solent under reduced pressure and filtration, the aqueous polymer solution was dialyzed and freeze–dried as in the preceding experiment, to give water–soluble **7** (x/y = 3.7) in 35% yield; η_{inh}, 11 mL g^{-1}. Anal. found: C/N, 2.9. Calcd. for **7** (x/y = 3.7): C/N, 3.0. ^{1}H NMR/δ: 7.3 (5H, arom.), 3.8 − 1.5 (14H, CH_2).

In an analogous reaction performed with **1** (x/y = 6.2), conjugate **7** (x/y = 6.2) was obtained in 29% yield; η_{inh}, 13 mL g^{-1}. ^{1}H NMR/δ: 7.3 (5H, arom.), 3.7 − 1.6 (19H, CH_2).

<u>Polymer–Bound cis–Dichloroplatinum(II) Complexes **8,9**</u>. Copolyamide **4a** (x/y = 2.34; 0.25 mmol), dissolved in H_2O (3mL), was added rapidly to the stirred solution of K_2PtCl_4 (0.30 mmol) in H_2O (3 mL). The pH was adjusted to 6.5 (1M HCl), and the solution was stirred in the dark for 20 h at room temperature and 5 h at 50°C, care being taken to maintain the pH at 6.0 – 6.5 throughout the reaction period. A slight precipitate, which had developed initially, for the most part redissolved at the higher reaction temperature. the filtered solution was dialyzed (12000 – 14000 molecular–mass cut–off) against several stationary charges of deionized H_2O and, after acidification to pH2 (1M HCl), was freeze–dried. This afforded beige–colored conjugate **9** as a water–soluble solid in 59% yield; η_{inh}, 12 mL g^{-1}. Anal. found: Pt, 18.1%. Calcd. for **8** (x/y = 2.34): Pt, 18.9%. IR/cm^{-1}: 320, 315 m (ν(Pt–Cl)). From the combined outer solutions collected in the dialysis step, volume reduction under reduced pressure, redialysis (3500 – 4000 molecular–mass cut–off), and freeze–drying provided another portion (19%) of low–molecular–mass conjugate **8**; η_{inh}, 4 mL g^{-1}. Anal. found: Pt, 18.7%. IR/cm^{-1}: 321, 315 m.

In an analogous fashion, reaction of the tetrachloroplatinate salt (0.30 mmol) with copolyamide **4c** (x/y = 3.0; 0.25 mmol) gave conjugate **9** (x/y = 3.0) in 52% yield; η_{inh}, 8 mL g^{-1}. Anal. found: Pt, 18.5%. Calcd. for **9** (x/y = 3.0): Pt, 18.3%. IR/cm^{-1}: 321, 317 m (ν(Pt–Cl)). The second portion was collected in 13% yield: η_{inh}, 4 mL g^{-1}. Anal. found: Pt, 17.9%. IR/cm^{-1}: 319, 312 m.

<u>Polymer–Bound Ferrocenylpropanoic Acid **10**</u>. For the preparation of 3–ferrocenylpropanoic acid N–hydroxysuccinimide ester, the free acid (**26**) (1.7 mmol) was dissolved in tetrahydrofuran (THF; 3 mL) together with N–hydroxysuccinimide (1.8 mmol). To the cooled (0°C) and stirred solution was added, dropwise, a solution of DCC (1.9 mmol) in the same solvent (3 mL). The mixture was stirred for 2 h at 0°C and 3 h at room temperature. The orange suspension was filtered for removal of the formed urea derivative and the residue thoroughly washed with THF. The combined filtrate and washings containing the active hydroxysuccinimide ester, after volume reduction to 5 – 6 mL, were added dropwise to the precooled (0°C) solution of copolyamide **4d** (x/y = 1.5; 1.3 mmol) in H_2O (5 mL). The mixture was stirred for 20 h at ambient temperature, NEt$_3$ (1.7 mmol) was added and stirring continued for another 8 h. Removal of the THF content under reduced pressure, addition of H_2O (10 mL) and filtration gave a reddish–brown solution, from which, by dialysis (3500 – 4000 molecular–mass cut–off) and freeze–drying, a brownish, water–soluble solid was obtained in 68% yield; η_{inh}, 8 mL g^{-1}. Anal. found: Fe, 5.9. Calcd. for **10** (x/y = 1.5; x/z = 2.3, corresponding to 65% substitution of primary amine functions): Fe, 5.9. IR/cm^{-1}: 820 m (δCH ⊥, ferrocene); 500, 480 m (ν_{21}, ν_{11}, ferrocene).

Acknowledgment

Financial support by the Chemical Division, Shell South Africa (Pty.) Ltd., is gratefully acknowledged. The authors are also indebted to Messrs. K.O. Sekonyela and A. Stephanou for experimental assistance, to Mrs. Lorraine Meredith for the proficient typing of the manuscript, and to Mrs. Margaret Crabb for her able assistance with the art work.

Literature Cited
1. Huntley Blair, A.; Ghose, T.I. *J. Immunol. Methods* 1983, *59*, 129.
2. Poznansky, M.J.; Juliano, R.L. *Pharmacol. Revs.* 1984, *36*, 277.
3. Duncan, R.; Kopecek, J. *Advan. Polym. Sci.* 1984, *57*, 51.

4. Dorn, K.; Hoerpel, G.; Ringsdorf, H. In *Bioactive Polymeric Systems*; Gebelein C.G.; Carraher, Jr., C.E., Eds.; Plenum, New York, 1985, Chap. 19.
5. Duncan, R. In *Controlled Drug Delivery*, 2nd Edn.; Robinson, J.R.; Lee, V.H.L., Eds.; Dekker, New York, 1987, Chap. 14.
6. Hodnett, E.M. *Polymer News* 1983, *8*, 323.
7. Pratten, M.K.; Cable, H.C.; Ringsdorf, H.; Lloyd, J.B. *Biochim. Biophys. Acta* 1982, *719*, 424.
8. Moroson, H.; Rotman, M. In *Polyelectrolytes and Their Applications*; Rembaum, A.; Sélégny, E., Eds.; Reidel Publishing Co., Dordrecht, 1975, p. 187.
9. Pethig, R.; Gascoyne, P.R.C.; McLaughlin, J.A.; Szent–Györgyi, A. *Proc. Natl. Acad. Sci. USA* 1984, *81*, 2088.
10. Neri, P.; Antoni, G. *Macromol. Synt.* 1982, *8*, 25.
11. Kovacs, J.; Könyves, *Naturwiss.* 1954, *41*, 333.
12. Vlasák, J., Rypáček, F.; Drobník, J., Saudek, V. *J. Polym. Sci. Polym. Symp.* 1979, *66*, 59.
13. Drobník, J.; Saudek, V.; Vlasák, J.; Kálal, J. *J. Polym. Sci. Polym. Symp.* 1979, *66*, 65.
14. Drobník, J.; Kálal, J.; Dobrowska, L.; Praus, R.; Váchová, M.; Elis, J. *J. Polym. Sci. Polym. Symp.* 1979, *66*, 75.
15. Rypáček, F.; Cífková, I.; Drobník, J. *Physiol. Bohem.* 1984, *33*, 481.
16. Rypáček, F.; Drobník, J.; Kálal, J. *Ann. N.Y. Acad. Sci.* 1985, *446*, 258.
17. Kőpf–Maier, P.; Kőpf, H.; Neuse, E.W. *J. Cancer Res. Clin. Oncol.* 1984, *108*, 336.
18. Kőpf–Maier, P. *Z. Naturforsch.* 1985, *40c*, 843.
19. Clarke, M.J. *ACS Symp. Ser.* 1983, *209*, 335.
20. Sava, G.; Zorzet, S.; Perissin, L.; Mestroni, G.; Zassinovich, G.; Bontempi, A. *Inorg. Chim. Acta* 1987, *137*, 69; and refs. therein.
21. Dabrowiak, J.C.; Bradner, W.T. In *Progress in Medicinal Chemistry*, Vol. 24; Ellis, G.P.; West, G.B., Eds.; Elsevier, New York, 1987, Chap. 4; and refs. therein.
22. Kőpf–Maier, P. *Naturwiss.* 1987, *74*, 374.
23. Ward, S.G.; Taylor, R.C. In *Metal–Based Anti–Tumor Drugs*; Gielen, M.F., Ed.; Freund, London, 1988, page 1; and refs. therein.
24. Crowe, A.J.; Smith, P.J.; Atassi, G. *Inorg Chim. Acta.* 1984, *93*, 179.
25. Bodanszky, M.; Ondetti, M.A.; Birkhimer, C.A.; Thomas, P.L. *J. Am. Chem. Soc.* 1954, *86*, 4452.
26. Dormond, A.; Décombe, J. *Bull. Soc. Chim. Fr.* 1968, 3673.

RECEIVED June 4, 1990

Chapter 26

Hydrosoluble Polymeric Drug Carriers Derived from Citric Acid and L-Lysine

J. Huguet, M. Boustta, and M. Vert

L.S.M., URA Centre Nationale de Recherche Scientifique 500,
INSA de Rouen, BP 08, 76131 Mont-Saint-Aignan Cedex, France

New functional polyamides with carboxyl alcohol pendent
groups were synthesized from two metabolites, namely citric
acid and L-lysine. These water-soluble macromolecules were
obtained by step-growth polymerization using the interfacial
method applied to protected L-lysine and citric acid, the
latter being in its diacyl chloride form. It is shown that a
side reaction occured during the polycondensation with the
formation of intramolecular cyclic imide groups. Chemical
modification due to the cleavage of the protecting groups and
the hydrolysis of the imide rings are discussed on the basis
of IR and ^{13}CNMR spectra. The imide rings were used
advantageously to couple low molecular primary amine
compounds. Resulting macromolecular conjugates were soluble
in water at neutral pH regardless of the degree of
substitution.

The interest of bioresorbable polymers is growing very fast in the field of
therapy for both surgery (1) and drug delivery (2). Among the various
systems which are presently studied to achieve controlled drug delivery
such as implants, micro and nanoparticles, polymerized liposomes,
hydrophobic microdomain-forming copolymers and macromolecular drugs or
prodrugs, the latter basically require water-soluble bioresorbable
macromolecules to act as drug carriers (3). By bioresorbable polymers, we
understand degradable synthetic polymeric systems which can yield low
molecular weights degradation products eliminated from the body through
natural pathways. One of our rationales for the synthesis of such polymers
aiming at biomedical or pharmacological applications is to start from
metabolites.

At the moment, the most popular bioresorbable polymers are the
aliphatic polyesters, especially those derived from lactic or glycolic
acids, two metabolites of the glycolic pathway. These polyesters are very
attractive compounds to achieve diffusion-controlled and / or

0097–6156/91/0467–0405$06.00/0
© 1991 American Chemical Society

degradation-controlled   drug   delivery   because   the   physical,
physico-chemical, mechanical and biological properties of resulting devices
can be adjusted through chirality and enantiomeric distributions or by
copolymerization to introduce achiral units as glycolic ones in polylactic
chains (4). However, none of the members of this family is water-soluble.

A few years ago, a route was found to make a water-soluble degradable
polyester derived from a metabolite of the Krebs cycle, namely malic acid.
This hydroxyacid metabolite is trifunctional and has to be monoprotected to
allow the synthesis of linear macromolecules. Accordingly, benzyl
malolactonate was synthesized and polymerized whereas Pd-charcoal catalyzed
hydrogenolysis   of   protecting   groups   yielded   poly(β-malic acid)
$-[-O-CH(COOH)-CO-]_n-$ whose polymer chains have racemic or optically active
repeating units with pendent carboxylic acid groups (4-6). It has been
shown that racemic poly(β-malic acid) is soluble in water regardless to pH,
is non toxic when given i.m. or i.v. to mice (7) and does degrade in vitro
to yield malic acid as ultimate degradation product (8). The coupling of
low molecular weights amine or alcohol compounds was shown feasible through
the pendent carboxylic acid group (9). However, a major drawback of such
coupling is that water-insolubility comes up rapidly when the resulting new
amide or ester pendent groups are hydrophobic. This occurs for no more than
20% substitution and then limit the possible load if water-solubility had
to be preserved (10).

The introduction of more than one hydrophilic ionic or non-ionic
groups per repeating unit can reasonably provide a means to retain
water-solubility even for systems with the attachment of one drug molecule
per repeat unit. A few years ago, we became interested in developping
linear water-soluble, polymeric drug carriers not only degradable and
derived from metabolites but also allowing high degree of hydrophobic
compound attachment with retention of water-solubility. One possibility was
offered by multifunctional hydroxyacids of the Krebs cycle and we selected
citric acid as a source of polymerizable monomers.

In a first approach, we tried to make multifunctional aliphatic
polyesters by ring opening polymerization of heterocyclic compounds derived
from citric acid. Two derivatives were investigated : β-citrolactone
dibenzyl ester, a β-lactone type monomer and citride tetrabenzyl ester, a
dioxane dione type monomer (the term citride being used by reference to
lactide and glycolide compounds). Unfortunately, these cyclic monomers
appeared as non polymerizable so far and thus we have looked for other ways
to make biodegradable polymers derived from citric acid (11).

As an alternative, we selected the route of step-growth
polymerization of the AA-BB type starting from citric acid as a diacid and
from L-lysine as a diamine comonomer. In this paper, we wish to report the
synthesis of the resulting new family of functional polyamides bearing
carboxyl and alcohol pendent groups (12).

Results and Discussion

Monomers synthesis

In order to achieve proper polycondensation, we have used partial
protection of the α-hydroxy part of the tetrafunctional citric acid and of
the carboxylic acid group of the trifunctional L-lysine to make both
compounds bifunctional.

The diacyl chloride type-AA monomer was obtained in two steps from citric acid or 2-hydroxy 1,2,3-propanetricarboxylic acid according to the following reactions :

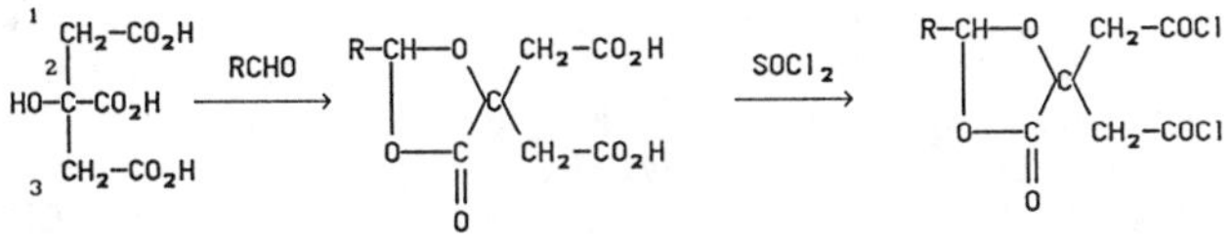

$$R = -CCl_3 \text{ citrochloral (13)} \qquad R = -CCl_3 \quad \underline{1}$$
$$\text{or } R = -C_6H_5 \text{ citrobenzal (14)} \qquad R = -C_6H_5 \quad \underline{2}$$

Trifunctional L-lysine was protected by esterification of the carboxylic group with benzylalcohol tobecome bifunctional. with p. toluene sulfonic acid to yield the L-lysine benzylester ditosylate $\underline{3}$ (15).

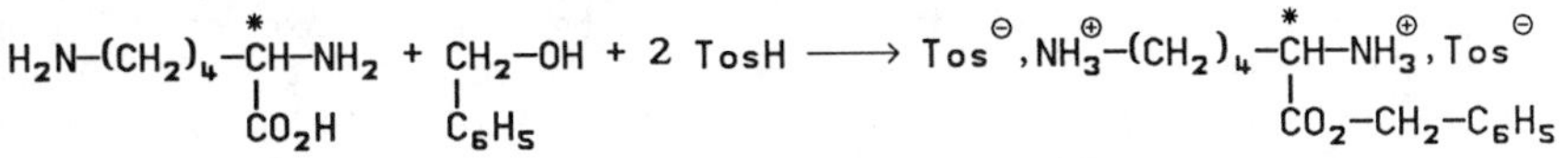

$$\underline{3}$$

L-lysine benzylester ditosylate

Interfacial polycondensation

Interfacial polycondensation of $\underline{1}$ or $\underline{2}$ with $\underline{3}$ were carried out in non miscible water/organic solvent mixtures under vigorous stirring. The reaction time was never more than 10 minutes.

The use of a dichloromethane / dichloro-1,2 ethane mixture as the solvent medium instead of benzene was ineffective to increase molecular weights. The highest values of GPC molecular weights ($M_{GPC} \simeq 20000$ in dioxane with respect to polystyrene standards) were obtained with an excess of 10 to 20 % of diacyl chlorides.

By reacting the protected diacyl chlorid with the protected diamine, the corresponding polyamide-type polymer was expected, i.e. a polyamide with the initial protecting groups, namely the oxolactonic group for citric units and the benzyl ester one for L-lysine units according to reaction I.

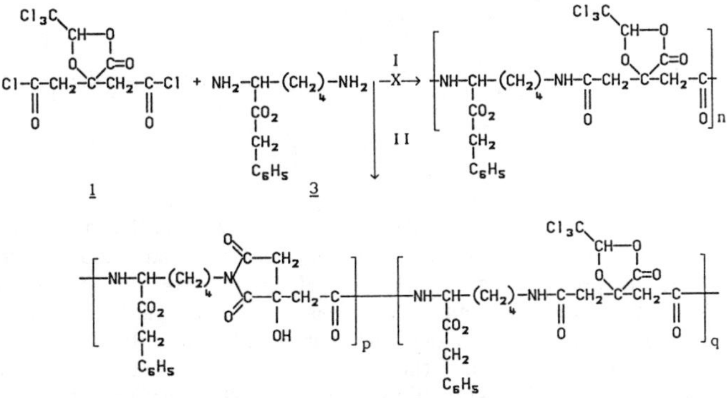

PLCAIp

Actually, we have found that interfacial polycondensation yielded a largely imidified hydroxylated polyamide imide polymer with a small content of residual oxolactonic groups. As the polycondensation reaction was carried out in a   non miscible alkaline water / organic solvents mixture, we concluded that oxolactonic ring was cleavable in this rather alkaline medium. The liberated hydroxy acid group was then available to react with newly formed amide bond to yield intramolecular imide group (reaction II).

The crude copolymers namely PLCAIp appeared as yellowish solids and were soluble in many usual organic solvents such as $CHCl_3$, acetone, dioxane, THF, DMF and insoluble in water, alcohols, benzene and diethylether.

The presence of imide rings was indentified by IR spectrometry. FTIR spectra of these crude polymers exhibited 6 bands in the 1500-1900 cm^{-1} zone (figure 1). The 3 bands located at 1550, 1660 and 1740 cm^{-1} were assigned to the >NH (amid II), >C=O (amid I) vibrations of amide groups and to the >C=O of benzyl ester group respectively. The band located at 1820 cm^{-1} characteristic of the v >C=O vibration of oxolactonic ring largely disappeared whereas two unexpected bands were detected at 1710 and 1780 cm^{-1}. These bands were assigned to the >C=O vibration of cyclic imide ring as refered to polysuccinimide (16,17). In the case of the crude polymer derived from citrochloral dichloride, the two bands corresponding to -CCl_3 group usually located at 820, 860 cm^{-1} were found very much decreased too.

Hydrogenolysis of protecting groups

In order to liberate the carboxyl protected groups of L-lysine units, we have used successfully the Pd-charcoal hydrogenolysis method. Hydrogenolysis carried out in N-methyl-pyrrolidone at 60°C, yielded the polyamide imide copolymers (PCLAI,H), bearing carboxyl and hydroxyl hydrophilic pendent groups as shown in reaction III.

Resulting PLCAI,H was found soluble in methanol, DMF, DMSO and insoluble in acetone, dioxane, ethers, chlorinated and aromatic organic solvents. In water, the solubility of PLCAI,H was found to depend on the proportion of imide groups whereas its Na salt form obtained by neutralization with sodium hydroxide was totally soluble.

The FTIR spectra of the copolymers PLCAI,H and PLCAI,Na in the 1500-1900 cm^{-1} zone are shown in figure 2. The benzyl ester band located at 1740 cm^{-1} is no longer present. We also found that the typical 1810 cm^{-1}

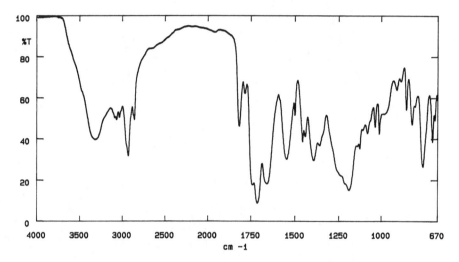

Figure 1.   FTIR   spectrum   of   PLCAIp   corresponding   to   interfacial polycondensation of citrochloral dichloride and protected L-lysine.

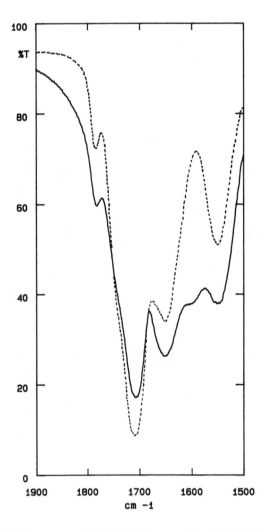

Figure 2. FTIR spectra of PLCAI,H (- - - -) and PLCA,Na (———).

band due to oxolactonic ring is no longer detectable in the IR spectrum whereas the two imide bands located at 1780 and 1710 cm^{-1} are still present. Therefore, we concluded that hydrogenolysis cleaved not only the benzyl ester protecting groups but also the residual oxolactonic ones. The IR bands of remaining imide groups were particularly visible in PLCAI,Na because of the shifting of carboxylic $>C=O$ vibration from 1715 cm^{-1} to 1600 cm^{-1} by neutralization with sodium hydroxide.

Hydrolysis of imide functions

It is well known that cyclic imides are stable in acidic medium but react in basic medium (18). The 100 % alkaline hydrolysis of cyclic imide groups present in PLCAI,H copolymers yielded the sodium salt of poly (L-lysinecitramide) (PLCA,Na), the initially desired multifunctional polyamide.

However, because of the detour through the imidified copolymers and because of the two possible ways to open the non-symmetric imide cycle, namely a and b, the actual PLCA chains include two types of L-lysine / citric acid enchainments, namely succinic and glutaric types corresponding respectively to (a) and (b) openings in reaction IV, with two different side chains, respectively carboxy and carboxymethylene.

PLCAI,H

(a)   (b)

IV

PLCA,Na

The acidic derivative of PLCA,Na (PLCA,H) was obtained by ion-exchange on a cationic resin in the H^+ form. PLCA,H appeared as a white solid which was soluble in organic solvents such as alcohols, DMF, DMSO and also in water regardless of pH. Compounds with molecular weights in the 20000-60000 daltons range (as determined by SALLS) can be easily achieved.

In figure 3, comparison is made of the IR spectra of the initially imidified copolymer (H form) and of a functional polyamide obtained after complete hydrolysis. The IR spectrum of the latter clearly shows 3 main bands in the 1500-1800 cm^{-1} region assigned to $>C=O$ carboxylic groups at 1730 cm^{-1} and to amide groups at 1650 cm^{-1} (amide I) and 1550 cm^{-1} (amide II), the two imide bands visible in the imidified copolymer spectrum having totally disappeared.

The ^{13}C NMR spectrum of the sodium salt form of the poly(lysine citramide) was recorded in D_2O (fig. 4). Seven signals were observed between 20 and 80 ppm and assigned to the various main chain carbon atoms. Five singulets were correlated to the five carbon atoms of lysine unit : C_6 for the asymetric methine at 55.3 ppm and C_7, C_8, C_9 and C_{10} for the four

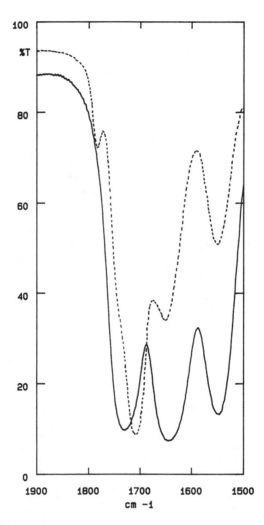

Figure 3. FTIR spectra of PLCAI,H (- - - -) and PLCA,H (———).

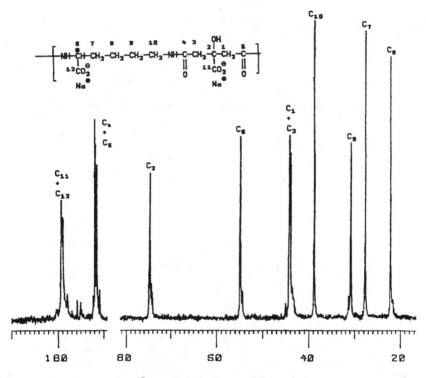

Figure 4. ^{13}C NMR spectrum of PLCA,Na in D_2O.

methylene carbon atoms at 28.1, 22.6, 31.3 and 39.2 ppm. The last two signals located at 44.5 and 74.7-75.2 ppm were respectively assigned to the C_2 quaternary and C_1-C_3 methylene carbon atoms of the citric unit.

Structured resonances were also detected and assigned respectively to carbonyl atoms C_4, C_5 of amide groups at 172-172.8 ppm and to side chains carbonyl atoms C_{11}, C_{12} at 176-177 ppm.

Self-coupling of amine-type compounds to PLCAI polymers

The coupling of amine-type compound can be basically achieved via two different routes : by using coupling reagents which can react with the pendent carboxylic acid groups of citric or/and L-lysine units or by self-coupling on the cyclic imide groups present in the polymeric intermediates.

We have tried to directly couple some primary amines such as 2-aminobutane, 1-phenylethylamine and amphetamine to PLCAI,H under the same conditions reported by Saudek et al. for polysuccinimide (18).

The coupling of amphetamine (Amp) to a 100 % imidified copolymer was carried out in hot DMF solutions with an excess of amine. Because of the two possibilities of ring opening, two isomeric amid-acid conjugates were actually obtained (reaction V).

The final amount of the covalently bound amino derivative was deduced from the ^{13}C NMR spectrum (fig. 5) of the sodium salt of PLCAAmp. From the two peaks corresponding respectively to the asymetric carton atoms of L-lysine unit (C_6 at 57 ppm) and of amphetamine (C_{13} at 49 ppm), we have found that 100 % amphetamine was covalently bound to polymer chains. In spite of this high degree of substitution which corresponds to one attached amphetamine per each citric unit, the resulting polymer chains were still soluble in water because of the remaining hydroxyl and carboxylate hydrophilic groups.

Conclusion

We have shown that it is possible to make highly hydrophilic water soluble drug carriers by polycondensation of metabolites such as citric acid and L-lysine after protection of some of the functional groups present in these molecules. In the course of polycondensation reaction, intramolecular imid groups are formed which can be used to couple directly primary amine compounds by avoiding the use of coupling reagents. High degree of coupling

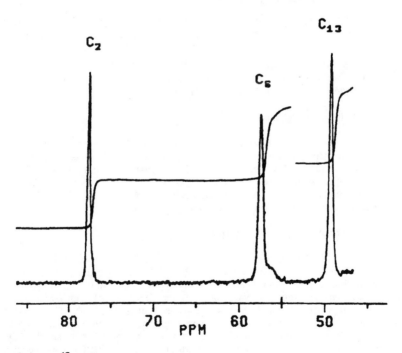

Figure 5. ^{13}C NMR spectrum in the 45–80 ppm range of amphetamine coupled to PLCAI in D_2O.

can be achieved without losing the water solubility. Now, it is attempted to know whether such polylysinecitramide which carries breakable bonds in the main chain can be biodegradable.

Experimental part

Typical interfacial polycondensation

L-lysine benzylic ester ditosylate (7.6 g), CO_3Na_2 (7.45 g) and dodecyl sodium sulfate (0.5 g) were dissolved in water-benzene mixture (60 cm^3, V/V, 66/33) in a Waring type mixer and stirred at 12-15000 t/mn. Citrobenzal dichloride (4.54 g) dissolved in benzene (40 cm^3) and CO_3Na_2 (8.2 g)dissolved in water (20 cm^3) were added simultaneously. First, 90% of these two solutions were introduced within 6 minutes. The reaction was allowed to stir for 2 minutes and then, the remaining 10% were added within two min.. The polymer precipitated as a yellow solid during the reaction. The medium was then acidified with dilute HCl. The polymer was washed with water and dried under vacuum at 50°C. 5.1 g of PLCAIp were recovered (yield ≈ 83%).

Catalytic hydrogenolysis of PLCAIp

PLCAIp (5 g) dissolved in N-methyl pyrrolidone and 10% Pd-charcoal catalyst (1 g) was allowed to stir under hydrogen atmosphere at 60°C until no more H_2 was absorbed. After 24 h., the catalyst was filtrated and the solvent evaporated to yield a slightly colored paste. The crude compound was dissolved in methanol, precipitated by ether and dried. 4.1 g of PLCAI,H (white solid) were recovered (yield ≈ 84%).

Hydrolysis of imide functions

PLCAI,H (4 g) were dissolved in aqueous 0.7 N NaOH (180 cm^3). The mixture was allowed to stir for 45 min. at room temperature. The basic solution was then diluted with water and dialyzed against neutral water through semipermeable dialysis tube (Wiskase type - cut off 6-8000). The final solution was concentrated, filtrated through a 0.45 μm Millipore filter and finally freeze-dried. 3.1 g of PLCA,Na were recovered (yield ≈ 60%).

Coupling of amphetamine to PLCAI,H

PLCAI,H (1.5 g) (100% imide rings with respect to citric acid repeating units), amphetamine (19 cm^3) and DMF (2 cm^3) were heated for 3 days at 60°C. The amine excess was evaporated under vacuum. After solubilizing the resulting polymer in aqueous NaCl M (100 cm^{-3}), the solution was dialyzed against neutral water through a Viskase tube (cut off 3500). The polymeric solution was then concentrated and percolated through an exchange cationic resin (Na$^+$ form), and then, freeze-dried. A brownish solid (1.6 g) containing 100% of coupled amine (as evaluated by ^{13}C NMR) was recovered (yield ≈ 65%).

IR spectra

IR spectra were recorded with a PERKIN-ELMER 1760 FTIR.

Molecular weights determination

Molecular weights of PLCAIp in dioxane were evaluated by GPC using a Waters apparatus equipped with μ-styragel columns and refractive index detection. Data are expressed with respect to GPC peaks in reference to polystyrene standards. For PLCA,H examples, molecular weights were evaluated by laser light scattering by using a Malvern SALLS diffusiometer.

NMR spectra

62.8 Mhz ^{13}C NMR spectra were recorded at 30°C by using a BRUCKER WH250 (ESSO Research Center-Mont-Saint-Aignan - FRANCE).

Acknowledgements

Authors are indebted to Mr. Plaindoux and the Esso Co. for their helpful contribution to the recording of ^{13}C NMR spectra.

References

1   Vert, M.; Christel, P.; Chabot, F.; Leray, J. In Macromolecular Biomaterials; Hastings, G.W.; Ducheyne, P., Eds.; CRC Press: Boca Raton, 1984; p 117.
2   Vert, M. In Polymers in Controlled Drug Delivery; Illum, L.; Davis, S.S., Eds.; Wright: Bristol, 1987; p 117.
3   Vert, M. In Therapeutic Drug carrier Systems; Bruck, S.D.,Ed.; CRC Press: Boca Raton, 1986; p 291.
4   Vert, M.; Lenz, R.W. Polym. Prep. 1979, 20, 608.
5   Guerin, P.; Vert, M.; Braud, C.; Lenz, R.W. Polym. Bull.1985, 14, 187.
6   Arnold, S.C.; Lenz, C.W. Makromol. Chem., Macromol.Symp. 1986, 6, 285.
7   Braud, C.; Bunel,C.; Vert, M.; Bouffard, Ph.; Clabaut, M.; Delpech,B. Proc. IUPAC 24th Int. Symp. Macromolecules, Amherst, 1982, p 384.
8   Braud, C.; Bunel, C.; Vert, M. Polym. Bull. 1985, 13, 293.
9   Bunel, C.; Vert, M. IUPAC 24th Microsymp., Prague, 1983, 52.
10 Braud, C.; Vert, M. In Polymers as Biomaterials; Shalaby, S.W., Hoffman, A.S.; Ratner, B.D.; Horbett, T.A., Eds.; Plenum Press: New-York, 1984; P 1.
11 Boustta, M. Ph. D; Thesis, Rouen University, France, 1988.
12 Boustta, M.; Huguet, J.; Vert, M. Br. Fr. PV 88-02956.
13 Boesenken, J. Versl. Akad. Wetenshap. 1926, 35, 1084.
14 Eggerer, H., Liebigs Ann. Chem. 1963, 666, 192.
15 Izumiya, N.; Makisumi, S. J. Chem. Soc. Japan, 1957, 78, 662.
16 Adler, A.J.; Fasman, G.D.; Blout, E.R. J. Am. Chem. Soc. 1963, 85, 90.
17 Rodriguez-Galan, A.; Munoz-Guerra, S.; Subirana-Buichong, J.A.; Sekiguchi, H. Makromol. Chem., Macrom. Symp., 1986, 6, 277.
18 Kovacs, J.; Kovacs, H.N.; Konyves, I.; Czaszar, J.; Vajda, T.; Mix, H. J. Org. Chem., 1961, 26, 84.
19 Vlasak, J.; Rypacek, F.; Drobnik, J.; Saudek, V. J. Polym. Sci., Pol. Symp. 1979, 66, 59.

RECEIVED June 4, 1990

Chapter 27

New Polyethylene Glycols for Biomedical Applications

J. Milton Harris[1], Julian M. Dust[1], R. Andrew McGill[1],
Patricia A. Harris[1], Michael J. Edgell[1], Reza M. Sedaghat-Herati[1],
Laurel J. Karr[2], and Donna L. Donnelly[2]

[1]Department of Chemistry, University of Alabama, Huntsville, AL 35899
[2]Space Science Laboratory, NASA–Marshall Space Flight Center,
Huntsville, AL 35812

Polyethylene glycols (PEGs) are neutral, water soluble polymers possessing an impressive array of biomedical and biotechnical applications. In the current article we describe work in our laboratories on synthesis and characterization of new PEG aldehydes. These novel PEGs have the advantages of selective reactivity toward amino groups and stability in an aqueous environment. The reactivity of two new aldehydes toward simple amines, proteins, and aminated surfaces is described. In addition we have examined some recent chemistry on amination of glass surfaces by silanization with different silanes under a variety of conditions. This work shows that monoethoxyaminosilanes, while giving relatively low percentages of nitrogen on the surface, yield more reactive amino groups on the surface than do triethoxy- and diethoxy-silanes.

Polyethylene glycols (PEGs) are neutral, hydrophilic polymers that are also soluble in a variety of organic solvents. Once in aqueous solution they are heavily hydrated, highly mobile and exclude other polymers (including proteins and nucleic acids). As a consequence PEGs are nontoxic, nonimmunogenic, and capable of forming aqueous two-phase systems with a variety of polymers. Molecules coupled to PEG become nontoxic, nonimmunogenic, soluble in water and many organic solvents, and surfaces modified by PEG attachment become hydrophilic and protein rejecting(1). These properties have led to a variety of biotechnical and biomedical applications including: aqueous two-phase partitioning (2-4), protein (i.e., enzyme, antibody, antigen) immobilization (5-9), drug modification (10-13), and preparation of protein rejecting surfaces (14-17). In addition, PEG-coated surfaces can be used to control wetting and electroosmosis (18,19).
    Preparation of new activated PEGs has become central to many studies on PEG applications(1). Typically, useful PEG derivatives contain at

0097–6156/91/0467–0418$06.00/0
© 1991 American Chemical Society

least one electrophilic center available for reaction with nucleophilic centers of biomolecules (e.g., lysine, cysteine and like residues of proteins) or surfaces (e.g., aminated glass). A variety of active PEG derivatives have been prepared, yet there is continued interest in synthesis of new derivatives possessing properties not now available. A property of particular interest is increased selectivity, including relative inertness toward water and high reactivity toward desired functional groups. As an example, PEG tresylate, a much used derivative (1, 20-22), is reactive toward a variety of nucleophiles including water, while PEG aldehydes (described below) are inert toward water and react primarily with amines. Inertness toward water is desired, not only because of efficiency of storage, preparation, and application, but also because it permits stepwise linkage, in aqueous media, of molecules to surfaces.

Another highly desirable property is ease of analysis. The literature of PEG applications is replete with studies in which the PEG derivatives used were either poorly characterized or uncharacterized. While this approach can be successful, lack of analysis can also make it extremely difficult to decipher the problems that are almost invariably encountered in any research project. Presumably, this situation results because polymer derivatives can be difficult to characterize. The active end groups that are generally of interest constitute such a small mass percentage of the polymer that small amounts of impurities can be misleading.

For example, consider the cyanuric-chloride derivative of the monomethyl ether of PEG 5000 (i.e., M-PEG 5000) (22-24). This derivative is frequently analyzed by hydrolysis to produce free chloride which is then measured by titration (22-24). However, a simple calculation shows that only 2% contamination with free cyanuric chloride would lead one to the assumption that the "derivative" is 90% substituted when in fact it is unsubstituted. Obviously, analysis for free cyanuric chloride is required (HPLC analysis is an excellent method in this case). Similarly, it can be difficult to determine the MW of this cyanuric-chloride derivative because the end groups can affect retention times on a size exclusion column that has been calibrated with PEG MW standards. We have found that proton NMR characterization (in deuterated dimethylsulfoxide solvent) is easy to apply and can reveal MW, purity, and percent modification in a single experiment for some derivatives; obviously, this capability is highly desirable. This work is described below.

Work in our laboratories involves: (1) synthesis and characterization of new PEG derivatives, (2) examination of reactivity of new PEGs with proteins and surfaces, (3) surface activation, (4) characterization of modified surfaces, and (5) applications of these PEG derivatives and modified surfaces in biotechnology. In the present manuscript we briefly review the following subjects: (1) proton-NMR characterization of PEG derivatives; (2) synthesis, characterization and reactivity of PEG aldehydes; and (3) amination of glass by silanization (preliminary to surface binding of PEG).

Proton-NMR Characterization of PEG Derivatives

As noted above, it is frequently necessary to apply several analytical techniques to characterize new PEG derivatives. In principal proton-NMR offers the possibility of rapid determination of impurity concentration, percentage end-group conversion, and MW. To illustrate the utility of this method we describe here determination of the percent end-group conversion in the preparation of PEG tosylate and determination of MW of some PEG standards (25).

Proton-NMR characterization of PEGs is complicated by two factors. First, the backbone peaks (from -O-CH$_2$CH$_2$-) are relatively large, and thus they can mask the smaller peaks of interest. For example, the backbone peak for M-PEG 2000 is 180 times larger than the hydroxyl peak and covers almost the entire region between 3 and 4 ppm when it is enlarged enough to permit ready viewing of the other peaks of interest (such as hydroxyl and methoxy). If the smaller peaks fall outside this 3-4 ppm region, however, the spectrum can be used.

A second problem is that the key absorption of the terminal hydroxyl group (-CH$_2$-OH) is generally shifted and broadened by variation in concentration of PEG, water, or impurities. However, we have recently shown that this second problem is eliminated when spectra are determined in deuterated DMSO; in this solvent the hydroxyl group shows up as a sharp triplet at 4.56 ($\pm$ 0.02) ppm (J = 5.2 Hz). Quantification of this resonance permits determination of degree of end group modification and of MW.

To test this NMR approach we prepared a sample of PEG tosylate (PEG-(OTs)$_2$; MW 1500 g/mol) of 97% conversion (based on titration and sulfur analysis) and mixed it with unmodified PEG 1500 to give a series of samples of varying tosylate/hydroxyl ratio. The percent conversion in these synthetic mixtures was then determined by NMR by measuring the area of the hydroxyl peak at 4.56 ppm and comparing it to the area of the tosylate aryl absorbance at 7.84 ppm (this absorbance is the downfield half of the aryl AB quartet). The results of this analysis are presented in Table I. As can be seen from the data, the NMR results are reasonable, with the method generally underestimating the percent substitution.

Table I. Comparison of degrees of substitution determined by hydrolysis-titration and by NMR (eq. 1) for synthetic mixtures of PEG-(OTs)$_2$ and PEG (both 1500 daltons)

Titration, %	NMR, %	% Difference
97.2 ± 0.1	100 ± 0	-2.8 ± 0.1
87.6 ± 0.3	91.0 ± 5.1	-3.4 ± 5.1
78.1 ± 0.3	66.3 ± 3.9	11.8 ± 3.9
67.3 ± 0.2	58.2 ± 4.5	9.1 ± 4.5
56.7 ± 1.9	49.4 ± 6.4	7.3 ± 6.7
48.6 ± 0.2	41.6 ± 2.9	7.0 ± 2.9
38.3 ± 0.7	31.9 ± 1.1	6.4 ± 1.3
30.2 ± 0.2	23.3 ± 0.9	6.9 ± 0.9
21.7 ± 0.7	15.0 ± 2.6	6.7 ± 2.8

As an additional test we have examined the ability of the NMR method to determine PEG number-average MWs by comparing the area of the PEG hydroxyl resonance (at 4.56 ppm) to that for the backbone (-O-CH$_2$CH$_2$-) absorption (at 3.51 ppm) for a series of commercial PEG standards. In this application, more inherent error would be anticipated since the method relies on comparison of the small end-group peak to that for the much larger backbone. The MW results are shown in Table II.

Table II. NMR determination of MW for some PEG MW standards [a]

NMR[b] (g/mol)	Standard[c] (g/mol)
155 (± 21)	150 (1.00)
245 (± 23)	202 (1.03)
592 (± 37)	586 (1.05)
1030 (± 230)	960 (1.07)
1508 (± 258)	1470 (1.05)
4495 (± 532)	4250 (1.03)
6968 (± 1871)	12600 (1.05)

[a]Dried in DMSO-d$_6$ solution with 3Å molecular sieves. [b]Average of three spectral determinations, estimated errors in parentheses (standard deviations). [c]M_w/M_n is listed in the parentheses as given by the suppliers, except in the case of the 150 g/mol sample which is triethylene glycol monomethyl ether.

The limit of practical MW determination of unsubstituted PEGs by this NMR method is reached close to 5000 daltons; at 12,600 daltons large errors appear, Table II. Errors in the integrals, upon which the method rests, arise from overlap of the small -OH peak with the broad shoulder of the large backbone peak and from interference from spinning side bands.

Although the present NMR method is apparently limited to lower MW PEGs, it is suitable for many applications. It is a primary method that is rapid and easy to apply. Most importantly the method can be applied to a large variety of derivatives for which valid MW standards are not available.

The NMR method can also be used to detect impurities. For example, proton NMR spectra for PEG tresylate and tosylate show separate peaks for the corresponding acids and acid chlorides. Similar impurity peaks can be found for reagents used to prepare many PEG derivatives.

Finally, it is important to note that there are derivatives which do not have appropriate proton absorptions that can be used to determine MW and percentage end-group conversion; an example is the cyanuric-chloride derivative. In our own case, we now consider the inapplicably of proton NMR sufficient to rule out utilization of a derivative.

Synthesis. Characterization and Reactivity of PEG Aldehydes

From our viewpoint, the ideal PEG derivative will be: (1) reactive with nucleophilic groups (typically amino) on proteins and surfaces; (2) heterobifunctional so that different chemistries can be performed on the two termini; (3) stable in aqueous media and on the shelf; (4) easily prepared and characterized; and (5) capable of coupling to proteins without reducing protein activity. Many of these properties are potentially possessed by PEG aldehydes. In particular, aldehydes are in equilibrium with their hydrates in water, and will react in this medium with amines to form imines (which can be reduced in situ with sodium cyanoborohydride). The availability of aqueous coupling chemistry has several benefits. For example, surfaces can be modified that either dissolve or soften deleteriously in organic solvents. Secondly, PEG aldehydes can potentially be used for tethering molecules to surfaces in water. To do this, the PEG can be first coupled to the surface, in water, and then reacted, in a second step still in water, to some other molecule such as a protein. More typical activated PEGs can not be used in this fashion since they react with water too rapidly to permit the two-step process. A second potential benefit is that the aldehyde reacts only with amino groups and not with other nucleophilic centers.

    As a consequence, we have actively pursued synthesis of PEG aldehydes. In our first work on these derivatives we prepared PEG acetaldehyde (PEG-O-CH$_2$CHO) (26,27). This compound has proven useful in some cases (15,26), but we experienced difficulty in achieving reproducible results with it. Closer examination showed that the compound is very sensitive to decomposition in base, presumably by aldol condensation (27). If aldol condensation is the decomposition route then more desirable derivatives would be the benzaldehyde derivative and the propionaldehyde derivative. Here we describe the synthesis and application of these derivatives.

PEG Benzaldehyde. We have prepared PEG benzaldehyde (PEG-O-C$_6$H$_4$-CHO, 1) by reacting the alkoxide of 4-hydroxybenzaldehyde with PEG tosylate in toluene. First, the potassium alkoxide is formed from the phenol by addition of potassium methoxide in methanol, and the methanol subsequently evaporated. PEG tosylate is then added and the reaction heated to reflux overnight. Workup required filtration to remove the potassium tosylate salt, concentration of the solution and precipitation by addition to an excess of cold ether. Reprecipitation from dichloromethane by addition of ether did not improve the appearance of the beige product, and proton NMR showed the presence of 4-hydroxybenzaldehyde. This impurity was removed by solution of the product in aqueous sodium carbonate and extraction with dichloromethane.

    This procedure was based on an earlier report of Bayer and coworkers (28) in which an ethanolic solution of 4-hydroxybenzaldehyde, PEG chloride and sodium ethoxide were sealed in a glass tube and heated to 130°C for 15 hours. The toluene procedure avoids use of sealed tubes, and formation of the potassium aryloxide in a separate step has the advantage of quantifying the amount of aryloxide present.

Proton NMR in DMSO shows the usual PEG backbone peak at 3.51 ppm plus resonances at: 9.87 ppm (1H, s, -CHO), 7.86 ppm (2H, d, Ar-H), 7.14 ppm (2H, d, Ar-H), and 4.22 ppm (2H, t, CH_2-O-Ar). Free 4-hydroxybenzaldehyde has corresponding, distinct resonances at 10.8 ppm (Ar-OH), 9.77 ppm, 7.74 ppm, and 6.92 ppm, making detection of impurity straightforward. Interestingly, it is more difficult to detect impurity with carbon-13 NMR, the method of characterization used by Bayer (28). UV-vis spectroscopy was also used to demonstrate the absence of free aldehyde in the polymer product. The UV-vis spectra of PEG-benzaldehyde and 4-hydroxybenzaldehyde are quite different in aqueous base, with the former showing a peak at 274 nm with a shoulder at 320 nm, and the latter showing peaks at 240 and 344 nm and a shoulder at 304 nm. The absence of 4-hydroxybenzaldehyde in our product was confirmed by HPLC analysis on a TSK 2000 size exclusion column. Proton NMR comparison of the PEG-OH peak at 4.56 ppm with the benzaldehyde peak at 9.87 ppm permited quantification of percent modification, which was typically around 80%. Note again that our NMR method typically underestimates the degree of substitution (Table I).

Next we examined the chemistry of PEG-benzaldehyde by reacting the compound with methylamine, aminopropyl glass, and alkaline phosphatase. Reaction with methylamine was done in a saturated solution of methylamine in deuterated DMSO at room temperature and monitored by proton NMR. The imine peak (-CH=N-Me) showed up, as expected, as a singlet at 8.25 ppm. This reaction was rather slow and required almost 12 hours to reach completion (although this imine is stable and requires no reduction to avoid reversal) (27). Thus the reaction with larger amines (such as proteins and aminated surfaces) was expected to be slow at room temperature. While this proved not to be a problem for high temperature reaction with surfaces, it did prove to be a problem with proteins. An indication of the rate-retarding effect of MW is given by the complete reaction in one hour of 4-methoxybenzaldehyde with methylamine in ethanol (27).

Reaction of PEG-benzaldehyde (MW 4000 g/mol) with aminated glass was examined also. The glass used was aminated porous glass (preparation described in next section) with 300 nm (7 m^2/g) or 50 nm (44 m^2/g) pore sizes. The density of amino groups (from titration) was shown to be approximately 0.40 nm^2/amino group. Reaction was conducted at 60°C in aqueous solution and followed by HPLC (using a UV-vis detector). Complete reaction was obtained within 18-24 hours to give a packing density of one PEG molecule per 3.6 nm^2 for 300nm glass and one per 4.9 nm^2 for 50nm glass. This is a densely packed PEG layer, as shown by the fact that free PEG of this MW in solution has an excluded volume comparable to that for a protein of 60,000 g/mol (29).

Finally, to test reactivity of the PEG-benzaldehyde with proteins, we reacted M-PEG-benzaldehyde (monomethyl ether, 5000 g/mol) with rabbit IgG at a variety of pHs, temperatures and concentrations. Slow reaction was observed under the following forcing conditions: pH 9-10, 20 mg aldehyde to 10 mg protein, at 37°C. Milder conditions showed no reaction. Reaction was demonstrated by fluorescamine assay (30) of the number of lysines modified and by the shifting of protein partitioning in an aqueous polymer two-phase system. Fluorescamine assay showed the product to be only 8% modified. Under similar conditions with typical activated PEGs (tresylates, cyanuric chloride, etc.) high degrees of modification approaching 60% would be observed under these conditions. Partitioning was studied in a two phase system consisting of 8% dextran T40 (40,000 g/mol), 6% PEG 8000, 0.15 M NaCl, 0.010 M sodium phosphate, pH 7.1, and 0.02% sodium azide. In this system, unmodified protein partitioned about 90% to the bottom, dextran-rich phase, and the PEG-modified protein partitioned 23-28% to the top phase.

To summarize this section, PEG-benzaldehyde can be prepared in high yield and readily analyzed to show percent conversion and absence of impurity. The compound is relatively unreactive, stable in water, reacts efficiently with aminated glass, and reacts only slightly with protein IgG. The compound would therefore appear to be of little utility for protein modification, unless a low degree of modification is desired.

PEG Propionaldehyde. To obtain a more reactive PEG aldehyde, which is resistant to aldol decomposition, we have prepared PEG-propionaldehyde (PEG-CH$_2$CH$_2$CHO, 2). Preparation of this compound proved rather difficult, with several promising routes proving to be failures; for example, reaction of the diethyl acetal of 3-chloropropionaldehyde with PEG alkoxide gave some of the desired product, but primarily gave elimination. An effective route is as follows (in DMSO):

HS-CH$_2$CH$_2$CH$_2$-SH + Cl-CH$_2$CH$_2$CH(OEt)$_2$ ---------->

$$\text{1. NaOMe/MeOH/PEG-OTs}$$
> HS-CH$_2$CH$_2$CH$_2$-S-CH$_2$CH$_2$CH(OEt)$_2$------------------------------------->
$$\text{2. acid}$$

> PEG-S-CH$_2$CH$_2$CH$_2$-S-CH$_2$CH$_2$CHO

Proton NMR is again useful for identifying compound 2. The intermediate acetal gives the following spectrum (in DMSO): 1.11 ppm, 6H, t, methyls; 1.77 ppm, 4H, m, -CH$_2$-CH$_2$-CH$_2$-; 2.54 ppm, m, 6H, -CH$_2$S-; 3.50 ppm, -O-CH$_2$-; 4.55 ppm, t, 1H, acetal CH. And the final product gives the spectrum (in DMSO): 1.78 ppm, t, 2H, -CH$_2$-CH$_2$-CH$_2$-; 2.54 ppm, 8H, m, CH$_2$S, 2.73 ppm, t, 2H, CH$_2$-CHO; 3.51 ppm, m, -O-CH$_2$-, backbone, 9.64 ppm, 1H, s, -CHO.

In sharp contrast to PEG-acetaldehyde, compound 2 is stable in the presence of mild base, and, in contrast to PEG-benzaldehyde, compound 2 is highly reactive in forming the imine from methyl amine. Reaction of 2 with methylamine in DMSO at room temperature is complete (as shown by proton NMR) within 5 minutes. Imine formation is reversible in this case, and sodium cyanoborohydride should be added to give reductive amination.

Compound 2 appears to be ideal for protein modification. Reductive amination proceeds rapidly under a variety of conditions. In a typical preparation we react about five mg of 2 for every mg of protein, along with a slight excess of cyanoborohydride, for one hour at room temperature in pH 9 borate buffer, followed by dialysis against phosphate-bufferred saline; high pH is utilized to avoid reduction of aldehyde. Under these conditions we find that 30-40% of the available lysines are modified, and we find that partitioning of the modified protein in a two-phase system is dramatically shifted. Two examples follow. The antibody against alkaline phosphatase (anti-alkaline phosphatase) is 38% modified under these conditions, and the modified protein partitions (in a system consisting of 8% dextran T-40, 6% PEG 8000, 0.15 M NaCl, 0.010M sodium phosphate, pH 7.2) 62% to the top phase; unmodified protein partitions 90% to the bottom phase. Similarly, the antibody against human red blood cells (anti-human RBCs) is 28% modified under these conditions, and the modified protein shifts RBC partition (in a system consisting of 4.55% dextran T-500, 3.86% PEG 8000, 0.15 M NaCl, 0.010 M sodium phosphate, pH 7.2) 43% to the top phase; free RBCs partition to the interface.

So in summary, PEG-propionaldehyde **2** is much more reactive toward amines than PEG-benzaldehyde **1**, stable in water, stable to base, and reactive with lysine groups of proteins (i.e., reductive amination). As such it should constitute a valuable new PEG derivative. We are currently examining use of this compound for reaction with surface-bound amines.

Silanization of Silica

As mentioned in the introduction, surfaces modified with PEG (and related polymers) have a variety of important applications. To prepare these surfaces, it is essential as a first step that one be able to "activate" the surface by introducing covalently-bound functional groups onto it. Here we examine some recent work from our laboratories on amination of silica by silanization with aminopropylethoxysilanes.

Silanization of silica is an important procedure for attaching groups to the silica surface (31,32) and it is one which many workers, including ourselves (7,18,33), have used for covalently linking amino groups to a variety of silicas including porous glass, quartz capillaries, optical fibers, and quartz slides. This surface amination procedure is frequently performed with 3-aminopropyltriethoxysilane (TES) by a variety of routes (including reaction in gas, aqueous, and organic phases) (31,32). Despite the great deal of activity in this area, problems remain with the silanization procedure (34). In our own work we have occasionally observed irregular surface coverage with this reagent. As has been pointed out (31,32,34), the di- or tri-functional silanes are sterically restricted as regards bonding with three silanols, they leave residual hydroxyl groups, and they are susceptible to polymerization reactions that can produce thick multilayers. Possibly the problems we have encountered could derive from removal of thick, polymerized patches during the exhaustive cleaning procedure to which we submit our samples. Kirkland et al. have recently shown that monoethoxy-silanes, applied in anhydrous hydrocarbons, give nonpolymerized monolayers that are resistant to acid degradation (34), and Jonsson et al. have shown that gas-phase silanization with TES produces monolayers not found with other application methods (35).

Since this amination procedure is so important to our work with polymer-modified glass, we have utilized X-ray photoelectron spectroscopy

(XPS or ESCA) to compare the surfaces produced by glass silanization with mono-, di-, and tri-alkoxy aminopropylsilanes under a variety of conditions. Following these initial experiments we examined the chemical availability of surface amino groups for the aminated silicas by following the extent of imine formation upon reaction with benzaldehyde. The silanes used in these experiments were: 3-aminopropyltriethoxysilane (TES - for triethoxy-silane), 3-aminopropylmethyldiethoxysilane (DES - for diethoxysilane), and 3-aminopropyldimethylethoxysilane (MES - for monoethoxysilane).

The glass used in the XPS experiments consisted of 1 cm^2 quartz slides and that used in the reactivity experiments was porous glass with 300 nm^2 pore size and a surface area of 7 m^2/g. The glass was cleaned by boiling in 30% aqueous hydrogen peroxide and then washed thoroughly with water and dilute hydrochloric acid. After silanization the slides were washed in turn with toluene, methylene chloride, ethanol, acetone, and water. The glass was then refluxed with acetone and methylene chloride and examined by GCMS to insure that no silane was present in the washes. Finally the treated glass was dried at 100°C under vacuum and stored in a vacuum desiccator until ready for use.

The clean quartz slides were subjected to amination conditions shown in Table III and then examined by XPS (using a Perkin Elmer 5400 system). XPS analysis provides an elemental analysis of the surface (approximately to a depth of 5 nm). We have found that it is most illustrative in this case to focus on the amount of nitrogen on the various surfaces, and we have given these data in Table III.

Table III. Percentage nitrogen (± 0.1%) on quartz slides (1 cm^2) as determined by XPS (1 mm diameter spot size) as a function of reaction conditions and silanating reagent

		%N (TES)		%N (DES)		%N (MES)	
medium	temp(°C)	dry	wet	dry	wet	dry	wet
toluene	111	1.2	3.2	-	-	-	-
m-xylene	128	1.9	4.1	-	0.9	-	-
mesitylene	163	4.2	5.3	-	0.9	0.7	0.9
water	100	-	1.3	-	<0.5	-	-
water(cure150°C)	100	-	1.4	-	-	-	0.6
gas phase	110	1.2	-	-	-	-	-
gas phase	160	1.7	-	-	-	-	-

It is clear from Table III that coatings derived from TES have a higher percentage surface nitrogen. While it might appear that TES is therefore producing better surfaces with more available amino groups for immobilization, we believe this is not the case but rather that polymerization is occurring in many cases with TES. This polymerization results in buried amino groups which are unavailable for reaction but lead to high apparent percentages of surface nitrogen. This conclusion is supported by earlier experiments from other groups (34,35) and it is supported by the reactivity analysis, discussed below, showing that TES coatings in fact do contain fewer available, reactive amino groups. Additionally, we have performed XPS measurements at several spots on the slides of Table III (0.6 mm diameter spots, 1.5 mm apart), and we find that the TES coatings from wet solutions are highly variable over the surface, while the MES slides from all conditions are uniform. An example is given in Table IV.

Table IV. XPS results from sampling of five spots (0.6 mm diamter) on TES-coated and MES-coated quartz slides prepared in refluxing dry or wet (saturated) xylene or mesitylene

spot#	TES-coated % N(dry xylene)	TES-coated %N (wetxylene)	MES-coated %N(wet mesitylene)
1	1.5	3.0	1.0
2	1.8	4.4	0.8
3	1.5	1.8	0.7
4	1.1	2.2	0.7
5	1.1	trace	0.9

As shown in Table IV, the nitrogen percentage from the TES-coated slide obtained from wet xylene is highly variable, indicating varying degrees of polymerization and "patchiness" on the surface. These experiments indicate to us that nitrogen percentages greater than 2.0% (as determined by XPS) result from polymerization. Note also from Table III that the TES surface from dry mesitylene also gives a high percentage of nitrogen. We feel this shows that polymerization can also be achieved in supposedly anhydrous solvents (presumably containing very small amounts of water) if the reaction temperature is high enough. Earlier gas phase work by Jonsson showed that TES applied in the gas phase gives monolayers (35), and this is confirmed by a low %N (<2%) as determined by XPS.

Thus to summarize the XPS experiments, we find low nitrogen percentages, consistent with monolayers for MES and DES and also for TES when TES is applied in very dry organics, water, or gas phase. Also, examination of varying locations on the surfaces shows that there is a correlation between high nitrogen percentage (above 2%) and variability of % nitrogen with location on the surface.

These initial results were further elucidated by determining the availability of reactive amino groups on the surfaces. This was done by measuring benzaldehyde depletion from aqueous suspensions of aminated glass by monitoring the benzaldehyde UV absorbance at 250 nm. Invariably we found the following order of benzaldehyde uptake for aminated glasses prepared under similar conditions but with the different silanes: MES-glass > DES-glass > TES-glass. For example, three samples of aminated porous glass were prepared by treatment with aqueous solutions of MES, DES, and TES for 12 hours at 100°C to give surfaces with approximately 1% surface nitrogen (see Table III). These glasses were then suspended in an aqueous, buffered solution (pH 10) of benzaldehyde sufficiently dilute so that reaction of the benzaldehyde produced a significant change in UV absorbance, and the absorbance monitored until it leveled out and remained stable. Normalization of the results (i.e., absorbance change / number of grams of glass) gave the following relative values of absorbance change per gram of glass: MES/DES/TES = 13.6/0.9/0.5. Note that these are relative values and have no absolute meaning outside this comparative experiment.

From this depletion experiment it appears that MES produces significantly more available amino groups even when the total amount of nitrogen on the surface is approximately the same. From this result we conclude that the di- and tri-alkoxy silanes are always giving some polymerization, which somehow restricts access to the amino groups or which leads to unstable linkages at this pH.

Finally, the aminated glass prepared from treatment with MES was reacted with an excess of benzaldehyde (i.e., too concentrated to follow with UV) in pH 9.2 borate buffer and the decrease in concentration measured by HPLC. From the amount of benzaldehyde depleted from solution we calculate that there was one available amino group per 0.40 nm^2 on the surface. This number compares favorably with a reported value of 0.20 nm^2 per silanol group on the surface of hydrated quartz (34).

Acknowledgment

The authors gratefully acknowledge the financial support of this work by The National Aeronautics and Space Administration and The National Institutes of Health (NIH GMS-40111-02).

Literature Cited

1. J.M. Harris, J. Macromol. Sci.-Rev. Macromol. Chem. Phys., C25, 325-373 (1985).
2. P. A. Albertsson, "Partition of Cell Particles and Macromolecules," Wiley Interscience, New York, 3rd ed., 1986.
3. H. Walter, D. E. Brooks, and D. Fisher (Editors), "Partitioning in Aqueous Two-Phase Systems," Academic Press, Orlando, FL, 1985.
4. D. Fisher and I. A. Sutherland (Editors), "Separations Using Aqueous Phase Systems: Applications in Cell Biology and Biotechnology", Plenum, London, 1989.
5. N. B. Graham in "Hydrogels in Medicine and Pharmacy," Vol. II, N. A. Pappas, ed., CRC Press; Boca Raton, FL, 1988, p. 95.
6. S. Zalipsky, F. Alericia and G. Barany, in "Peptides: Structure and Function," 9th American Peptide Symposium, C. M. Deber, V. J. Hruby and K. D. Kopple, 1985, p. 257.
7. K. Yoshinaga and J. M. Harris, J. Bioact. Comp. Polym., 1, 17-24 (1989).
8. H. A. Jacobs, T. Okano, and S. W. Kim, J. Biomed. Mat. Res., 23, 611 (1989).
9. M.-B. Stark and J. K. Holmberg, Biotech. Bioeng., 34, 942 (1989)..
10. S. Zalipsky, C. Gilon, and A. Zilkha, Eur. Polym. J., 19, 1177 (1983).

11. E. G. MacEwen, R. Rosenthal, R. Matus, A. T. Viau and A. Abuchowski, Cancer, **59**, 2011 (1987).
12. N. V. Katre, M. J. Knauf, and W. J. Laird, Proc. Natl. Acad. Sci. USA, **84**, 1487 (1987).
13. T. Suzuki, N. Kanbara, T. Tomono, N. Hayashi, and I. Shinohara, Biochim. Biophys. Acta, **788**, 248 (1984).
14. J. D. Andrade, S. Nagaoka, S. Cooper, T. Okano, S. W. Kim, J. Am. Soc. Artif. Intern. Organs, **10**, 75 (1987).
15. C.-G. Golander and E. Kiss, J. Colloid Interface Sci., **121**, 240 (1988).
16. E. W. Merrill, E. W. Salzman, J. Am. Soc. Artif. Intern. Organs, **6**, 60 (1983).
17. W. R. Gombotz, W. Guanghui, and A. S. Hoffman, J. Apply. Polym. Sci., **37**, 91 (1989).
18. B. J. Herren, S. G. Shafer, J. M. Van Alstine, J. M. Harris, and R. S. Snyder, J. Colloid Interface Sci., **51**, 46-55 (1987).
19. J. M. Harris, D. E. Brooks, J. F. Boyce, R. S. Snyder, and J. M. Van Alstine, in "Dynamic Aspects of Polymer Surfaces," J. D. Andrade, Ed., Plenum, 1988, pp 111-119.
20. K. Nilsson and K. Mosbach, Methods in Enzymology, **104**, 56 (1984).
21. C. Delgado, G. E. Francis, and D. Fisher, in "Separations Using Aqueous Phase Systems," D. Fisher and I. A. Sutherland, Eds., Plenum, London, 1989, pp. 211-213.
22. K. Yoshinaga, S. G. Shafer, and J. M. Harris, J. Bioact. Compatible Polym., **2**, 49-56 (1987).
23. S. G. Shafer and J. M. Harris, J. Polym. Sci. Polym. Chem. Edn., **24**, 375-378 (1986).
24. A. Abuchowski, T. van Es, N. C. Palczuk and F. F. Davis, J. Biol. Chem., **252**, 3578 (1977).
25. A detailed description of this method is given in: J. M. Dust, Z. Fang, and J. M. Harris, Macromolecules, 0000 (1990).
26. J.M. Harris, M. Yalpani, J.M. Van Alstine, E.C. Struck, M.G. Case, M.S. Paley, and D.E. Brooks, J. Poly. Sci., Poly. Chem. Ed., **22**, 341 (1984).
27. M. S. Paley and J. M. Harris, J. Polym. Sci. Polym. Chem. Edn., **25**, 2447-2454 (1987).
28. E. Bayer, H. Zheng, K. Albert and K. Geckeler, Polym. Bull., **10**, 102 (1983).
29. D. Knoll and J. Hermans, J. Biol. Chem., **258**, 5710 (1983).
30. S. J. Stocks, A. J. Jones, C. W. Ramey and D. E. Brooks, Anal. Biochem., **154**, 232 (1986).
31. "Silanes, Surfaces, and Interfaces," D. E. Leyden, Ed., Gordon and Breach, New York, 1986.
32. E. P. Plueddemann, "Silane Coupling Agents," Plenum, New York, 1982.
33. S. M. Klainer and J. M. Harris, SPIE Proceedings, **906**, 139-147 (1988).
34. J. J. Kirkland, J. L. Glajch, and R. D. Farlee, Anal. Chem., **61**, 2 (1989).
35. U. Jonsson, G. Olofsson, M. Malmqvist, and I. Ronnberg, Thin Solid Films, **124**, 117 (1985).

RECEIVED July 2, 1990

Chapter 28

Biomedical Applications of High-Purity Chitosan

Physical, Chemical, and Bioactive Properties

Paul A. Sandford[1] and Arild Steinnes[2]

[1]Protan, Inc., Bio Applications, P.O. Box 1632, Woodinville, WA 98072
[2]Protan A/S, Bio Applications, P.O. Box 420, N–3002, Drammen, Norway

Since chitosan, partially deacetylated
chitin, is a natural biopolymer composed
primarily of 2-amino-2-deoxy-D-glucose
(D-glucosamine) and 2-acetamido-2-deoxy-D-
glucose (N-acetyl-D-glucosamine), two
common constituents of the body, and since
chitosan has unique solution, chemical,
physical, and biological properties,
chitosan is being used in a wide variety
of biomedical applications. Many
applications use chitosan's ability to form
films, fibers and gels. Chitosan also has
lubricating and viscosifying properties.
Many biomedical applications (eg. hemostatic
agent, wound dressings, bacteriostatic agent,
contact lens, anticholesteremic agents,
spermicide) are based on chitosan's bioactive
properties, in particular its ability to
bind and activate mammalian and microbial
cells. Chitosan in liquid, bead, film,
fiber and powder forms will be discussed.

Structure of Chitin and Chitosan

Chitosan, the partially deacetylated form of chitin, is a
natural biopolymer composed of the two common sugars,
D-glucosamine and N-acetyl-D-glucosamine, both of which
are constituents of the body (1-10).
     In Figure 1, note that chitin and chitosan, like
cellulose, are beta 1-->4 linked biopolymers and
traditionally commercial products have been composed of
~80% D-glucosamine and 20% N-acetyl-D-glucosamine.
Chitin, next to cellulose, is nature's second most
plentiful natural biopolymer.

0097–6156/91/0467–0430$06.00/0
© 1991 American Chemical Society

Commercial Sources

Protan, currently the only US manufacturer of chitin or chitosan, extracts chitin from the outer shell of the crustaceans, shrimp (Pandulus borealis) and crab (Cancer magister) (11-16) in its new facilities in Raymond, Washington. Figures 2 and 3 illustrate the basic steps in the isolation of chitin and the production of chitosan.

PRONOVA chitosan is the name of Protan's high purity chitosan. SEACURE is a high quality chitosan used for many cosmetic applications (3). PROFLOC chitosan is used primarily as a cationic flocculent in waste water clarification applications (17).

Prior to Protan's recent installation of a large scale purification facility, chitosan for most medical and cosmetic applications had to be purified by the user. The vast interest in chitosan for medical, cosmetic and bioreactor applications has led to Protan's commercial development of high purity grades of chitosan (18).

Protan has just brought on stream a large production facility where many natural biopolymers including chitosan, alginate, hyaluronic acid, agarose., and laminarin, are being purified on a large scale (19,20). However, only purified chitosan will be discussed here.

Chitosan Purification Scheme

Currently, the main purification step starts with making a solution of "regular chitosan" in an organic acid and removing insolubles, protein, low molecular weight materials, pyrogens and toxic metals by filtration and other proprietary methods (Figure 4). The goal is to provide a commercial product of highest purity that can be readily "filter sterilized". Extensive analyses are performed to guarantee the material is suitable for biomedical applications. Figure 5 is a typical "Certificate of Analysis" for a purified chitosan glutamate.

Chitosan's Key Properties Relating to Biomedical Applications

Cationic properties--Chitosan is a linear polymer of mainly anhydro glucosamine (see Figure 1). At pHs below ~6.5 chitosan in solution it carries a positive (+) charge. Many of chitosans uses depend on its "cationic nature" (Table I).

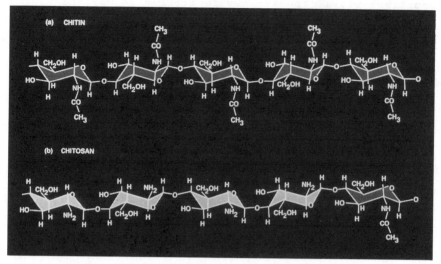

Figure 1.   Structure of chitin and chitosan.

SHRIMP/CRAB --------> SHRIMP/CRAB SHELLS --------> CHITIN

       1) Cook   1) Grind
       2) Peel   2) Deproteinize
       3) Can/freeze meat   3) Demineralize
                                   4) Rinse
                                   5) Dry

Figure 2.   Chitin extraction process.

CHITIN -------> CHITOSAN ----------> CHITOSAN
            (free amine)   (ammonium salt)

1) NaOH   1) Dissolve in acid
2) Rinse   2) Filter
3) Dry   3) Spray/freeze dry

Figure 3.   Chitosan manufacturing scheme.

```
Chitosan ------>Bioreactor ----> In vitro ---->In vivo
                grade             grade         grade

* filtration              * remove low molecular weight
                            materials
* remove pyrogens

* remove protein          * remove toxic metals
```

Figure 4.   Key parameters in purifying chitosan.

A N A L Y T I C A L   C E R T I F I C A T E

PRODUCT NAME: PRONOVA BLV CHITOSAN
                                 GLUTAMATE

PRODUCT CODE: 18060101

LOT NUMBER   808-572-01

COMMON NAME: CHITOSAN GLUTAMATE,
                                 LOW VISCOSITY

PURITY: BIORERACTOR GRADE
                                   -THROUGH 5 MICRON FILTER
                                   -SPRAY DRIED

VISCOSITY: 440 CPS, 2% SOLUTION

PH: 5.0, 2% SOLUTION

LOSS ON DRYING: 5.2%

PROTEIN CONTENT: 0.16%

ELEMENTAL ANALYSIS:

```
            Ca          75 ppm
            Mg           4  "
            Cu           5  "
            Zn          20  "
            Si           2  "
            Sr           3  "
            Fe          80  "
            P        <0.01  "
            Mn       <0.01  "
            Pb           1  "
            As          26 ppb
            Hg           9  "
```

Protan A/S, Biopolymer Division, Drammen
June 1, 1989

Figure 5.   Analytical Certificate for PRONOVA
chitosan glutamate.

Table I.   Chitosan's Cationic Properties
----

* Linear polyelectrolyte
* High charge (+) density
* Adheres to negatively charged materials
  -excellent flocculent
  -binds to many cell surfaces
  -substantive to hair and skin
  -interacts with polyanions (eg. alginate, DNA,
  lipopolysaccharide)
  -forms ionotropic gels with polyanions (eg.
  polyphosphate, alginate)
  -forms membranes with negatively charged polymers

Since most living tissues have a "negative (-)
charge", chitosan being positively charged is attracted
to most tissues, skin, bone and hair. The outer surfaces
of most microbes are also negatively charged. Thus,
binding of chitosan to living cells of all types is a
very important property in the use of chitosan in
biomedical applications.

Being a linear polyelectrolyte, chitosan has both
reactive amino groups and hydroxyl groups that can be
used to chemically alter its physical and solution
properties (Table II). There are many interesting
chitosan derivatives (4-10) but these will not discussed
here.

Table II. Chitosan's Chemical Properties
----

* Linear polyamine (poly-D-glucosamine)
* Reactive amino groups (-NH^2) available
* Reactive hydroxyl groups (-C^{3}OH, -C^{6}OH) available
* Chelates many transitional metal ions

Biomedical applications of chitosan are also
dependent on its many and useful biological properties
(Table III). First, it is a natural and biodegradable
polymer that is biocompatible with most living systems.
This biocompatibility is now being exploited with the
introduction of sophisticated methods to purify chitosan
to very high qualities. Thus, by removing toxic and
contaminating bioburden materials such as protein, heavy
metals, and pyrogens, biocompatibility and safety are
assured.

Chitosan's key biological properties are quite
extensive with many major medical companies pursuing the
safety and efficacy of chitosan in a variety of forms.
From a biological standpoint, the major reasons chitosan
is being researched are shown in Table III.

Table III. Chitosan's Biological Properties (21,22)

----

* Biocompatible
  -natural polymer
  -biodegradable to normal body constituents
   (glucosamine, N-acetylglucosamine)
  -safe, non-toxic
* Binds to mammalian and microbial cells aggressively
* Regenerative effect on connective gum tissue
* Accelerates the formation of osteoblasts
  responsible for bone formation
* Hemostatic
* Bacteriostatic
* Fungistatic
* Spermicidal
* Antitumor
* Anticholesteremic
* Accelerates bone formation
* Central nervous system depressant
* Immunoadjuvant

Chitosan's Many Useful Forms

Another important factor why chitosan is used in such a
large variety of biomedical applications is the large
variety of useful physical forms (Table IV).

Powder forms--Chitosan, both in the "free amine" and
"ammonium salt" forms, can be used directly, but usually
they are dissolved and used commercially as aqueous
solutions. Chitosan is available in laboratory to bulk
quantities.

Aqueous forms of chitosan--Solutions of ultrapure
chitosan (Figure 6) form crystal clear solutions that can
be used "as is" or as the starting point for making
films, fibers, pastes, salves, gels or beads. Solutions,
depending on the molecular weight of the chitosan and its
salt form, can have viscosities ranging from that of a
gel to "watery" solutions.

Chitosan films--Chitosan films can be made readily by
casting a solution of chitosan on a flat surface and
merely allowing it to air dry into a tough, flexible film
(Figure 7). Chitosan films in both the "ammonium salt"
and "free amine" forms can be made. At least two

Figure 6.   Ultrapure chitosan powder forms
crystal clear solutions.

Figure 7.   Chitosan readily forms tough, flexible
films by merely air drying aqueous solutions.

companies are exploring the use of films for contact lens
materials and for wound dressings.

Table IV.   Physical Forms of Chitin/Chitosan in
           Biomedical Applications

Form of Material	Application (7,21,22)
Solution/gel	Bacteriostatic agent Fungistatic agent Hemostatic agent Cosmetics Flocculating agent Coating agent
Gel/paste	Drug delivery vehicle Spermicide Immobilization/encapsulation agent
Beads	Immobilizing enzymes/cells Protein separations
Film/Membrane	Dialysis membrane Contact lens Wound dressing Encapsulation of mammalian cells
Sponge	Mucosal hemostatic agent Wound dressings
Fiber	Suture Wound dressings Membrane Drug delivery (hollow fiber)

[This table is Adapted from Table 1, R. Olsen et. al.,
Biomedical Appl. of Chitin and Its Derivatives, 4th
Inter. Conf. Chitin and Chitosan, 1988]

Chitosan shaped objects--Chitosan, being cationic (+
charged) can interact with negatively (-) charged
polymers such as alginate or hyaluronate. Chitosan will
react with alginate to form a chitosan-alginate membrane
suitable for encapsulation live cells (23) and to form

capsules (24). If chitosan is precipitated under mild alkaline conditions, gels and shaped objects can be made (25).
        Chitosan beads can be made easily in a variety of porosities and sizes, and types (cross linked). Currently, Fuji Spinning and Protan are introducing chitosan beads suitable for drug delivery, enzyme and cell immobilization (26,27) (Figure 8). Figure 9 is SEM picture of a mammalian cell culture (L-929) adhering to a chitosan bead as a microcarrier.

        Figure 10 is another illustration of using chitosan to immobilize enzymes. In this case, glucose isomerase was immobilized on chitosan and then the beads were packed into columns for making high fructose corn syrup.

        Chitosan fibers--Chitosan can be made into strong flexible fibers that are biodegradable by spinning from aqueous solutions (4,28). Chitosan forms beautiful silk-like fibers (Figures 11 and 12). Chitosan fibers are tough enough to be knitted into pure chitosan cloth (Figure 13). There is great potential in chitosan fibers as wound dressings and sutures.

        Chitosan sponge--By freeze drying chitosan solutions, a spongy chitosan sheet can be obtained (Figure 14).

        Chitosan/Chitin paper--Chitin or chitosan fibers can be made into a non-woven matrix. Squid pen chitin can be made into a paper like material (29) (Figure 15).

Chitosan as an anticholesteremic agent

Chitosan, when ingested by rats, is as effective as cholestyramine, a drug currently on the market, in lowering cholesterol and lipid levels (30). Chitosan, besides having a cholesterol lowering effect, it has many useful physical properties that are useful in the delivery of drugs (19).

Conclusion

The physical, mechanical, chemical, bioactive properties, and commercial availability of chitosan has made chitosan a very interesting biomaterial and the leading candidate in many biomedical applications.

Figure 8.   Fuji Spinning chitosan beads have many
uses (eg. drug delivery, column chromatography,
enzyme/cell immobilization).

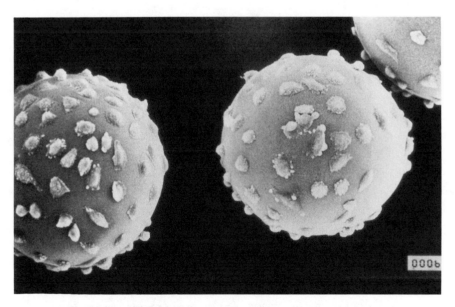

Figure 9.   Scanning electron micrograph of
mammalian cells (L-929) immobilized on chitosan
bead as microcarrier.

Figure 10.   Chitosan beads on which glucose
isomerase has been immobilized (TAKA-SWEET, Miles
Laboratories) for the production of high-fructose
corn syrups.

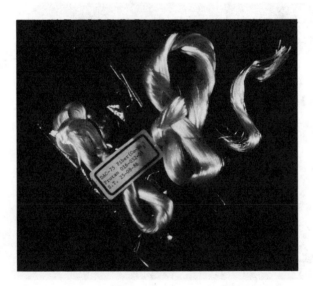

Figure 11.   Pure, 100% chitosan fibers made from
"crab shell chitosan" (courtesy of Prof.   S.
Tokura, Hokkaido University, Japan) have a
silk-like appearance.

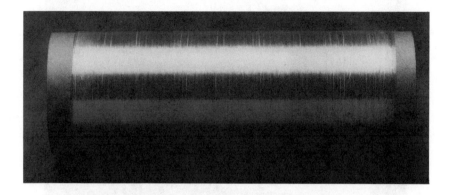

Figure 12.   Pure, 100% chitosan fibers made by
Fuji Spinning, Japan, have potential as
biodegradable biomaterial in biomedical
applications.

Figure 13.   100% "Chitosan fabric" made by
weaving chitosan fibers.

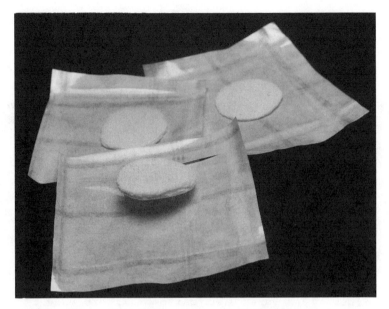

Figure 14. Chitin sponge, made by freeze drying
squid pen chitin dispersions, may have
application as a biomaterial (courtesy of Nippon
Suisan Co., Ltd., Tokyo, Japan).

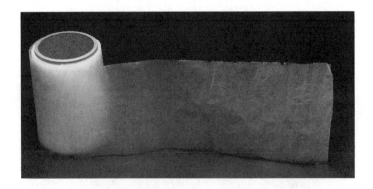

Figure 15. Chitin paper made from squid pen
chitin (courtesy of Prof. S. Tokura, Hokkaido
University, Japan).

Literature Cited

1. Braconnot, H. A. Chi. Phys. 1811, 79, 265-304.

2. Odier, A. Mem. Soc. Hist. Nat. Paris, 1823, 29-42.

3. Sandford, P. A. In Chitin and Chitosan: Sources,
    Chemistry, Biochemistry, Physical Properties and
    Applications; Skjak-Braek; Anthonsen, T.; Sandford,
    P. A. Eds.; Elsevier Applied Science; New York,
    1989.

4. Proceedings of the First International Conference on
    Chitin/Chitosan; Muzzarelli, R. A. A.; Pariser, E.
    R., Eds.; MIT Sea Grant Report MITSG 78-7: Cambridge,
    MA, 1978.

5. Chitin and Chitosan: Proceedings of the Second
    Conference on Chitin and Chitosan; Hirano, S.;
    Tokura, S., Eds.; The Japanese Society of Chitin and
    Chitosan: Tottori 680, Japan, 1982.

6. Chitin in Nature and Technology: Proceedings of the
    Third International Conference on Chitin and
    Chitosan; Muzzarelli, R.; Jeuniaux, C.; Gooday, W.
    G., Eds.; Pleum Press: New York, 1985.

7. Chitin and Chitosan: Sources, Chemistry,
    Biochemistry, Physical Properties and Applications:
    Proceedings from the 4th International Conference on
    Chitin and Chitosan; Skjak-Braek, G.; Anthonsen, T.;
    Sandford, P., Eds.; Elsevier Applied Science: New
    York, 1989.

8. Muzzarelli, R. A. A. Chitin; Pergamon Press: New
    York, 1977.

9. Chitin, Chitosan, and Related Enzymes; Zikakis, J. P.,
    Ed.; Academic Press, San Diego, CA, 1984.

10. Pariser, E. R.; Lombardi, D. P. Chitin Sourcebook;
    A Guide to the Research Literature; John Wile &
    Sons: New York, 1989.

11. Peniston, Q. P.; Johnson, E. L.   U.S. Patent 3 533
    940, 1970.

12. Peniston, Q. P.; Johnson, E. L. U.S. Patent 3 862
    122, 1975.

13. Peniston, Q. P.; Johnson, E. L. U.S. Patent 4 066
    735, 1978.

14. Peniston, Q. P.; Johnson, E. L. U.S. Patent 4 159
    932, 1979.

15. Peniston, Q. P.; Johnson, E. L. U.S. Patent 4 195
     175, 1980.

16. Peniston, Q. P.; Johnson, E. L. U.S. Patent 4 199
     496, 1980.

17. PROFLOC--Natural Cationic Polymer for Recovering
     Valuable By-Products from Food Process Wastes;
     Protan, Inc., Commack, New York, 1987.

18. Purified Chitosan-PROTASAN; Protan, Inc.: Commack,
     New York, 1988.

19. Sandford, P. A., Proc. of the 16th Inter. Symp. on
     Controlled Release of Bioactive Materials, 1989, p
     454.

20. PRONOVA-The New Generation of Polysaccharides;
     Protan, Inc., Commack, New York, 1989.

21. Olsen, R.; Schwartzmiller, D.; Weppner, W.; Winandy,
     R. In Chitin and Chitosan: Sources, Chemistry,
     Biochemistry, Physical Properties and Applications;
     Skjak-Braek, G.; Anthonsen, T.; Sandford, P. A.
     Eds.; Elsevier Applied Science: New York, 1989.

22. Chitin and Chitosan: Specialty Biopolymers for
     Foods, Medicine, and Industry; Technical Insights,
     Inc., Englewood, New Jersey, 1989.

23. McKnight, C. A.; Ku, A.; Goosen, M. F. A.; Sun, D.;
     Penny, C. J. Bioact. Biocompat.Polymer 1988, 3,
     334-355.

24. Daly, M. M.; Keown, R. W.; Knorr, D. W. U. S. Patent
     4 808 707, 1989.

25. Chitosan for Cell Immobilization; Protan Inc.:
     Commack, New York, 1987.

26. FUJIBO Chitosan Beads; Fuji Spinning Co. Ltd.,
     Tokyo, Japan, 1987.

27. Chitopearl: Enzyme Immobilizing Carrier/Protein
     Adsorbing Agent; Fuji Spinning Co. Ltd., Tokyo,
     1988.

28. Tokura, S.;Nishi, N.; Nishimura, S. Sen-I Gakkaishi
     1983, 12, T-507-511.

29.   Takai, M.; Shimizu, Y.; Hayashi, J.; Tokura, S.;
      Uraki, Y. Ogawa, M. 1990, Chem. Abstr. Cell-41,
      199th Am. Chem. Soc. Nat. Meet., April
      22-27, Boston, MA.

30.   Jennings, C. D.; Boleyn, K.; Bridges, S. R.; Wood,
      P. J.; Anderson, J. W.; Proc. of the Soc. for Exp.
      Biol. and Med. 1988, 189, 13-20.

RECEIVED July 2, 1990

Chapter 29

Viscosity Behavior and Oil Recovery Properties of Interacting Polymers

John K. Borchardt[1]

Research Center, Halliburton Services, P.O. Drawer 1431,
Duncan, OK 73533

Certain combinations of polymers have been found which
exhibit substantially increased solution viscosity as
compared to solutions of the individual polymers at the
same total polymer concentration. Such combinations
include poly(styrene sulfonate) and either xanthan gum
or hydroxyethyl cellulose, poly(vinyl sulfonate) and
xanthan gum, a quaternary ammonium salt modified guar
and either hydroxypropyl guar or hydroxyethyl cellulose,
and a sulfonated guar and either hydroxyethyl cellulose
or carboxymethylhydroxyethyl cellulose. Another polymer
system of interest is a blend of $\leq 5\%$ hydrolyzed polyacrylamide and a quaternary ammonium salt organic polymer (QASP). The QASP functions as a salt reducing
polyacrylamide solution viscosity. As the QASP adsorbs
on silicaceous mineral surfaces, solution viscosity
increases. The result of solution viscosity and oil
recovery experiments are reported.

The ability of small concentrations of hydrophobically modified
water-soluble polymers ($\underline{1}$) such as alkylhydroxyethyl celluloses ($\underline{2}$)
and associative polymers such as poly(acrylamide-co-N-
alkylacrylamide) ($\underline{3}$) to increase water viscosity has led to
proposals for their use in various applications such as latex
paints and enhanced oil recovery ($\underline{4},\underline{5}$). However, these polymers
are relatively expensive. Less expensive combinations of polymers
have been discovered which exhibit high viscosity values ($\underline{6},\underline{7}$).
One such combination that has been known for some time is that of
xanthan gum and locust bean gum ($\underline{8},\underline{9}$).

---

[1]Current address: Shell Development Company, P.O. Box 1380, Houston, TX 77251–1380

0097–6156/91/0467–0446$06.00/0
© 1991 American Chemical Society

Polymers used as mobility control agents must fulfill several requirements. These are detailed in reference 4. One critical property is shear behavior. Ideally, a polymer will reversibly shear thin under high shear conditions such as found during mixing and pumping the polymer down a wellbore. Flow from the wellbore into and through the formation is radial. So flow rates initially are quite high but decline exponentially with increasing distance from the injection wellbore. As the polymer solution moves very near the production well, flow again becomes more rapid and shear rates increase. Ideally, a polymer system would exhibit low viscosity under the relatively high shear conditions near the injection and production wells and high viscosity in the bulk of the reservoir (low shear conditions). High aqueous solution viscosity is required in the reservoir to efficiently displace the oil (4).

Two types of polymer systems are discussed herein. The first type is interacting polymers which do not contain hydrophobic groups. These exhibit the desired properties of high viscosity at low concentration under low shear conditions and reversible shear thinning under high shear conditions. The second type is $\leq$ 5 mole percent hydrolyzed polyacrylamides in combination with certain organic quaternary ammonium salt polymers. The quaternary ammonium salt polymers act as salts. Polyacrylamide solution viscosity is sharply reduced in saline solution (4) The low viscosity solutions are more easily mixed and injected into the oil-bearing formation. The quaternary ammonium salt polymer is rapidly adsorbed on mineral surfaces thereby reducing salinity. This increases polyacrylamide solution viscosity within the reservoir.

Experimental Procedures

Solutions of the individual polymers were prepared and aged 24 hours before combining them to produce solutions of polymer blends. These in turn were aged 24 hours before determining solution viscosity using a Brookfield Model LVT viscometer. All comparisons were made at similar (usually equal) shear rates. Viscosity values for identical solutions in different data sets may differ because they were determined at different shear rates.

Oil recovery experiments were performed at 23°C. using 4 ft. heavy-walled glass tubes (inside diameter 3.1 cm). These were packed with a 70-170 U.S. mesh sand and had a pore volume of ca. 140 cc. Injection pressure was one pound per square inch (psi). The permeability to fresh water (10.5-13.5 darcies) was determined and the column then saturated with a 40 cps oil. Then 300 cc fresh water was injected to model secondary oil recovery followed by injection of 140 cc polymer solution and finally 250 cc fresh water. Fluid flow rate through the test column was 1.1-1.7 ft./day. Sand pack columns were used for a single experiment and then discarded.

Oil recovery experiments utilizing partially hydrolyzed polyacrylamides and quaternary ammonium salt organic polymers were also performed in Berea sandstone cores at 46°C (115°F) at an

applied pressure of 100 psi. The oil used in oil recovery
experiments utilizing these two types of polymers had a Brookfield
viscosity of 13.5 centipoise (cps) at a shear rate of 6.60 sec^{-1}
and contained 0.3% by weight asphaltenes and 2.7% by weight
paraffins. The cores were hydrated in aqueous 5% sodium chloride
and then saturated with oil. During oil recovery experiments,
aqueous fluid was injected until no oil was produced in three
successive 25 cc aliquots.

Smectite clay untreated with any polymers was the substrate
used in X-ray diffraction studies. The total amount of polymer
used in each test was 150% of the theoretical amount which would
produce two layers of adsorbed organic polymer between smectite
clay platelets. Clay - polymer solution slurries were prepared and
transferred to glass microscope slides. The slides were heated to
drive off unbound water and the samples placed in the X-ray
diffractometer for analysis.

Results

Poly(styrene sulfonate). Poly(styrene sulfonate), PSS, did not
substantially increase the viscosity of fresh water solutions, even
when the molecular weight was increased to 6 x 10^6 (entries 1-3,
Table I). PSS did not appear to be a shear thinning polymer
(entries 3 and 4, Table I).

The remaining entries in Table I indicate that the presence of
1000-4000 ppm PSS substantially increased the viscosity of 1500-
5000ppm xanthan solutions. Xanthan solution viscosity increased
significantly with increasing PSS molecular weight (Figure 1). The
effect of PSS concentration on xanthan solution viscosity was
dependent on molecular weight of the synthetic polymer (Figure 2).
For the lower molecular weight PSS, solution viscosity increased
more rapidly with increasing PSS concentration.

Other polysaccharides did not always behave similarly to
xanthan gum when in the presence of PSS. Hydroxyethyl cellulose
solution viscosity was not substantially increased by the presence
of a PSS having a molecular weight of 7 x 10^4 daltons (PSS-3, Table
II). Only when the PSS molecular weight was increased to 6 x 10^6
was a substantial viscosity increase observed.

The fresh water solution viscosity values of guar and
hydroxypropyl guar were not substantially increased by the presence
of PSS having a molecular weight of 70 x 10^4 daltons (Table III).

Other Sulfonate Polymers. These did not consistently behave in a
similar manner to poly(styrene sulfonate). Neither a hydrolyzed
copolymer of styrene sulfonate and maleic anhydride, HPSMA, or
poly(vinyl sulfonate) alone are water thickeners (Table IV). The
presence of HPSMA resulted in a modest increase in xanthan solution
viscosity (Table IV). The first three entries of Table IV indicate
that the effect of PSS on xanthan solution viscosity was
significantly greater than that of HPSMA.

Table I. EFFECT OF Poly(styrene sulfonate) (PSS) Molecular Weight
on Viscosity of Xanthan Gum Solutions

PSS Molecular Weight	Concentration (ppm) PSS	Xanthan	Shear Rate (sec^{-1})	Solution Viscosity (cps)
70,000	1500	0	39.6	1.2
500,000	1500	0	39.6	1.2
6,000,000	1000	0	79.2	15.5
6,000,000	1000	0	6.60	5.4
----	0	1500	6.36	132
6,000,000	1500	1500	6.36	678
----	0	5000	6.36	594
70,000	1000	5000	6.36	678
500,000	995	5000	6.36	809
6,000,000	995	5000	6.36	890
70,000	1990	5000	2.54	2,062
6,000,000	1990	5000	2.54	2,355
70,000	995	5000	2.54	1,888
70,000	1990	5000	2.54	2,062
70,000	3979	5000	2.54	2,170
500,000	995	5000	6.36	809
500,000	1990	5000	6.36	915
500,000	3979	5000	6.36	2,035
6,000,000	995	5000	6.36	890
6,000,000	1990	5000	2.54	2,355
6,000,000	3979	5000	1.26	5,700

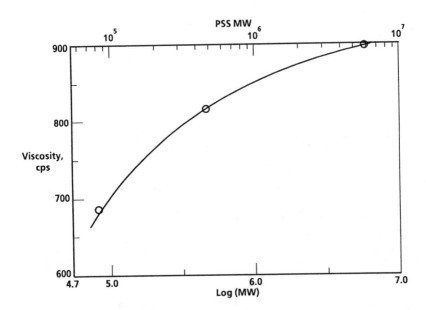

Figure 1. Effect of Poly(styrene sulfonate) (PSS) Molecular
Weight on Xanthan Solution Viscosity

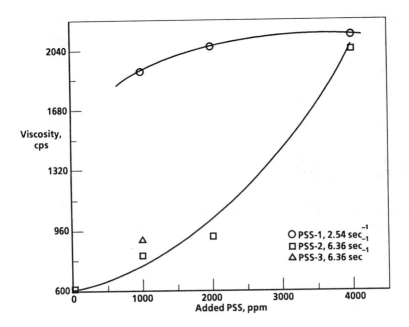

Figure 2. Effect of Poly(styrene sulfonate) (PSS) on Xanthan
Solution Viscosity

Table II. Effect of PSS Molecular Weight On Viscosity
of HEC Solutions

PSS Molecular Weight	Concentration (ppm) PSS	HEC	Shear Rate (sec^{-1})	Viscosity (cps)
70,000	1500	0	39.6	1.2
6,000,000	1000	0	79.2	15.5
6,000,000	1000	0	6.60	5.4
---	0	5000	6.36	244
70,000	995	5000	6.36	230
70,000	1990	5000	6.36	244
70,000	3979	5000	6.36	260
---	0	5000	1.32	640
70,000	1174	5000	1.32	705
---	0	500	6.60	3.6
6,000,000	500	500	6.60	12.0

Table III. Effect of PSS Molecular Weight on Viscosity
of Guar and Hydroxypropyl Guar (HPG) Solutions

PSS Molecular Weight	Concentration (ppm) PSS	Polymer	Shear Rate (sec^{-1})	Viscosity (cps)
70,000	1500	0	39.6	1.2
6,000,000	1000	0	79.2	15.5
6,000,000	1000	0	6.60	5.4
---	0	5000 G	6.36	340
70,000	995	5000 G	6.36	355
70,000	1990	5000 G	6.36	362
70,000	3979	5000 G	6.36	365
---	0	5000 HPG	2.64	208
70,000	995	5000 HPG	6.36	215
70,000	1990	5000 HPG	6.36	218
70,000	3979	5000 HPG	6.36	218
---	0	500 HPG	6.60	3.6
6,000,000	500	500 HPG	6.60	12.0

Table IV. Effect of Sulfonate Polymers on Viscosity of
5,000 ppm Xanthan Gum Solutions

Sulfonate Polymer	Concentration		Viscosity[a]
	Sulfonate (ppm)	Xanthan (ppm)	(cps)
---	0	5000	1648
PSS-3	1990	5000	2170^b
PSS-3	3979	5000	5700^c
HPSMA	1000	0	1
HPSMA	2000	0	1
HPSMA	4000	0	1
HPSMA	1000	5000	2930
HPSMA	2000	5000	3895
HPSMA	4000	5000	3877
PVS	10,000	0	5
PVS	10,000	5000	3116

a. Shear rate = 1.27 sec^{-1} unless otherwise noted.
b. Shear rate = 1.45 sec^{-1}.
c. Shear rate = 1.26 sec^{-1}.

The presence of poly(vinyl sulfonate), PVS, resulted in a substantial increase in xanthan solution viscosity (Table IV). However, a large PVS concentration, 10,000 ppp. was used.

Solute Effects. Salinity effects on solution viscosity are critical since fresh water is often unavailable to dissolve polymers. Increasing the solvent salinity to 9.75 pounds per gallon NaCl reduced the solution viscosity of a xanthan/PSS blend by 10.8% while not having a significant effect on xanthan solution viscosity (Table V).

The results summarized in Table VI indicate that the presence of PSS-3 substantially increased xanthan solution viscosity in acetic acid. The effect of PSS-3 on xanthan solution viscosity in hydrochloric acid solution was dependent on HCl concentration. Significant viscosity increases were observed in 3.1% HCl but not in 5% HCl (Table VI).

Measuring Viscosity Effects. The effect of additives on polymer solution viscosity may be quantified using the following equation (10):

$$\log \ \mu_B = X \ (\log \ \mu_1) + (1-X) \ \log \ \mu_2$$

wherein: μ_B = solution viscosity of the polymer blend
μ_1 = solution viscosity of component 1
μ_2 = solution viscosity of component 2
X^2 = weight fraction of component 1

The polymer concentrations of solutions of the individual polymers and the total polymer concentration in the polymer blend solutions were the same. The calculated viscosity values referred to in Tables VII and VIII were calculated using the above equation. When the calculated viscosity is significantly less than the observed viscosity, polymer interactions gave rise to an increased solution viscosity.

Charged Polysaccharides. Ionic polysaccharides have a substantial effect on the solution viscosity of polysaccharides. Unlike the synthetic sulfonate polymers previously discussed, the charged polysaccharides studied are themselves water thickeners.

Cationic guar (CG) contains trimethylammonium groups with a degree of substitution of 0.13-0.17. The presence of CG resulted in very substantial increases in the viscosity of fresh water solutions of hydroxyethyl cellulose and a hydroxypropyl guar having a molar substitution of hydroxypropyl groups of 0.35-0.45, HPG-1 (Table VII). At a higher M.S., 1.8, HPG-2, the observed viscosity was less than the calculated value. Using the higher purity hydroxypropyl guar, HPG-3 with an M.S. of 1.8, a significant viscosity increase was observed.

Sulfonated guar had a D.S. of 0.10. Sulfonated guar blends with hydroxyethyl cellulose (HEC) and carboxymethylhydroxyethyl cellulose exhibited substantially higher solution viscosity than

Table V. Effect of Solvent On Viscosity Of
Xanthan Gum/PSS-3 Solutions

Concentration (ppm)		Solvent	Solution Viscosity
PSS	Xanthan		(cps)
0	5000	pH 10.8 D.I. H_2O	594
995	5000	pH 10.8 D.I. H_2O	890
0	5000	pH 9.5, 9.75 ppg NaCl	592
995	5000	pH 9.5, 9.75 ppg NaCl	794

Table VI. Effect of Solvent on Viscosity of
Xanthan Gum/PSS-3 Solutions in Acid

Concentration (ppm)		Solvent	Solution Viscosity
PSS	Xanthan		(cps)
0	5000	13% acetic acid	220
498	5000	13% acetic acid	340
995	5000	13% acetic acid	460
0	5000	5% acetic acid	220
498	5000	5% acetic acid	308
995	5000	5% acetic acid	494
0	5000	5% HCl	192
995	5000	5% HCl	184
1990	5000	5% HCl	178
0^a	5987	3.1% HCl	891
3587^a	5987	3.1% HCl	1220
3587^b	5987	3.1% HCl	999

a. Shear rate = 6.30 sec^{-1}. Viscosity determined after
stirring solution for 5 minutes.
b. Shear rate = 6.30 sec^{-1}. Viscosity determined after
shearing solution for 20 minutes.

Table VII. Effect of Cationic Guar on Polysaccharide
Solution Viscosity

Concentration (ppm)		Shear Rate	Viscosity (cps)	
CG	Polysaccharide	(sec^{-1})	observed	calculated
5000 CG-1	0	2.52	702	---
---	5000 HEC	2.52	422	---
2500 CG-1	2500 HEC	2.52	6355	544
5000 CG-2	0	6.36	220	---
---	5000 HPG-1	6.36	450	---
2500 CG-2	2500 HPG-1	6.36	690	315
---	5000 HPG-2	6.36	140	---
2500 CG-2	2500 HPG-2	6.36	85	176
---	5000 HPG-3	6.30	3540	---
2500 CG-2	2500 HPG-3	6.30	2180	882
2500 CG-2	0	6.36	58	---
---	2500 HPG-3	6.36	330	---
1250 CG-2	1250 HPG-3	6.36	370	138

Table VIII. Effect of Sulfonated Guar (SG) on
Solution Viscosity of Cellulose Derivatives

Concentration (ppm)		Shear Rate	Viscosity (cps)	
SG	Polymer	(sec^{-1})	observed	calculated
5000	0	6.30	690	---
0	5000 HEC	6.30	270	---
5000	5000 HEC	6.30	3776	432
2500	0	6.36	61	---
0	2500 HEC	6.36	34	---
2500	2500 HEC	6.36	366	46
0	5000 CMHEC	6.30	216	---
5000	5000 CMHEC	6.30	3464	386
2500	0	6.36	41	---
0	2500 CMHEC	6.36	61	---
2500	2500 CMHEC	6.36	460	50

the calculated values (Table VIII). This indicated that polymer
interaction gave rise to a substantial viscosity increase.
Temperature effects are summarized in Figures 3 and 4. The
viscosity of CG/HEC blends did not decrease more rapidly with
increasing temperature than did the solution viscosity of the
individual polymer solutions. The viscosity of an SG/HEC solution
is not decrease more rapidly with increasing temperature than did
HEC solution viscosity.

Quaternary Ammonium Salt Organic Polymers. These polymers
(abbreviated QASP) functioned in the same manner as inorganic salts
in reducing the viscosity of polyacrylamides. This can be seen in
the results for a ≤5 mole percent hydrolyzed polyacrylamide
summarized in Table IX below:

Table IX.   Effect of 1,5-dimethyl-1,5-diazaundecamethylene
            polymethobromide on Solution Viscosity of Hydroxyethyl
            Cellulose and Polyacrylamide

Concentration of Cationic Polymer (ppm)	Hydroxyethyl Cellulose cps	% initial viscosity	Polyacrylamide cps	% initial viscosity
0	77.8	100.0	18.1	100.0
2000	----	----	9.3	51.4
4000	81.5	106.1	8.0	44.4
8000	81.2	105.7	7.6	41.8

Initial viscosity was the viscosity of the fresh water polymer
solution prior to the addition of any quaternary ammonium salt
polymer. Shear rate was constant in these experiments. The QASP
had the structure:

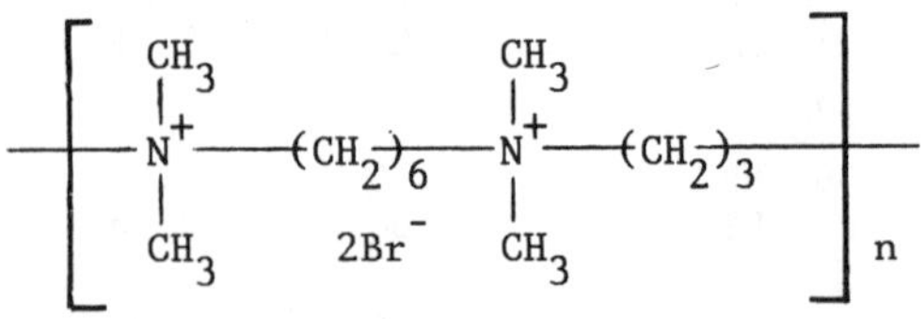

$$\left[\begin{array}{c} CH_3 \\ | \\ -N^+---(CH_2)_6-N^+---(CH_2)_3- \\ | \quad\quad\quad\quad | \\ CH_3 \quad 2Br^- \quad CH_3 \end{array} \right]_n$$

The molecular weight of the ≤5% hydrolyzed polyacrylamide was 12-15
x 10^6.
      The viscosity of the 4000 ppm hydroxyethyl cellulose solution
remained unchanged even when the QASP concentration was twice that
of the polysaccharide. In contrast, the presence of 2000 ppm QASP
(37% by weight of the polyacrylamide) in a 5400 ppm polyacrylamide
solution resulted in a viscosity decline of nearly 50%.
      Consistent results were observed in an aqueous solvent
containing 198,305 ppm total dissolved solids including 20,459 ppm
Ca^{+2}, Mg^{+2}, and $Fe^{+2,+3}$ ions. When polymers containing quaternary
nitrogen atoms in the polymer backbone or in hetercyclic rings such

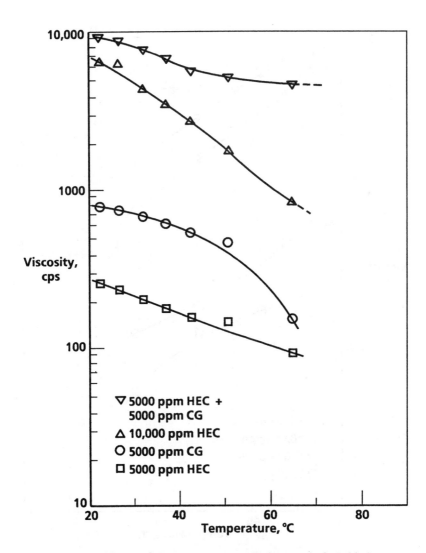

Figure 3. Effect of Temperature on Hydroxyethyl Cellulose (HEC) and Cationic Guar (CG) Solution Viscosity

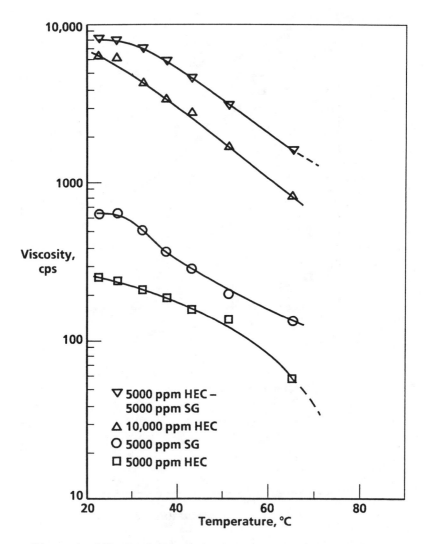

Figure 4. Effect of Temperature on Hydroxyethyl Cellulose
(HEC) and Sulfonated Guar (SG) Solution Viscosity

as poly(diallyldimethylammoium chloride) were mixed with the
essentially unhydrolyzed polyacrylamide, no solids formation was
observed in this brine after 24 hours. In contrast, when a polymer
containing quaternary nitrogen atoms in long flexible sidechains
such as poly(methacrylamidopropyltrimethylammonium chloride) was
mixed with the same polyacrylamide, a fine white precipitate was
observed three hours after mixing. Concentration of each polymer
after mixing was 4000 ppm.

Oil Recovery Experiments

A blend of poly(styrene sulfonate) and xanthan gum was used in oil
recovery experiments performed in a four foot sand pack (see
above). Results are summarized in Table X.

Table X. Oil Recovery Experiments Using Four Foot Sand Packs

Polymer	Polymer Concentration (ppm)	Oil Recovery, % OOIP	ORIP
None	0	64.6	
PSS-1	1000	1.7	4.5
Xanthan	1000	17.4	32.5
Xanthan + PSS-1	500 + 500	15.8	38.3

Polymer concentrations and flow rate (1.1-1.7 ft./day) are
representative of those used in enhanced oil recovery. OOIP refers
to the volume of 40 cps mineral oil in place prior to injection of
any aqueous fluid. ORIP refers to oil remaining in place after
water injection (modeling secondary oil recovery or waterflooding)
and prior to polymer solution injection.
     Results summarized in Table X indicate that, although the
poly(styrene sulfonate) solution recovered only a small amount of
additional oil, the presence of poly(styrene sulfonate) in a dilute
xanthan gum solution resulted in a small but significant increase
in oil recovery as compared to the xanthan solution itself.
     Similar experiments were performed (Table XI) using a blend of
quaternary ammonium salt organic polymer and a copolymer of
acrylamide (95 mole percent) and dimethylaminoethyl methacrylate
sulfuric acid salt. Test column permeability was 10.5 - 13.5
darcies. The polymer solvent and water used to waterflood the test
column were fresh water. The QASP was poly(dimethylamine-co-
epichlorohydrin). Unless otherwise noted, oil and polymer solution
viscosity were determined at a shear rate of 6.60 sec^{-1}. The first
experiment using 500 ppm of the acrylamide copolymer can be viewed
as a control experiment. Excellent oil recovery was obtained.
Although the presence of 500 ppm of quaternary ammonium salt
polymer greatly reduced viscosity of the injected polymer solution,
instead of the expected decrease in oil recovery, an increase was
observed (test 2 below).

Table XI. Oil Recovery Experiments Using Blends of a
Polyacrylamide Copolymer and a Quaternary Ammonium Salt
Organic Polymer

Test	Concentration (ppm)		Solution	Oil Recovery
	Flood Polymer	QASP	Viscosity (cps)	% OOIP
1	500	0	37.0	90.1
2	500	500	8.2	96.0
3	250	0	14.0	93.4
4	250	250	3.0	99.2
5	125	0	8.0 3.3[a]	82.4
6	125	125	2.6[a]	97.8

a. Determined at a shear rate of 36.71 sec^{-1}.

Similar results were observed in tests 3 and 4 and tests 5 and
6 (Table XI) at lower polymer concentrations.

Berea core flood test results were performed at $46^\circ C$ ($115^\circ F$)
using the same polymers and stock tank oil as used in tests 1-6
described above. Results are summarized in Table XII.

Test 7 was a control experiment in which the acrylamide
copolymer concentration was much greater than that of the crude oil
giving rise to a favorable mobility ratio. When the flood polymer
solution contained quaternary ammonium salt polymer (test 8), the
viscosity was significantly less than that of the oil. Despite
this, the oil recovery was significantly higher. Tests 9-12
indicate that similar results were observed at lower polymer
concentrations.

Discussion

The blends described in Tables I-VIII and Figures 1-4 were highly
shear thinning. This can clearly be seen by comparing the
viscosity of 5000 ppm xanthan gum solutions containing ca. 1000 ppm
poly(styrene sulfonate) having a molecular weight of 70,000 daltons
designated PSS-2 (Table I). As the shear rate decreased from 6.36
to 2.54 sec^{-1}, the Brookfield viscosity increased by a factor of
2.8. In tests of 5000 ppm hydroxyethyl cellulose solutions, the
PSS-2 concentration was not held quite constant, varying from 995
ppm to 1174 ppm (Table II). However, other data summarized in
Table II indicates that this small change in PSS-2 concentration
cannot account for the viscosity increase observed in decreasing
the shear rate from 6.36 to 1.32 sec^{-1}. The viscosity increased by
a factor of 3.1 on this decrease in shear.

The polymer blend shear thinning was reversible. This was
indicated by sequential viscosity measurements under low shear,
then high shear, and finally the original low shear rate.
Reversible shear thinning would tend to preclude the use of these

Table XII. Berea Core Flood Experiments. Effect of
poly(dimethylamine-co-epichlorohydrin) on Oil Recovery
Obtained Using poly(acrylamide-co-dimethylaminoethyl
methacrylate sulfuric acid salt)[a]

Test	Core Permeability (md)	Concentration (ppm) polymer 1	polymer 2	Solution Viscosity (cps)	Oil Recovery (% OOIP)
7	6.1	1000	0	71	70.0
8	1.9	1000	990	9	76.2
9	31.2	500	0	35	89.9
10	55.7	500	500	8.3	100.0
11	40.0	125	0	8.0	75.3
12	57.2	125	490	8.0	83.9

a. 95 mole % acrylamide/5 mole % dimethylaminoethyl
methacrylate sulfuric acid salt. Designated polymer 1.
Poly(dimethylamine-co-epichlorohydrin) designated polymer 2.
md = millidarcies
See Experimental Section for details of the test procedures
and conditions.

polymer blends under high shear oil field conditions such as hydraulic fracturing fluids. However, these blends could be used as low fluid loss wellbore fluids. They may also have some utility as drilling fluids and in gravel packing methods of sand control.

The results of the oil recovery experiments indicate that these polymer blends have potential application in polymer flooding, particularly in graded viscosity banks. The shear thinning characteristic of these polymer blends would actually be an advantage in polymer flooding. In the high shear region around the injection well bore, the effective viscosity of the injected polymer blend would be relatively low. At low viscosity, there is less resistance to flow and it is easier to inject large volumes of fluid. Since the rate of oil recovery in a well engineered project is directly related to the rate of fluid injection (at injection pressures below the fracture or parting pressure of the reservoir), a high injection rate is most desirable. Viscosity would then increase with decreasing shear as the polymer solution penetrates more deeply into the formation.

Polymer solution viscosity is directly related to oil recovery. The volumetric sweep efficiency (defined as the fraction of the oil reservoir volume contacted by the injected fluid) and thus oil recovery is related to the mobility of the injected fluid and the crude oil in the reservoir rock. This mobility ratio, M, is defined as the mobility of the injected aqueous fluid in the highly flooded low oil saturation zone, m_w divided by the mobility of the oil in the oil-bearing portions of the reservoir, m_o (11,12). This relationship may be expressed by the equation:

$$M = m_w/m_o = (k_{rw}/\mu_w)/(k_{ro}/\mu_o)$$

wherein k_{rw} and k_{ro} represent the relative permeability to water and oil respectively and μ_w and μ_o represent the viscosity of the aqueous and oil phases respectively.

The volumetric sweep efficiency increases with decreasing mobility ratio and thus with increasing injected polymer solution viscosity. So the polymer blends such as detailed in Tables I-VIII have the desirable properties of behaving as low viscosity fluids to increase injection rates and as high viscosity fluids to invade a larger fraction of the reservoir and recover more oil.

Quaternary Ammonium Salt - Polyacrylamide Blends. For $\leq 5\%$ hydrolyzed plyacrylamides, blends with QASP polymers also exhibit a low viscosity in the near injection well bore region where a low viscosity is desirable in maintaining high injection rates without exceeding the fracture or parting pressure of the formation. Certain QASP polymers are known to adsorb rapidly on clay and other silicaceous mineral surfaces (13). These quaternary ammonium salt polymers generally desorb very slowly. Thus adsorption soon decreases the salinity of the polyacrylamide solution and the viscosity increases within the reservoir where a higher viscosity contributes to a more favorable mobility ratio, increased volumetric sweep efficiency, and thus, improved oil recovery.

The QASP blends with <5% hydrolyzed polyacrylamide offer an additional advantage; the quaternary ammonium salt polymer can be chosen from the polymers known to function as swelling clay stabilizers (13). These QASP polymers are known to control permeability reductions caused by clay swelling and consequent mineral fine particle migration. This aids in maintaining high fluid injection rates.

Calcium ions are known to increase the adsorption of partially hydrolyzed polyacrylamide 14). Three mechanisms have been proposed for this behavior: reduction in electrostatic repulsion of anionic groups due to charge screening by calcium ions, specific interactions of calcium with polyacrylamide decreasing its charge and affinity for water, and fixation of calcium on mineral surfaces reducing the negative surface charge and providing new adsorption sites for polyacrylamide.

Oil recovery results summarized in Tables XI and XII imply that such mechanisms are not causing increased polyacrylamide adsorption from quaternary ammonium salt organic polymer - polyacrylamide blends. In addition, X-ray diffraction studies of smectite clay indicated that a quaternary ammonium salt organic polymer, poly(dimethylamine-co-epichlorohydrin), abbreviated poly(DMA-co-EPI), preferentially adsorbed on smectite clay in the presence of a <5% hydrolyzed polyacrylamide. The results summarized in Table XIII indicate no evidence for the wider smectite basal spacing characteristic of polyacrylamide adsorption:

Table XIII.   X-Ray Diffraction Analyses of Polymer Treated Smectite

Polymer	Concentration (ppm)	Smectite Clay Basal Spacing (A)
Polyacrylamide	5000	15.0
poly(DMA-co-EPI)	5000	14.2 ± 0.2
Polyacrylamide +	5000	
poly(DMA-co-EPI)	5002	14.1

Nature Of Polymer Interactions. Presently the nature of the polymer interactions giving rise to the large viscosity increases detailed in Tables I-VIII are not well understood. Certainly, no hydrophobic groups are present in the polymers to give rise to hydrophobic domains and polymer association proposed for hydrophobically modified water-soluble polymers 1-3. The nature of the interactions of the quaternary ammonium salt polymers with polyacrylamides and the effect of this interaction on solution viscosity appears to be similar to that of inorganic salts. The interaction of the cationic atoms or sites with the anionic or polar groups of polyacrylamide lessens electrostatic repulsive interactions of these groups tightening the polymer coil and reducing the hydrodynamic volume and thus the solution viscosity (4).

Mixing experiments indicate that only when there is a very
high degree of steric hindrance around the quaternary nitrogen
atom, as when the nitrogen atom is located in or held close to the
polymer backbone, are solutions of the QASP and the $\leq$5% hydrolyzed
polyacrylamide compatible, i.e., no precipitate formation occurs
for 24 hours after mixing. When the quaternary nitrogen atom is
located in a long flexible sidechain, steric hindrance about the
quaternary nitrogen atom is much less. Sufficient interaction with
carboxylate groups of the $\leq$5% hydrolyzed polyacrylamide takes place
to result in fairly rapid (<24 hour) precipitate formation.

Conclusions

Various combinations of polymers which do not contain hydrophobic
groups can give rise to substantially increased aqueous fluid
viscosity. These polymer blends undergo reversible shear thinning.
This viscosity cannot be rationalized in terms of increased total
polymer concentration. The chain interactions giving rise to the
increased solution viscosity are not well understand but are
different in kind from those of hydrophobically associating
polymers.
    Polyquaternary ammonium salt polymers function in a similar
manner as inorganic salts in reducing the viscosity of <5%
hydrolyzed polyacrylamide.
    The viscosity behavior of these two types of polymer blends
can be utilized to provide improved mobility control solutions for
polymer flooding enhanced oil recovery. Polymer blends that
produce increased solution viscosity under low shear conditions may
also be useful in various types of well bore fluids.

Legend of Symbols

PSS-1   sodium poly(styrene sulfonate), 6,000,000 daltons
PSS-2   sodium poly(styrene sulfonate), 500,000 daltons
PSS-2   sodium poly(styrene sulfonate), 70,000 daltons
HPSMA   hydrolyzed poly(styrene sulfonate-co-maleic anhydride) PVS
poly(vinyl sulfonate)
HEC   hydroxyethyl cellulose, M.S. 2.5
CMHEC   carboxymethylhydroxyethyl cellulose, D.S. 0.4, M.S. 2.4
HPG-1   hydroxypropyl guar, M.S. 0.35-0.45
HPG-2   hydroxypropyl guar, M.S. 1.8
HPG-3   hydroxypropyl guar, purified, M.S. 1.8
SG   sulfonated guar, D.S. 0.10
CG   cationic $(N(CH_3)_3^+$ guar, D.S. 0.17, 0.13 (two different
          samples
QASP   quaternary ammonium salt organic polymer
cps   centipoise
A   Angstrom
ppm   parts per million by weight
OOIP   oil originally in place after saturating the sand pack or
          core with oil
ORIP   oil remaining in place after injecting water through the

sand pack or core. ORIP = OOIP - oil recovered by water injection.

Literature Cited

1. Glass, J.E., Ed. Polymers in Aqueous Media: Performance Through Association; American Chemical Society: Washington, D.C., 1989, pp. 317-549.
2. Landoll, L.M. Netherlands Patent 80 03 241, 1980.
3. Bock, J.; Valint, P.L. U.S. Patent 4 730 028, 1988.
4. Borchardt, J.K. In Oil-Field Chemistry: Enhanced Recovery and Production Stimulation; Borchardt, J.K. and Yen, T.F., Eds.; American Chemical Society, Washington, D.C. 1989, pp. 27-32.
5. Evani, S.; Rose, G.D. Polym. Mater. Sci. Eng., 1987, 57, 477-81.
6. Borchardt, J.K. U.S. 4 508 629, 1985.
7. Borchardt, J.K. U.S. 4 524 003, 1985.
8. Kovacs, P.; Food Technol., 1973, 27, 26.
9. Rocks, J.K.; Food Technol., 1971, 25, 22.
10. DeMartino, R.N. U.S. Patent 4 172 055, 1979.
11. Crafts, B.C.; Hawkins, M.F. Applied Petroleum Reservoir Engineering, Prentice-Hall, New York, 1959.
12. Craig, Jr., F.F. The Reservoir Engineering Aspects of Water Flooding, Society of Petroleum Engineers, Dallas, Monograph No. 3, 1971, pp. 45-47.
13. Borchardt, J.K. In Oil-Field Chemistry: Enhanced Recovery and Production Stimulation; Borchardt, J.K. and Yen, T.F., Eds.; American Chemical Society, Washington, D.C. 1989, pp. 204-21.
14. Lee, L.T.; Lecourtier, J.; Chauveteau, G. In Oil-Field Chemistry: Enhanced Recovery and Production Stimulation; Borchardt, J.K. and Yen, T.F., Eds.; American Chemical Society, Washington, D.C. 1989, pp. 225-40.

RECEIVED April 9, 1990

ADVANCES IN LESS CONVENTIONAL SYSTEMS

Chapter 30

Hydrophilic–Hydrophobic Domain Polymer Systems

Kishore R. Shah

Ethicon, Inc., Somerville, NJ 08876

It is the objective of this paper to review the chemistry of hydphobically modified water soluble and other hydrophilic polymer systems, and their biocompatibility for potential use in biomedical and pharmaceutical applications. The systems reviewed include block copolymers, graft copolymers, polymer blends, and networks which exhibit hydrophilic/hydrophobic microphase domain structures. Enhanced biocompatibility and mechanical strength appear to be a general characteristic of such systems as compared to single phase hydrophilic polymers.

Water soluble polymers play a very important role in biological systems and are of considerable interest in industrial, biomedical, and pharmaceutical applications. It is the purpose of this paper to review the chemistry of hydrophobically modified water soluble and other hydrophilic polymer systems, and their biocompatibility for potential use in biomedical and pharmaceutical applications. The hydrophobically modified polymer systems include block copolymers, graft copolymers, polymer blends, and networks which exhibit hydrophilic/hydrophobic microphase domain structures.

Covalent and ionically crosslinked networks of hydrophilic polymers, commonly referred to as hydrogels in their hydrated state, have been known for a long time and have found a number of biomedical applications, such as soft contact lenses, wound management, and controlled drug delivery. The methods of preparation of these hydrogels are subject to the constraints imposed by their thermosetting nature, and consequently they do not enjoy the benefits of thermoplastic processing. In addition, such hydrogels are mechanically weak, which further limits their usefulness. On the other hand, the hydrophobic/hydrophilic domain polymer systems which hydrate to form hydrogels are of special interest on account of their unique morphological features which greatly influence their properties. The hydrophobic domains in such systems behave as thermally labile pseudocrosslinks. Their two-phase hydrophobic/hydrophilic nature is also responsible for their enhanced

0097–6156/91/0467–0468$06.00/0
© 1991 American Chemical Society

biocompatibility and superior mechanical strength as compared to that of covalently crosslinked hydrophilic polymers. Few such systems have been reported to date, and the ones reported are of relatively recent origin. Some of these polymer systems will be discussed here.

BLOCK COPOLYMERS

Okano, et al. (1,2) have prepared A-B-A type block copolymers of 2-hydroxyethyl methacrylate, HEMA (A) and styrene, St (B) by condensation of amino-semitelechelic-oligo HEMA (Prepolymer A) with isocyanate-telechelic-oligo-St (Prepolymer B) under the condition of $[NH_2]/[NCO] = 1.2$ (synthetic scheme I). The Prepolymer A, in turn was prepared by free radical initiated oligomerization of HEMA in the presence of 2-aminoethanethiol, used as a chain transfer agent. The Prepolymer B was prepared by photooligomerization of styrene initiated by p,p'-diisocyanate diphenyl disulfide.

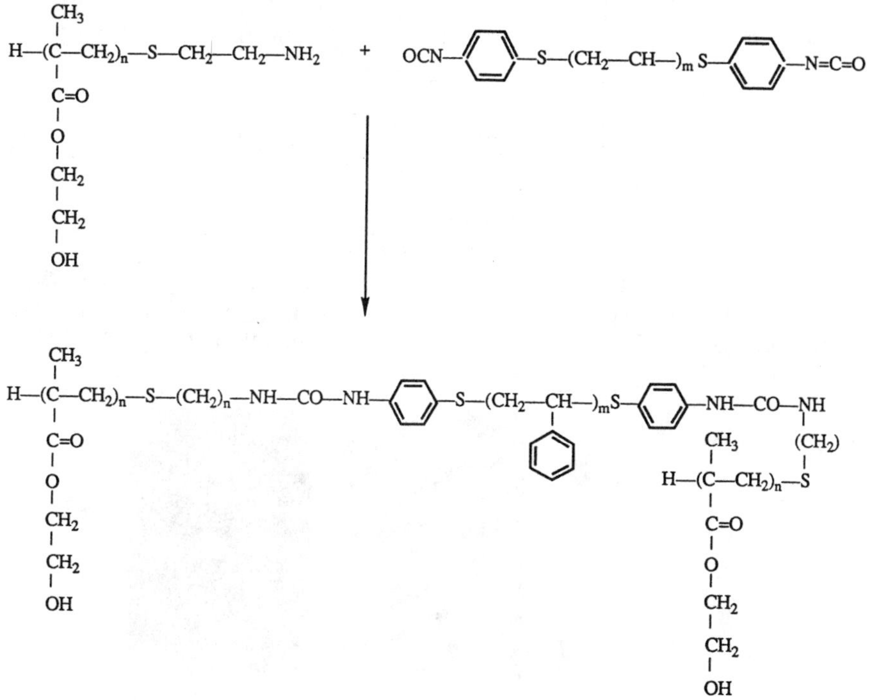

Scheme I

The wettability (Figure 1) of the block copolymers, as measured by cosine of the contact angle of water on the copolymer film, increased with an increase in the mole fraction of HEMA. However, the block copolymer was less wettable than a random copolymer of similar composition because of the presence of hydrophobic domains of polystyrene, which are clearly seen in the transmission electron micrographs (Figure 2). As one would expect, PHEMA-St block copolymers show spherical domains at high and low HEMA weight fractions. Whereas, the 60:40 PHEMA-St block copolymer exhibits lamellar morphology.

Hemocompatibility is an important attribute required of materials for many biomedical applications. Hemocompatibility is also one of the more stringent

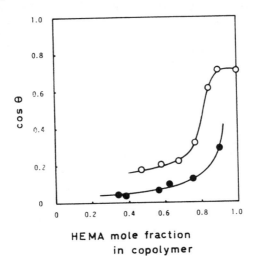

HEMA mole fraction
in copolymer

Figure 1: Relation between wettability and copolymer composition: (●) PHEMA-St ABA-type block copolymer system; (○) PHEMA-St co-oligomer system. (Reprinted with permission from ref. 1. Copyright 1978 Wiley.)

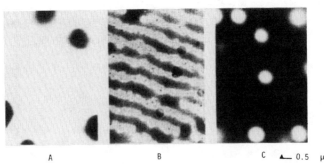

A   B   C   ▲— 0.5 μ —▲

Figure 2: Electron micrographs of PHEMA-St ABA-type block copolymer films cast from DMF at 40 °C and having HEMA mole fractions of (A) 0.347, (B) 0.608, and (C) 0.884. (Reprinted with permission from Ref. 2. Copyright 1981 Wiley.)

criteria for biological acceptance. It is of interest to note that copolymers having a balance of both hydrophilic and hydrophobic microphase domains were the most blood compatible. Hemocompatibility of the block copolymers was evaluated by their interaction with canine blood, which is known for its ease of coagulation. Minimum platelet adhesion was achieved for the block copolymer having lamellar morphology (0.6 mole fraction HEMA) (Table I).   In general, the block copolymers, having both hydrophilic and hydrophobic domains, were more resistant to platelet adhesion than either homopolymers of HEMA and styrene or their random copolymers.

TABLE I. Platelet Adhesion and Aggregation on the Surface
of PHEMA-St Block Copolymer

Samples	HEMA Mole Fraction in Copolymer	Adhesion (%)	Morphology of Platelets
Polystyrene	0	33.8	Aggregation
ABA-type block copolymer	0.347	19.7	Round
ABA-type block copolymer	0.608	11.8	Round
ABA-type block copolymer	0.884	17.6	Aggregation
PolyHEMA	1	26.8	Aggregation
Random copolymer	0.640	38.5	Aggregation
Glass beads	0	34.7	Spread out

Further evidence of antithrombogenicity of the block copolymer was lack of platelet aggregation and their round morphology, similar to that of native form. Platelet adhesion and aggregation in blood are believed to take place on a layer of proteins which are initially adsorbed on the material surface. Selective protein adsorption on the hydrophobic and hydrophilic domains has been observed (3).   The organized surface protein layers on the domains of these block copolymers are believed to be responsible for their improved hemocompatibility.

The chemistry, properties, and biomedical applications of multiblock copolymers, having alternating sequences of hydrophobic polyacrylonitrile and hydrophilic derivatives of acrylic acid, has been recently reviewed by Stoy (4).

The polyacrylonitrile sequences in the block copolymers, HYPAN, form crystalline domains, and being incompatible they phase separate from the amorphous hydrophilic sequences (Figure 3). The existence of microphase separated crystalline domains was detected by x-ray diffraction (5).

Controlled hydrolysis of polyacrylonitrile (PAN), the starting material for the synthesis of HYPAN copolymers, in a homogenous medium yields PAN-polyacrylamide multiblock copolymer. The mechanism of the multiblock copolymer formation first involves formation of a restricted number of amido groups amidst the PAN chains (6). Further hydrolysis involves only those nitrile groups that have an already formed amido group in their proximity, thus resulting in the formation of polyacrylamide blocks separated by PAN blocks. Next, acid catalyzed cyclization at elevated temperatures in an inert atmosphere results in the formation of the precursor, containing glutarimide units, to the HYPAN copolymers (7). The hydrophilic sequences (soft block) of the HYPAN copolymers are formed by ring opening of glutarimide units of the precursor copolymer by a base. The chemical composition of the soft blocks can vary depending upon the base employed for opening the glutarimide ring (synthetic scheme II). Depending on the character of the resulting amorphous soft blocks, HYPAN copolymers can be neutral, cationic, anionic, or amphoteric.

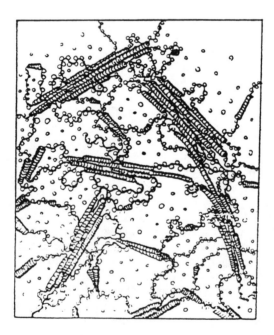

Figure 3: Schematic representation of HYPAN having permanent crystalline network. The PAN blocks (⟨αcoccoo⟩) separate in the presence of water from the hydrated "soft blocks" (⟨ᵃᵥᵃᵥᵃᵥ⟩) to form crystalline clusters. (Reprinted with permission from Ref. 4. Copyright 1989 Technomic.)

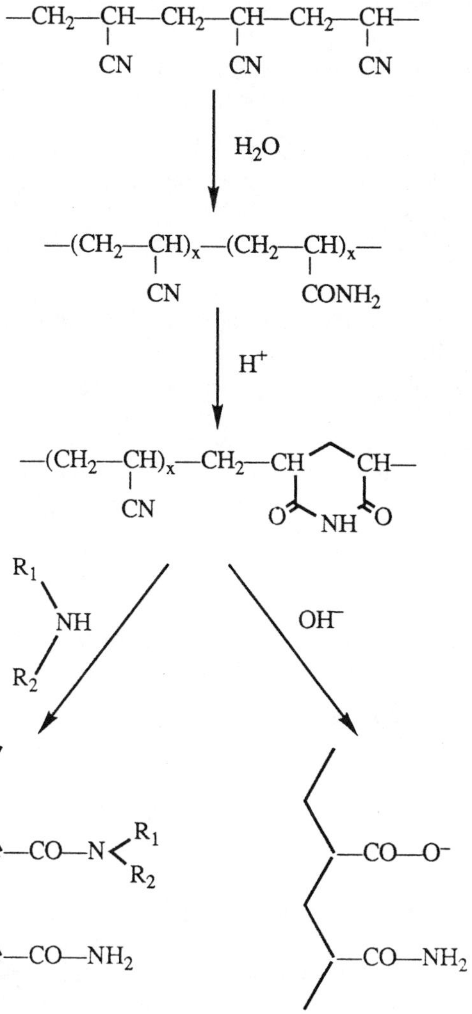

$$-CH_2-CH-CH_2-CH-CH_2-CH- \atop \quad\;\; CN \qquad\quad CN \qquad\quad CN$$

$$\downarrow H_2O$$

$$-(CH_2-CH)_x-(CH_2-CH)_x- \atop \qquad\;\; CN \qquad\qquad CONH_2$$

$$\downarrow H^+$$

$$-(CH_2-CH)_x-CH_2-CH\quad CH- \atop \qquad\;\; CN \qquad\qquad O^{\diagup}{}_{NH}{}^{\diagdown}O$$

$$R_1 \atop {}^{\diagdown}NH \atop R_2{}^{\diagup} \qquad\qquad\qquad OH^-$$

$$-CO-N{<}^{R_1}_{R_2} \qquad\qquad\qquad -CO-O^-$$

$$-CO-NH_2 \qquad\qquad\qquad -CO-NH_2$$

Scheme II

HYPAN copolymers are soluble in solvents, such as dimethyl sulfoxide and 50-55% aqueous sodium thiocyanate, which are thermodynamically good solvents for both PAN and hydrophilic soft blocks. On account of the high crystalline melting point of PAN ($\sim$325°C) and its low decomposition temperature ($\sim$220°C) solvent plasticization of HYPAN is required for thermoplastic melt processing. The properties of HYPAN copolymers are determined by the character of the crystalline networks, the composition of the amorphous phase, and the ratio of the two phases. The HYPAN copolymers, having substantial proportions of the soft blocks, are water absorbent and form hydrogels upon hydration. The crystalline PAN domains confer high mechanical strength to the HYPAN hydrogels. Good biocompatibility of these hydrogels was manifested by low protein adsorption and the formation of a thin stable, fibrous capsule around implanted HYPAN (8).

Living cationic polymerization initiated by HI/I_2 system has been used by Sawamoto, et al., (9) to synthesize amphiphilic block copolymers of vinyl ethers (synthetic scheme III). The living nature of this polymerization system stems from the stability of the propagating species consisting of a terminal carbon-iodine bond associated with molecular iodine (shown below). This stability allows the growing end to elude not only chain transfer and termination, but side reactions with polar groups in monomers and polymers. Therefore, this cationic polymerization process permits the synthesis of block copolymers of vinyl ether derivatives by sequential monomer addition.

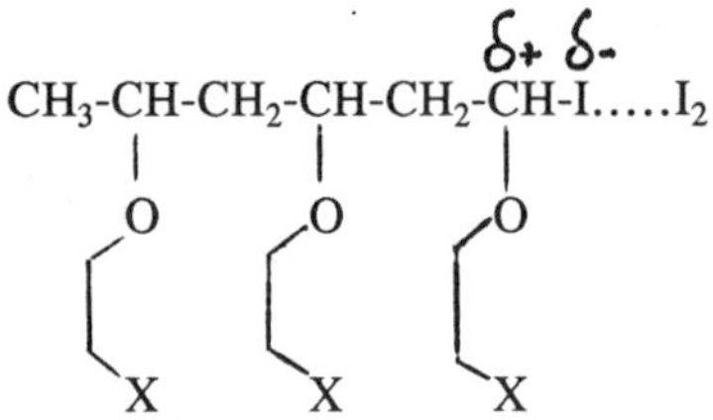

$$\overset{\delta+\ \delta-}{CH_3\text{-}CH\text{-}CH_2\text{-}CH\text{-}CH_2\text{-}CH\text{-}I.....I_2}$$

This synthetic method yields block copolymers, which are free from contaminating homopolymers, having a narrow molecular weight distribution (Mw/Mn $\le$ 1.1), and controlled composition. The amphiphilic block copolymers (VEC 1, VEC 2,& VEC 3) shown in the Scheme III were water soluble and exhibited surface active properties. However, their solid state microstructure has not been reported. Such copolymers having sufficiently long segmented block lengths would be expected to show microphase separation in the solid state.

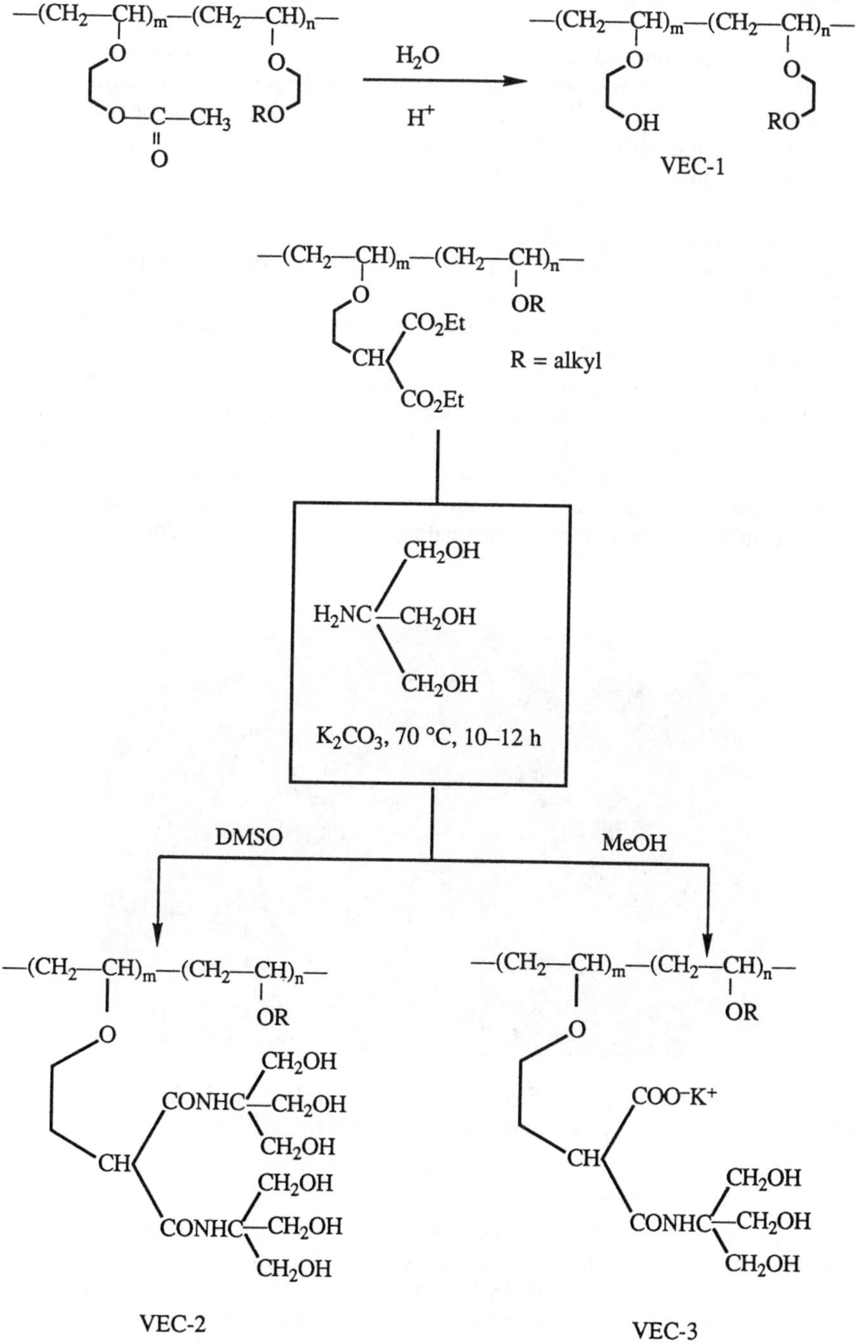
—(CH2—CH)m—(CH2—CH)n—
O
O
O—C—CH3
O
RO
H2O
H+
OH
RO
VEC-1
OR
CO2Et
CH
CO2Et
R = alkyl
CH2OH
H2NC—CH2OH
CH2OH
K2CO3, 70 °C, 10–12 h
DMSO
MeOH
CONHC—CH2OH
COO−K+
VEC-2
VEC-3

Scheme III

GRAFT COPOLYMERS

Free radical initiated solution copolymerization of poly(methyl methacrylate) (PMMA) macromonomer with HEMA has been used to prepare PHEMA-PMMA graft copolymers (10). The preparation of PMMA macromonomer, which has been described by Palit and Mandal (11), involved acylation of hydroxy terminated PMMA with methacryloyl chloride.

Transmission electron microscopy of the PHEMA-PMMA graft copolymer, having the PMMA content of 29 wt %, indicated a microphase separated structure (Figure 4), the size of the dispersed PMMA phase being 200-300 Å. Thromboresistance of the graft copolymers was evaluated employing canine blood and expressed as its relative thrombogenicity (RT), at 9 minutes of blood contact time, with respect to silastic. By definition RT is 0 for silastic and 100 for glass. The negative values of RT indicate thromboresistance better than that of silastic. As in the case of PHEMA-St block copolymers, the results (Table II) indicated that the microphase separated PHEMA-PMMA graft copolymers, having hydrophobic domains, were much more hemocompatible than the corresponding homopolymers, their random copolymers, and silastic.

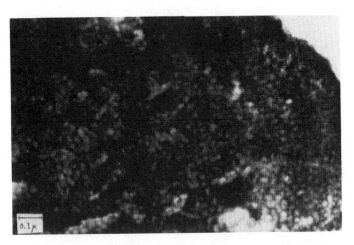

Figure 4: Electron micrograph of a thin film of PHEMA-MMA graft copolymer. (Reprinted with permission from Ref. 10. Copyright 1977 Wiley.)

Onishi has reported polymerization of methyl methacrylate onto dextran polymer chains (Mw, 60,000) in the presence of ceric ions to form graft copolymers having hydrophilic-hydrophobic domain structures (11). In vitro studies, on the rate of thrombus formation on the surface of copolymer in canine blood, have further confirmed improved hemocompatibility of such two-phase copolymers

TABLE II. In vitro Thrombogenicity of
PHEMA/PMMA Graft-Type Copolymer

PMMA Mn	MMA Content (wt %)	Water Uptake (%)	RT
2 x 10⁴	9	52	-27
	42	29	-4
5 x 10⁴	12	74	-52
	35	37	55
HEMA/MMA random copolymer	44	24	86
PHEMA (crosslinked)		66	0
PMMA		0	73
Polytetrafluorethylene		0	31
Silastic		0	0

Milkovich has described the preparation of hydrophobic macromonomer, by living anionic polymerization (Scheme IV), and its subsequent free radical copolymerization with a hydrophilic monomer resulting in the formation of a hydrophilic polymer having hydrophobic polymer chains grafted onto it (12). Such graft copolymers were microphase separated, thermoplastic, and absorbed water to form hydrogels. Thus, for example, styrene was polymerized with sec-butyl lithium to form a living polystyrene carbanion, which was then "capped" with ethylene oxide. The end-capped polymeric alkoxide anion, is more stable and therefore, less susceptible to undesirable side reactions with terminating agents than the polystyrene carbanion. Nucleophilic displacement of chlorine of methacryoyl chloride by the alkoxide anion yielded polystyrene macromonomer (Mn, 13,400, Mw/Mn. = 1.05).

Free radical initiated solution copolymerization of the polystyrene macromonomer and N,N-dimethylacrylamide (2:3 weight ratio) yielded a graft copolymer, which swelled in water to form a transparent hydrogel having 69% equilibrium water content. Unusually high strength (tensile strength 6.0 MPa and elongation of 100%) of the fully hydrated copolymer indicated that the hydrophobic domains acted as physical crosslinks to the normally water soluble poly(N,N-dimethylacrylamide).

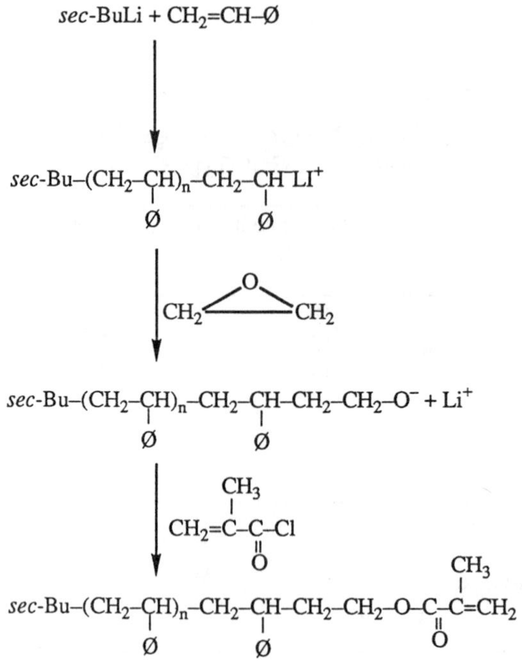

sec-BuLi + CH$_2$=CH–Ø

$\downarrow$

sec-Bu–(CH$_2$–CH)$_n$–CH$_2$–CH$^-$Li$^+$
          | |
          Ø   Ø

$\left|\right.$ CH$_2\overset{O}{\underset{}{\triangle}}CH_2$

$\downarrow$

sec-Bu–(CH$_2$–CH)$_n$–CH$_2$–CH–CH$_2$–CH$_2$–O$^-$ + Li$^+$
          | |
          Ø   Ø

$\left|\right.$ CH$_2$=C–C–Cl with CH$_3$ above and O below

$\downarrow$

sec-Bu–(CH$_2$–CH)$_n$–CH$_2$–CH–CH$_2$–CH$_2$–O–C–C=CH$_2$
          | | ‖ |
          Ø   Ø   O CH$_3$

PS macromonomer

Scheme IV

An alternate method described for the preparation of the polystyrene macromonomer is coupling of carboxy-terminated polystyrene, obtained by free radical polymerization of styrene using monoiodoacetic acid as a chain transfer agent, with glycidyl methacrylate (13). However, unlike anionic polymerization, this method does not yield macromonomer having a narrow molecular weight distribution.

POLYMER BLENDS

Shah (14,15) has reported an approach of attaining microphase separation in blends of two polymers via interpolymer hydrogen bonding interactions. In this system, one of the two polymeric components is either a homopolymer or a copolymer of an N-vinyl lactam, and the other component is an acrylate or a related copolymer containing a small proportion of acidic groups. Interpolymer hydrogen bonding interactions (Figure 5) between the acidic groups and the lactam carbonyl groups make the blends compatible on an optical scale. An interplay of the thermodynamic forces of phase separation, an inherent tendency in a mixture of two dissimilar polymers, and the extent of interpolymer

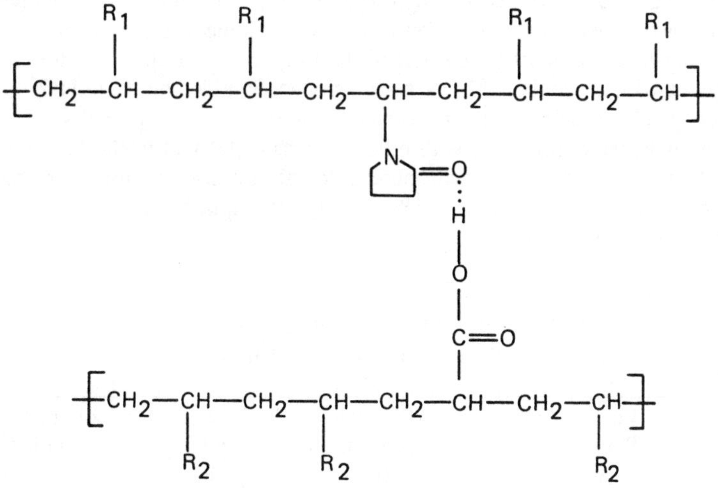

Figure 5: Interpolymer hydrogen bond formation

interactions result in microphase separation. The size of the phase domains is generally less than 1000 Å.

For example, blends of water soluble poly(N-vinyl 2-pyrrolidone) (PVP) and a water insoluble poly(n-butyl methacrylate-co-methacrylamide-co-acrylic acid) (Terpolymer) are optically transparent in both dry and fully hydrated state. The mixtures of these two hydrophilic and hydrophobic polymers were organic solvent soluble, water swellable but water insoluble, and amenable to processing by thermoplastic methods. The equilibrium water content of the fully hydrated blends varied inversely with the proportion of the water insoluble copolymer in the blend (Figure 6). The hydrated network was stable over a pH range of 1 to 11 and in salt solutions. This stability is explained by the hypothesis that the two polymeric phases in the blend are not pure, and that the segments of the water soluble polymer are trapped in the domains of the water insoluble polymer.

High degree of tissue and blood compatibility (16) of these polymer blend hydrogels make them potentially useful in diverse biomedical applications. Stress-strain measurements (Table III) of the fully hydrated blends of the Terpolymer with a copolymer of N,N-dimethylacrylamide and N-vinyl 2-pyrrolidone reflect their excellent mechanical properties. These results are consistent with the observation with other systems (4,12) that in the fully hydrated state the hydrophilic-hydrophobic domain network exhibits very much greater mechanical strength than the covalently crosslinked network of hydrophilic polymers.

Table III. Mechanical Properties of
Polymer Blend Hydrogels

Terpolymer in Blend Wt. %	Equilibrium Water Content Wt. %	Ultimate Tensile Strength MPa	Young's Modulus MPa	Ultimate Elongation %
20	71	0.428	0.794	336
40	55	1.787	9.232	273
60	40	4.878	33.30	238

An interesting property of the hydrogel forming polymer blends was their ability to form an adhesive bond to certain polymeric materials [e.g., ethylene-acrylic acid copolymer, poly(vinyl chloride), and cellulose acetate] having functional groups capable of forming interactions with N-vinyl lactam groups of the polymer blend (17). The adhesive bond between the substrate and the polymer blend remained strong even upon equilibrium hydration of the blend. This property enables the preparation of hydrogel coatings on plastic surfaces or hydrogel/plastic laminates by a simple process of solution coating or heat lamination.

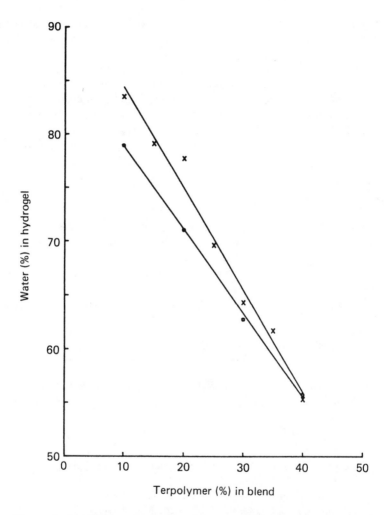

Figure 6: Equilibrium hydration of the PVP-Terpolymer (x) and the
DMA/VP copolymer-Terpolymer (0) blends (Terpolymer
= 73% BMA, 22% MAA and 5% AA; DMA/VP
copolymer = 68% DMA and 32% VP)

NETWORKS

Chen, et al. (18) have reported new amphiphilic network by free radical copolymerization of hydrophobic methacryloyl capped polyisobutylenes (MA-PIB-MA) with hydrophilic 2-(dimethylamino)ethyl methacrylate (Figure 7). Two MA-PIB-MAs were prepared with Mn = 4920 and 10,200, and two series of networks were prepared with MA-PIB-MA contents between 48 and 71.5%. The water swelling of the networks ranged from 170% to 20% with increasing MA-PIB-MA content. A phase separated domain structure in these networks was evidenced by the existence of two glass transition temperatures indicated by differential scanning calorimetry. As with the block and graft copolymer systems, the most biocompatibility was found for network hydrogels having a balance of hydrophilic and hydrophobic properties (19).

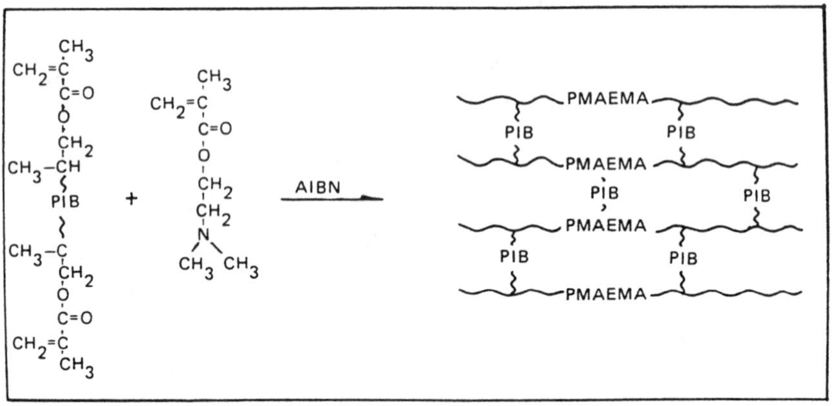

Figure 7: Copolymerization of MA-PIB-MA with DMAEMA and the resulting network. (Reprinted with permission from Ref. 19. Copyright 1989 Wiley.)

Literature Cited
1. T. Okano, M. Katayama, and I. Shinohara, J. Appl. Polym. Sci., 22, 369(1978).
2. T. Okano, S. Nishiyama, I. Shinohara, T. Akaike, Y. Sakurai, K. Kataoka, and T. Tsuruta, J. Biomed. Mat. Rsearch, 15, 393 (1981).
3. T. Okano, S. Nishiyama, I. Shinohara, T. Akaike, and Y. Sakurai, Polym. J., 10, 23(1978).
4. V. A. Stoy, J. Biomat. Appl., 3, 552(1989).
5. J. Soler and J. Baldrian, Angew. Makromol. Chemie, 49, 49(1976).
6. V. A. Stoy, O. Wichterle, and A. Stoy, U. S. Patent 4,095,877 (June 20, 1978).
7. V. A. Stoy, U.S. Patent 4,369,294 (Jan. 18, 1983).
8. M. Chvapil, P. R. Weinstein, R. L. Misiorowski, D. Tellers, L. Rankin, and V. A. Stoy, J. Biomed. Mat. Res., 18, 757(1984).

9. M. Sawamoto, S. Aoshima, and T.Higashimura, Makromol. Chem, Macromol. Symp., 13/14, 513(1988).
10. T. Nakashima, K. Takakura, and Y. Komoto, J. Biomed. Mat. Res., 11, 7(1977).
11. Y. Onishi, S.Maruno, S. Kamiya, K. Nishioka, Y. Kikuchi, and T. Hanafusa, in "Contemporary Topics in Polymer Science", W. J. Bailey and T. Tsuruta, Ed., Plenum Publ. Co., N.Y., 1984, p. 149.
12. R. Milkovich and M. T. Chiang, U.S.Patent 4,085, 168 (April 18, 1978).
13. Y. Yamashita, K. Ito, H. Mizuno, and K. Okada, Polymer J., 14, 255 (1982).
14. K. R. Shah, Polymer, 28, 1212(1987).
15. K. R. Shah, in "Contemporary Topics in Polymer Science, Vol. 6", B. M. Culbertson, Ed., Plenum Publ. Co., N.Y., 1989, p. 473.
16. K. R. Shah, Unpublished results.
17. K. R. Shah, U.S. Patent 4,369,229 (Jan. 18, 1983).
18. D. Chen, J. P. Kennedy, and A. J. Allen, J. Macromol. Sci.-Chem., A25, 389(1988).
19. D. Chen, J. P. Kennedy, M. K. Kory, and D. L. Ely, J. Biomed. Mat. Res., 23, 1327(1989).

RECEIVED July 24, 1990

Chapter 31

Enzyme-Degradable Hydrogels

Properties Associated with Albumin-Cross-Linked Polyvinylpyrrolidone Hydrogels

Waleed S. W. Shalaby[1], William E. Blevins[2], and Kinam Park[1]

[1]School of Pharmacy and [2]Department of Veterinary Clinical Sciences, Purdue University, West Lafayette, IN 47907

The enzyme-degradable properties of albumin-crosslinked hydrogels have been studied both in-vitro and in-vivo. The kinetics of gel swelling and enzymatic degradation were dependent on the functionality of albumin used as a crosslinker. A predominance of either surface or bulk degradation was observed depending on the the degree of albumin functionality. As the degree of functionalization increased, transition from surface degradation to bulk degradation occurred. The swelling and degradation behavior of hydrogels in the canine stomach was non-invasively examined by radiographic and ultrasonographic imaging. The potential use of enzyme-digestible hydrogels as platforms for long-term oral drug delivery is discussed.

Biodegradable hydrogels can be described as water-swellable, crosslinked polymers that undergo a net dissolution process in various environments. Gel degradation can result in the formation of water soluble, oligomeric or polymeric fragments. During dissolution, these fragments may remain within the gel network until complete gel disruption occurs or diffuse away from the gel surface. For hydrogel networks prepared through chemical crosslinking, gel degradation can be achieved by cleavage of either the crosslinking agent or the polymer backbone. Hydrogel degradation may occur at either the gel surface or within the matrix. Surface degradation is characterized by a localized dissolution process occurring at the gel surface. Bulk degradation, however, is defined as a homogeneous dissolution process that occurs throughout the gel network.

To date, the most common mechanisms of gel degradation are simple hydrolysis and enzyme-catalyzed hydrolysis. Early investigators examined the hydrolysis of N,N'-methylene-bisacrylamide crosslinks and the release of macromolecules from 1-vinylpyrrolidone hydrogels (1, 2). It was found that gel dissolution could be efficiently controlled by the content of the crosslinking agent. In contrast to gel dissolution via the crosslinking agent, hydrolysis of the polymer backbone has been studied using hydrogels prepared from unsaturated polyester prepolymers (3, 4). Some hydrogels that degrade through the polymer backbone have also been used to coat multifilament sutures and ligatures (5). A simple hydrolysis method for solubilizing hydrogels may be useful in a variety of applications.

Enzyme-catalyzed hydrolysis can be utilized to degrade both natural and synthetic networks. Dickinson et al. (6, 7) studied diaminododecane-crosslinked poly (2-hydroxyethyl-L-glutamine) hydrogels as subcutaneous implants. Their results suggested that bulk degradation occurred during the first two weeks of implantation and was attributed to the presence of proteolytic enzymes, pronase and papain. In a related work, Pangburn et al. (8) examined the degradation by lysozyme of glutaraldehyde-crosslinked, deacetylated chitin hydrogels. The use of enzyme-catalyzed hydrolysis for cleaving hydrogel crosslinks was first investigated by Ulbrich et al. (9). The use of oligopeptide crosslinks was found to be quite useful in controlling gel degradation by chymotrypsin. Both the degree of crosslinking and the length of oligopeptide sequences could be manipulated to alter the solubilization kinetics of the gels. More recently, we have developed biodegradable hydrogels using modified albumin as a crosslinking agent (10). Albumin-crosslinked hydrogels were found to possess unique dissolution characteristics (Shalaby, W.S.W.; Park, K. Pharm. Res., in press.). Currently, our research efforts are being focussed on the study of these systems for their potential use as long-term oral drug delivery systems. We will discuss, in this paper, the swelling and degradation properties associated with albumin-crosslinked hydrogels and their preliminary development as platforms for oral drug delivery.

Experimental Methods

Functionalization of Albumin. Human serum albumin (Sigma, fraction V) was alkylated using glycidyl acrylate (Aldrich) as described previously (10). A 5% (w/v) solution of albumin was prepared in phosphate-buffered saline solution. While stirring at room temperature, the glycidyl acrylate was added directly to the solution. Typically, 200 µl of glycidyl acrylate was added to 5 ml of the albumin solution. The degree of albumin alkylation was measured using the amine group titration method developed by Snyder and Sobocinski (11).

Hydrogel Preparation. Albumin-crosslinked polyvinylpyrrolidone hydrogels were prepared in distilled deionized water using 1-vinyl-2-pyrrolidinone (VP, Aldrich) at 40% (w/v), 2,2-azobis(2-methyl-propionitrile) (ABMP, Eastman Kodak) as initiator at 1% (w/w) of the monomer, and functionalized albumin as crosslinking agent. The concentration of functionalized albumin was varied from 4.5% (w/w) to 8% (w/w) of the monomer, while the degree of albumin alkylation ranged from 8% to 90%. The monomeric solution was prepared by first solubilizing the ABMP in previously de-gassed VP followed by the addition of functionalized albumin. Polymerization was carried out for 12 h at 60 °C. Following gelation, the hydrogels were cut into discs and washed extensively with distilled deionized water over a 3 day period. The gels were then cut into smaller discs (0.5 cm diameter x 0.3 cm thickness) and dried for 24 h at room temperature and for 12 h at 60 °C.

Equilibrium Swelling Properties. Dried gels were equilibrated in pepsin-free simulated gastric fluid for 32 h at 37 °C. The equilibrium swelling ratio (Q_{eq}) was calculated by dividing the fully swollen gel weight by the dried gel weight.

Enzyme-Digestion of Hydrogels. To characterize the degradable properties of these systems, dried gels were swollen in pepsin-containing simulated gastric fluid at 37°C. The concentration of pepsin was 250 unit/ml. We had chosen 250 unit/ml since at lower concentrations gel degradation was found to be concentration dependent. During the uptake of penetrants, the weight of the swollen gels was monitored over time. Here Q represents the swelling ratio during dynamic swelling. To compare the effects of pepsin, Q was also calculated for identical gels which were swollen in pepsin-free simulated gastric fluid.

In-Vivo Experiment. Albumin-crosslinked hydrogels were prepared as mentioned above. The content of functionalized albumin was 10% (w/w) of the monomer while the degree of alkylation was 90%. Once polymerization was complete the gels were cut into cylinders (1.3 cm length x 1.0 cm diameter). The gels were purified, dried, and then swollen for 32 h at 37 °C in a 4% (v/v) solution of diatrizoate meglumine/sodium diatrizoate (Gastrografin (GG), Squibb Diagnostics). The GG-loaded gels were then air dried for 24 at room temperature and oven dried at 37 °C for at least one week. Just prior to administering the gel to the dog, the gel was pre-swollen in a 4% (v/v) GG solution until the swollen dimensions were 1.6 cm in length by 1.4 cm in diameter. This pre-swelling step was intended to impart lubricity to the gel surface to ensure safe transit to the stomach. Before each experiment, the dog was fasted for at least 15 h and radiographed just prior to gel administration to ensure the absence of food. To provide a suitable swelling environment, 400 ml of water was pre-administered to the dog using a stomach tube and syringe. Immediately following water instillation, the gel was orally adminstered. At every half-hour following gel administration, 380 ml of water was administered to the dog to maintain the water content in the stomach. Radiographic and ultrasonographic imaging were used to monitor the gel. The experiment was carried out for 8 h.

Results and Discussion

Functionalization of Albumin. The degree of albumin alkylation was proportional to the reaction time. Samples reacted for periods of 0.5 h to 49 h resulted in a degree of alkylation ranging from 8% to 90% of the total available amine groups. Table I illustrates the relationship between the degree of alkylation and reaction time.

Table I.   Relationship Between Reaction Time and
Degree of Albumin Alkylation

Reaction Time (h)	Degree of Alkylation (%)
0.5	8
1	11.7
2	12
5	15
12	27
19	50
26	63
49	90

Equilibrium Swelling Properties. Albumin with as low as 8% alkylation was able to function as a crosslinking agent and form hydrogels. Equilibrium swelling was found to be dependent on both the concentration of functionalized albumin and the degree of albumin alkylation (Figure 1). As the degree of alkylation increased up to 27%, Q_{eq} decreased almost linearly. Little differences, however, were observed at a degree of alkylation greater than 27%. Q_{eq} was also found to decrease by increasing the concentration of functionalized albumin.

Enzyme-Digestion of Hydrogels. The effects of pepsin on gels prepared with albumin having 8% to 15% alkylation is shown in Figure 2. At times exceeding 1 h, the swelling ratios for the gels in the pepsin-containing medium became substantially lower than the gels in the pepsin-free medium. This trend in swelling was also consistent when the content of functionalized albumin was varied. During degradation, it was observed that the actual size of the gels was reduced over time while the integrity of the network was relatively unaltered. This allowed us to calculate swelling ratios as low as 2 below which the gels were too small to be retrieved. The results of this study indicate that the gels were solubilized predominantly by surface degradation. We refer to a predominance of surface degradation since the presence of bulk degradation cannot be excluded; however, when the degree of albumin alkylation is less than 15%, bulk degradation appears to be negligible compared to surface degradation.

When the degree of albumin alkylation was varied from 27% to 90%, gels were more resistant to pepsin degradation. This resistance was directly proportional to the degree of albumin alkylation. Comparison of the gel degradation profiles with similar gels swollen in pepsin-free simulated gastric fluids shows that the dynamic swelling ratios were distinctly different at times greater than 8 h (Figure 3). Since the greater degree of swelling was observed with degrading gels as opposed to non-degrading gels, degradation was the result of net bulk degradation. Although it is possible to lose polymer chains from the gel surface during bulk degradation, this loss of polymer was believed to be minimal; however, further characterization is required.

Mechanisms of Surface and Bulk Degradation. The existence of the two degrading mechanisms which can be effectively controlled by varying the degree of albumin alkylation is a unique phenomenon. As the degree of alkylation increases from 15% to 27%, a transition from net surface degradation to net bulk degradation occurs. In general, albumin-crosslinked hydrogels become more impervious to degradation as the degree of alkylation is increased. The mechanism controlling the presence of either net surface or net bulk degradation is dependent on two factors. The first factor is the modification of albumin by glycidyl acrylate. We have shown previously that modified albumin is more resistant to enzymatic degradation than control albumin (Shalaby, W.S.W; Park, K. Pharm. Res., in press.). The second factor is the degree to which each albumin molecule is incorporated or crosslinked to form a network. When a dried gel begins to swell in the presence of pepsin, it is conceivable that two penetration fronts are established: the first being the pepsin-free swelling front and the second the pepsin-containing degrading front. Initial gel swelling, regardless of the albumin alkylation, can be attributed to the penetration of the pepsin-free swelling front since mesh size restrictions would limit penetration of the pepsin-containing degrading front. The mechanism of degradation is therefore controlled by the hydrogel's response to the degrading front. For gels which are only moderately crosslinked, that is, having a degree of albumin alkylation of 15% or less, penetration of the degrading front will be localized at the gel surface since the digestion of albumin and subsequent loss of polymer chains occurs very efficiently. When the degree of albumin alkylation is 27% or greater, the functionalized albumin is digested more slowly. As a result, the

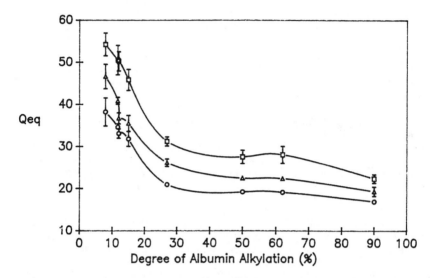

Figure 1   Equilibrium swelling ratio (Q_{eq}) as a function of the degree of albumin alkylation.   The concentration of functionalized albumin was 4.5% (□), 6% (△), and 8% (○).   The absence of error bar indicates that the standard deviation of the data is smaller than the size of the symbol. (n=3).

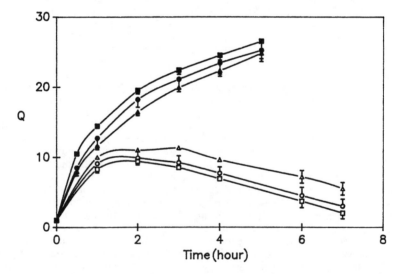

Figure 2   Dynamic swelling of albumin-crosslinked hydrogels containing 8% functionalized albumin as a function of time.   The degree of albumin alkylation was 8% (□), 12% (○), and 15% (△) in the presence of pepsin, and 8% (■), 12% (●), and 15% (▲) in the absence of pepsin.   The absence of error bar indicates that the standard deviation of the data is smaller than the size of the symbol. (n=4).

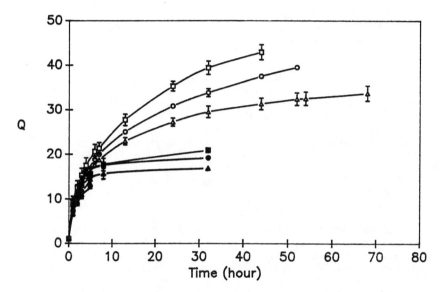

Figure 3   Dynamic swelling of albumin-crosslinked hydrogels containing 8% functionalized albumin as a function of time. The degree of albumin alkylation was 27% (□), 50% (○), and 90% (△) in the presence of pepsin, and 27% (■), 50% (●), and 90% (▲) in the absence of pepsin. The absence of error bar indicates that the standard deviation of the data is smaller than the size of the symbol. (n=4)

degrading front can penetrate into the gel with minimal loss of polymer chains at the gel surface.

The digestibility of albumin-crosslinked hydrogels can be conceptualized by considering the average size and mobility of peptide segments that exist between polymer chains on each albumin molecule. The increase in the degree of albumin alkylation produces a greater reduction in both the size and mobility of these segments since albumin becomes more incorporated within the network with increasing numbers of vinylic pendant groups. By reducing the size and mobility of these segments, steric hindrance at the enzyme active sites arises, penetration of the degrading front occurs, and net bulk degradation results. Conversely, by reducing the degree of alkylation, larger and more mobile segments are produced. The increase in size and mobility enhances the segment's susceptibility to degradation by pepsin. Above a critical segment size and mobility, the rapid digestion of the peptide segments result in net surface degradation. The data presented in this paper supports the above hypothesis; however, further characterization is required.

In-Vivo Experiment. Proteolytic enzymes present in the gastrointestinal tract can be used to degrade albumin-crosslinked hydrogels. Both the mechanism of degradation and the rate of degradation may have a profound influence on the drug release kinetics as well as the gel's gastrointestinal residence time. Hydrogels that could be selectively retained in the stomach to provide sustained drug absorption are promising for long-term oral drug delivery. For long-term gastric retention, gels must swell rapidly to prevent entrance into the intestine, maintain sufficient integrity to overcome disruption by contractile waves, and biodegrade to simplify removal. With these requirements in mind, we studied the in-vivo behavior of hydrogels with net bulk degrading properties.

The location, movements, deformation, and swelling of hydrogels could be monitored in real-time using radiographic and ultrasonographic imaging. Typical radiographic images taken in the stomach of a dog can be seen in Figure 4. Although it may be difficult to see, a contrast in intensity was observed in the original radiographs between the central and peripheral regions of the gel. This contrast was indicative of solvent penetration. The hyperboloidal geometry of the swelling gel has been discussed by Davidson and Peppas (12). Ultrasound images taken after 6 h and 30 min following gel administration (Figure 5) showed a 1.0 cm increase in gel length. During the course of this study, only a moderate increase in swelling was observed. This behaviour indicated that the concentration of pepsin present in the stomach was much lower than the concentration used in in-vitro experiments. The lower pepsin concentration ultimately reduced the enhanced swelling effects expected from bulk degradation. For future experiments, the exact determination of pepsin acitivity in both the fasted and fed state of a canine's stomach will be useful. In the presence of large volumes of fluids, gastric emptying of the gel did not occur during the 8 h experiment. Based on these results, future work will be focussed on using hydrogels that are more susceptible to pepsin in an effort to improve gel swelling and to better understand gel retention in the canine stomach.

Summary

The use of modified albumin as a multifunctional crosslinker can produce hydrogels with a wide variety of swelling and degradation properties. The potential use of these systems for long-term oral drug delivery is promising. With additional modifications, hydrogels that undergo net bulk degradation may be retained in the stomach for long periods of time until enzymatic digestion leads to its solubilization. In the future, this approach may be useful in the delivery of both hydrophobic and possibly hydrophilic drugs. Although our current emaphasis has been placed on net bulk degrading networks to achieve gastric retention, hydrogels that can undergo net surface

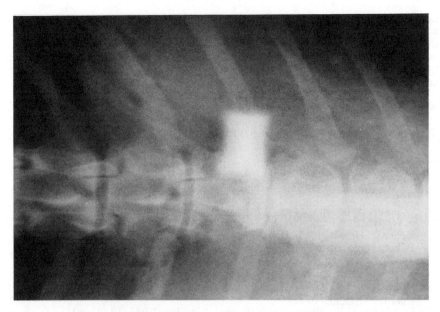

Figure 4   Radiographic image of albumin-crosslinked PVP hydrogel in the stomach of a dog. The image was taken after 1 h following gel administration.

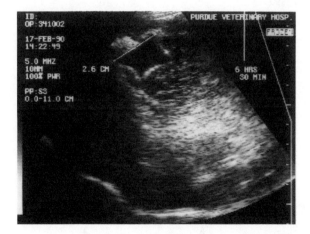

Figure 5   Ultrasonographic image of albumin-crosslinked PVP hydrogel in the stomach of a dog. The image was taken after 6 h and 30 min following gel administration. The length of the gel was determined to be 2.6 cm.

degradation in the gasrtrointestinal tract could also provide unique therapeutic advantages. Since the presence of proteolytic enzymes will produce hydrogel degradation, the application of these systems as implant devices may also be considered. In short, the data presented in this paper suggest that albumin-crosslinked hydrogels represent a promising new alternative to the study and application of water-swellable, yet biodegradable polymer networks.

Acknowledgments

This study was supported in part by ICI Pharmaceuticals Group.

Literature Cited

1. Torchilin, V.P.; Tishenko, E.G.; Smirnov, V.N.; Chason, E.I. J. Biomed. Mat. Res. 1977, 11, 223.
2. Heller, J; Baker, R.W.; In Controlled Release of Bioactive Materials; Baker, R.W., Ed.; Academic: New York, 1980; Chapter 1.
3. Heller, J; Helwing, R.E.; Baker, R.W.; Tuttle, M.E. Biomaterials 1983, 4, 262.
4. Baker, R.W.; Tuttle, M.E.; Helwing, R. Pharm. Technol. 1984, 8 (2), 26, 28, 30.
5. Rosati, L.; Casey, D.J. Polym. Mater. Sci. Eng. 1988, 59, 516.
6. Dickinson, H.R.; Hiltner, A.; Gibbons, D.F.; Anderson, J.M. J. Biomed. Mat. Res. 1981, 15 (4), 577.
7. Dickinson, H.R.; Hiltner, A. J. Biomed. Mat. Res. 1981, 15 (4), 591.
8. Pangburn, S.H.; Trescony, P.V.; Heller, J. Biomaterials 1982, 3 (2), 105.
9. Ulbrich, K.; Strohalm, J.; Kopecek, J. Biomaterials 1982, 3, 150.
10. Park, K. Biomaterials 1988 9 (5), 435.
11. Snyder, S.L.; Sobocinski, P.Z. Anal. Biochem. 1975, 64, 284.
12. Davidson, G.W.R.; Peppas, N.A. J. Controlled Release 1986, 3, 243.

RECEIVED June 4, 1990

Chapter 32

Comparison of Solution and Solid-State Structures of Sodium Hyaluronan by ^{13}C NMR Spectroscopy

Joan Feder-Davis, Daniel M. Hittner, and Mary K. Cowman

Department of Chemistry, Polytechnic University, 333 Jay Street, Brooklyn, NY 11201

Solid state CPMAS and solution ^{13}C-NMR spectra of Streptococcal sodium hyaluronan indicate that conformational changes occur upon dissolution in aqueous solvents. The solid state sample was significantly crystalline; X-ray powder diffraction analysis indicated that it was in a four-fold helical form. There were substantial differences between the solid state NMR spectrum, obtained at 75 MHz, and that of the same sample dissolved in physiological saline solution. The spectral changes are attributed at least in part to a change in the orientation of the acetamido substituent group, and are compatible with additional changes in the dihedral angles about the glycosidic linkages.

Hyaluronan (HA) is a connective tissue polysaccharide with multiple biological functions and a growing commercial importance. HA contributes to the shock absorbing properties of the eye vitreous and the joint synovial fluid, participates in the organization of other macromolecular species in the extracellular matrix, and exerts specific biological effects via cell surface interactions (1-4). Current or potential biomedical applications include uses in ophthalmic surgery, arthritis treatment, wound healing, and as a coating or component of implant materials (5). It is also used as a component of cosmetic formulations (6).

Solutions of HA are highly viscoelastic (7). This property is of particular importance for its mechanical role in the eye and joint, and for many of its commercial applications. We have been involved in efforts to understand and modulate the physical properties of HA solutions, by investigation of the conformation and self-association of HA chains in solution.

HA (Figure 1) is composed of alternating N-acetyl-β-D-glucosamine (GlcNAc) and β-D-glucuronic acid (GlcUA) residues linked at the 1,3 and 1,4 positions, respectively (8). In the solid state, HA has been observed to adopt two-, three-, and four-fold single helices, as well as a four-fold double helix (9-14). The helical form depends on the counterion type and extent of sample hydration. Each of these conformations is stabilized by intrachain hydrogen bonds across the glycosidic linkages, and interchain hydrogen bonds and cation bridges between adjacent helices. The

0097–6156/91/0467–0493$06.00/0

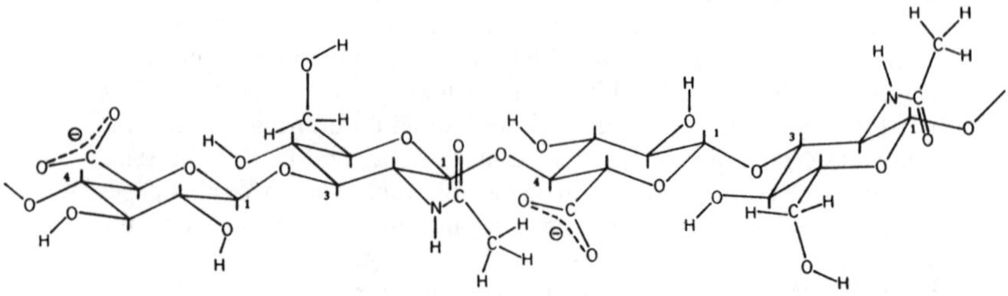

GlcUA   GlcNAc   GlcUA   GlcNAc

Figure 1. Hyaluronan structure. The repeat unit is a disaccharide.

solution conformation is less well established. At physiological pH and ionic strength, HA behaves as a worm-like coil (15) . The moderate degree of chain rigidity has been attributed by Scott (16) to the persistence in aqueous solution of multiple intrachain hydrogen bonds. Another source of rigidity is interchain association. Self-association of HA chains occurs in physiological NaCl solutions (17-19). The chain-chain interactions appear to be relatively weak and dynamic, but may contribute significantly to solution viscoelasticity.

In an effort to further investigate the solution conformation of HA, we have obtained the ^{13}C-NMR spectrum of a solid sample of NaHA, and compared it to the ^{13}C spectrum of the same sample in physiological aqueous solution. Winter et al. (20) reported ^{13}C-NMR studies of a related connective tissue polysaccharide, dermatan sulfate, which showed little evidence for conformational change between solution and solid state. For NaHA, we observe marked differences in the spectral properties upon dissolution. These spectral differences are consistent with loss of at least some inter- and intra-chain hydrogen bonds as well as changes in chain conformation.

Materials and Methods

Sodium hyaluronan, isolated from Streptococcus zooepidemicus, was obtained from Sigma Chemical Company (sample code H-9390, lot 27F-0399). The weight-average molecular weight was determined to be approximately 2.1×10^6 by agarose gel electrophoresis (Lee & Cowman, unpublished). It was stored at 4-8°C under desiccation, and used without further treatment for solid state NMR and X-ray powder diffraction experiments. For solution NMR, the NaHA was dissolved at 4-8°C, at a concentration of 8 mg/mL in H_2O/D_2O (4/1) containing 0.15 M NaCl, 0.34 mM NaH_2PO_4, 1.5 mM Na_2HPO_4, pH 7.1. The solution was transferred to a 10 mm NMR tube and allowed to equilibrate at room temperature prior to analysis.

The X-ray powder pattern was acquired using a Philips XRG-3100 X-ray generator with nickel-filtered copper radiation, operated at 50 kV, 35 mA, with a scanning rate of 0.02°/sec. ^{13}C-NMR spectra were obtained at 21°C using a General Electric GN-300 wide-bore spectrometer operated at 75.57 MHz. For the solid state cross-polarization magic angle spinning (CPMAS) spectrum, a Chemagnetics probe with zirconia rotor containing 200 mg of NaHA was used. The contact time was 2.5 ms, acquisition time 102 ms, and the pulse repetition time was 9 s. The spinning rate was 5 kHz. Approximately 8000 transients were accumulated over a spectral width of 20 kHz using 4K data points. For the solution spectrum, the pulse angle was 30°, and the acquisition time 0.49 s, with no delay between acquisitions. Approximately 150,000 transients were acquired using a spectral width of 16.7 kHz and 16K data points. Line broadening of 10 Hz was used for the solid state spectrum, and 4 Hz for the solution spectrum. No internal reference was used, but the methyl carbon peak was set to 25.18 ppm downfield of DSS (sodium 2,2-dimethyl-2-silapentane-5-sulfonate), based on extensive previous studies of NaHA fragments and polymer in solutions containing acetone as secondary reference (Hittner and Cowman, unpublished).

Results and Discussion

X-ray Analysis. The dry bacterial NaHA sample had a granular appearance, and was found to be significantly crystalline by x-ray diffraction analysis. The powder diffraction pattern yielded five readily measurable maxima (Table I). These were compared to diffraction data published by Arnott and co-workers (9,10,14) for the three different

Table I. Identification of Strep NaHA Conformation:
Comparison of X-Ray Powder Pattern With Published Fiber Patterns

Powder Pattern -This Work		Four-fold Tetragonal Na$^+$ (Ref 9,14)			Four-fold Orthorhombic Na$^+$ (Ref 9,14)			Three-fold Hexagonal Na$^+$ with Ca^{++} (Ref 10)		
2θ[a]	I$_o$[c]	2θ[b]	hkl	I$_o$[d]	2θ[b]	hkl	I$_o$[d]	2θ[b]	hkl	I$_o$[d]
10.3	MS	10.4	102	MS	9.3	102	M	8.7	100	MS
					10.4	012	S	9.3	101	MS
12.6	S	12.9	111	S	12.9	112	S	12.8	103	W
14.6	W	13.7	112	M	14.2	113	W	15.1	110	S
None Observed		16.4	114	M	15.6	201	S	16.4	112	S
					15.8	114	S			
17.8	S	18.1	201	S	17.8	210	S	17.5	200	S
		18.7	202	S	18.0	211	S	18.6	202	S
					18.6	212	VS			
19.6	VS	20.2	211	VS	19.6	023	S	20.7	106	M
		20.8	212	S	19.7	121	S			
					20.2	122	S			
Possibility of weak reflections		22.2	205	S	22.2	124	S	23.2	210	VS
		25.5	220	M	24.9	310	S	24.1	212	S

[a] Scattering angle (degrees).

[b] Calculated from published unit cells.

[c] Intensity observed. **W**, weak; **M**, medium; **MS**, medium strong; **S**, strong; **VS**, very strong.

[d] Based on published structure amplitudes.

helical forms adopted by NaHA in oriented films. The existence of a very strong reflection for our sample at a scattering angle of 19.6° favors the tetragonal or orthorhombic four-fold forms. Furthermore, the hexagonal three-fold form would be characterized by strong reflections near 23-24°, which are not observed in the NaHA sample studied here. The tetragonal form is most consistent with the absence of an observed reflection at 15-16°. All other reflections observed are in agreement with expectations for the tetragonal four-fold helical form. This helical form is stabilized by two hydrogen bonds per disaccharide repeat unit: 1) from the GlcNAc acetamido N-H to the GlcUA carboxylate across the β-1,4 linkage, and 2) from the GlcNAc C4-OH to the GlcUA ring oxygen across the β-1,3 linkage (9,14).

^{13}C-NMR Analysis. The solution and solid state ^{13}C spectra of bacterial NaHA are compared in Figure 2. The solution spectrum is in good agreement with previously published data (21). Chemical shifts and assignments of the observed resonances are given in Table II. The assignments are taken from the work of Bociek et al.(21). We have found (Hittner, Feder-Davis and Cowman, unpublished) that the ^{13}C chemical shifts observed for neutral salt solutions of NaHA are independent of sample molecular weight, down to at least the hexasaccharide level (disregarding resonances from sugars at the chain ends).

Table II. [13]C Chemical Shifts For NaHA
In Solution and Solid State

Assignment	Chemical Shift (ppm)	
	Solution	Solid State (CPMAS)
CO-GlcNAc	177.6	(177.7, 176.7, 175.1)
COO-GlcUA	176.7	(177.7, 176.7, 175.1)
C1-GlcUA	105.8	(103.6, 102.0, 100.9)
C1-GlcNAc	103.1	(103.6, 102.0, 100.9)
C3-GlcNAc	85.2	(88.7, 86.7, 85.2)
C4-GlcUA	82.6	(88.7, 86.7, 85.2)
C5-GlcUA	78.9	80.2
C5-GlcNAc	78.1	(77.5 - 78.1)
C3-GlcUA	76.3	76.2
C2-GlcUA	75.2	75.1
C4-GlcNAc	71.1	72.3
C6-GlcNAc	63.3	(62-68)
C2-GlcNAc	57.0	56.5
CH$_3$-GlcNAc	25.2	(26.2, 25.5, 24.9, 24.2)

The solid state ^{13}C-CPMAS spectrum shows several significant differences relative to the solution spectrum. The resonance for C6 of GlcNAc is very broad and appears to be shifted downfield in the solid state spectrum. It should be noted that the hydroxymethyl group participates in hydrogen bonds linking adjacent helices in the tetragonal four-fold form (14). These hydrogen bonds would be lost upon dissolution. (As described above, weak chain-chain association is known to exist in solutions of

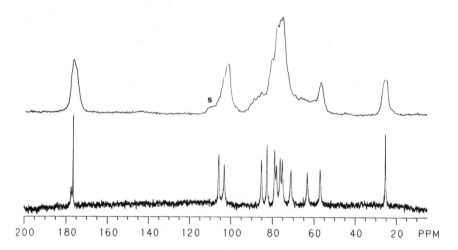

Figure 2. 75 MHz ^{13}C-NMR spectra of Streptococcal sodium hyaluronan. Top: solid state CPMAS spectrum. The peak labelled (s) is a spinning sideband. Bottom: solution spectrum. The solvent was 0.15 M NaCl, phosphate buffered to pH 7.1.

NaHA, but has proven difficult to detect by ^{13}C-NMR [Hittner and Cowman, unpublished]).

Dissolution of NaHA is also known to disrupt the intrachain hydrogen bond linking the acetamido N-H (amide proton) and carboxylate groups. Circular dichroism studies gave the first indication for significant change in the acetamido group environment upon transition from solid state to solution (22). The amide proton was subsequently observed directly by ^{1}H-NMR in studies of aqueous solutions of HA fragments (23). The coupling constant $^3J_{HNCH}$ indicates that the amide proton is nearly trans to H2. More recent ^{1}H-NOESY studies by Lerner and co-workers (personal communication) show that the amide proton is closer to the GlcUA Hl than to the GlcNAc Hl, and cannot be directly hydrogen bonded to the carboxylate. Heatley and Scott (24) have recently proposed the existence of a hydrogen bond from the amide proton to the carboxylate via a bound water molecule. This proposal must be examined by further experiments. In either case, the results indicate that a change in the acetamido group orientation occurs upon transfer from the solid to the solution state.

We have observed only subtle changes in the ^{13}C resonances assigned to the acetamido and carboxylate groups. The methyl carbon shows several environments in the solid state. There are similarly several environments for the acetamido carbonyl and/or carboxylate carbons, at least one of which moves downfield in solution. This may primarily reflect cation binding to the carboxylate in the solid state. It may, however, also reflect the altered carboxylate environment in the presence of the hydrogen bond. The existence of a different carboxylate environment in the solid state is further supported by the observation of an upfield shift for C5-GlcUA upon transfer to solution.

In going from the solid to the solution state, a loss of the acetamido-to-carboxylate hydrogen bond across the β-1,4 linkage may be expected to result in a change in the lowest energy conformers about that linkage. The Cl resonance of GlcNAc appears to occur at a more downfield position in aqueous solution than in the solid state. (While it is possible that only the Cl of GlcUA is at a substantially different chemical shift in solution vs solid, we favor an assignment scheme in which both Cl resonances are moved. This interpretation is based on studies of HA hexasaccharide in dimethyl sulfoxide solution, wherein the acetamido-to-carboxylate hydrogen bond is known to exist. [Feder-Davis and Cowman, unpublished]). The environment of the C4 resonance of GlcUA is also clearly different in solution vs solid, being shifted upfield upon dissolution. Thus a change in dihedral angle about the β-1,4 linkage is compatible with our data.

The β-1,3 linkage is stabilized in the solid state by a hydrogen bond from C4-OH of GlcNAc to the GlcUA ring oxygen. This hydrogen bond may also be disrupted in aqueous solution. We observe a moderate change in the C4-GlcNAc resonance position. We also observe a significant change in the Cl-GlcUA chemical shift (more downfield in solution), and possibly the C3-GlcNAc resonance (more upfield in solution, assuming that both C3-GlcNAc and C4-GlcUA resonances move). Thus a change in the conformation about the β-1,3 linkage is compatible with our data.

An alternative explanation for the changes in chemical shift of the Cl-GlcUA resonance exists, which may not require substantial change in the β-1,3 linkage geometry. Bock et al. (25) have correlated the ^{13}C chemical shifts of Cl carbons in disaccharides with interproton distances across the glycosidic linkage. Downfield shifts occur with shorter interproton distances. We observe a downfield shift for Cl- GlcUA upon change from the solid to the solution state. This may result primarily from the closer approach of the amide proton to Hl of GlcUA in solution, as opposed to its more distant position in the solid state.

Conclusions

The viscoelasticity of NaHA solutions reflects both interchain association and inherent chain stiffness. The stiffness of NaHA chains arises largely from the limited conformational freedom about the glycosidic linkages (26). Whether any of the intrachain hydrogen bonds suggested by Scott (16,24) also exist in aqueous solution (and contribute to chain rigidity) remains open to question. Competition by solvent molecules should diminish their frequency of occurrence. It is clear from the present work that the rigid helical structure found in the solid state is altered in aqueous solution, with the rupture of at least one intrachain hydrogen bond.

Acknowledgments

We thank Dr. L. Lerner and Dr. W.T. Winter for providing helpful discussions.

Literature Cited

1.   Balazs, E.A.; Denlinger, J.L. In The Eye, Vol. IA. Vegetative Physiology and Biochemistry; Davson, H., Ed.; 3rd Ed., Academic: New York, 1984; pp 533-589.
2.   Heinegård, D.; Sommarin, Y. Meth. Enzymol. 1987, 144D, 305-319.
3.   Laurent, T.C.; Fraser, J.R.E. In Functions of the Proteoglycans; Ciba Foundation Symposium No. 124; Wiley: Chichester, 1986; pp 9-29.
4.   Toole, B.P. Conn. Tiss. Res. 1982, 10, 93-100.
5.   Van Brunt, J. Bio/Technology 1986, 4, 780-782.
6.   Balazs, E.A.; Band, P. Cosmet. Toiletries 1984, 99, 65-72.
7.   Gibbs, D.A.; Merrill, E.W.; Smith, K.A.; Balazs, E.A. Biopolymers 1968, 6, 777-791.
8.   Meyer, K. Federation Proc. Fed. Amer. Soc. Exp. Biol. 1958, 17, 1075-1077.
9.   Guss, J.M.; Hukins, D.W.L.; Smith, P.J.C.; Winter, W.T.; Arnott, S.; Moorhouse, R.; Rees, D.A. J.Mol. Biol. 1975, 95, 359-384.
10. Winter, W.T.; Smith, P.J.C.; Arnott, S. J. Mol. Biol. 1975, 99, 219-235.
11. Sheehan, J.K.; Gardner, K.H.; Atkins, E.D.T. J.Mol.Biol. 1977, 117, 113-135.
12. Winter, W.T.; Arnott, S. J. Mol. Biol. 1977, 117, 761-784.
13. Sheehan, J.K.; Atkins, E.D.T. Int. J. Biol. Macromol. 1983, 5, 215-221.
14. Mitra, A.K.; Raghunathan, S.; Sheehan, J.K.; Arnott, S. J. Mol. Biol. 1983, 169, 829-859.
15. Cleland, R.L. Biopolymers 1984, 23, 647-666.
16. Scott, J.E. In The Biology of Hyaluronan; Ciba Foundation Symposium No. 143; Wiley: Chichester, 1989; pp 6-20.
17. Welsh, E.J.; Rees, D.A.; Morris, E.R.; Madden, J.K.; J. Mol. Biol. 1980, 138, 375-382.
18. Sheehan, J.K.; Arundel, C.; Phelps, C.F. Int. J. Biol. Macromol. 1983, 5, 222-228.
19. Turner, R.E.; Lin, P.; Cowman, M.K. Arch. Biochem. Biophys. 1988, 265, 484-495.
20. Winter, W.T.; Taylor, M.G.; Stevens, E.S.; Morris, E.R.; Rees, D.A. Biochem. Biophys. Res. Commun. 1986, 137, 87-93.
21. Bociek, S.M.; Darke, A.H.; Welti, D.; Rees, D.A. Eur. J. Biochem. 1980, 109, 447-456.
22. Buffington, L.A.; Pysh, E.S.; Chakrabarti, B.; Balazs, E.A. J. Am. Chem. Soc. 1977, 99, 1730-1734.
23. Cowman, M.K.; Cozart, D.; Nakanishi, K.; Balazs, E.A. Arch. Biochem. Biophys. 1984, 230, 203-212.

24. Heatley, F.; Scott, J.E. Biochem. J. 1988, 254, 489-493.
25. Bock, K.; Brignole, A.; Sigurskjold, B.W. J. Chem. Soc., Perkin Trans. II 1986, 1711-1713.
26. Potenzone, R.; Hopfinger, A.J. Polymer J. 1978, 10, 181-199.

RECEIVED June 4, 1990

Chapter 33

Thermoreversible Gels

Shalaby W. Shalaby

Bioengineering Department, Clemson University, Clemson, SC 29034-0905

Consistent with the theme of this book, the present chapter should address water-soluble, thermoreversible gels. However, due to the great importance of the non-aqueous gels and availability of significant theoretical treatments on these systems, discussion of non-aqueous thermoreversible gels is included as part of the chapter.

Definition, Major Types and Theoretical Aspects

Different types of linear, long-chain polymers which can form gel or network structures have been recognized over 25 years (Flory, 1974; Mandelkern, 1964; Russo, 1987). For copolymers which undergo gelation by crystallization from dilute solutions, the formation of three-dimensional network was suggested, where primary molecules of finite molecular weight are reversibly bound by crystallites and hence non-covalently crosslinked (Flory, 1974; Mandelkern 1979; Takahashi, Sakai and Kato, 1980). When gelation occurs as a result of crystallization from dilute or moderately dilute solutions, a very fluid solution is converted to a rigid medium of infinite viscosity and the polymer system pervades the entire volume (Domszy, et. al, 1986). These gels are formed by crystallization upon cooling and revert to fluid solutions or melts upon heating and hence denoted as thermoreversible gels (Domszy, et. al, 1986). Formation of thermoreversible gels by this mechanism was described not only for copolymers but also for a number of homopolymers (Domszy, et. al, 1986). The latter include poly(4-methylpentene) (Charlet, Phuong-Nguyen and Delmas, 1984), isotactic polystyrene (Guenet, Lotz and Wittman 1985) and poly(vinyl chloride) (Guerrero and Keller, 1981). In a recent study of thermoreversible gelation of linear poly(ethylene) and ethylene copolymers, it was found that the thermodynamic properties of the gels are continuous from the homopolymers to the highly branched copolymers (Domszy, et. al., 1986). At the same time, major changes in the supermolecular structure were observed and noted to reflect the gradual degradation of the lamellar crystallite structure, responsible for the quasi-crosslinking in the gel. Domszy, et. al., (1986) proposed that the principles governing the gel formation of the above ethylene-based polymer should be applicable to all types of crystallizable homopolymers and copolymers. In more recent studies on the molecular structure of isotactic polystyrene, thermoreversible gels, it was noted that the chain conformation in the gel (present as solvated helices) is totally different from what is in the semicrystalline form (Kline, Brulet and Guenet, 1990).

So far the discussion has been limited to reversible gel formation of crystallizable polymers from dilute non-aqueous solutions. However, gel formation of amorphous polymers such as atactic polystyrene is also achievable from dilute

· 0097–6156/91/0467–0502$06.00/0

non-aqueous solutions (Tanaka, 1989; Wellinghoff and Shaw, 1979). Tanaka (1989) introduced a simple model to describe observed coexistance of gelation and phase mixing of non-crystallizable atactic polystyrene in carbon disulfide. The model demonstrates that multiple-equilibria conditions for molecular clustering caused by physical crosslinking can describe the characteristic features of the temperature-concentration phase diagram. Predominantly syndiotactic poly(methyl methacrylate) (s-PMMA) is noted for its ability to form thermoreversible gels by cooling its solutions in any of the several organic solvents which display strong interaction with this polymer (Konnecke and Rehage, 1983; Spevacek, et. al, 1982). Until recently, the exact mechanism of thermoreversible gelation in solutions of s-PMMA was not fully understood. However, in a report by Berghams and coworkers (1987) it was proposed, based on extensive experimental data, that the thermoreversible gelation of s-PMMA in o-xylene proceeds by a two-step mechanism, different from the usual crystallization-gelation discussed above. This entails a fast conformational change followed by a slow intermolecular association. It was also noted that a minimum degree of stereoregularity, corresponding to a minimum average sequence length, which is characteristic of certain degree of cooperativeness in the conformational change, is needed. A relation between this minimum length and the gelation temperature was, thus, documented.

Thermoreversible gels originating from the non-aqueous polymer solutions, discussed above, are formed by cooling these solutions to allow phase-separation and the establishment of the physical network structures. Upon dissolving polymers in organic solvents, a small positive change in enthalpy is encountered, which is counterbalanced by large positive gain in entropy and hence the process of dissolution can be realized. For aqueous polymer solutions, the opposite is often observed. Subsequently, the process of gelation through phase separation takes place as the solution temperature is raised to a critical value, the "lower critical solution temperature" or LCST (Franks, 1983; Hoffman, 1987). Such phase separation is also an entropy-driven process where the release of structured bound-water from the hydrophobic groups along the polymer main chain, is the major contributing thermodynamic force. An important group of water-soluble, non-ionic polymers, which form thermoreversible gels having a wide range of LCST's, are based on N-alkyl acrylamide homopolymers and copolymers with or without basic or acidic comonomers (Hoffman, 1987; Russo, 1987). Of these polymers, poly(N-isopropylacrylamide) (PolyNiPAAm) has received the attention of most investigators (Heskins and Guillet, 1986, Hoffman, 1987, Kubota and Fujishige, 1990; Winnik, Ringsdorf and Winzmer, 1990). A major reason for the great interest in PolyNiPAAm, is having an LCST around 32°C (Heskins and Guillet, 1986), which allows one to incorporate proteins, with denaturing, in aqueous systems of this polymer. Having an LCST approaching that of body temperature, made PolyNiPAAm also an attractive system for studies pertinent to pharmaceutical and biomedical applications. (Dong and Hoffman, 1986; Hoffman 1987). In their study of PolyNiPPAm in aqueous solutions and the effect of various molecular parameters on the solution properties Kubota and coworkers (1990) have noted that the polymer molecule behaves like a flexible coil and the unique thermal behavior of its solution, in the concentration range of 1 wt. %, should be attributed mainly to the hydrated structure. In a note by Winnik and coworkers (1990) methanol-water was described as a co-nonsolvent system for PolyNiPPAm.

A second important group of non-ionic, water-soluble polymers which can form thermoreversible gels, are the triblock copolymers made of poly(ethylene oxide) (PEO) and poly(propylene oxide) (PO) with the crystallizable PEO as the terminal blocks (Molyneux, 1982); Prasad, et. al., 1979). Depending on the length of the individual blocks and the total average molecular weight of the PEO-PO-PEO chains, these polymers can form thermoreversible gels with a range of LCST's (Prasad, et. al., 1979).

Technological Aspects

This section outlines advances in materials development approaches to control the gelation processes and new applications.

A. Advances in Materials & Properties - Recent information pertinent to the development of new thermoreversible gel systems are disclosed, mostly, in the patent literature. Among the interesting systems are those (a) made, primarily, of agar, glycerol and certain electrolytes and geared for use as impression material in dentistry (Skalska, 1984), (b) based on 1,3-glucan polysaccharide (Provonchee, and Renn, 1987) and (c) prepared from partially hydrolyzed poly(acrylonitrile) (Stoy, et. al., 1979). In a report by Taylor and Kolesinski (1986) poly(trimethylamine-p-vinylbenzimide) was noted as a polymer whose aqueous solutions show the properties of negative thixotropy and thermoreversible gelation. In addition to the search for new materials, a number of attemps have been made, recently, to modify the performance of known thermoreversible gels, which included use of gamma radiations to achieve partial crosslinking in the PEO-PO-PEO triblock system (Attwood, Tait and Collett, 1987).

Among the recent studies on the properties of thermoreversible hydrogels which may have broad implications on future developments and applications of these systems are those of Stoks and coworkers (1988) on the gelation of poly(vinyl alcohol) in ethylene glycol. Similar to other non-aqueous systems, the gelation that occurred on cooling resulted in a liquid-liquid demixing followed by a crystallization of the polymer in the concentrated domains. Rheological studies of these gels showed that under no circumstances was a real network formed. In the study of the thermoreversible gelation of atactic polystyrene in cyclohexanol by Hikmet, Collister and Keller (1988), matrix inversion from a solvated rubbery to a glassy matrix with increasing polymer concentration was demonstrated, with bicontinuous network structures in the intermediate stages, both by inspection of macroscopic consistency and by SEM study of the morphology. Glass transition, crystallization and thermoreversible gelation in ternary poly(propylene oxide) solutions and relationship to asymmetric membrane formation was studied by Burghardt, Yilmaz & McHugh (1987). The thermally reversible sol-gel transition in newly synthesized poly(organophasphazes), containing OEt, NHEt., and OH side groups , were studied, recently, by Tanigami and coworkers (1989). The phase separation in aqueous system was attributed to the balance of the hydrophilic and hydrophobic groups in the polymer and gelation was noted to occur by hydrogen bonding among the side groups.

B. Control of the gelation process & new applications - The general phenomenon of reversible gel formation, which is associated with phase transition, usually takes place in response to infinitesimal changes in temperature, solvent composition, pH, ionic composition and small electric fields. (Amiya and Tanaka, 1987; Hirokawa and Tanaka, 1984, Tanaka, et. al, 1982). The phase transition can be accompanied by a reversible, discontinuous volume change. Due to the technological importance of large volume changes during the formation of gels other than the thermally-activated types, this section will also address athermally activated reversible gels. Among the existing or potential applications of reversible gels are those pertinent to their use as sensors, chemomechanical transducers, switches, display units, controllable drug delivery systems and selective pumps (Mamada, et. al., 1990 and Tanaka, 1978). In some of these applications it is desirable to have light-induced phase transitions, where diffusion controlled variables (pH, temperature and electric field) may compromise the response time during gel formation (Mamada, et. al., 1990). In response to this need Mamada and coworkers (1990) directed their efforts to the study of photo-induced phase transition of gels and reported the results of the first successful observation of ultraviolet-induced phase transition in PolyNiPAAm in presence of a photosensitive molecule, bis(4-(dimethylamino)phenyl)(4-vinylphenyl)methyl leucocyanide.

The chemistry thermoreversible gels based on homopolymers and copolymers of N-isopropylacrylamide and their biomedical and pharmaceutical applications have been studied extensively by A. S. Hoffman and his coworkers (Hoffman, 1987). Dong and Hoffman (1988) have shown, for the first time, that an enzyme can be immobilized within a thermally reversible hydrogel and the enzyme activity is "shut off" when the gel temperature is raised above its LCST. The enzyme-gel can regain its activity below the LCST. Hoffman, Afrassiabi and Dong (1986) have shown that these hydrogels can be used, broadly, for the delivery and selective removal of substances from aqueous solutions.

Recent applications of thermoreversible gels include (a) the diffusion controlled delivery of a peptide drug (rat atrial natriuretic factor) in PEO-PO-PEO (or Poloxamer) (Juhasz, et. al., 1989) and (b) the immobilization of Arthrobacter simplex cells and comparative effects of organic solvent and polymeric surfactant on steroid conversion (Park and Hoffman, 1989).

Thermoreversible Gels Based on Proteins and Polysaccharides

Formation of thermoreversible gels of proteins and polysaccharides is covered extensively in the old & recent literatures, (Glass, 1986, Molyneaux, 1982; Russo, 1987). Among the most addressed systems are collagen as a protein and agarose as a polysaccharide. Additional treatment of the subject in this chapter is, thus, limited to relatively new findings pertinent to polymeric systems of growing importance.

In their study of chitosan, Hayes and Davies (1978) proposed that in aqueous solutions, the polymer normally exists in a random coil configuration. However, in the presence of oxalic acid the chitosan chains from double helices, thus creating physical crosslinks which eventually lead to gel formation. In contrast, no chitosan gels could be formed with other di- and poly-basic acids such as boric, maliec, malic, succinic, phosphoric and citric acids (Yamaguchi, et. al., 1978). The chemical structures of various carrageenans, an important group of polysaccharides, and the mechanism of reversible gelation were studied by Ter Meer and Burchard, (1983). Starch hydrolysis products with low level of debranching, (e.g. low debranched maltodextrins) were shown to form thermally reversible gels (Bulpin, Cutler and Dea, 1984). Kyu and Mukherjee (1988) studied the kinetics of phase separation in a liquid crystalline solution of hydroxypropyl-cellulose gel.

Literature Cited

1. Amiya, T. and Tanaka, T., Macromolecules, **20**, 1162(1987).
2. Attwood, D., Tait, C. J. and Collet, J. H., A.C.S. Symp Ser. **348**, 128(1987).
3. Berghams, H., Donkers, A., Frenay, L., Stoks, W., DeSchryver, F.E., Moldenaers, P. and Mewis, J., Polymer, **28**, 98(1987).
4. Bulpin, P.V., Cutler, A.N. and Dea, I.C.M., Gums Stab. Food Ind.Appl. Hydrocolloids, Proc. Int. Conf., (G.O. Phillips, D.J.Wedlock and P.A. Williams, Eds.) Pergamon, Oxford, U.K. 1984, p:475.
5. Burghardt, W.R., Yilmaz, L. and McHugh, A.J., Polymer 28, 2085(1987).
6. Charlet, G., Phuong-Nguyen, H. and Delmas, G., Macromolecules, **17**:1200(1984).
7. Domszy, R.C., Alamo, R., Edwards, C.O. and Mandelkern, L., Macromolecules **19**, 310(1986).
8. Dong, L.C. and Hoffman, A.S., J. Contr. Rel. **4**, 223(1986).
9. Flory, P. J., Discuss. Faraday Soc. **57**, 7(1974).
10. Frank, F. in "Chemistry and Technology of Water-Soluble Polymers" (Finch, C.A., Ed.) Plenum Press, New York 1983, Chap. 9.
11. Fujishige, S., Kubota, K. and Ando, I., J. Phys. Chem. **93**, 3311(1989).
12. Glass, J. E., Ed. "Water-Soluble Polymers - Beauty with Performance", ACS Adv. Chem. Series, Vol. **213**, ACS-Washington, D.C., 1986.
13. Guenet, J.M., Lotz, B. and Wittman, J.C., Macromolecules **18**, 420(1985).
14. Guerrero, S.J. and Keller, A., J. Macromol. Sci., Phys. **B20**,167(1981).
15. Hayes, E.R. and Davies, D.H., Proc.Int.Conf.Chitin and Chitosan, M.I.T., Cambridge, MA, 1978; Chem. Abstr. **91**, 20930k(1979).
16. Heskins, M. and Guillet, J.E., J. Macromol. Sci. Chem., **A-2**, 1441(1986).
17. Hikmet, R.M., Callister, S. and Keller, A., Polymer, **29**, 1378(1988).
18. Hirokawa, Y. & Tanaka, T., J. Chem. Phys. **81**, 6379(1984).
19. Hoffman, A.S., J. Contr. Rel. **6**, 297(1987).

20. Hoffman, A.S., Afrassiabi, A. and Dong, L.C., J. Contr. Rel., **4**, 213(1986).
21. Juhasz, J., Lenaerts, V., Raymond, P. and Ong, H., Biomaterials **10**, 265(1989).
22. Klein, M., Brulet, A. and Guenet, J.-M., Macromolecules **23**, 540(1990).
23. Konnecke, K. and Rehage, G., Makromol. Chem. **184**, 2679(1983).
24. Kubota, K., Fujishige, S. and Ando, I., Polymer J., **22**, 15(1990).
25. Kyu, T. and Mukherjee, P., Liq. Cryst. **3**, 631(1988).
26. Mamada, A., Tanaka, T., Kungwatchakun, D. and Irie, M., Macromolecules, **23**, 1517(1990).
27. Mandelkern, L. "Crystallization of Polymers" McGraw Hill, New York, 1964 pages 113 and 308.
28. Mandelkern, L., Discuss. Faraday Soc. **68**, 685(1979).
29. Molyneaux, P. "Water-Soluble Polymers: Properties and Behavior", CRC Press, Vol. I, p.58, Vol. II, p.86, New York, 1982.
30. Park, T.G. and Hoffman, A.S., Biotechnol. Lett. **11**, 17(1989).
31. Prasad, K.N., Luong, T.T., Florence, A.T., Paris, J., Vaution, C., Seiller, M. and Puisieux, F., J. Colloid Interface Sci., **69**, 225(1979).
32. Provonchee, R.B. and Renn, D.W., Ger. Offen. (To FMC)3,621,303 (1987); Chem. Abstr. **106**, 86565e(1987).
33. Russo, P.S., Ed., "Reversible Polymeric Gels & Related Systems", ACS Symp. Ser. **350**, Washington, D.C., 1987.
34. Skalska, A. Czech. Pat., 225,302(1984); Chem. Abstr. **104**,75083d(1986).
35. Spevacek, J., Schneider, B., Bohdanecky, M. and Sikora, A.J., J. Polym. Sci., Polym. Phys. Edit. **20**, 1623(1982).
36. Stoks, W., Berghmans, H., Moldenaers, P. and Mewis, J., Br. Polym. J. **20**, 361(1988).
37. Stoy, V., Stoy, A., Zima, J. and Ilavsky, M., Czech. Pat, 178,638(1979); Chem Abstr. **92**, 239899d(1980).
38. Takahashi, A., Sakai, M. & Kato, T., Polym. J. (Tokyo) **12**, 335(1980).
39. Tanaka, F., Macromolecules, **22**, 1988(1989).
40. Tanaka, T., Phys. Rev. Lett. **40**, 820(1978).
41. Tanaka, T., Nishio, I., Sun, S-T. and Ueno-Nishio, S., Science **218**, 467(1982).
42. Tanigami, T., Ono, T., Suda, N., Sakamaki, Y., Yamaura, K. and Matsuzawa, S., Macromolecules **22**, 1397(1989).
43. Taylor, L.D. and Kolesinski, H.S., J. Polym. Sci., Part C:Polym.Lett. **24**, 287(1986).
44. Ter Meer, H.U. and Burchard, W., Org. Coat. Appl. Polym. Sci. Proc., **48**, 338(1983).
45. Wellinghoff, S., Shaw, J. & Baer, A., Macromolecules, **12**, 932(1979).
46. Winnik, F.M., Ringsdorf, H. & Winzmer, J., Macromolecules **23**, 2415(1990).
47. Yamaguchi, R., Hirano, S., Arai, Y. and Ito, T., Agric. Biol. Chem. **42**, 1981(1978).

RECEIVED September 21, 1990

Author Index

Ander, Paul, 202
Audebert, R., 218
Blevins, William E., 484
Borchardt, John K., 446
Boustta, M., 405
Butler, George B., 25,151,159,175
Clark, Mark D., 291
Cowman, Mary K., 493
Cummins, Dosha F., 338
Da, A-H, 159
Daly, William H., 189
Do, Choon H., 151
Donnelly, Donna L., 418
Dust, Julian M., 418
Edgell, Michael J., 418
Ezzell, Stephen A., 130
Feder-Davis, Joan, 493
Frank, Curtis W., 303
Gopalkrishnan, Sridhar, 175
Hamielec, A. E., 82,105
Harris, J. Milton, 418
Harris, Patricia A., 418
Hemker, David J., 303
Hester, Roger D., 276,320
Hittner, Daniel M., 493
Hogen-Esch, Thieo E., 159,175
Hoyle, Charles E., 291
Huguet, J., 405
Hunkeler, D., 82,105
Iliopoulos, I., 218

Karr, Laurel J., 418
Lee, Soo, 189
Leung, Sau-Hung S., 350
McCormick, Charles L.,
        2,119,130,291,320,338
McGill, R. Andrew, 418
Mettille, Michael J., 276
Middleton, John C., 338
Molyneux, Philip, 232
Morgan, Sarah E., 320
Neuse, Eberhard W., 394
Oyama, Hideko T., 303
Park, Kinam, 484
Perlwitz, Axel G., 394
Regelson, William, 367
Robinson, Joseph R., 350
Salazar, Luis C., 119
Sandford, Paul A., 430
Schild, Howard G., 249,261
Schulz, D. N., 57
Sedaghat-Herati, Reza M., 418
Shah, Kishore R., 74,468
Shalaby, Shalaby W., 74,502
Shalaby, Waleed S. W., 484
Steinnes, Arild, 430
Vert, M., 405
Wang, T. K., 218
Zhang, Nai Zheng, 25,175
Zhang, Y-X, 159

Affiliation Index

Clemson University, 74,502
Columbia Research Laboratories, 350
Ethicon, Inc., 74,468
Exxon Research and Engineering
  Company, 57
Halliburton Services, 446
INSA de Rouen, 405
Louisiana State University, 189
McMaster University, 82,105
Macrophile Associates, 232
Medical College of Virginia, 367
NASA–Marshall Space Flight Center, 418
Polytechnic University, 493
Protan A/S (Norway), 430
Protan, Inc. (USA), 430

Purdue University, 484
Seton Hall University, 202
Stanford University, 303
Université Pierre et Marie Curie (Paris), 218
University of Alabama—Huntsville, 418
University of Florida—Gainesville,
  25,151,159,175
University of Massachusetts—Amherst,
  249,261
University of Southern California, 159
University of Southern Mississippi,
  2,119,130,276,291,320,338
University of Wisconsin—Madison, 350
University of the Witwatersrand
  (South Africa), 394

Subject Index

A

Acidic fibroblast growth factor, effect of
  heparin, 377–378
Acrylamide, mechanism and kinetics of
  persulfate-initiated polymerization,
  82–101
Acrylamide–3-(2-acrylamido-2-
  methylpropyl-dimethylammonio)-1-
  propanesulfonate copolymers
  added electrolytes, effect, 124,126f,127
  characterization, 122
  compositions, 123t,124,125f
  macromolecular structure, 122,123t
  microstructure, 124t
  molecular weight, 123t
  monomer synthesis, 120,121f
  pH, effects, 127,128f
  polymer concentration, effect, 127,128f
  polymerization, degree, 123t
  reaction parameters for
    copolymerization, 122t
  reactivity ratios, 123,124t,125f
  synthesis, 120,122
  viscosity measurements, 122

Acrylamide-based polyelectrolytes,
  hydrophobically modified, See
  Hydrophobically modified acrylamide-
  based polyelectrolytes
Acrylamide–n-decylacrylamide copolymer
  system, effect of concentration on
  apparent viscosity, 130,131f
Acrylamide polymerization
  inverse emulsion polymerization,
    64,65f,66
  inverse microemulsion polymerization, 66
  precipitation polymerization, 67
  solution polymerization, 60–63t
Acrylamide polymerization using persulfate
  initiation
  cage-effect theory, 87–88
  complex theory, 87–88
  inhibition, 85
  initiation mechanism
    elucidation, 91–98,100
    generalization to other water-soluble
      monomers, 98–99,100t,101
  kinetics, 85–91
  radical addition mechanism, 85
  redox pairs, use, 83,84t

Acrylamide polymerization using persulfate
  initiation—*Continued*
reduction oxidation, 83
solvent(s), effects, 83,85
solvent-transfer theory, 85
Acrylamide polymerization using persulfate
  initiation at high monomer concentration
comparison of kinetic model to
  experimental data, 112,114–117*f*
consumption rate of initiator, effect on
  initial monomer concentration,
  112,117*f*
conversion time and weight-average
  molecular weight conversion data and
  model predictions, 112,114–116*f*
experimental conditions, 109,110*t*
experimental method of polymerization,
  107,108*f*
experimental procedure, 107
limiting conversions, 109,110*t*,111*f*
molecular-weight determination, 106–107
parameter estimation, 109,112,113*t*
rate order vs. monomer
  concentration, 109
residual monomer concentration
  determination, 106
thermal and monomer-enhanced
  decomposition of potassium persulfate,
  relative magnitudes, 112,117*f*
Acrylamide polymer synthesis,
  See Acrylamide polymerization
Acrylamide–sulfobetaine amphoteric
  monomer copolymers, *See*
  Acrylamide–3-(2-acrylamido-2-methyl-
  propyldimethylammonio)-1-propane-
  sulfonate copolymers
3-(Acrylamido)-3-
  methylbutyltrimethylammonium
  chloride, synthesis, 35
2-(Acrylamido)-2-methylpropanesulfonic
  acid, synthesis, 44
3-(2-Acrylamido-2-methylpropyldimethyl-
  ammonio)-1-propanesulfonate–acrylamide
  copolymers, *See* Acrylamide–3-[2-(acrylamido)-2-methylpropyldimethylammonio]-
  1-propanesulfonate copolymers
Aggregation
diffusion-limited cluster–cluster
  aggregation model, 315–317
process, 315

Albumin–cross-linked polyvinylpyrrolidone
  hydrogels
albumin functionalization, 486*t*
bulk degradation mechanism, 487,490
enzyme digestion, 487,488–489*f*
equilibrium swelling properties,
  485,487,488*f*
hydrogel synthesis, 485
in vivo experiment, 486,490,491*f*
procedure for albumin
  functionalization, 485
procedure for enzyme digestion of
  hydrogels, 486
radiographic image, 490,491*f*
surface-degradation mechanism, 487,490
ultrasonographic image, 490,491*f*
Alginate, 76–77
Algorithm development for molecular-
  weight-distribution determination
choice of function-fit method,
  279–280,281*f*,282
regression algorithm, 284
series use for integration, 282–284
Aliphatic polyesters, application, 405–406
Alkyldiallylamine
characterization procedure, 153
^{13}C-NMR spectra of copolymers, 153,154*f*
^{13}C-NMR spectrum, 153*f*
cyclocopolymerization
  procedure, 153
  reaction, 152
homocyclopolymerization, 152
polymeric micelles, formation, 157*f*
side-chain crystallization of
  copolymers, 155,156*f*
synthesis, 152
Alkyldiallylmethylammonium chloride
characterization procedure, 153
^{13}C-NMR spectra of copolymers,
  153*f*,154,155–156*f*
cyclocopolymerization
  procedure, 153
  reaction, 152
homocyclopolymerization, 152
polymeric micelles, formation, 157*f*
synthesis, 152
Amine-functionalized water-soluble
  polyamides as drug carriers
anchoring reactions, 399–400
experimental procedure, 400–403

Amine-functionalized water-soluble
  polyamides as drug carriers—*Continued*
  structural prerequisites, 395–396
  synthetic approaches, 396–399
Amine imides, 46–47
6-Amino-6-deoxycellulose, synthesis, 190
6-Amino-6-deoxycellulose acetate
  6-azido-6-deoxycellulose acetate,
    synthesis, 191
  azido groups, reduction to amino
    groups, 193,196
  cellulose–*g*-peptide graft
    copolymers, use in synthesis, 196–197
  ^{13}C-NMR spectra of cellulose acetate
    precursors, 193,195*f*
  experimental procedure, 190–191,192*t*
  graft copolymers
    copolymerization procedure, 191,192*t*
    synthetic scheme, 191,193
  ^{1}H-NMR spectra of *g*-peptide graft
    copolymer, 196,198*f*
  IR spectra
    cellulose acetate precursors, 193,194*f*
    g-peptide graft copolymer, 196,199*f*
    synthesis, 191
  6-*O*-tosylcellulose acetate, synthesis, 191
Ammonium imides, 46–47
Amphiphilic associating polymers, use in
  thickening of aqueous solutions, 218–219
Amphiphilic polyelectrolytes containing
  aromatic chromophores, 291–292
Amphoteric polymers
  aminimide synthesis, 46–47
  commercial applications, 21
  examples, 17,20*f*,21
  polyampholytic synthesis, 47
  properties, 21
  quaternary ammonium carboxylate
    synthesis, 46
Angiogenesis, effect of heparin, 384
Anionic dyes, clinical applications, 370
Anionic monomers, application of
  persulfate-initiated polymerization,
  99,100*t*,101
Anionic polyelectrolytes, synthesis, 17
Anionic polymers
  polyphosphonic acid and salt synthesis,
    45–46
  polysulfonic acid and salt synthesis,
    43–45

Antibody localization, polyanionic
  polymers, 375
Antigen localization, polyanionic
  polymers, 375
Antipolyelectrolytic effect, description, 119
Antiviral action of polyanions, 371–373
Aqueous solution(s), roles of molecular
  structure and solvation on drag
  reduction, 320–335
Aqueous solution behavior, hydrophobically
  modified poly(acrylic acid), 218–231
Aqueous solution properties of
  naphthalene-pendent acrylic copolymers
  added urea concentration, effect,
    294,296*f*
  copolymer structure, 293
  diamine copolymers
    solution pH, effect, 300,301*f*
    structure, 297
  experimental materials, 292–293
  fluorescent spectroscopy, 293
  hydrophobic methyl groups, effect, 300,301*f*
  polymer characterization, 293
  polymer concentration, effect,
    294,297,298*f*
  solution pH, effect, 294,296*f*
  solution pH for diamine copolymers,
    effect, 300,301*f*
  steady-state fluorescent spectra, 294,295*f*
  steady-state spectra of diamine
    copolymers, 297,298–299*f*,300
  structure
    copolymer, 293
    diamine copolymers, 297
  viscosity measurements, 293
Aryl cyclic sulfonium zwitterions,
  synthesis, 41
ASi-containing polyacrylamides, synthesis
  and properties, 165*t*,166
Associative copolymer systems
  behavior, role of liaisons, 239–245
  free energies of hydrophobic bonding
    between alkyl chains,
    242,243*f*,244*t*,245
Associative polymers
  thickening of solutions, use, 218
  water viscosity, increase, 446
Atherosclerosis, role of heparin, 383–384
6-Azido-6-deoxycellulose acetate,
  synthesis, 189–190

B

Basic fibroblast growth factor, inhibition by heparin, 378
Beclomethasone dipropionate, nasal delivery, 356
γ-Benzyl-α,L-glutamate n-carboxyanhydride cellulose copolymer, synthesis, 196–197,198–199f
  experimental procedure, 190–191,192t
  graft copolymerization procedure, 191,192t
  synthetic scheme, 191,193
Bioadhesive drug delivery
  advantages, 350–351
  bioadhesive parameters, 351,353
  buccal drug delivery, 357–358
  cervical and vaginal drug delivery, 350
  colon and rectal drug delivery, 359
  components, 351,352f
  epithelium, 354–355
  mucus layer, 353–354
  nasal drug delivery, 356
  ocular drug delivery, 355–356
  stomach and intestinal drug delivery, 358–359
  strategies, 355–360
Bioadhesive parameters
  chain segment mobility, 353
  charges, 351,353
  hydration and expanded nature of polymer, 353
Bioadhesive screening techniques, 351
Biodegradable hydrogels, 484
Biopolymers, water soluble, See Water-soluble biopolymers
Bioresorbable polymers, 405–406
2,3-O-Bis(phenylcarbamoyl)-6-azido-6-deoxycellulose, synthesis, 196
Block copolymers
  antithrombogenicity, 469
  electron micrographs, 469,470f
  hemocompatibility, 470,471t
  platelet adhesion and aggregation on surface, 471t
  properties, 474
  schematic representation, 471,472f
  synthesis, 469,472,475
  wettability, 469,470f
Brain, role of polyanionic polymers, 380

Buccal drug delivery, 357–358
Bulk degradation, definition, 484
Bulk polymerization of n-vinylpyrrolidone
  reactivity ratios, 69t
  typical recipe, 69

C

Cage-effect theory, description, 87–88
Calcium metabolism, role of heparin, 380–381
Carbetimer, antitumor application, 370–371
Cat-Floc, synthesis, 27
Cathepsin, vascular effects, 376
Cationic antibiotics, detoxification, 369–370
Cationic guar, effect on solution viscosity of polysaccharides, 453,455t,456,457–458f
Cationic monomers, application of persulfate-initiated polymerization, 99,100t,101
Cationic polyelectrolytes, synthesis, 17,19f
Cationic polymer synthesis
  oxonium polymer, 43
  phosphonium polymer, 41,42f,43
  quaternary ammonium polymer, 27–40
  sulfonium polymer, 40–41
Cationic quaternary ammonium monomers, 34
Cell membrane transduction, role of heparin, 381
Cellulose, chemical modification for increased water solubility, 75
Cervical drug delivery, 360
Chain-growth polymerization, synthesis of water-soluble copolymers, 2,4
Charge-density parameter, theory for polyelectrolytes, 203–204
Charged polymers
  amphoteric polymers, 17,20f,21
  classification, 17
  examples
    anionic monomers for polyelectrolytic synthesis, 17,19f
    quaternary ammonium salts for cationic polyelectrolytic synthesis, 17,19f
  properties, effect of structure, 17
Charged polysaccharides, effect on solution viscosity of polysaccharides, 453,455t,456,457–458f

Charge-transfer interactions, liaisons, 234
Chelation, role of heparin, 381
Chemical modification of natural polymers
  alginate, 76–77
  cellulose, 75
  chitosan, 74–75
  hyaluronic acid, 77
  pectin, 76
  proteins, 77–78
  starch, 75–76
Chitin
  extraction process, 431,432*f*
  structure, 430,432*f*
Chitosan
  biological properties, 434,435*t*
  cationic properties, 431,434*t*
  certificate of analysis for purified
    materials, 431,433*f*
  chemical modification for increased
    water solubility, 74–75
  chemical properties, 434*t*
  commercial sources, 431
  fibers, 438,440–441*f*
  films, 435,436*f*,437
  forms
    aqueous, 435,436*f*
    physical, 435,437*t*,438
    powder, 435
    shaped, 437–438,439–440*f*
  key properties relating to biomedical
    applications, 431,434–435*t*
  manufacturing scheme, 431,432*f*
  paper, 438,442*f*
  purification scheme, 431,433*f*
  sponge, 438,442*f*
  structure, 430,432*f*
Cluster–cluster aggregation,
  description, 315
Colon drug delivery, factors affecting
  residence time, 359
Comblike cyclopolymers of
  alkyldiallylamines and
  alkyldiallylmethylammonium chlorides,
  synthesis and characterization, 151–157
Complex theory
  application to persulfate-initiated
    acrylamide polymerization, 88–90
  description, 87–88
  evaluation for acrylamide
    polymerization, 88

Complexation reactions between synthetic
  polymers, objectives of
  investigations, 303
Condinine, polymerization, 39
Conformational behavior, effect of extra
  liaisons, 237–239
Connective tissue, role of polyanionic
  polymers, 379–380
Contact, definition, 233
Coordination-complex forces, liaisons, 234
Copoly(α,β-D,L-aspartamides), synthesis,
  401–402
Copoly(imide–amides), synthesis, 396–397
Cosolute-binding systems, role of liaisons
  in behavior, 244,246
Coulombic forces, liaisons, 234
Cross-link, definition, 233
Cyclopolymerization, description, 27

D

Derivatized polysaccharides, examples, 16
Detoxification, use of polyanionic
  polymers, 369–370
Dextran sulfate, oral absorption, 385
Diacetone acrylamide, 33
n-[(Dialkylamino)alkyl]acrylamides,
  synthesis, 35
Dialkyldiallylammonium salts, synthesis of
  water-soluble cyclopolymers, 151–152
Dialkyl(2-methylene-3-butenyl)sulfonium
  chloride, synthesis of monomers and
  polymers, 40–41
Diallyldecylmethylammonium chloride
  hydrophobically associating copolymers
  applications, 188
  Brookfield viscosity measurement, 179
  ^{13}C-NMR spectrum, effect of solvent,
    182*f*,183
  copolymerization procedure, 178,179*t*
  copolymerization with diallyldimethyl-
    ammonium chloride, 183*t*
  experimental materials, 176
  NMR measurements, 179–180
  shear rate, effect on viscosity vs.
    concentration, 185,186–187*f*,188
  viscosity of copolymer
    comonomer content, effect, 185*f*
    concentration effects, 183,184*f*,185

Diallylmethyl(β-propionamido)ammonium chloride monomer and homopolymers, 31–32

Diffusion, study of Na$^+$ ion interactions with polyions in aqueous salt-free solutions by diffusion, 202–216

Diffusion coefficient
calculation, 203
relationship to molecular weight, 278

Diffusion-limited aggregation, description, 315

Diffusion-limited cluster–cluster aggregation, model, 315–317

[β-(Dimethylamino)ethyl]acrylamide, 33

p-[(Dimethylamino)ethyl]styrene, 32

[(Dimethylamino)methyl]acrylamide, 32–33

p-[(Dimethylamino)methyl]styrene, 32

(γ-Dimethylaminopropyl)acrylamide, 33

Dimethyldiallylammonium chloride, 27–28,29f

Dimethyldiallylammonium chloride polymer
electroconductivity, 50,51t
fractionation and characterization by size exclusion chromatography, 30,31f
intrinsic viscosity vs. ionic strength, 29,30f
intrinsic viscosity vs. temperature and molecular weight, 28,29f
^{14}N-NMR spectrum, 28,29f
structure, 27
synthesis, 27

n-[1,1-Dimethyl-3-(dimethylamino)propyl]-acrylamide, 33

DNAs, molecular structure, 15

Drag reduction
classification of models, 320–321
coil size vs. drag reduction efficiency, 321
discovery, 320
influencing factors, 320

Drag reduction in aqueous solutions
associations, effects, 334,335f
composition, effect, 323,328f,329
experimental materials, 321
experimental results, agreement with theoretical models, 334
friction factor vs. Reynolds number, plot, 323,326f
hydrodynamic volume changes, effect, 329,332–333f
measurements, 321–322
molecular weight, effect, 329,330–331f

Drag reduction in aqueous solutions—
Continued
percent drag reduction, definition, 323
percent drag reduction values, 323,327t
percent drag reduction vs. polymer volume fraction, 323,326f
polymer characterization, 321
solvent ordering, effect, 334,335f
tailored copolymers, 322–323,324–325t
universal calibration for diverse polymer types, 334,335f

Drug and enzyme conjugates, examples of use with polyanionic polymers, 368–369

Drug carriers, amine-functionalized, water-soluble polyamides, 394–403

Drug delivery, bioadhesive, See Bioadhesive drug delivery

Dynamic light scattering for molecular-weight-distribution determination of water-soluble macromolecules
algorithm development, 279–284
data collection, 285,287
data processing, 285–287
experimental procedure, 284–287
polyacrylamide molecular-weight distribution, 288,289f
polyethylene oxide molecular-weight distribution, 288,289f
sample preparation, 284–285
series approximations, 288
terminology, 277
theory, 277–279

E

Electroconductivity in electrophotography, use of water-soluble polymers, 50,51t,f

Encephalomyocarditis viruses, effect of pyrans, 373

Endothelial cell growth factor, effect of heparin, 377–378

Enhanced oil recovery, use of water-soluble polymers, 175

Enzyme(s), applications for chemically modified water-soluble compounds, 77–78

Enzyme-catalyzed hydrolysis, network degradation, 485

Epidermal growth factor, applications for
chemically modified water-soluble
compounds, 77–78
Epithelium, bioadhesion, 354–355
Ethylene oxide polymers, classes, 26

F

Fibronectin, description, 354–355
First-order normalized autocorrelation
function, definition, 277–278
Flocculation, use of water-soluble
polymers, 48f
Fluorescent probes of complexation between
poly(n-isopropylacrylamide) and sodium
n-dodecyl sulfate, comparison, 261–273
Fluorocarbon surfactants, hydrophobic
bonding to β-cyclodextrins, 159–160
Fractal dimensionality, description, 316–317
Free-radical polymerization
kinetics scheme, practical
consequences, 57–58,59t
reactions, 57–58
Friction factor, definition, 322
Friend leukemia, protection by pyran, 372
FX–13-containing polyacrylamide
Brookfield viscosities vs. mole percent
comonomer, 161–162,164f
synthesis, 161,162t
FX–14-containing polyacrylamide
Brookfield viscosities vs. mole percent
comonomer, 161–162,164f
synthesis, 161,162t

G

Gastric and intestinal mucosa, role of
heparinoids and heparin, 384–385
Gastrointestinal tract, motility
pattern, 358
Gel degradation, mechanisms, 484
Generalized exponential distribution
adjustable parameters, 280,281f
function, 279–280
integrability, 280
molecular-weight distribution,
effect of choice of function, 280
Glycocalyx, description, 354

Graft copolymers
electron micrograph, 476f
synthesis, 474,477–478
thrombogenicity, 476,477t

H

Heparanase, inhibition, 371
Heparan sulfate, role in brain, 380
Heparan sulfate proteoglycans, role in
brain, 380
Heparin
affinity for superoxide dismutase, 368
angiogenesis, effect, 384
detoxification of cationic antibiotics,
369–370
glomerular effects, 376–377
growth, effect, 377–378
inhibition of heparanase, 371
oral absorption, 385
proteinase inhibition, 375–376
role(s)
atherosclerosis, 383–384
brain, 380
cell membrane transduction, 381
connective tissue, 379–380
gastric and intestinal mucosa, 384–385
immunity, 374
inflammation, 383
lipids, 383
structure, 385
transcription inhibition, 382
tumor inhibition, 378
vascular effects, 376
Heparinoids
gastric and intestinal mucosa, role, 384–385
immunity, role, 374
vascular effects, 376
Heparin sulfate, antigen or antibody
localization, 375
Heterochain-type polymers, use as drug
carriers, 394–395
^{1}H-NMR characterization of polyethylene
glycol derivatives
degrees of substitution, comparison
with those from titration, 420t
end-group conversion,
determination, 420
impurity detection, 421

1H-NMR characterization of polyethylene
glycol derivatives—*Continued*
molecular-weight determination, 421*t*
problems, 420
Homogeneous precipitation, example, 237
Homopolymer solutions, role of liaisons in
behavior, 234–236
Hyaluronan
biomedical applications, 493
description, 493
function, 493
structural comparisons, 495–499
structure, 493,494*f*,495
viscoelasticity, 493
Hyaluronic acid, applications for
chemically modified water-soluble
compounds, 77
Hydrodynamic radius, definition, 279
Hydrodynamic volume
definition, 4
dimension of macromolecular chain in
solution vs. repeating unit
structure, 4–5,6–7*f*
methods for increasing, 5,8*f*
Hydrogels, 468
Hydrogen bond(s), liaisons, 234
Hydrogen bonding, prerequisites, 304
Hydrophilic–hydrophobic domain polymer
systems
advantages, 468–469
block copolymers, 469–475
graft copolymers, 474,476*f*,477*t*,478
networks, 482*f*
polymer blends, 478,479*f*,480*t*,481*f*
Hydrophobically associating ionic
copolymers, typical hydrophobic
association, 175–176
Hydrophobically associating ionic
copolymers of methyldiallyl-(1,1-
dihydropentadecafluorooctoxyethyl)-
ammonium chloride, effect of perfluoro
carbon groups on water solubility, 176–188
Hydrophobically associating
polyacrylamides, synthesis, 67,68*t*
Hydrophobically associating polymers
apparent viscosity vs. concentration,
339,340*f*
hydrocarbon replacement by
perfluoroalkyl groups, 159
proposed model, 339,340*f*

Hydrophobically associating systems, role
of liaisons in behavior, 242
Hydrophobically modified acrylamide-based
polyelectrolytes, 338–347
Hydrophobically modified
hydroxyethylcellulose, role of
liaisons in behavior, 240,241*f*
Hydrophobically modified poly(acrylic
acid)
aqueous solution behavior
alkyl chain content and length,
effect, 221,222*f*
comparison to hydrophobically modified
nonionic polymers, 228–229
ionic strength, effect, 221,222*f*,228–229
pH, effect, 223,224*f*
shear rate, effect, 223,224*f*,229–230
experimental apparatus, 220
experimental conditions, 220–221
experimental materials, 219
mixed aggregates, formation, 229
modification of poly(acrylic acid),
procedure, 219,220*t*
solubilization of hydrophobic additives,
226,227*f*,230
surfactants, effect, 223,225*f*,226
viscosity vs. interchain association, 226,228
Hydrophobically modified polymer(s), 21–22
Hydrophobically modified polymer systems,
examples, 468
Hydrophobically modified water-soluble
polymers
applications, 130
examples, 130
water viscosity, increase, 446
Hydrophobic bond, description of term, 338
Hydrophobic bonding, free energies between
alkyl chains of associative copolymer
systems, 242,243*f*,244*t*,245
Hydrophobic effects on complexation and
aggregation in water-soluble polymers
chain configuration, 308,311
complexation measurements
fluorescent measurements, 306,307*f*
interpretation, 308,310*f*,311–312
pH measurements, 308,309*f*
complexation mechanism, experimental
support, 311–312
diffusion-limited cluster–cluster
aggregation model, 315–317

Hydrophobic effects on complexation and aggregation in water-soluble polymers—*Continued*
dynamic-light-scattering measurements of aggregation, 312,313–314f,315
experimental procedure, 305–306
Hydrophobic interactions, liaisons, 234
Hydrosoluble polymeric drug carriers derived from citric acid and L-lysine
catalytic hydrogenolysis, procedure, 416
coupling of amine-type compounds, procedure, 416
hydrogenolysis of protecting groups, 408,410f,411
hydrolysis of imide functions, 411,412–413f,414,416
interfacial polycondensation, 407–408,409f,416
IR spectra procedure, 417
molecular-weight determination, 417
monomer synthesis, 406–407
self-coupling of amine-type compounds, 414,415f
Hypercoils, formation, 291

I

Icebergs, description, 339
Immunity, role of polyanionic polymers, 374
Indifferent solvent, definition, 234
Inflammation, role of heparin, 383
Initiation mechanism for persulfate-initiated acrylamide polymerization
chain initiation, 95
complex–cage equivalence, 91
historically used mechanism, 91–94
initiator reactions, 94
propagation, 95
proposed mechanism, 94–97
rate equation, derived, 97–98,100t
swollen cage formation and decomposition, 94–95
termination, 95
transfer to monomer, 95
Inorganic polymers, 16–17
Insulin, nasal delivery, 356
Intensity–time autocorrelation function, definition, 277

Interacting polymers, viscosity behavior and oil-recovery properties, 446–464
Intestinal drug delivery, description, 358–359
Intraparticle interference factor approximation, 279
occurrence, 278–279
Intrinsic viscosity, measurement, 9,11
Inverse emulsion polymerization of acrylamide
advantages, 64,66
schematic diagram, 64,65f
typical recipe, 64
Inverse microemulsion polymerization of acrylamide, 66
Ionenes, synthesis, 35–36
Ionomers, description, 218
Isotactic polyacrylamide, synthesis, 83

K

Kinetics, free-radical polymerization, 57–58,59t
Kinetics of persulfate-initiated acrylamide polymerization
cage-effect theory, 87–88
complex theory, 87–88
rate dependence, 85,86t
solvent-transfer theory, 85–86

L

Large-scale aggregates, occurrence, 303–304
Laurylacrylate-containing polyacrylamides, synthesis and properties, 161,163t
Lectins, description, 355
Liaisons
associative copolymer systems, effect on behavior, 239–245
bimolecular formation, 238–239
characteristics, 233
charge-transfer interactions, 234
conformational behavior, effect of extra liaisons, 237–239
coordination-complex forces, 234
cosolute-binding systems, effect on behavior, 244,246
Coulombic forces, 234
description, 232–233

Liaisons—*Continued*
equilibrium constant for simple
homopolymer solutions, 235–236
estimation of average number in simple
homopolymer solutions, 235
Flory–Huggins interaction parameter, 236
hydrogen bonds, 234
hydrophobic interactions, 234
interaction forces, types, 233–234
simple homopolymer solutions, effect
on behavior, 234–236
solubility behavior, effect of extra
liaisons, 237
spectroscopic behavior, effect of
extra liaisons, 237
unimolecular formation, 238
van der Waals forces, 234
Linear crystalline polyethylenimine, 37
Lipids, role of heparin, 383
Lower critical solution temperature probes
of poly(n-isopropylacrylamide)
cloud-point measurements, 253,255f
experimental materials, 250–251
fluorescent probes, 254,256,257f,258
hydrogen bonding, effect, 249–250
measurements, 252–253
microcalorimetry, 253–254,255f
nonradiative energy transfer, 258–259
polymer synthesis, 253
sample preparation, 252
synthesis of fluorene-labeled
poly(n-isopropylacrylamide), 251–252
synthesis of poly(n-
isopropylacrylamide), 251
thermal response in aqueous media for
practical applications, use, 250
Lymphocyte mobilization, effect of
polyanionic polymers, 373–374

M

Macromolecular systems, importance of
molecular-weight characterization, 276
2-(Methacryloyloxy)ethyl]sulfonate,
structure, 44
[3-(Methacryloyloxy)-2-hydroxypropyl]-
sulfonate
structure, 45
synthesis, 44

Methacryloylurea derivatives containing
quaternary ammonium groups,
synthesis, 38
Methyldiallyl-(1,1-dihydropentadecafluoro-
octoxyethyl)ammonium chloride (FX–15)
hydrophobically associating copolymers
applications, 188
Brookfield viscosity measurement, 179
Br vs. F group, effect on
polymerization, 180f
^{13}C-NMR spectrum, effect of solvent,
181–182f
copolymerization with diallyldimethylammonium chloride, 183t
experimental materials, 176
homopolymerization procedure,
178t,179
NMR measurements, 179–180
synthesis
2-(bromoethyl)-1,1-dihydro-
pentadecafluoro-n-octyl ether,
176–177
diallyl-(1,1-dihydropenta-
decafluorooctoxyethyl)amine, 177
FX–15, 177–178
viscosity of copolymers
comonomer content, effect, 185f
concentration, effect, 183,184f,185
viscosity vs. concentration, effect
of shear rate, 185,186–187f,188
n-Methylolacrylamide, 32–33
Microemulsion, definition, 66
Mobility of injected fluid and crude oil
in reservoir rock, definition, 462
Molecular studies, role in drag reduction
in aqueous solutions, 320–335
Molecular weight, determination,
106–107
Molecular-weight characterization of
macromolecular systems, 276–277
Molecular-weight distribution,
determination for water-soluble
macromolecules by dynamic light
scattering, 276–289
Molecular weight maximum,
determination, 280
Mucus layer
charge, effect on bioadhesion, 354
presence, 353–354
thickness, effect on bioadhesion, 354

N

Naphthalene-pendent acrylic copolymers, photophysical and rheological studies of aqueous solution properties, 291–301
Nasal drug delivery, 356
Natural polymers, chemical modification, 74–78
Neocarzinostatin, 369
Networks of polymers, copolymerization, 482f
Nonionic polymers
  applications, 17
  repeating units, 17,18f
Nonpolyelectrolytic water-soluble polymers, 25–26
Nonradiative energy transfer between species in solution, investigation, 258–259
N-substituted α,β-D,L-aspartamide polymers, synthesis, 396–399
Nucleic acids, 15
Nucleus, role of polyanionic polymers, 381–382

O

Ocular drug delivery, 355–356
Oil-recovery properties of interacting polymers
  copolymer–crude oil ratio, effect, 460,461t
  experimental results, 459t
  quaternary ammonium salt organic polymer, effect, 459,460t
Organic polymers, functional groups that impart water solubility, 25,26t
Osteoporosis, role of heparin, 380–381
Oxonium monomers and polymers, 43

P

Particle-cluster aggregation, description, 315
Pectins, applications for chemically modified water-soluble compounds, 76
Peptide graft copolymers, synthesis using soluble aminodeoxycellulose acetate, 188–199
Percent drag reduction, definition, 323

Persulfate-initiated polymerization of acrylamide, mechanism and kinetics, 82–101
Persulfate-initiated polymerization of acrylamide at high monomer concentration
  comparison of kinetic model to experimental data, 112,114–117f
  conversion time and weight-average molecular weight conversion data and effect of consumption rate of initiator on initial monomer concentration, 112,117f
  experimental conditions, 109,110t
  experimental procedure, 107
  limiting conversions, 109,110t,111f
  model predictions, 112,114–116f
  molecular-weight determination, 106–107
  parameter estimation, 109,112,113t
  polymerization, experimental method, 107,108f
  rate order vs. monomer concentration, 109
  relative magnitudes of thermal and monomer-enhanced decomposition of potassium persulfate, 112,117f
  residual monomer concentration determination, 106
Phosphonated polyethylene, synthesis, 45
Phosphonium monomers and polymers, viscosity data, 41,42f,43
Piloplex, function, 356
Polyacrylamide(s)
  applications, 82
  degradation, 82
  intrinsic viscosity, effect of time, 83
  molecular-weight distribution, 288,289f
  poly(vinyl alcohol) synthesis, 26
  synthesis, 25–26
Polyacrylamide–perfluorocarbon group containing water-soluble polymers
  Brookfield viscosities, 161–166
  Brookfield viscosity–comonomer content profiles, 168,172
  characterization, 166t,167f
  diffusion coefficient vs. polymer concentration, 166,167f,172–173
  experimental procedure, 160–161
  NaCl concentration, effect on Brookfield viscosity, 168,169f

Polyacrylamide–perfluorocarbon group
    containing water-soluble polymers—
    Continued
nonionic surfactants, effect on
    Brookfield viscosity, 168,171*f*,173
rheology, 168,169–171*f*
shear, effect on Brookfield viscosity,
    168,169*f*
structures of comonomers, 160
surfactant concentration, effect on
    Brookfield viscosity, 168,170*f*
temperature, effect on Brookfield
    viscosity, 168,170*f*
viscosity vs. concentration, 161,164*f*
Poly(acrylic acid)
hydrophobically modified, *See*
    Hydrophobically modified
    poly(acrylic acid)
synthesis, 26
Polyampholytes, 47
Polyampholytic polymers, 119–120
Polyanionic polymers
angiogenesis, effect, 384
antigen or antibody localization, 375
antitumor application, 370–371
antiviral activity, 371–373
applications
    anionic dyes, 370
    drug carriers and enhancers, 368
detoxification, use, 369–370
drug and enzyme conjugates, use, 368–369
glomerular effects, 376–377
growth, effect, 377–378
heparanase inhibition, 371
immunoadjuvancy, 371–373
interest, 367
lymphocyte mobilization, 373–374
oral absorption, 385
proteinase inhibition, 375–376
reviews, 367–368
role(s)
    atherosclerosis, 383–384
    brain, 380
    cell membrane transduction, 381
    chelation, 381
    connective tissue, 379–380
    gastric and intestinal mucosa, 384–385
    immunity, 374–375
    inflammation, 383
    lipids, 383

Polyanionic polymers
role(s)—*Continued*
    nucleus, 381–382
    osteoporosis, 380–381
    respiration, 385
    structure, 384
    transcription inhibition, 382
    tumor inhibition, 378–379
    tumor necrosis factor, isolation, 379
    vascular effects, 376
Poly(α,β-D,L-aspartamides)
side chain modification procedure, 402
synthesis, 401
Poly(α,β-D,L-asparthydrazide-*co*-D,L-
    ornithine, synthesis, 400–401
Poly(diallyldimethylammonium chloride),
    cyclocopolymerization reaction, 152
Polyelectrolytic water-soluble polymers
amphoteric polymer synthesis, 46–47
anionic polymer synthesis, 43–46
cationic polymer synthesis, 27–43
Polyethylene glycol(s)
analysis, 419
biotechnical and biomedical
    applications, 418
properties, 418
selectivity, 419
solubility, 418
superoxide dismutase, use
    as carrier, 368
Polyethylene glycol aldehydes, 422–425
Polyethylene glycol benzaldehyde
characterization, 423
reactivity, 423–424
synthesis, 422
Polyethylene glycol derivatives
^{1}H-NMR characterization, 419,420–421*t*
properties, ideal, 422
Polyethylene glycol propionaldehyde,
    424–425
Polyethylene oxide, molecular-weight
    distribution, 288,289*f*
Polyethylene sulfonate, antitumor
    application, 370–371
Polyiminodiacetamides, synthesis,
    396–399
Poly[imino-2,1-propyleneoxy-poly(ethylene-
    oxy)-1,2-propyleneiminocarbonyl-
    methyleneiminomethylenecarbonyl],
    synthesis, 401

Poly(*n*-isopropylacrylamide)
  applications for complexation with
    surfactants, 262
  probes of lower critical solution
    temperature, 249–259
  structure, 249,261
Poly(*n*-isopropylacrylamide)–sodium
    n-dodecyl sulfate complexes
  bound probes of critical micelle
    concentration, 269,270*f*,272*t*
  characterization, 264,265*t*
  cloud points, 265,267
  critical micelle concentration vs. lower
    critical solution temperature, 272*t*
  experimental materials, 262–263
  free probes of critical micelle
    concentration, 267,269,270*f*
  lower critical solution temperatures,
    272,273*f*
  measurements, 264
  modeling of complexation, 269,271–272
  polymer synthesis, 264
  sample preparation, 263–264
  sodium dodecyl sulfate, effect on
    lower critical solution temperature,
    265,266*f*,267,268*f*
  structures, 264,266*f*
  synthesis, 263
Poly(β-malic acid), synthesis, 406
Polymer behavior, effect of extra liaisons, 237
Polymer blends
  equilibrium hydration, 480,481*f*
  interpolymer hydrogen bond formation,
    478,479*f*,480
  mechanical properties, 480*t*
Polymer-bound aminosalicylic acid,
    synthesis, 402
Polymer-bound *cis*-dichloroplatinum(II)
    complexes, synthesis, 403
Polymer-bound drugs, smooth solubility as
    prerequisite of efficacious
    administration, 394–395
Polymer-bound ferrocenylpropanoic acid,
    synthesis, 403
Polymer-bound phenylacetic acid,
    synthesis, 402
Polymer complexation
  degree of acid dissociation, effect, 305
  molecular weight, effect, 305
  study methods, 304–305

Polymerization, acrylamide, 83,84*t*,85
Polymer solution viscosity, vs. oil
    recovery, 462
Polymers used as mobility control agents, 447
Poly(methacrylic acid), synthesis, 26
Poly(methyl vinyl ether), synthesis, 26
Polypeptides, 15–16
Polyphosphonic acids and salts,
    synthesis, 45
Polysaccharides, 16
Poly(styrene sulfonate) blends
  applications, 462
  oil-recovery properties, 459,460–461*t*
  viscosity behavior, 448–459
Polysulfonic acids and salts, synthesis,
    43–45
Poly(vinyl acetate-*co*-vinyl alcohol), role
    of liaisons in behavior, 239–240,241*f*
Poly(vinyl alcohol), synthesis, 26
Precipitation polymerization of
    acrylamide, 67
Primary recombination, definition, 88
Primary structure, influencing factors,
    2,3*f*
Proteins
  applications for chemically modified
    water-soluble compounds, 77–78
  commercial applications, 16
  molecular structure, 15
  properties, effect of structure, 15–16
  synthesis, 16
Pyran
  antiviral action, 371–372
  immunity, role, 374
  vaccine adjuvancy, 371
Pyran copolymer
  antitumor application, 370–371
  superoxide dismutase, use as carrier, 368
Pyrene, solubilization, 226,227*f*,230
Pyrene-labeled polyacrylamides
  absorption spectrum, 141,142*f*
  characterization methods, 132
  decay profiles, 146,148–149*f*
  excimer–monomer intensity ratios vs.
    concentration, 141,145*t*,146
  excitation spectra, 146,147*f*
  experimental materials, 132
  fluorescence emission spectra,
    141,142–143*f*
  instrumentation, 132

Pyrene-labeled polyacrylamides—*Continued*
model-compound synthesis, 135–136,138
monomer synthesis, 132–135
photophysical characterization, 141–149
rheological characterization, 138,141
solution-polymerized microstructure,
141,144*f*
solution techniques for copolymerization,
135,137–138,139*f*
surfactant-polymerized microstructure,
141,143*f*
surfactant technique for
copolymerization, 135–136,140*f*

Q

Quaternary ammonium carboxylates, 46
Quaternary ammonium polymers, synthesis,
27–40
Quaternary ammonium salt organic polymers,
effect on solution viscosity of
polysaccharides, 456*t*,459
Quaternary ammonium salt–polyacrylamide
blends
oil-recovery properties, 463*t*
polymer interactions giving rise
to large viscosity increases, 463–464
viscosity behavior, 462–463
Quaternary structure, influencing factors, 2

R

Reaction-limited aggregation,
description, 315
Rectal drug delivery, factors affecting
residence time, 359
Reduction activation, development, 83
Residual monomer concentration,
determination, 106
Respiration, role of polyanionic polymers, 385
Reynolds number, definition, 322
RF–4 and RF–8 containing polyacrylamides,
synthesis and properties, 163,165*t*
Rheological properties of hydrophobically
modified acrylamide-based
polyelectrolytes
apparent viscosity vs. polymer
concentration, 342,343–345*f*

Rheological properties of hydrophobically
modified acrylamide-based
polyelectrolytes—*Continued*
conceptual model, 342
electrostatic repulsions, shielding by
added electrolyte, 346,347*f*
experimental procedure, 339,341*f*,342
monomer synthesis, 339,341*f*
polymer synthesis, 339
solution studies, 342,343–345*f*
terpolymers, comparison,
342,346,347*f*
viscometric measurement, 339,342
Rheology, 11,13*f*
RNAs, molecular structure, 15

S

Secondary recombination, definition, 88
Secondary structure, influencing factors, 2
Second-order temporal correlation
function, determination, 305–306
Sedimentation of suspended polymers, use
of water-soluble polymers, 49*f*,50
Sensitivity, definition, 43
Serpins, 375–376
Silanization of silica
importance, 425
percent N vs. reaction conditions and
silanating reagent, 426–427*t*
problems, 425
reactive amino groups, availability on
surfaces, 427–428
Silica, silanization, 425
Size exclusion chromatography, advantages
and disadvantages for molecular-weight-
distribution determination, 277
Sludge dewatering, use of water-soluble
polymers, 48,49*f*
Sodium 2-(acrylamido)-2-methylpropane-
sulfonate, 44,45
Sodium cromoglycate, nasal delivery, 356
Sodium ion interactions with polyions in
aqueous salt-free solutions by diffusion
additivity rule for Na$^+$ ion diffusion,
211,214,216
average tracer diffusion coefficient
values of different normalities, 206*t*,207
charge-density parameter, 203–204

Sodium ion interactions with polyions in
    aqueous salt-free solutions by diffusion—
    Continued
charge fraction vs. inverse of charge
    density parameter, 213–214,215*f*
condensed Na$^+$ ions, fraction in
    solution, 213*t*,214,215*f*,216
diffusion measurement procedure, 202–203
Na$^+$ ion diffusion additivity
    rule, test, 211,212*f*
Na$^+$ ion diffusion coefficient vs.
    equivalent concentration, 204,205*f*
Na$^+$ ion diffusion ratios, 210*t*,211
tracer diffusion coefficient, comparison of
    experimental vs. theoretical values,
    204*t*,206
tracer diffusion coefficient vs.
    concentration parameter, 211*t*,212*f*,213
tracer diffusion coefficient vs.
    equivalent concentration and nature
    of polyelectrolyte, 204,205*f*
tracer diffusion coefficient vs. inverse of
    charge density parameter, 207–208,209*f*
tracer diffusion coefficient vs. normality
    of sodium carboxymethylcellulose,
    208,209*f*,210
Soluble aminodeoxycellulose acetate,
    synthesis of peptide graft copolymers,
    189–199
Solution polymerization of acrylamide
kinetic parameters, 60*t*,61*t*,62*t*
pH, effect on kinetic parameters, 60,61*t*
rate constants, 60*t*
reactivity ratios, effect of medium, 62,63*t*
solvent polarity, effect on kinetic
    parameters, 61,62*t*
typical recipe, 63
Solution polymerization of
    n-vinylpyrrolidone, 70,71*f*
Solvation, role in drag reduction in
    aqueous solutions, 320–335
Solvation of polymers, role of water, 9,10*f*
Solvent(s), effect on acrylamide
    polymerization, 83,85
Solvent-transfer theory, description, 85
Starch, applications for chemically
    modified water-soluble compounds,
    75–76
Step-growth polymerization, synthesis of
    water-soluble copolymers, 2,4

Stomach drug delivery, description, 358–359
Structural comparisons of hyaluronan by
    ^{13}C-NMR spectroscopy
^{13}C-NMR analysis, 497*t*,498*f*,499
    experimental materials, 495
    experimental procedure, 495
    X-ray analysis, 495,496*t*,497
Structural design, water-soluble
    copolymers, 2–23
Sulfobetaine amphoteric monomer–
    acrylamide copolymers, *See*
    Acrylamide–3-(2-acryl-amido-2-
    methylpropyldimethylammonio)-1-
    propanesulfonate copolymers
Sulfonated guar, effect on solution
    viscosity of polysaccharides,
    453,455*t*,456,457–458*f*
Sulfonium monomers and polymers,
    synthesis, 40–41
Superoxide dismutase
    affinity of heparin, 368
    pyran copolymer, use as carrier, 368
Suramin, clinical applications, 370
Suspension polymerization of
    n-vinylpyrrolidone, typical recipe, 72
Synthetic methods, synthesis of
    water-soluble copolymers, 2,4
Synthetic water-soluble polymers
    charged polymers, 17,19–20*f*,21
    hydrophobically modified polymers, 21–22
    nonionic polymers, 17,18*f*

T

Tertiary recombination, definition, 88
Tertiary structure, influencing factors, 2
Thermoreversible gels
    advances, 504
    applications, 504–505
    definition, 502
    formation mechanism, 502
    formation of gels based on proteins and
        polysaccharides, 505
    gelation process, control, 504
    theory, 502–503
    types, 502–503
Transcription, inhibition by heparin, 382
Trimethyl(*p*-vinylbenzyl)ammonium
    chloride, 32

Tumor inhibition, effect of polyanionic polymers, 378–379
Tumor necrosis factor, isolation by heparinoids, 379
Type IV collagen, function, 379

V

Vacuum filtration, use of water-soluble polymers, 50f
Vaginal drug delivery, 360
van der Waals forces, liaisons, 234
n-Vinylacetamide, 40
Vinylimidazolium sulfobetaine, structure, 44
2-Vinylpyridine, polymerization, 37–38
4-Vinylpyridine, polymerization, 37–38
n-Vinylpyrrolidone polymer synthesis
    bulk polymerization, 69t
    solution polymerization, 70,71f
    suspension polymerization, 72
Vinyl-type polymers, use as drug carriers, 394
Viscosity
    concentration, effect, 9,11,12f
    determination of onset of associations, 11,12f
    measurement, 9,11
Viscosity behavior of interacting polymers
    charged polysaccharides, effect, 453,455t,456,457–458f
    poly(styrene sulfonate), effect, 448,449t,450f,451t
    quaternary ammonium salt organic polymers, effect, 456t,459
    shear thinning, reversibility, 460,462
    solute, effect, 453,454t
    sulfonate polymers, effect, 448,452t,453
    viscosity effects, measurement, 453,455t

W

Water, role in solvation of polymers, 9,10f
Water solubility, effect of functional groups, 2,3f
Water-soluble biopolymers
    derivatized polysaccharides, 16
    examples, 11,14f,15
Water-soluble biopolymers—Continued
    nucleic acids, 15
    polypeptides, 15–16
    polysaccharides, 16
    proteins, 15–16
Water-soluble copolymers
    biopolymers, 11,14f,15–16
    hydrodynamic volume, 4–8
    inorganic polymers, 16–17
    rheology, 9,11,12f
    solvation, role of water, 9,10f
    structural design, 2–23
    synthetic methods, 2,4
    synthetic polymers, 17–22
    viscosity, 9,11,12f
Water-soluble monomers, application of persulfate-initiated polymerization, 98–99,100t,101
Water-soluble polymer(s)
    applications, 48–50f,51t,f,175
    functional groups imparting water solubility, 2,3f
    hydrophobic effects on complexation and aggregation, 303–317
    importance in biological systems, 468
    liaisons, role in behavior, 232–246
    molecular-weight-distribution determination by dynamic light scattering, 276–289
    nonpolyelectrolytic synthesis, 25–26
    polyelectrolytic synthesis, 27–47
    primary structure, factors affecting, 2,3f
    quaternary structure, factors affecting, 2
    secondary structure, factors affecting, 2
    tertiary structure, factors affecting, 2
    uses
        electroconductivity, 50,51t,f
        flocculation, 48f
        sedimentation of suspended polymers, 49f,50
        sludge dewatering, 48,49f
        vacuum filtration, 50f
Water-soluble polymer synthesis
    acrylamide, 60–68
    kinetics of free radical polymerization, 57–58,59t
    n-vinylpyrrolidone, 69t,70,71f,72
Weight-average molecular weight, determination, 280

Production: Kurt Schaub
Indexing: Deborah Steiner
Acquisition: Robin Giroux
Cover design: Tina Mion

Printed and bound by Maple Press, York, PA

Paper meets minimum requirements of American National Standard
for Information Sciences—Permanence of Paper for Printed Library
Materials, ANSI Z39.48–1984 ∞

Other ACS Books

Chemical Structure Software for Personal Computers
Edited by Daniel E. Meyer, Wendy A. Warr, and Richard A. Love
ACS Professional Reference Book; 107 pp;
clothbound, ISBN 0–8412–1538–3; paperback, ISBN 0–8412–1539–1

Personal Computers for Scientists: A Byte at a Time
By Glenn I. Ouchi
276 pp; clothbound, ISBN 0–8412–1000–4; paperback, ISBN 0–8412–1001–2

Biotechnology and Materials Science: Chemistry for the Future
Edited by Mary L. Good
160 pp; clothbound, ISBN 0–8412–1472–7; paperback, ISBN 0–8412–1473–5

Polymeric Materials: Chemistry for the Future
By Joseph Alper and Gordon L. Nelson
110 pp; clothbound, ISBN 0–8412–1622–3; paperback, ISBN 0–8412–1613–4

The Language of Biotechnology: A Dictionary of Terms
By John M. Walker and Michael Cox
ACS Professional Reference Book; 256 pp;
clothbound, ISBN 0–8412–1489–1; paperback, ISBN 0–8412–1490–5

Cancer: The Outlaw Cell, Second Edition
Edited by Richard E. LaFond
274 pp; clothbound, ISBN 0–8412–1419–0; paperback, ISBN 0–8412–1420–4

Practical Statistics for the Physical Sciences
By Larry L. Havlicek
ACS Professional Reference Book; 198 pp; clothbound; ISBN 0–8412–1453–0

The Basics of Technical Communicating
By B. Edward Cain
ACS Professional Reference Book; 198 pp;
clothbound, ISBN 0–8412–1451–4; paperback, ISBN 0–8412–1452–2

The ACS Style Guide: A Manual for Authors and Editors
Edited by Janet S. Dodd
264 pp; clothbound, ISBN 0–8412–0917–0; paperback, ISBN 0–8412–0943–X

Chemistry and Crime: From Sherlock Holmes to Today's Courtroom
Edited by Samuel M. Gerber
135 pp; clothbound, ISBN 0–8412–0784–4; paperback, ISBN 0–8412–0785–2

For further information and a free catalog of ACS books, contact:
American Chemical Society
Distribution Office, Department 225
1155 16th Street, NW, Washington, DC 20036
Telephone 800–227–5558

Highlights from ACS Books

Bestsellers from ACS Books

The ACS Style Guide: A Manual for Authors and Editors
Edited by Janet S. Dodd
264 pp; clothbound, ISBN 0–8412–0917–0; paperback, ISBN 0–8412–0943–X

Chemical Activities and Chemical Activities: Teacher Edition
By Christie L. Borgford and Lee R. Summerlin
330 pp; spiralbound, ISBN 0–8412–1417–4; teacher ed. ISBN 0–8412–1416–6

Chemical Demonstrations: A Sourcebook for Teachers,
Volumes 1 and 2, Second Edition
Volume 1 by Lee R. Summerlin and James L. Ealy, Jr.;
Vol. 1, 198 pp; spiralbound, ISBN 0–8412–1481–6;
Volume 2 by Lee R. Summerlin, Christie L. Borgford, and Julie B. Ealy
Vol. 2, 234 pp; spiralbound, ISBN 0–8412–1535–9

Writing the Laboratory Notebook
By Howard M. Kanare
145 pp; clothbound, ISBN 0–8412–0906–5; paperback, ISBN 0–8412–0933–2

Developing a Chemical Hygiene Plan
By Jay A. Young, Warren K. Kingsley, and George H. Wahl, Jr.
paperback, ISBN 0–8412–1876–5

Introduction to Microwave Sample Preparation: Theory and Practice
Edited by H. M. Kingston and Lois B. Jassie
263 pp; clothbound, ISBN 0–8412–1450–6

Principles of Environmental Sampling
Edited by Lawrence H. Keith
ACS Professional Reference Book; 458 pp;
clothbound; ISBN 0–8412–1173–6; paperback, ISBN 0–8412–1437–9

Biotechnology and Materials Science: Chemistry for the Future
Edited by Mary L. Good (Jacqueline K. Barton, Associate Editor)
135 pp; clothbound, ISBN 0–8412–1472–7; paperback, ISBN 0–8412–1473–5

Personal Computers for Scientists: A Byte at a Time
By Glenn I. Ouchi
276 pp; clothbound, ISBN 0–8412–1000–4; paperback, ISBN 0–8412–1001–2

Polymers in Aqueous Media: Performance Through Association
Edited by J. Edward Glass
Advances in Chemistry Series 223; 575 pp;
clothbound, ISBN 0–8412–1548–0

For further information and a free catalog of ACS books, contact:
American Chemical Society
Distribution Office, Department 225
1155 16th Street, NW, Washington, DC 20036
Telephone 800–227–5558